Circuits and Signals:
An Introduction
To Linear and Interface Circuits

Circuits and Signals:
An Introduction
To Linear and Interface Circuits

Roland E. Thomas
Mission Research Corporation

Albert J. Rosa
U. S. Air Force Academy

John Wiley & Sons
New York   Chichester   Brisbane   Toronto   Singapore

Cover design by Jim Massey

Library of Congress Cataloging in Publication Data:

Thomas, Roland E., 1930–
Circuits and signals.

 Includes index.
  1. Interface circuits. 2. Linear integrated
circuits. I. Rosa, Albert J., 1942–   . II. Title.

TK7868.I58T53 1984   621.381′73   83-21588
ISBN 0–471–89560-1

Printed in the United States of America

10 9 8 7 6 5 4 3 2 1

To our wives, Juanita and Kathleen

Preface

This book evolved from teaching an introductory circuits course over several years. In writing it, our objective was to merge the essentials of linear circuit theory with the imperatives of integrated circuit technology. An electronic world dominated by chips is a world dominated by digital systems. As a result, the applications of classical linear circuits have been driven to the periphery of the system—its interfaces.

The digital domain moreover is dominated by design, rather than by analysis considerations. In this regard it is important to distinguish between the design *of* integrated circuits and design *with* integrated circuits. The former (design of) requires the ability to deal with large-scale circuits, with an attendant emphasis on the concepts needed to understand computer-aided circuit analysis and design. The latter (design with) basically implies the interconnection of standard, commercial assemblies. Design in this context normally involves only relatively simple circuits that are invariably located at the interfaces between chips.

The purpose of this book is to develop the viewpoint and skills that ultimately will lead to an ability to design *with* integrated circuits. The focus of the book is perhaps best captured in its subtitle—*An Introduction to Linear and Interface Circuits*. In pursuit of this focus there is an early introduction to the concept of a circuit interface as a pair of terminals at which certain signal conditions may be prescribed or observed. There is also an integrated treatment of the premier linear interface device—the OP AMP. Over the past decade or so the OP AMP has revolutionized the way in which electrical engineers design linear circuits. Its inclusion in a book that first teaches them how to think about linear circuits is clearly necessary.

The content is packaged in three blocks. Block I (Chapters 1–5) treats most of the classical topics of circuit theory in the context of resistance circuits and introduces the OP AMP as a circuit element. A central theme of this block is that circuit response is the result of constraints that arise from two sources: circuit connections and circuit devices. This theme reappears in several places and is crucial to the development in later chapters. Block II (Chapters 6–10) introduces time-varying waveforms and energy storage elements, or memory elements as we call them. This is followed by an introduction to the classical methods of solving first- and second-order linear differential equations, and an axiomatic introduction to the Laplace transformation. This block culminates with the concept of transforming the circuit into the *s*-domain and viewing the circuit response in terms of transforms rather than waveforms. Block III (Chapters 11–13) presents three important applications of the *s*-domain approach: step response (transients), sinusoidal response (phasors), and frequency response (filters). This area makes extensive use of the pole-zero approach to explain the results.

An important organizational feature of the book is the use of Bloom's taxonomy[1] to provide a framework of educational objectives and supporting homework. *En route Objectives* are explicitly listed at the end of each chapter. The *Exercises* related to the en route objectives are designed to test mastery at the knowledge, comprehension, and application levels of the taxonomy. The *Problems* at the end of each chapter require the integration or extension of several en route objectives and, as such, often test mastery beyond the application level. *Block Objectives* listed at the beginning of each block require the integration of en route objectives across several chapters and provide opportunities to encourage creativity. The problems supporting the block objectives are given in Appendix A and are designed to test mastery at the analysis, design (synthesis), and evaluation levels of the taxonomy. This framework of objectives allows the development of well-organized courses in which students need never ask what they are expected to know.

The book is intended to provide the foundations for subsequent courses in electronics and in systems, and it contains sufficient material for a two-semester sequence if all en route and block objectives are used. Two-quarter and one-semester treatments are possible by the proper selection of en route objectives and reduction of the expected level of achievement in the block objectives. Some examples of en route objective selections are shown in Table 1, included as part of this preface. This basic coverage can then be followed by different application emphasis, as suggested in Table 2, which follows Table 1. These examples illustrate that it is possible to maintain an unambiguous framework of objectives for the student. A

	Basic En route Objectives	
Chapter	**One Semester**	**Two Quarters**
1	All	All
2	All	All
3	3-2, 3-3, 3-5	All
4	4-1, 4-3, 4-4	4-1, 4-2, 4-3, 4-4
5	5-1	5-1
6	6-1	6-1, 6-2
7	7-1	7-1, 7-2, 7-4
8	8-1	8-1
9	9-1, 9-2, 9-3	9-1, 9-2, 9-3
10	10-1, 10-3	10-1, 10-3, 10-4

TABLE 1
Suggested Basic En route Objective Selections for Varying Course Lengths

[1]"Taxonomy of Educational Objectives—The Classification of Educational Goals," *Handbook I, Cognitive Domain,* Benjamin S. Bloom (editor), Longman Inc., New York, 1956.

Application Emphasis En route Objectives				
Chapter	**General**	**Systems**	**Power**	**Filters**
11	11-1, 11-2	11-1 thru 11-4	11-1	11-1
12	12-1, 12-2	12-1	12-1 thru 12-5	12-1
13	13-2, 13-3	13-1, 13-2		13-1 thru 13-5

TABLE 2
Possible Application Alternatives

more detailed presentation of various alternatives is given in the accompanying *Teacher's Manual,* including a one-semester course for nonelectrical engineers that emphasizes OP AMP interfacing in instrumentation systems.

We are deeply indebted to the many Air Force Academy cadets who suffered through the ever-changing pattern of the thoughts recorded here. Special thanks are due the faculty members who contributed to this evolution, and especially to M. J. O'Brien, W. H. Block, A. C. Dwelis, C. J. Corley, J. J. Connery, Jr., S. L. Hammond, M. Alexander, W. C. Hobart, Jr., P. L. Sisson, W. D. Wilson, and J. E. Erickson. We also are indebted to C. V. Stewart for his meticulous review of earlier versions of this material, and to K. A. Rosa, who typed the final manuscript with efficiency and enthusiasm. Finally, we gratefully acknowledge the support of our wives, Juanita and Kathleen, who never ceased to encourage us in this work.

Colorado Springs, Colorado   **Roland E. Thomas**
July 1983   **Albert J. Rosa**

Contents

BLOCK III
Applications

APPENDICES

Circuits and Signals:
An Introduction
To Linear and Interface Circuits

BLOCK I
MEMORYLESS CIRCUIT

BLOCK OBJECTIVES

ANALYSIS

Given a memoryless circuit with prescribed input signals, determine prescribed output signals or input-output relationships using any analysis technique.

DESIGN

Devise a memoryless passive or active circuit or modify an existing circuit to obtain a specified output signal for given input signals or to implement a given input-output relationship.

EVALUATION

Given two or more circuits that perform the same signal-processing function, select the best circuit on the basis of given criteria such as performance, cost, parts count, power dissipation, and simplicity.

Chapter 1
Introduction

A Brief History
About This Book
Symbols and Units

Electrical networks are processors of electrical signals and the function to be performed is specified in terms of the input and output signals.

John G. Linvill

Electrical engineering involves a diversity of fast-changing topics. Yet the discipline is unified by an unchanging interest in two basic commodities—energy and information. The purpose of this chapter is to offer some background that can help the reader use this unifying concept to integrate the technical matters discussed in succeeding chapters. We begin with a brief historical sketch of the events leading up to the emergence of information systems as the primary field of electrical engineering. The next section uses the concept of signal processing to identify the broad outline of the scope of our study, and discusses how this text is structured to help the student develop the engineering abilities needed to analyze, design, and evaluate signal-processing circuits. The final section gives some of the standard notation used throughout the book.

1-1   A BRIEF HISTORY

The widespread use of electricity in our modern world is based on several of its unique properties. Electric energy can be transmitted to distant points almost instantaneously, and there efficiently converted into other forms of energy such as mechanical, chemical, thermal, and acoustical. In an electric power system, some form of stored energy is converted into an electrical form and transferred to consumers or "loads," where it is converted into the form required by the application. Much of the work of our industrial society is accomplished through the generation, transfer, and conversion of large blocks of electric energy.

Electric energy also can be transferred in very tiny, carefully controlled amounts. The transfer of energy can take the form of telephone conversations or of numbers exchanged between computers. Often electricity is the most effective means of controlling the flow of other forms of energy. In such applications the ability of electricity to convey information is the primary consideration. The use of electricity as a medium for information processing and interchange has become the dominant concern of electrical engineering in modern times.

A brief survey of the history of electrical engineering may provide the perspective to see that modern technology is the product of human experience in the fairly recent past. Today's technology may make more sense if one has even a very rudimentary concept of how it got to be that way. Perhaps one can also gain some humility in realizing that pioneers in technology cleared the paths through the wilderness that have led this generation to the electrical age.

The study of history can be organized in a variety of ways, with the most obvious a chronological development. Most other engineering fields have their historical roots in arts and crafts that go back to antiquity. In contrast, the chronology of electrical engineering covers a relatively brief time span since essentially all applications of electrical phenomena have been developed since the turn of the nineteenth century. Three major themes thread through this time span. The fundamental understanding provided by **electrical science** led to the development of engineering systems of two basic types—**energy systems** and **information systems.** In tracing these themes we will encounter the major historical personages whose contributions seem (at least to the authors) to have had the greatest impact. Electrical engineering, perhaps more than any other technological discipline, recognizes these individuals by the units and terminology it uses. The volt, ampere, ohm, hertz, and farad all honor people, as do such terms as the Edison effect, the Scott connection, the Nyquist criterion, the Bode plot, and the Shannon–Hartley law.

Electrical Science

The momentous discovery of Alessandro Volta in 1800 marked the first milestone in the electrical age. Volta found that by alternating disks of dissimilar metals, separated by acid-moistened pieces of paper, he could produce a continuous flow of electricity. The "Voltaic pile," or battery, provided a steady flow of electric charge that allowed experimenters to explore the properties of electricity in a repeatable and scientific way.

In 1820 the Danish scientist Hans Christian Oersted observed that a magnetic compass was influenced by the flow of electricity in a wire. Oersted's experiments showed that electric currents produce magnetic fields, and for the first time there was evidence of a connection between electricity and magnetism. By 1825 André Marie Ampere had formulated the quantitative relationships involved, and had also become the first to recognize the difference between electric current and voltage.

Early investigators were puzzled by the paradox that although electricity produced magnetism, apparently magnetism did not produce electricity. However, by 1831 Michael Faraday had found that a changing magnetic field would indeed produce a flow of electricity, an effect uncovered at about the same time by Joseph Henry. Faraday's discovery of magnetic induction provided the impetus for the initial development of the electric generator and telegraph.

By mid-century, Gustav Robert Kirchhoff had formulated the laws that govern the behavior of electric circuits. In 1873, after many years of study, James Clerk Maxwell published his classic paper *Electricity and Magnetism,* which unified all knowledge of electricity through a set of relationships that have come to be known as **Maxwell's equations.** Maxwell predicted electromagnetic waves as early as 1864, although it was not until 1887 that Heinrich Hertz experimentally verified this prediction. With the discovery of cathode rays and the electron in the last decade of the nineteenth century, most of the fundamental knowledge of electricity was at hand.

Energy Systems

The discovery of magnetic induction by Michael Faraday (1831) provided a means for converting mechanical energy into electrical form, and this triggered the development of electric generators, or dynamos as they were called. For nearly fifty years progress was rather slow since the scientific knowledge had not yet been translated into engineering design know-how. The early machines produced direct current (dc). Their performance was gradually improved through a series of trial-and-error developments until, by 1880, there were a number of isolated dc systems providing energy for illumination, traction, and electrotype applications.

In 1882 Thomas Alva Edison patented his incandescent lamp. The wide acceptance of this lamp by the public led to a rapid increase in the demand for electric power. But Edison did more than simply create a practical incandescent lamp. He and his associates produced dynamos, distribution lines, switches, fuses, sockets—in other words, all of the apparatus required in an economical working power system. The famous Pearl Street Station in New York City was the first of the power systems installed by

Edison. By 1884 there were 20 such stations operating around the United States.

The period of cut-and-try design of dc dynamos came to a close in the 1880s, as men trained in science and engineering were attracted to the growing field of energy systems. John Hopkinson, professor of electrical engineering at the University of London, redesigned Edison's dynamos and nearly doubled their capacity. In 1886 Hopkinson published a classic paper, *Dynamoelectric Machines,* which provided a mathematical description of the performance of a dc generator. Machine designers were then able to predict performance from design calculations. By the last decade of the nineteenth century, dc power system technology was fairly well advanced.

At about the same time a competing power system loomed on the horizon. The development of the power transformer in 1882, and Nicola Tesla's induction motor in 1887, provided the last missing links in the alternating-current (ac) power system. By 1895 Charles Steinmetz had published a series of classical papers that provided the mathematical foundation for the design of ac systems. The competition between the older dc and the newer ac systems developed into a heated controversy in the early 1890s. In the United States the controversy hinged on the system for a large power plant to be installed at Niagara Falls, N.Y. The decision to use an ac system was the precursor of a trend that led to the very large, highly interconnected ac power systems of today.

Information Systems

Faraday's discovery also triggered the development of the electric telegraph. Early systems were produced by Schilling (1832), Gauss–Weber (1833), and Wheatstone–Cooke (1837). The first commercial telegraph system in the United States was installed by Samuel F. B. Morse between Baltimore, Md., and Washington, D.C., in 1843. By the end of the Civil War, the Western Union telegraph system spanned the continent. The expansion of land-line telegraphy was accompanied by the development of underwater cables, or submarine cables as they were called. The first of these was laid across the English Channel in 1850. A proposed submarine cable across the Atlantic presented early telegraph engineers with many challenges. Sir William Thomson (Lord Kelvin) provided the theoretical analysis showing that telegraph signals could indeed be transmitted via cable over the distances involved. Kelvin also invented a sensitive meter to detect the very weak signals at the receiving end of submarine cables. In 1857 the first attempt was made to lay such a cable. However, it was not until 1866, after several failures, that the project was finally brought to a successful completion. By 1902 submarine cables circled the globe.

The first practical acoustical transducer was patented by Alexander Graham Bell in 1875, although there were precursors of this invention in the earlier works of Philipp Reis. The first telephone system based on

Bell's patents was installed in New Haven, Conn., in 1878, and involved a total of but eight lines and 24 customers. As demand for telephone service expanded, a major bottleneck developed in the manual telephone exchanges that interconnected users. The automatic step-by-step switching station, originally developed in 1887, was improved over the years and eventually replaced manual operation. The theoretical contributions of Oliver Heaviside, Michael Pupin, and George Campbell in the closing years of the nineteenth century led to the transmission-line loading coil. This development improved long-line telephone performance, and allowed service to be extended to distances of up to 1500 miles. Further extension of long-distance operation had to await the development of electronic amplifiers.

Hertz's classical experiments (1887) verifying Maxwell's prediction led to the development of wireless telegraphy. Guglielmo Marconi first demonstrated wireless operation in 1895 and took out patents on a complete wireless system two years later. Oliver Lodge greatly improved the selectivity of wireless receivers by perfecting this coil-condenser tuning method in 1898. For nearly 20 years shipboard telegraphy was the main use of Marconi's system. The ramming of a passenger steamer off Nantucket in 1909 dramatically demonstrated that wireless could greatly increase safety at sea. In that same year the United States passed a law requiring wireless on all its passenger ships.

The electronic age dawned with the discovery of the electron by Joseph John Thomson in 1897. In 1904 John Ambrose Fleming produced a signal-processing device based on the flow of electrons, the vacuum diode. Curiously, Edison had produced a crude vacuum diode in the process of perfecting his incandescent lamp. He described the device in a paper presented at the inaugural meeting of the American Institute of Electrical Engineers in 1884, but unfortunately neither Edison nor anyone in attendance at that meeting recognized the importance of his results. Lee De Forest added a third electrode in 1906 to produce the vacuum triode, or audion as he called it. This invention made possible the development of electronic signal amplifiers. New advancements followed this breakthrough with the perfection of the electronic oscillator, the superheterodyne receiver (1918), and wireless transmission of voice signals. The first commercial radio broadcast station (KDKA Pittsburgh) went into operation in 1920.

The two decades between 1920 and 1940 were one of the most remarkable periods in the history of electrical engineering. The central engineering problem of the time was the development of economical long-distance telephone service on a rather wide scale. In the process, the engineers of this era also formulated much of the basic theory and techniques of information systems. The contributions of George Campbell, Otto Zobel, Ronald Foster, Sidney Darlington, Harry Nyquist, Hendrik Bode, and others laid the foundation of what is now called circuit and filter theory. The negative feedback amplifier was invented by Harold Black in 1934. Feedback theory was further advanced by Bode and Nyquist to provide a mathematical basis for the practical design of high-quality amplifiers. The systematic study of electrical noise, both as a physical phenomenon and as an information-limiting factor, was begun by J. B. Johnson (1927) and Nyquist (1928). Many signal-processing techniques were first proposed during this time, including single-sideband modulation (John Carson), frequency modulation (Edwin Armstrong), signal sampling (Harry Nyquist), and pulse code modulation (Alec Reeves, 1937). Finally, the early formulations of information theory by Nyquist and R. V. L. Hartley date from this period. It remained for Claude Shannon to unify the theory of information with his classic paper *The Mathematical Theory of Communication*.

The theory and techniques from the telephone industry were to find wide application during World War II. Developments in radar, feedback control systems, electronic computers, and radio navigations systems were rapidly advanced as a result of the imperatives of the time. These tremendous advances in technology, largely cloaked in secrecy during the war years, burst upon the scene in the postwar era. With the invention of transistors in the late 1940s, solid-state electronics emerged to enter a period of rapid and continuing growth. By 1950 the center of gravity of electrical engineering had shifted from energy and power systems to electronics and information processing.

The advent of solid-state electronics revolutionized signal processing, and the transistor rapidly replaced the vacuum tube in most applications. Beginning in about 1960, discrete transistors were in turn replaced by integrated circuits (ICs) or "chips" as they are commonly called. By 1970 large-scale integrated (LSI) circuits involving many thousands of transistors were commercially available in various forms, most notably as a computer on a "chip," or microprocessor. The size, cost, and reliability advantages of integrated circuits are so overwhelming that this technology now dominates in almost all signal-processing applications. As a result, a major concern of today's circuit design engineer is the interconnection and interfacing of standard IC assemblies.

Table 1-1 gives a brief chronology of the three historical themes in this discussion. This listing presents some of the more important discoveries, events, and classic papers leading up to the emergence of information systems as the dominant field in electrical engineering. A chronology cannot indicate the relative importance and interrelationship of events, but does provide, in skeleton form, a broad outline of the major developments. Additional **historical notes** are included at appropriate places in succeeding chapters.

1-2   ABOUT THIS BOOK

The basic purpose of this book is to introduce circuit analysis. Although circuit analysis is one of the fundamental tools of electrical engineering,

Electrical Science	Energy Systems	Information Systems
1800 Volta announces the voltaic pile	1809 Davy produces the arc light	1832 Schilling telegraph
1820 Oersted discovers electromagnetism	1825 Sturgeon produces electromagnets	1843 Morse installs telegraph from Baltimore to Washington
1825 Ampere presents important results to the French Academy of Science	1832 Dal Negro and Pixie produce primitive electric generators	1850 First submarine cable across English Channel
1826 Ohm publishes his treatise	1832 Sturgeon develops magnetic motor	1856 Kelvin develops theory of submarine cables
1831 Faraday and Henry discover electromagnetic induction	1836 Daniel develops an improved battery	1857 First attempt to lay an Atlantic cable
1845 Kirchhoff's results first published	1860 Pacinotti invents generator using electromagnets	1861 Western Union establishes transcontinental telegraph
1864 Maxwell predicts electromagnetic waves	1867 Seimans, Varley, and Wheatstone produce self-excited dynamos	1863 Reis reproduces sounds electrically
1873 Rowland describes magnetic saturation and permeability	1871 Gramme invents the slotted armature	1875 Bell patents telephone
1873 Maxwell publishes *Electricity and Magnetism*	1882 Edison patents his incandescent lamp	1895 Marconi transmits wireless telegraphy 1 km
1887 Hertz produces electromagnetic waves	1882 Edison installs the Pearl St. Station	1898 Lodge perfects coil-condensor tuning
1892 Heaviside publishes *Electrical Papers*	1882 Gauland and Gibbs produce ac power transformers	1901 Marconi transmits across Atlantic
1897 J. J. Thomson discovers the electron	1884 American Institute of Electrical Engineers established	1904 Fleming invents the vacuum diode
	1886 Hopkinson publishes *Dynamoelectric Machines*	1906 De Forest invents vacuum triode
	1887 Tesla patents polyphase ac systems	1914 First transcontinental telephone channel
	1888 Tesla announces the ac induction motor	1917 Campbell develops the electric wave filter
	1888 Parson installs first steam-turbine-driven ac generator	1918 Armstrong perfects the superheterodyne receiver
	1893 Steinmetz publishes *On the Law of Hysteresis*	1920 First commercial AM radio station KDKA Pittsburgh
	1894 Steinmetz publishes *Complex Quantities and Their Use in Electrical Engineering*	1924 Nyquist publishes *Certain Factors Affecting Telegraph Speed*
	1895 AC power system installed at Niagara Falls	1928 Hartley publishes *Transmission of Information*
	1895 Steinmetz publishes *Theory of the General AC Transformer*	1934 Black publishes *Stablized Feedback Amplifiers*
	1918 Fortesque develops method of symmetrical components	1934 Hazen publishes *The Theory of Servomechanisms*
		1936 Armstrong describes frequency modulation
		1939 –45 Rapid development of communication, control, computer, and radar systems
		1947 Bardeen, Brattain, and Shockley invent the transistor
		1948 Shannon publishes *The Mathematical Theory of Communication*
		1957 "Planar" silicon technology emerges
		1958 Kilby develops the integrated circuit
		1963 Hofstein and Heiman invent MOS transistor
		1969 Hoff designs the first microprocessor and large-scale integration (LSI) is born

TABLE 1-1
A Chronology of Electrical Engineering

it is important to understand at the outset that circuits do not exist solely for analysis purposes. Circuits exist and are worthy of study because they process signals. For the present we can define a circuit as an interconnection of electric devices, and a signal as a time-varying electrical quantity whose purpose is to carry information or energy. Circuit analysis tells us how the signal is altered in its passage from circuit input to circuit output. Thus in our view circuits and signals do not exist in isolation, but are intertwined and inseparable.

The subtitle of this book suggests that we are primarily interested in linear circuits. In this context a linear circuit is one in which the output signals are proportional to the input signals. This important principle, called superposition, is introduced in Chapter 3, and is an extremely useful conceptual tool. Many circuits are essentially linear within a useful range of signal levels. When driven outside this range they become nonlinear so that superposition no longer applies. We shall deal briefly on occasion with such nonlinear circuits to contrast their behavior, but our primary attention will be focused on circuits operating within their linear range.

The subtitle also indicates that our study includes interface circuits. This terminology has been spawned by integrated-circuit (IC) technology. With this technology most of the circuit interconnections are inaccessible and the basic signal-processing function has been predetermined by the manufacturer. The problem of using ICs is a matter of interconnecting large circuits at the few accessible connections in such a way that they are compatible. This process often involves a relatively small interface circuit whose purpose is to change the signal level or format to ensure compatibility. For the purposes of this book we define an interface as a pair of accessible terminals at which signal characteristics may be observed or specified. An interface circuit then is one that is intentionally interjected at an interface to ensure that the appropriate signal conditions exist.

This book has been structured with a well-defined set of objectives and related homework that can be used to demonstrate the achievement of those objectives. The overarching goals, called **block objectives,** are listed at the beginning of each of the three parts of the book. The corresponding homework problems are in Appendix A. Block objectives are defined at three levels: analysis, design, and evaluation. In terms of signal processing, analysis means that the input signals and the circuit are known, and the goal is to predict the output signals. The compelling feature of analysis is that a unique solution always exists. The analysis process will occupy the bulk of our attention since it is the foundation for understanding the interaction of signals and circuits. The design process involves given input and output signals, and requires the identification of a circuit that will perform this prescribed signal-processing function. Design is the ultimate goal of engineering, and hence the ultimate demonstration of understanding. But in contrast to analysis, a design situation may have no solutions, or a number of different solutions. This latter possibility leads to the evaluation objective. Given several circuits that perform the same basic function, we must evaluate the alternatives using some criterion and ultimately make a choice. In reality the engineer's role involves all three processes: analysis, design, and evaluation. It is our hope that this book will help prepare you for that role by stating these goals explicitly and providing homework problems that test your mastery of these goals.

Although the block objectives represent the ultimate goals of this book, their achievement is build upon a structured set of smaller goals called **en route objectives.** These smaller milestones are listed at the end of each chapter, together with related exercises designed to test your understanding of basic concepts and techniques. Although they are listed at the end of the chapter, the en route objectives are an integral part of the chapter and are keyed to one or more sections of the text. You should rely heavily on these objectives in using this book. Collectively the en route objectives represent the basic knowledge and understanding needed to achieve the block objectives. Following the en route objectives in each chapter are a number of **problems.** These problems are designed to encompass several en route objectives and offer the ability to practice mastery of broader concepts before attempting the even more global block objectives.

1-3   SYMBOLS AND UNITS

Like all disciplines, electrical engineering has its own terminology and symbology. The symbols used to represent some of the more important physical quantities and their units are listed in Table 1-2. It is not our purpose to define these quantities here, or to offer this list as an item for memorization. Rather the purpose of this table is simply to list in one place all of the quantitites commonly used in this book.

Numerical values encountered in electrical engineering range over many orders of magnitude. Consequently the system of standard decimal prefixes in Table 1-3 is used. These prefixes on the unit abbreviation of a quantity indicate the power of 10 that is applied to the numerical value of the quantity.

The following examples illustrate the use of the standard decimal prefixes on the unit abbreviation symbol:

$$1 \text{ M}\Omega = 1 \text{ megohm} = 1 \times 10^6 \text{ ohms}$$
$$5 \text{ kW} = 5 \text{ kilowatts} = 5 \times 10^3 \text{ watts}$$
$$3.6 \text{ ms} = 3.6 \text{ milliseconds} = 3.6 \times 10^{-3} \text{ seconds}$$
$$0.3 \text{ }\mu\text{F} = 0.3 \text{ microfarads} = 0.3 \times 10^{-6} \text{ farads}$$

Incidentally, this illustrates the use of coding in information processing. Clearly the symbols 1 MΩ contain the same data as the set of symbols 1,000,000 ohms. That is, they contain the same data once you understand the code in Tables 1-2 and 1-3.

Quantity	Symbol	Unit	Abbreviation of Unit
Time	t	second	s
Frequency	f	hertz	Hz
Radian frequency	ω	radian/second	rad/s
Phase angle	θ, ϕ	deg or radian	° or rad
Energy	w	joule	J
Power	p	watt	W
Charge	q	coulomb	C
Current	i	ampere	A
Electric field	$\mathcal{E}$	volt/meter	V/m
Voltage	v	volt	V
Impedance	Z	ohm	Ω
Admittance	Y	mho	℧
Resistance	R	ohm	Ω
Conductance	G	mho	℧
Reactance	X	ohm	Ω
Susceptance	B	mho	℧
Inductance	L, M	henry	H
Capacitance	C	farad	F
Flux	ϕ	weber	Wb
Flux linkage	λ	weber	Wb

TABLE 1-2
Some Important Quantities, Their Symbols and Unit Abbreviations

Multiplier	Prefix	Abbreviation
10^{12}	tera	T
10^{9}	giga	G
10^{6}	mega	M
10^{3}	kilo	k
10^{-1}	deci	d
10^{-2}	centi	c
10^{-3}	milli	m
10^{-6}	micro	μ
10^{-9}	nano	n
10^{-12}	pico	p
10^{-15}	femto	f

TABLE 1-3
Standard Decimal Prefixes

EN ROUTE OBJECTIVES
AND RELATED EXERCISES

1-1 HISTORICAL ROOTS (SECS. 1-1 to 1-3)

The student will be able to describe the origins and development of electrical engineering.

Exercises

1-1-1 Explain the difference between electrical science, energy systems, and information systems.

1-1-2 Refer to the chronology in Table 1-1 and select the one event you feel most affected technology as you know it today. Explain your choice.

1-2 ELECTRICAL QUANTITIES (SEC. 1-3)

Given an electrical quantity described in terms of words, scientific notation, or decimal prefix notation, convert the quantity to an alternate description.

Exercises

1-2-1 Write the following in symbolic form:
  (a) four megahertz
  (b) sixteen nanoseconds
  (c) point forty-seven microfarads
  (d) one hundred kilohms
  (e) five millihenries

1-2-2 Write out in words the following symbols:
  (a) 0.005 Trad/s
  (b) 2 J
  (c) 6.02 nV
  (d) 10 dB
  (e) 102 fA

Chapter 2
Basic Circuit Concepts

"The electromotive action appears in two sorts of effects that I think ought to be distinguished by precise definitions. I will call the first electric tension, the second electric current."

André Marie Ampere

We begin our study of circuits by studying the parameters that will become second nature to us as engineers. A number of physical devices and their ideal models are introduced and applied. Kirchhoff's laws are the constraints on interconnections of elements. The application of Kirchhoff's laws through loop and node equations results in useful analysis tools such as voltage and current division. The concept of equivalent circuits is developed and applied to solve analysis problems efficiently.

2-1 CIRCUIT VARIABLES

The underlying physical quantities in the study of electronic systems are two basic variables—**charge** and **energy.** Of the two, charge is electric in character. The concept of electric charge explains the very strong electrical forces that occur in nature. To explain both attraction and repulsion, we say that there are two kinds of charge—positive and negative. Like charges repel whereas unlike charges attract. The symbol q is used to represent charge. In the MKS (meter-kilogram-second) system charge is measured in coulombs (abbreviated C). The smallest quantity of charge in nature is an electron's charge ($q_E = 1.6 \times 10^{-19}$ C).

Electric charge is a rather cumbersome variable to work with in practice. Moreover, in most situations the charges are moving, and so we find it more convenient to measure the amount of charge passing a given point per unit time. To do this, we define a signal variable i called **current** as follows:

$$i = \frac{dq}{dt} \tag{2-1}$$

Current is a measure of the flow of electric charge. It is the time rate of change of charge passing a given point. The physical dimensions of current are coulombs per second. In the MKS system, the unit of current is the **ampere** (abbreviated A). That is,

1 coulomb/second = 1 ampere

Since there are two types of electric charge, there is a bookkeeping problem associated with the direction of current flow. In electrical engineering it is customary to define the direction of current as the direction of the net flow of positive charges.

The concept of **voltage,** a second signal variable, is associated with the change in energy that would be experienced by a charge as it passes through a circuit. The symbol w is commonly used to represent energy. In the MKS system, energy carries the units of joules (abbreviated J). If a small charge dq were to experience a change in energy dw in passing from point A to point B, then the voltage v between A and B would be defined as the change in energy per unit charge or

$$v = \frac{dw}{dq} \tag{2-2}$$

Historical Note

Most of the basic quantities in electrical engineering are named for early contributors to the understanding of electricity. Charles Augustin de Coulomb (1736–1806) was a prominent French army officer who invented a very delicate torsion balance. In using his balance to measure electrical forces, he discovered the law now known as Coulomb's law.

Historical Note

André Marie Ampere (1775–1836) was born in Lyons, France, on January 20, 1775. A natural genius with an affinity for the mathematical sciences, by age 12 he had mastered all the mathematics then known. He studied physics and founded the science of electrodynamics—now called electromagnetics. By 1801 he had become a professor of both chemistry and physics at the University of Bourg. He later moved to the École Polytechnique in Paris, where he served as a professor of mathematics and was elected to the French Academy of Sciences. Unlike many of his contemporaries, he was not a methodical experimenter, but relentlessly pursued an idea to the bitter end. When he received word that Oersted had discovered that a magnetic needle is deflected by a current-carrying wire, he immediately postulated a theory regarding the interaction of current and magnetism. Within a week he had published the first of several papers explaining his theory. From this work the law now known as Ampere's law was defined. This law states that the line integral of the magnetic field around an arbitrarily chosen path is proportional to the net current enclosed by that path. Ampere was also the first to recognize the difference between voltage and current, and discussed that fact in the same paper. The quotation at the beginning of this chapter was obtained from that source. Ampere also developed the galvanometer, a tool to measure current and voltage, which is a forerunner of today's ammeter and voltmeter. André Marie Ampere is recognized as one of the greatest electrical scientists of all times.

Voltage does not depend on the path followed by the charge dq in moving from point A to point B. Furthermore, there can be a voltage between two points even if there is no charge motion, since voltage is a measure of how much energy would be involved if a charge dq were moved. The dimensions of voltage would be joules per coulomb. In the MKS system, the unit of voltage is the **volt** (abbreviated V). That is,

$$1 \text{ joule/coulomb} = 1 \text{ volt}$$

Historical Note

Count Alessandro Guiseppe Antonio Anastasio Volta (1745–1827) was born in Como, a small town in what is now northern Italy. A member of a noble family, he pursued his interests in electrochemistry. At the age of 24 he invented the electrophorus, the first machine ever to generate electricity, and during that same year, was appointed to the chair of physics at the University of Pavia. A year later, when his friend Luigi Galvani found that electricity could be produced by two dissimilar metals imbedded in a frog's muscle, Volta argued that the animal tissue was not necessary. This sparked considerable controversy, which lasted until 1800, when Volta demonstrated the first battery. When he showed his invention to Napoleon, the emperor was so impressed that he made Volta a count and senator of the Kingdom of Lombardy (as that part of Italy was then called). Some fourteen years later the emperor of Austria appointed him director of the philosophical faculty at the University of Padua. Volta's invention, the battery, aided the future contributions of such scientists as Ohm and Kirchhoff. As a tribute, the volt was named in his honor.

A third signal variable is **power,** which is defined as the time rate of change of energy.

$$p = \frac{dw}{dt} \tag{2-3}$$

The dimensions of power would be joules per second, which in the MKS system is called a **watt** (abbreviated W). In electrical situations, it is useful to have power expressed in terms of current and voltage. This is done by writing Eq. 2-3 as

$$p = \frac{dw}{dq}\frac{dq}{dt}$$

Using Eqs. 2-1 and 2-2, we obtain

$$p = vi \tag{2-4}$$

Thus the electric power associated with a situation is determined by the product of current and voltage.

Example 2-1

In a TV picture tube, an electron beam carrying 10^{14} electrons per second is accelerated by a voltage of 50 kV. Let us determine the power in the beam. First, the current is the rate of charge flow. The net current in this case is q_E times the rate of electron flow dn_E/dt. Hence

$$i = q_E \frac{dn_E}{dt} = (1.6 \times 10^{-19})(10^{14})$$
$$= 1.6 \times 10^{-5} \text{ A}$$

Therefore, the beam power is

$$p = vi = (50 \times 10^3)(1.6 \times 10^{-5}) = 0.8 \text{ W}$$

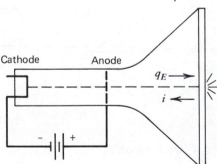

FIGURE 2-1
A cathode-ray tube (CRT).

Figure 2-2 shows the interrelation between the two basic variables, charge and energy, and the three signal variables, current, voltage, and power. Charge and energy, like mass, length, and time, are taken as basic assumed concepts.

While the roots of electrical engineering are intimately entwined with the basic variables, the engineer only occasionally becomes concerned about them, but works instead with the signal variables. The reason for

---

Historical Note

James Watt (1736–1819) was born in Greenock, Scotland. He studied for the trade of mathematical instrument maker in London, and by 1757 had become the instrument maker for the University of Glasgow. Watt became famous for his work with steam power, and in 1774 he invented his own steam engine. He never really worked with electricity, but his work with steam engines firmed his relationship with mechanical power, although in those days the principal measure was in horsepower (hp). The unit of mechanical power was later refined to include a smaller unit as well—that is, 1 newton acting through 1 meter caused 1 joule of work to be done. A joule per second was a new unit of power given the name of ''watt'' (W) in his honor (1 hp = 746 W). When it was later shown that 1 joule was the energy required to transport 1 colomb between two points having a potential difference of 1 volt, it was realized that the mechanical unit and electrical unit of power were the same.

---

this choice of working parameters is that current and voltage are relatively easy to measure, and therefore convenient to use to represent data. Since the processing of data is the fundamental reason for studying electronic systems, it follows that current and voltage are our working variables.

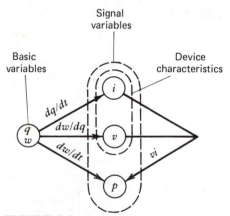

FIGURE 2-2
Flow diagram of circuit variables.

Up to this point in the text, we have said that a signal can be either a current or a voltage, and we have not made any great distinction between them. From this point forward, it is essential that the reader recognize that current and voltage are not the same thing. Current is a measure of the time rate of charge passing a point. Since it is a measure of flow, we think of current as a **through** variable. Voltage is a measure of the net change in energy involved in moving a charge from one point to another. Note that the concept of voltage involves two points. Voltage is not measured at a point, but rather between two points. We therefore call voltage an **across** variable.

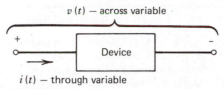

FIGURE 2-3
Assignment of reference direction to current and reference polarity to voltage in a two-terminal device.

Figure 2-3 shows a notation used for assigning reference directions to current and voltage. The reference mark for current (an arrow below the wire) does *not* indicate the actual direction of current flow. The actual direction of flow may be reversing a million times per second. However, when the actual direction coincides with the reference direction, we say that the current is positive. When the opposite occurs, we say that the current is negative. Thus, if the net flow of positive charge in Figure 2-3 is to the right, we say that the current $i(t)$ is positive. Conversely, if the current $i(t)$ is positive, then we know the net flow of positive charge is to the right.

Similarly, the voltage reference marks (+ and − symbols) do not imply that the potential at the + terminal is always higher than the potential at the − terminal. However, when this is true, we will say that the voltage across the device is positive. When the opposite is true, we say that the voltage is negative. The bracket in Figure 2-3 is used here to clarify the situation, and to remind us that voltage is an **across variable.** When we interconnect devices to form a circuit, it becomes impractical to draw all of the brackets that would be required.

2-2 ELEMENT CONSTRAINTS

The device shown as a rectangular box in Figure 2-3 is representative of a whole family of components that engineers connect together to form circuits. Since there are many different devices that possess different, and at times divergent, current–voltage characteristics, (usually abbreviated *i-v* characteristics), engineers have developed models representing the different devices. An electric device is a component that is viewed as an entity. To study a device is to study the *i-v* characteristics of that device. Most devices are nonlinear, that is, their *i-v* relationship is nonlinear. The study of circuits containing several nonlinear devices would require enormous amounts of effort to analyze if it were not for the model. A model, which usually takes the form of an equation or graphical relation, is a linearized version or approximation of the more complex nonlinear *i-v* relation. To distinguish between circuit devices (the real thing) and circuit models (an approximate stand-in), we call the model a circuit element. Thus circuit devices appear as hardware, and are listed in manufacturers' catalogs. Circuit elements, or models, appear in books on circuit analysis and electronics.

The first device in our study is the resistor shown in Figure 2-4*a*. The current–voltage characteristic of this device is quite nonlinear, as shown by the dashed line in Figure 2-4*b*. To write a reasonably accurate mathematical expression for this device would require a cubic equation. Such a complex relation would make analysis of circuits containing resistors very difficult.

If we look closely at the *i-v* relation, we can see that if we limit the range over which we will use the device, we can approximate, with only minimal error, its characteristics with a straight line. This limit is usually described as a power rating of the resistor and in general is not exceeded in designing circuits that employ resistors. In sum, if we stay within the device's prescribed limits, we can obtain our first circuit element, the linear resistor.

The equations that describe this element are

$$v = Ri \quad \text{or} \quad i = Gv \tag{2-5}$$

where

$$G = 1/R \tag{2-6}$$

These equations are collectively known as Ohm's law. The parameter R is called resistance and has the unit ohms (Ω). The parameter G is called conductance, with the unit mho ($\mho$) (ohm spelled backward). The rela-

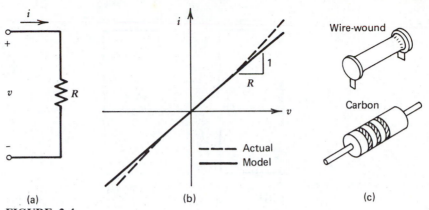

FIGURE 2-4
The linear resistor. (*a*) Circuit symbol. (*b*) *i-v* characteristics. (*c*) Actual resistors.

tionships expressed in these equations can also be presented graphically, as the solid line in Figure 2-4*b*. Such a graph is called an *i-v* characteristic curve. This model is linear and bilateral. In fact, the linear resistor model is so pervasive that we normally drop the adjective "linear" and simply say "resistor." A sketch of actual resistors is shown in Figure 2-4*c*. Most resistors use a color code to identify their value and their tolerance. This code is described in Appendix C.

The power associated with the resistor can be found from $p = vi$. Using Eqs. 2-5 to eliminate v from this relationship yields

$$p = Ri^2 \tag{2-7}$$

or using the same equations to eliminate i yields

$$p = Gv^2 = v^2/R \tag{2-8}$$

Since the parameter R is positive, these equations tell us that the power is always nonnegative. This means that the resistor always absorbs power.

Example 2-2

The linear description of a resistor applies as long as the voltage and current are within certain limits. For resistors, this limit is expressed as a power rating. Suppose we have $R = 47$ kΩ and a maximum power rating of 0.25 W. Let us determine the maximum current and voltage that can be applied to the resistor.

From

$$p = Ri^2$$

we obtain

$$I_{\text{MAX}} = \sqrt{P_{\text{MAX}}/R} = \sqrt{\frac{0.25}{47 \times 10^3}} = 2.31 \text{ mA}$$

Similarly, from $p = v^2/R$ we obtain

$$V_{\text{MAX}} = \sqrt{RP_{\text{MAX}}} = \sqrt{47 \times 10^3 \times 0.25} = 108 \text{ V}$$

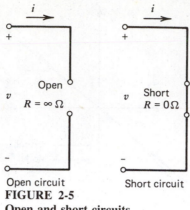

Open circuit   Short circuit
FIGURE 2-5
Open and short circuits.

The next two circuit elements occur so often in electrical engineering that they deserve a special introduction. Consider a resistor R with a voltage v applied across it. Let's calculate the current flowing through the resistor for different values of resistance. If $v = 10$ V and $R = 1\Omega$, we readily find, using Ohm's law, that $i = 10$ A. If we increase our resistance to $100\ \Omega$, we find that i has decreased to 0.1 A or 100 mA. If we continue to increase R to 1 MΩ, i becomes very small at 10 μA. Continuing this process, we arrive at a point at which R is very nearly infinite and i just about zero. When i is zero, we call the special value of resistance, that is, $R = \infty$, an open or an **open circuit.** Similarly, if we reduce R until it approaches zero, we find that the voltage is very near zero. When v is zero, we call the special value of resistance, that is, $R = 0$, a closed or **short circuit.** The circuit symbols for these two elements are shown in Figure 2-5. Elements in a circuit are considered to be connected by ideal wire, that is wire with zero resistance or short circuits.

Example 2-3

A switch occurs often in electrical engineering and is an element with which we have been familiar for a long time. Some people consider a switch a unique circuit element but actually, it is an exclusive combination of an open and a short circuit. Figure 2-6 shows the circuit symbol and i-v characteristics of a switch. When it is closed,

$$v = 0 \quad \text{and} \quad i = \text{any value}$$

and when it is open, $\hspace{4cm}$ (2-9)

$$i = 0 \quad \text{and} \quad v = \text{any value}$$

For the switch element, there really is no relationship between current and voltage. When the switch is closed, the voltage across the device is zero and the element will pass any current that may result. When it is open, no current flows and the element will accommodate any voltage across its terminals. The switch element does not involve any power since the product $vi = 0$ regardless of whether the switch is open or closed. This is, of course, only a model. The switch device has some limitations; among these are the maximum current it will carry when closed and the maximum voltage it will accommodate when open. The switch is activated (closed or opened) by some external influence, such as a mechanical motion, temperature, or pressure.

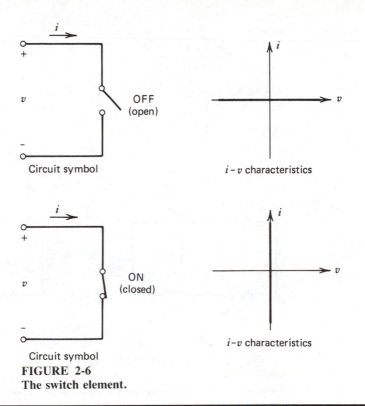

Circuit symbol $i-v$ characteristics

Circuit symbol

FIGURE 2-6
The switch element.

Example 2-4

An important device for analog-digital interface circuits is the analog switch. This device, shown in Figure 2-7, is an electronically controlled switch that changes the resistance of the switch from a few hundred ohms when the switch is ON to millions of megohms when it is OFF. There are several levels of models, each increasing in its ability to describe the switch's operation accurately and each increasing in complexity of use. Figure 2-7a shows a sketch of the actual device, and Figures 2-7b through 2-7d are various models of increasing complexity.

The switch operates as follows: When a suitable voltage, usually digital in nature, is applied to G the switch changes state. In the basic model the switch is either ON or OFF. In the intermediate model, the model most commonly used, the applied voltage switches the switch, but instead of being a pure conductor, the switch is lossy. When it is ON, a 200 Ω or so resistance is connected to the switch. Any current must flow through this resistance, causing joulean heating and subsequent loss of power to the rest of the circuit. Even when the switch is OFF, some small current still flows, although it may be insignificant. More advanced models consider other properties of the switch that cause it to behave differently than the basic or ideal model would suggest. This example illustrates how engineers can adapt various combinations of circuit elements to model other electric devices. It also suggests that no one model can serve all applications. The engineer usually will use the simplest model that will adequately represent the actual device in a particular application.

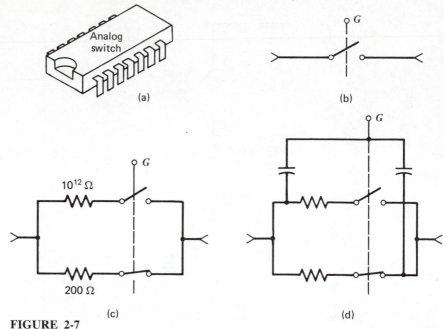

FIGURE 2-7
The analog switch. (*a*) Actual device. (*b*) Basic model. (*c*) Intermediate model. (*d*) Advanced model.

Electronic circuits require power to operate. In electronics, there are two types of power sources: voltage sources and current sources. In addition, each of these sources can be either constant, non-time-varying, or time-varying signal sources. The circuit symbols and the *i-v* characteristics of each of these four sources are shown in Figure 2-8. It is noted that there is no separate symbol for a constant current source, whereas there is a symbol for a constant voltage source (a battery). It should also be noted that while the constant voltage source can be only just that, the signal voltage source symbol can also be used to indicate a constant voltage source.

The element equations for the current source are

$$i = i_s \qquad v = \text{depends} \qquad \textbf{(2-10)}$$

In words, the current source supplies i_s amperes out of its + terminal and into its − terminal, and will furnish whatever voltage is required by the circuit to which it is connected. The signal voltage source is described by

$$v = v_s \qquad i = \text{depends} \qquad \textbf{(2-11)}$$

This means that the voltage source produces v_s volts across its terminals

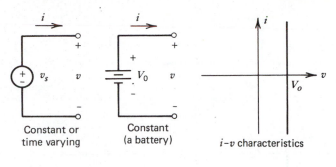

Voltage Sources

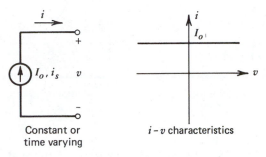

Current Source

FIGURE 2-8
Ideal sources.

and will supply whatever current may be required by the circuit to which it is connected. Sources are often called forcing functions and are said to force or drive the circuit.

This completes the initial entries in our catalog of circuit elements (device models). Later in this text, we develop models for the inductor and capacitor. Other devices will be introduced in other courses as we proceed in our study of systems. Thus the study of devices and device equations is a continuing process.

Example 2-5

Our use of circuit elements in analysis in reality is the use of an appropriate model of the actual device. A device may have several models that can be used, depending on the nature of the circuit application. For example, Figure 2-9 shows several possible models of a resistor that depend on the frequency of signals in the circuit. Similarly, Figure 2-9 shows practical models for the ideal voltage and current sources. These models are called practical because they more accurately represent the characteristics of signal sources as they exist in the real world. It is important to remember that resistors do not always behave as an ideal resistance. Conversely, resistance in a circuit model does not always come from a resistor. The resistance may be needed only to model the inherent characteristics of a device.

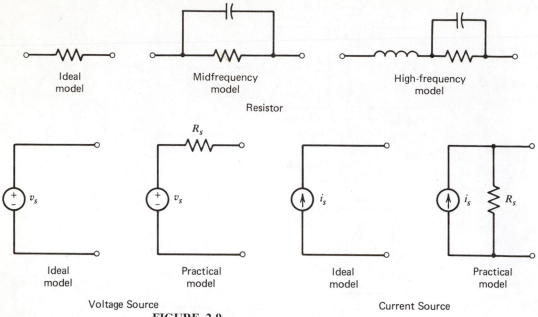

FIGURE 2-9
Examples of models and comparison between ideal and real sources.

2-3   CONNECTION CONSTRAINTS

In the previous section we dealt with individual devices and models. In this section we turn our attention to the interconnections of devices to form circuits. The laws governing circuit behavior are based on the meticulous work of the German scientist Gustav Kirchhoff (1824–1887). Kirchhoff's laws apply to circuits, and they tell us that the process of interconnecting devices to form a circuit forces the device currents and voltages to behave in certain ways. These constraints are based only on the circuit connections and not on the specific devices in the circuit. For this reason, we call the equations derived from Kirchhoff's laws **connection constraints.**

Our treatment of Kirchhoff's laws begins with some definitions. A **circuit** is any collection of devices connected at their terminals. A **node** is a point at which two or more devices are connected together.

In this text we indicate wires that are connected together (physically tied together electrically) as shown in Figure 2-10*a*. Sometimes wires are not connected together, but cross over or under each other. Since we are restricted to drawing these wires on a planar surface, we indicate crossovers as shown in Figure 2-10*b*. In engineering systems very often two or more circuits are connected together. The junction of these two circuits is called an **interface** and represents not only a physical electrical connection, but also a contractural boundary between the different manu-

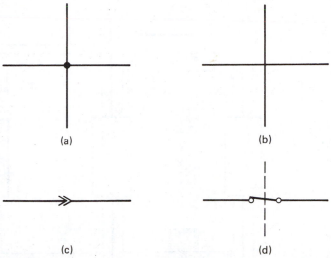

FIGURE 2-10
Types of connections. (*a*) An electrical connection. (*b*) A crossover or no electrical connection. (*c*) Jack or interface, an electrical connection. (*d*) Control line, a nonelectrical connection.

facturers of the individual circuits. Interface connections are electrical connections as in Figure 2-10*a*, but because of their unique purpose of representing divisions between circuits, **jack** or **interface** symbol is used as shown in Figure 2-10*c*. On certain occasions a control line is required to show a mechanical or other nonelectrical interdependency. Figure 2-10*d* indicates how this interdependency is shown in this text.

While it is customary to designate a juncture of two or more elements as a node, it is important to realize that a node is not confined to a point but includes all the wire from the point to each element. In the circuit of Figure 2-11 there are only three different nodes: Ⓐ, Ⓑ, and Ⓒ. Points ② and ③, for example, are really one and the same node Ⓑ.

Kirchhoff's first law, known as Kirchhoff's Current Law (KCL), states:

The algebraic sum of the currents entering a node is zero at every instant.

In forming the algebraic sum of currents, we must take into account the current reference directions associated with the devices. It the current reference direction is into the node, we assign a positive sign in the algebraic sum to the corresponding current. If the reference direction is away from the node, we assign a negative sign. In applying this convention to the nodes in Figure 2-11, we obtain the following KCL connection equations:

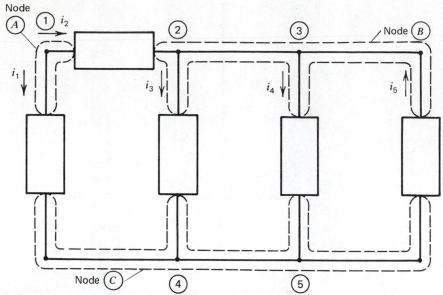

FIGURE 2-11
Node identification for application of Kirchhoff's current law.

Node Ⓐ	$- i_1 - i_2 = 0$
Node Ⓑ	$i_2 - i_3 - i_4 + i_5 = 0$

$$(2\text{-}12)$$

The KCL equation at node Ⓐ does not mean that all of the currents are negative. The minus signs in this equation simply mean that the reference direction for each current is directed away from node Ⓐ. Likewise the equation at node Ⓑ could be written as

$$i_3 + i_4 = i_2 + i_5 \qquad (2\text{-}13)$$

This form illustrates an alternate statement of KCL: The sum of the currents entering a node equals the sum of the currents leaving that node.

There are two signs associated with each current in the application of KCL. The first is the sign given to a current in writing a KCL connection equation. This sign is determined by the orientation of the current reference direction relative to a node. The second sign is determined by the actual direction of current flow relative to the reference direction. The next example illustrates these two signs.

Example 2-6

Suppose in Figure 2-11 we are given $i_1 = -4$ A, $i_3 = +1$ A, $i_4 = +2$ A. Then the first KCL equation in Eq. 2-12 yields

$$- i_1 - i_2 = - (-4) - i_2 = 0$$

The sign outside the parentheses comes from the KCL equation and the sign inside the parentheses comes from the actual direction of current flow. Solving this equation for the unknown current, we obtain

$$i_2 = +4 \text{ A}$$

The plus sign here indicates that the current i_2 actually flows to the right in Figure 2-11, that is, in the direction of its assigned reference direction. Using this current in the second KCL equation of Eq. 2-13, we write

$$i_5 = -i_2 + i_3 + i_4$$
$$= -(+4) + (+1) + (+2) = -1 \text{ A}$$

Again, the sign inside the parentheses is associated with the actual direction of current flow and the sign outside with the KCL connection equation. The minus sign means that the current i_5 actually flows in the direction opposite to its assigned reference direction. Given three currents in this case, we have determined all of the remaining currents in the circuit using only KCL.

The second of Kirchhoff's circuit laws, Kirchhoff's Voltage Law (KVL), states:

The algebraic sum of all of the voltages around a loop is zero at every instant.

A loop is a sequence of devices that forms a closed path. For example, two loops are shown in the circuit of Figure 2-12. In writing the algebraic sum of voltages, we must account for the assigned reference marks. When we traverse a loop, if we go from a "+" to "−" reference mark, we use a positive sign in the sum for the corresponding voltage. If the opposite occurs, then we use a minus sign. Traversing the two loops in Figure 2-12 in the indicated clockwise direction yields the following KVL connection equations:

Historical Note

The fabrication of electric circuits has undergone a revolution in recent times. Prior to 1960 circuit fabrication was dominated by discrete component technology. With this approach, the electric devices are fabricated by one manufacturer, who then distributes and sells them as separate entities. At some later time, the devices are connected to form a specific circuit, and this is usually accomplished by a different process and manufacturer. This approach is known as discrete component technology as the individual devices are identified separately in parts lists and catalogs. With IC technology, the devices are fabricated and interconnected on a single piece of semiconductor material in one continuous process. This is done by a single manufacturer, who then distributes and sells the completed circuit as a separate physical entity. Once the production process for a specific circuit is perfected, the manufacturer can produce thousands of identical duplicates of the circuit. The size, cost, and reliability advantages inherent in the process are so overwhelming that IC technology has become the dominant method of circuit fabrication. But regardless of the method of fabrication, the completed circuit is governed by the laws of Gustav Robert Kirchhoff (1824–1887). Kirchhoff's work closely followed the pioneering discoveries of Ampere and Ohm. He first published his results as an appendix to a paper in 1845. Kirchhoff was a 21-year-old student at the time.

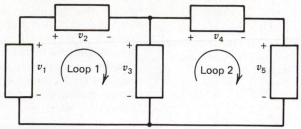

FIGURE 2-12
Circuit diagram showing two loops and five device voltages.

Loop 1 $\qquad\qquad -v_1 + v_2 + v_3 = 0$

Loop 2 $\qquad\qquad -v_3 + v_4 + v_5 = 0$ $\qquad$ **(2-14)**

There are two signs associated with each voltage. The first is the sign determined by the actual polarity of a voltage relative to its assigned reference polarity. The second is the sign given the voltage in a KVL connection equation. This second sign is determined by the reference polarity relative to the direction of traversing the loop. The following example illustrates these concepts.

Example 2-7

Suppose in Figure 2-12 we are given $v_1 = 5$ V, $v_2 = -3$ V, and $v_4 = 10$ V. The loop 1 KVL equation reads:

$$-v_1 + v_2 + v_3 = -(+5) + (-3) + (v_3) = 0$$

The sign outside the parentheses comes from the KVL equation and the sign inside from the actual polarity

of the voltage. This equation yields $v_3 = +8$ V. Using this value in the second KVL equation,

$$-v_3 + v_4 + v_5 = -(+8) + (+10) + (v_5) = 0$$

yields $v_5 = -2$ V. The minus sign here means that the actual polarity of v_5 is the opposite of its assigned reference polarity.

In our study of connection constraints, two unique connections occur so frequently in circuit analysis that they deserve special attention. Consider the circuit of Figure 2-13, in which the two elements 1 and 2 are

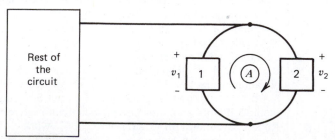

FIGURE 2-13
A parallel connection.

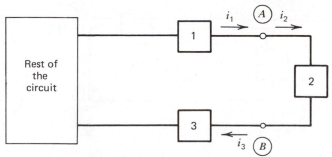

FIGURE 2-14
A series connection.

connected between two common nodes. As a consequence of this connection, KVL around the loop between the two elements yields the fact that the voltages across each of the two elements are equal—that is: KVL around loop Ⓐ

$$- v_1 + v_2 = 0 \qquad \text{or} \qquad v_1 = v_2$$

As a result of being connected between two common nodes, the two elements are said to be connected in **parallel.** This connection is not restricted to two elements; any number of elements connected between two common nodes are in parallel and, as a result, all the voltages across them are equal.

The circuit in Figure 2-14 shows a **series** connection. Two elements are said to be connected in series if they share one common node to which no other element is connected. In Figure 2-14 elements ☐1 and ☐2 are connected in series since only these two elements are connected at node Ⓐ. A consequence of the series connection is that the current through each element must be equal. Applying KCL at node Ⓐ yields

$$i_1 - i_2 = 0 \qquad \text{or} \qquad i_1 = i_2 \qquad\qquad \textbf{(2-15)}$$

Any number of elements may be connected in series, for example, element ☐3 in Figure 2-14 is connected in series with element ☐2 at node Ⓑ. As a result, $i_2 = i_3$ by KCL. Hence in this circuit, $i_1 = i_2 = i_3$ and we say that elements ☐1, ☐2, and ☐3 are connected in series and the same current flows through all three.

Example 2-8

For all of the circuits in Figure 2-15, identify the elements connected in parallel and in series.
In circuit $C1$ elements ☐1 and ☐2 are connected in series at node Ⓐ, while elements ☐3 and ☐4 are connected in parallel between nodes Ⓑ and Ⓒ. In circuit $C2$ elements ☐1 and ☐2 are connected in series at node Ⓐ,

as are elements ☐4 and ☐5 at node Ⓓ. There are no elements connected in parallel in circuit $C2$. In circuit $C3$ there are no elements connected in either series or parallel. It is important to realize that elements need not be connected in series or parallel, but may be connected in neither.

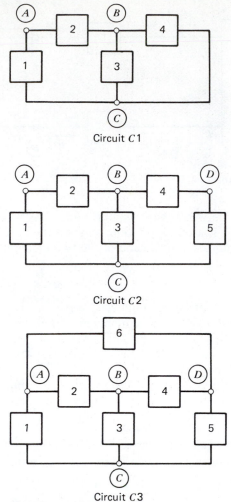

FIGURE 2-15
Circuits for understanding series and parallel connections.

2-4 COMBINED CONSTRAINTS

The usual goal of circuit analysis is to determine the currents or voltages at various places in a circuit. This analysis is based on sets of equations or constraints of two distinctly different types. The element constraints are based on the models of the specific devices connected in the circuit. The connection constraints are based on Kirchhoff's laws and the circuit connections. The element equations are independent of the circuit in

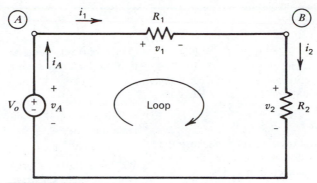

FIGURE 2-16
A circuit used to demonstrate Kirchhoff's voltage law.

which the device is connected. Likewise, the connection equations are independent of the specific devices that may be in the circuit. But taken together, the element and connection equations provide the data needed to analyze a circuit.

The circuit in Figure 2-16 can be used to illustrate the formulation of these equations. The element equations for this circuit are

$$v_A = V_O \qquad v_1 = R_1 i_1 \qquad v_2 = R_2 i_2 \qquad \textbf{(2-16)}$$

These equations describe the three devices in the circuit. They do not depend on how the devices are connected in the circuit. The connection equations are obtained from Kirchhoff's laws:

Node Ⓐ $\qquad\qquad\qquad\qquad i_A - i_1 = 0$

Node Ⓑ $\qquad\qquad\qquad\qquad i_1 - i_2 = 0$ $\qquad\qquad\textbf{(2-17)}$

Loop $\qquad\qquad\qquad -v_A + v_1 + v_2 = 0$

These equations are independent of the specific devices in the circuit. They depend only on the circuit connections.

In this example, there are six unknowns: three device currents and three device voltages. Taken together, the element and connection equations give us six equations. In general, for a network with N nodes and E two-terminal elements, we must write $N - 1$ KCL connection equations and $E - N + 1$ KVL connection equations, and we can write E element equations. The total number of equations thus generated is

KCL	$N - 1$
KVL	$E - N + 1$
Element	E
Total	$2E$

The grand total is then $2E$ combined connection and element equations, which is exactly the number of equations needed to solve for the voltage across and current through every element—a total of $2E$ unknowns.

Example 2-9

Suppose that we are given $V_o = 10$ V, $R_1 = 2000$ Ω, and $R_2 = 3000$ Ω for the circuit in Figure 2-16. Let us solve for the device currents and voltages. Substituting the element equations into the KVL connection equation produces

$$- V_o + R_1 i_1 + R_2 i_2 = 0$$

This equation can be used to solve for i_1 because the second KCL connection equation requires that $i_2 = i_1$, since R_1 and R_2 are connected in series.

$$i_1 = \frac{V_o}{R_1 + R_2} = \frac{10}{2000 + 3000} = 2 \text{ mA}$$

By finding this current, we have determined every device current, because the KCL connection equations collectively require that

$$i_A = i_1 = i_2$$

since all three elements are connected in series. Substituting all of the known values into the element equations gives

$$v_A = 10 \text{ V} \quad v_1 = R_1 i_1 = 4 \text{ V} \quad v_2 = R_2 i_2 = 6 \text{ V}$$

We have found every device voltage and current. Note the analysis strategy used here. We first determined all of the device currents and then found the voltages from these values using the element constraints.

Example 2-10

This example deals with the analysis of the source-resistor-switch circuit in Figure 2-17. Suppose that $V_o = 10$ V and $R_1 = 2000$ Ω. The connection equations for the circuit are

Node Ⓐ $i_A - i_1 = 0$

Node Ⓑ $i_1 - i_2 = 0$

Loop $- v_A + v_1 + v_2 = 0$

These connection equations are identical to those in Eqs. 2-17 for the circuit in Figure 2-16. The two circuits have different devices but the same connections. When the switch is open, $i_2 = 0$. The KCL connection equations require that all device currents in this circuit be equal; hence $i_1 = 0$ and $v_1 = R_1 i_1 = 0$. The KVL connection equation then yields $v_2 = v_A = 10$ V. Thus, when the switch is open, no current flows and all of the source voltage appears across the open switch. When the switch is closed, $v_2 = 0$ and the KVL connection equation gives

$$v_1 = V_o = 10 \text{ V}$$

Hence

$$i_1 = v_1/R_1 = 10/2000 = 5 \text{ mA}$$

Thus, when the switch is closed, all of the source voltage appears across the resistor that determines the current through the switch.

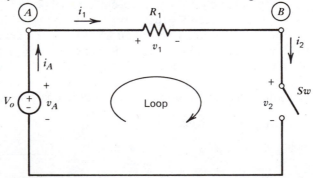

FIGURE 2-17
A switched series circuit.

2-5   EQUIVALENT CIRCUITS

Very often in the analysis of circuits, it is to the engineer's benefit if he or she can replace a portion of the circuit with a simpler equivalent one. The underlying basis for two circuits to be equivalent is contained in their *i-v* relation, as shown in the following definition:

Two circuits are said to be equivalent if they have identical i-v *characteristics at a specified pair of terminals.*

Consider the circuit shown in Figure 2-18*a*. Suppose that between a pair of terminals ⒶＡ and ⒷＢ there are two resistors connected in series. Our objective here is to try to simplify the circuit without altering the electrical behavior of the rest of the circuit.

A KVL equation around the loop from ⒶＡ to ⒷＢ is

$$v = v_1 + v_2 \qquad\qquad (2\text{-}18)$$

Since the two resistors are connected in series, the same current i flows through both. Hence by applying Ohm's law we get $v_1 = iR_1$ and $v_2 =$

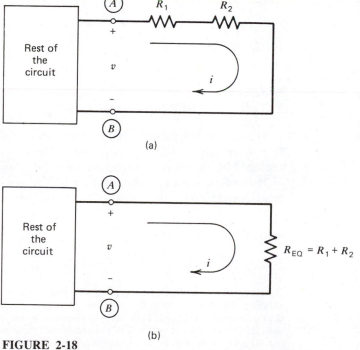

(a)

(b)

FIGURE 2-18
A series resistance circuit and its equivalent. (*a*) Original circuit. (*b*) Equivalent circuit.

iR_2. Substituting these relations into Eq. 2-18 and then simplifying yields

$$v = R_1 i + R_2 i = i(R_1 + R_2) \tag{2-19}$$

Thus finally we can write

$$v = iR_{EQ}$$

where

$$R_{EQ} = R_1 + R_2 \tag{2-20}$$

This result is significant. The response of the rest of the circuit would be unchanged if the two resistors R_1 and R_2 were replaced by a single equivalent resistor R_{EQ}, as shown in Figure 2-18b.

The dual[1] situation occurs in the parallel resistance circuit shown in Figure 2-19a. Suppose that between the two terminals Ⓐ and Ⓑ a pair of conductances is connected in parallel. Let's simplify our circuit in a similar fashion without altering its behavior.

A KCL equation at node Ⓐ yields

$$i = i_1 + i_2 \tag{2-21}$$

Since the conductances are connected in parallel, the same voltage v appears across both of them. Hence in applying Ohm's law we get $i_1 = vG_1$ and $i_2 = vG_2$. Substituting these relations into Eq. 2-21 and then simplifying yields

$$i = vG_1 + vG_2 = v(G_1 + G_2) \tag{2-22}$$

Finally we can write

$$i = vG_{EQ}$$

where

$$G_{EQ} = G_1 + G_2 \tag{2-23}$$

Hence the response of the rest of the circuit would be unchanged if the two parallel conductances connected between terminals Ⓐ and Ⓑ were replaced by an equivalent conductance G_{EQ}, as shown in Figure 2-19b.

It is often useful to rewrite Eq. 2-23 in terms of resistance R, since resistors are rarely described in terms of their conductance. That is,

$$\frac{1}{R_{EQ}} = \frac{1}{R_1} + \frac{1}{R_2} \tag{2-24}$$

$$R_{EQ} = \frac{R_1 R_2}{R_1 + R_2}$$

[1]In circuit analysis often two seemingly different circuits will have identical results upon solution, with the exception that the roles of voltage and current and resistance and conductance are interchanged. This repeated similarity is not coincidental but is only a part of the identical patterns of behavior of parameters in circuit analysis. This similarity is called the principle of **duality.** In later chapters we will see this principle exhibited in other parameters. For the experienced circuit analyst this principle helps in verifying results obtained and assists in quickly identifying errors in analysis.

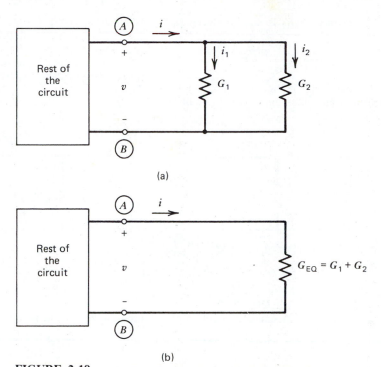

FIGURE 2-19
A parallel resistance circuit and its equivalent. (*a*) Original circuit. (*b*) Equivalent circuit.

A word of caution: The "product over sum" form in Eq. 2-24 only works for *two* resistors connected in parallel. If a circuit contains more than two resistors in parallel, we use the general result ($G_{EQ} = G_1 + G_2 + G_3 + \cdots$) implied by Eq. 2-23 to obtain the equivalent resistance.

Example 2-11

Find the equivalent resistance connected between terminals Ⓐ–Ⓑ, R_{EQ1}, and the equivalent resistance connected between terminals Ⓒ–Ⓓ, R_{EQ2}, for the circuit of Figure 2-20*a*.

We begin by noting that resistors R_2 and R_3 are in parallel since they share two common nodes. Applying Eq. 2-24, we obtain

$$R_2 \| R_3 = \frac{R_2 R_3}{R_2 + R_3}$$

As an interim step in simplifying our circuit, we can redraw it as shown in Figure 2-20*b*.

Up to this point in our analysis it did not matter whether we were finding R_{EQ1} or R_{EQ2}. To find the equivalent resistance between terminals Ⓐ and Ⓑ, we note that R_1 and the equivalent resistance we just found, $R_2 \| R_3$, are in series. Thus the total equivalent resistance R_{EQ1} between terminals Ⓐ and Ⓑ is

$$R_{EQ1} = R_1 + R_2 \| R_3$$

$$R_{EQ1} = R_1 + \frac{R_2 R_3}{R_2 + R_3}$$

Looking into terminals Ⓒ–Ⓓ, however, yields a different result. R_1 is *not* involved since there is an open circuit (an infinite resistance) between terminals Ⓐ

and $\circledB$. Therefore only $R_2\|R_3$ is seen between terminals $\circledC$ and $\circledD$.

$$R_{EQ2} = R_2\|R_3 = \frac{R_2 R_3}{R_2 + R_3}$$

From this simple example it should be clear that the equivalent circuit found can vary with the pair of terminals selected. The student should be able to show that the equivalent circuit for this circuit between terminals $\circledA$ and $\circledC$ is R_1, between $\circledB$ and $\circledD$ is zero, between $\circledA$ and $\circledD$ is the same as between $\circledA$ and $\circledB$, and between $\circledB$ and $\circledC$ is the same as between $\circledC$ and $\circledD$.

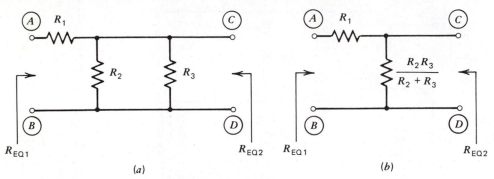

(a) (b)

FIGURE 2-20
Demonstration circuit for computing equivalent resistance.

Up to now we have considered sources as the ideal elements shown in Figure 2-8. Real sources rarely can be modeled in such an ideal manner. If we look back at Figure 2-9, we see a more realistic representation of a voltage source—an ideal voltage source in series with a resistor. Similarly, a more realistic representation of a current source is an ideal current source in parallel with a resistor.

In our study of equivalent circuits, suppose that either of the two nonideal sources can be connected between two terminals $\circledA$ and $\circledB$ as shown in Figure 2-21. A simultaneous analysis of both of these circuits will yield some interesting results.

Let's start by applying Kirchhoff's laws.

Circuit A

KVL
$V_S = v_R + v$

Circuit B

KCL
$I_S = i_R + i$

Applying Ohm's law:

Circuit A

$v_R = R_1 i$

Circuit B

$i_R = v/R_2$

By combining these results we see that as far as the rest of the circuit is concerned, the i-v characteristic equations between terminals $\circledA$ and $\circledB$ are

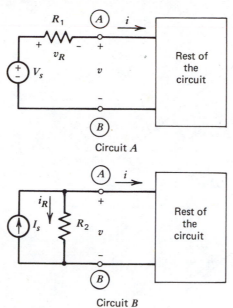

FIGURE 2-21
Source transformation equivalent circuits.

Circuit A

$$i = -\frac{v}{R_1} + \frac{V_s}{R_1}$$

Circuit B

$$i = -\frac{v}{R_2} + I_s$$

If we graph these two characteristic equations, we get the results shown in Figure 2-22. Clearly both circuits will have the same i-v characteristics (shown both graphically and in equation form) if

$$R_1 = R_2 = R$$

and

(2-25)

$$V_s = I_s R_2 = I_s R$$

What this analysis implies is that as far as the rest of the circuit is concerned, a real-world voltage source (ideal source V_s in series with resistor R) connected to terminals Ⓐ and Ⓑ can be replaced with an equivalent real-world current source (ideal source I_s in parallel with resistor R_2), provided that the conditions of Eq. 2-25 are met.

A word of caution: Equivalence does not mean identically equal. As far as the rest of the circuit is concerned, any electrical measurement made at terminals Ⓐ and Ⓑ produces the same results regardless of which source is connected. However, both circuits are not identical. Assume that the two circuits are connected to an open circuit. Circuit A has no current flowing while circuit B does, and is dissipating energy.

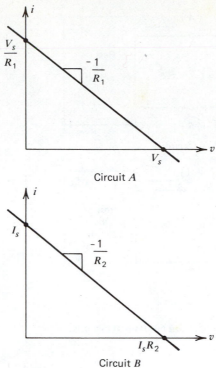

Circuit A

Circuit B

FIGURE 2-22
Source *i-v* characteristics.

Example 2-12

Convert the voltage source in Figure 2-23 into an equivalent current source. From Eq. 2-25 we have

$$R_1 = R_2 = R = 10 \ \Omega$$

and

$$I_S = V_S/R = 5 \text{ A}$$

Hence the equivalent current source circuit can be drawn as shown in Figure 2-23.

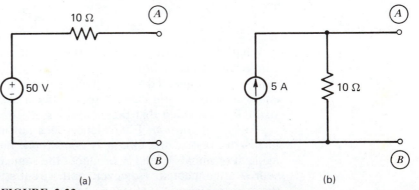

(a)

(b)

FIGURE 2-23
Example of source transformation. (*a*) Given source. (*b*) Equivalent source.

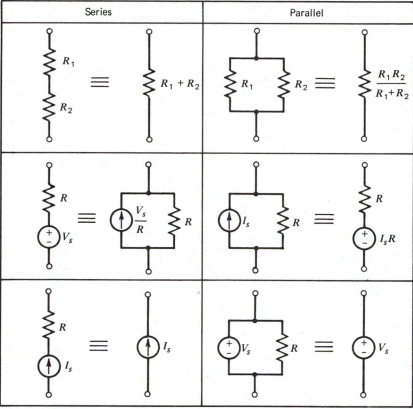

FIGURE 2-24
Some two-terminal equivalent circuits.

In Chapter 3 we shall see that entire circuits can be represented by the source models developed here. The equivalency of the two circuits shown here will work with those application as well.

Figure 2-24 summarizes some of the more common equivalent circuits we will encounter in linear circuit analysis.

2-6   VOLTAGE AND CURRENT DIVISION

No discussion of series and parallel circuits can be complete without the inclusion of voltage and current division. These two tools find wide application in the analysis of circuits.

Suppose we have a circuit consisting of a voltage source in series with three resistors as shown in Figure 2-25. Suppose that the voltage across R_2 is desired. The following analysis will yield the correct result, and a very useful analysis tool.

The application of KVL around the loop in Figure 2-25 yields

$$v_S = v_1 + v_2 + v_3$$

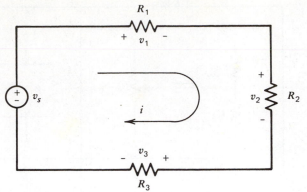

FIGURE 2-25
A voltage divider circuit.

Using Ohm's law and the fact that the same current i flows through all three resistors since they are connected in series, we find that

$$v_S = R_1 i + R_2 i + R_3 i$$

Solving for i,

$$i = \frac{v_S}{R_1 + R_2 + R_3}$$

Now to find v_2, we again use Ohm's law,

$$v_2 = iR_2$$

so that substituting for i, we obtain our result:

$$v_2 = \frac{v_S R_2}{R_1 + R_2 + R_3} \tag{2-26}$$

Finding the result for v_1 and v_3 is similar and we obtain

$$v_1 = \frac{v_S R_1}{R_1 + R_2 + R_3} \tag{2-27}$$

$$v_3 = \frac{v_S R_3}{R_1 + R_2 + R_3} \tag{2-28}$$

In looking over our results we observe an interesting phenomenon. In each case the voltage across the desired resistor is equal to the value of its resistance divided by the total resistance in the circuit times the total voltage across the series circuit. In other words, each resistor extracts its fraction of voltage so that the sum of the voltages across all the resistors in the loop equals the total voltage applied:

$$v_S = v_1 + v_2 + v_3$$

$$v_S = \frac{v_S R_1}{R_1 + R_2 + R_3} + \frac{v_S R_2}{R_1 + R_2 + R_3} + \frac{v_S R_3}{R_1 + R_2 + R_3}$$

$$v_S = v_S \frac{(R_1 + R_2 + R_3)}{R_1 + R_2 + R_3} = v_S \tag{2-29}$$

Once we recognize this phenomenon, we need not repeat the analysis! Whenever we have a number of resistors in series with a voltage source, we can immediately use the voltage division rule to find the desired voltages. Two examples will help clarify this concept.

Example 2-13

Find the voltage indicated in the circuit of Figure 2-26.

Using the voltage division rule, we see that

$$v_o = \frac{(35)30}{10 + 20 + 30 + 10} = 15 \text{ V}$$

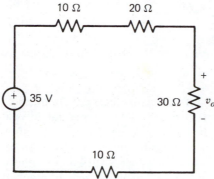

FIGURE 2-26
Example of a voltage divider.

Example 2-14

Use the voltage division rule to find the output voltage (V_o) of the circuit in Figure 2-27.

At first glance it appears that the voltage division rule does not apply since the resistors are not connected in series. However, the output of the circuit is an open circuit. Therefore the current through R_3 is zero, and consequently the voltage across it is also zero. The same current must flow through R_1 and R_2, and the voltage V_o also appears across R_2. In essence it is as if R_1 and R_2 were connected in series. Hence by voltage division,

$$V_O = \frac{V_S R_2}{R_1 + R_2}$$

The reader should carefully review the logic leading to this result since voltage division applications of this type occur frequently, particularly in the determination of what are called Thévenin equivalent circuits (Chapter 3).

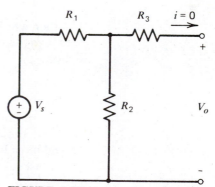

FIGURE 2-27
A "special" application of a voltage divider.

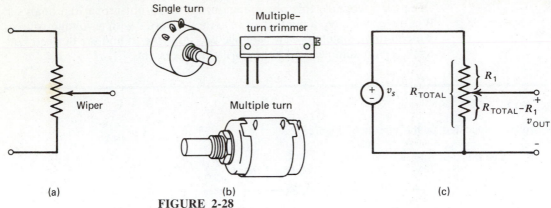

FIGURE 2-28
The potentiometer. (*a*) **Circuit symbol.** (*b*) **Actual potentiometers.** (*c*) **Application.**

An important device in electrical engineering is the **potentiometer.** This device makes use of voltage (**potential**) division to **meter** out a desired fraction of the input voltage. Figure 2-28 shows the circuit symbol of a potentiometer, a typical application, and sketches of actual potentiometers.

The voltage v_{OUT} can be varied by simply turning the knob on the potentiometer to move the wiper arm contact. By using the voltage divider rule, v_{OUT} can be found as

$$v_{OUT} = v_s \frac{R_{TOTAL} - R_1}{R_{TOTAL}} \qquad (2\text{-}30)$$

When we make R_1 zero by moving the wiper all the way to the top, we get

$$v_{OUT} = v_s \frac{R_{TOTAL} - 0}{R_{TOTAL}} = v_s \qquad (2\text{-}31)$$

In other words, the full voltage is available to the rest of the circuit. If we move the wiper all the way to the bottom, we make R_1 equal to R_{TOTAL}, hence

$$v_{OUT} = v_s \frac{R_{TOTAL} - R_{TOTAL}}{R_{TOTAL}} = 0 \qquad (2\text{-}32)$$

This is the other extreme. We can vary the output voltage all the way from v_S, the applied voltage, to zero. Halfway between, we intuitively should have half of the applied voltage; in checking we get, as expected,

$$v_{OUT} = v_s \frac{R_{TOTAL} - \frac{1}{2} R_{TOTAL}}{R_{TOTAL}} = \frac{v_s}{2} \qquad (2\text{-}33)$$

Applications for potentiometers are almost endless—control of volume, control of voltage output, and fine adjustments for alignment of circuits are just a few.

Example 2-15

A certain device requires 9 V to operate. It has an equivalent input resistance of 50 Ω. You have a 12-V source available that has a 10-Ω internal resistance. Design a suitable interface that will permit the 9-V device to operate properly. Figure 2-29 illustrates the task.

The problem is to provide exactly 9 V to the device. One way to solve the problem is to use voltage division. Consider the interface to be a series resistor R_1 as shown in Figure 2-29b. We can write a suitable voltage division equation as

$$9 \text{ V} = \frac{50 \ (12 \text{ V})}{50 + R_1 + 10}$$

which we can solve for R_1, as

$$R_1 = \frac{600 - 540}{9} = 6.67 \ \Omega$$

It should be noted that this solution is not unique. It is left up to the student to show that the interface shown in Figure 2-29c will also work.

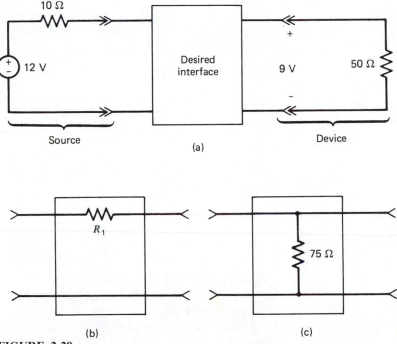

FIGURE 2-29
(*a*) An example of interface design. (*b*) A solution. (*c*) An alternate solution.

The dual of voltage division is current division. Since it is the dual, we deal with conductances and current sources connected in parallel.

For example, Figure 2-30 contains a typical circuit that lends itself to solution by current division. To analyze this circuit, we begin by applying KCL at node Ⓐ:

$$i_S = i_1 + i_2 + i_3$$

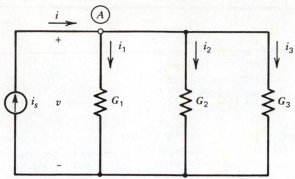

FIGURE 2-30
A current divider.

Using Ohm's law, and the fact that the same voltage v appears across all three conductances since they are connected in parallel,

$$i_S = vG_1 + vG_2 + vG_3$$

Solving for v, we find

$$v = \frac{i_S}{G_1 + G_2 + G_3}$$

Once we find v, we can find any current using Ohm's law:

$$i_1 = vG_1 = \frac{G_1 i_S}{G_1 + G_2 + G_3} \tag{2-34}$$

$$i_2 = vG_2 = \frac{G_2 i_S}{G_1 + G_2 + G_3} \tag{2-35}$$

$$i_3 = vG_3 = \frac{G_3 i_S}{G_1 + G_2 + G_3} \tag{2-36}$$

We see that the total current delivered to the parallel network is *divided* among the conductances G_1, G_2, and G_3 in proportion to their value compared with the total of the conductances in parallel.

Since we do not always work with conductance, it is useful to solve the current division rule for two resistors in parallel. Consider the circuit of Figure 2-31.

To find i_1, we use the current division rule as

$$i_1 = \frac{i_S G_1}{G_1 + G_2}$$

$$i_1 = \frac{i_S \, 1/R_1}{1/R_1 + 1/R_2}$$

$$i_1 = \frac{i_S \, 1/R_1}{(R_2 + R_1)/R_1 R_2} = \frac{i_S R_2}{R_1 + R_2} \tag{2-37}$$

Solving for i_2, we get

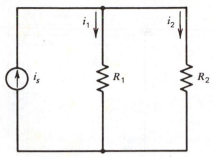

FIGURE 2-31
A two-element current divider.

$$i_2 = \frac{i_s G_2}{G_1 + G_2}$$

$$i_2 = \frac{i_s R_1}{R_1 + R_2} \qquad \textbf{(2-38)}$$

Looking at our results, we see that when we can reduce our parallel circuit to just two resistors in parallel, the desired current is equal to the resistance in the *nondesired* current path times the total current divided by the sum of the resistance in both paths. *Caution:* this technique will work only if we reduce the circuit to two paths, a desired path and a second path consisting of an equivalent resistance of all other paths.

Example 2-16

Find the currents i_x and i_o in Figure 2-32a. To find i_x, we reduce the circuit to two paths, the desired path and a path equivalent to all other paths as shown in Figure 2-32b.

Now we can use our rule:

$$i_x = \frac{6.67}{20 + 6.67} \times 5 = 1.25 \text{ A}$$

Similarly, to find i_o, we reduce the circuits to the two paths shown in Figure 2-32c.

$$i_o = \frac{5 + 5}{5 + 5 + 10} \times 5 = 2.5 \text{ A}$$

Although not asked for, the middle path has 1.25 A flowing through it. Adding all our currents, we find that they equal 5 A, the source current, as they should!

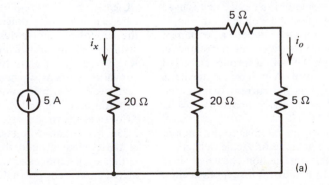

(a)

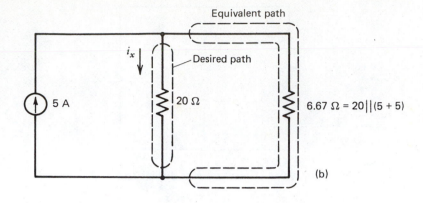

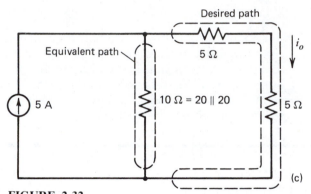

FIGURE 2-32
(*a*) A current divider problem. (*b*) Solution for i_x. (*c*) Solution for i_o.

Example 2-17

The D'Arsonval galvanometer is a device that is used to measure direct currents and voltages. In simple terms, a coil of wire is mounted between a permanent magnet in such a manner that it is free to rotate. As a current flows through the coil, a torque is created that causes the coil to turn. A pointer or needle is attached to the coil and as the current is increased the pointer deflection is linearly proportional to the current. See Figure 2-33*a*.

D'Arsonval movements are rated as to the amount of current necessary to achieve full-scale deflection. Depending on the actual construction, the usual range of current necessary to achieve full-scale deflection is 1 μA to 1 mA.

Clearly engineers require instruments that can measure a much wider range of values. To get around this limitation, meter designers employ a shunt resistance R_S, as shown in Figure 2-33*b*. Suppose it was desired to measure 10 A full scale, and the D'Arsonval movement we had available provided a full-scale reading with 10 μA. What shunt resistance will be required if the coil resistance is 20 Ω?

Consider the model of an ammeter shown in Figure 2-33*b*. Essentially the task at hand is to shunt most of the current around the actual meter movement, letting only (and exactly) 10 μA flow through the meter movement when 10 A is flowing through the newly designed meter. In Figure 2-33*b*, the actual current desired to be measured is indicated by i_M, the resistance of the meter movement R_M, and the full-scale current of the movement I_{FS}. R_S is the shunt resistance we must design for. Once we show the model, it should become clear that the problem can be solved using current division, that is,

$$I_{FS} = \frac{R_S I_M}{R_M + R_S}$$

$$10^{-5} = \frac{R_S (10)}{20 + R_S}$$

This can be solved for R_S as

$$R_S = 20 \ \mu\Omega$$

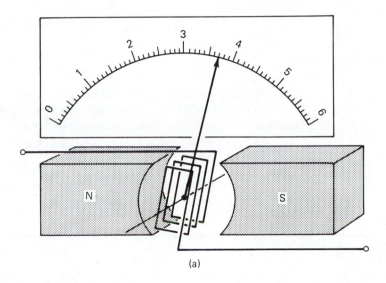

(a)

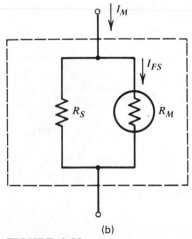

(b)

FIGURE 2-33
Simple representation of a D'Arsonval meter movement and model. (a) Representation of a D'Arsonval galvanometer. (b) Model of an ammeter.

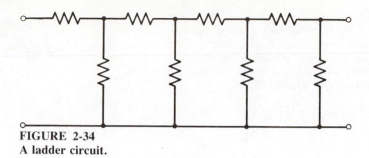

FIGURE 2-34
A ladder circuit.

2-7 CIRCUIT REDUCTION

The concepts of series/parallel equivalence, voltage/current division, and source transformations can be used to analyze ladder circuits of the type shown in Figure 2-34. The basic analysis strategy is to reduce the circuit to some simpler equivalent in which the desired output is easily found by voltage or current division, or even perhaps by Ohm's law. There is no fixed pattern to the reduction process, and much depends on the insight of the analyst. But in any case, with circuit reduction we work directly with the circuit model, and so the process gives us insight into circuit behavior. The next three examples serve to introduce circuit reduction.

Example 2-18

Let us determine the output voltage v_F and the input current i_S of the ladder circuit in Figure 2-35a. Figures 2-35b and 2-35c show the steps required to determine the equivalent resistance between the terminals Ⓐ and Ⓑ. We first determine the parallel equivalent resistance of the $2R$ and R resistors as

$$R_{EQ1} = \frac{(R)\,(2R)}{R + 2R} = 2R/3$$

This equivalent resistance is then combined with the series resistor R to obtain

$$R_{EQ2} = R + R_{EQ1} = R + 2R/3 = 5R/3$$

By alternate application of series/parallel equivalence, we have reduced the ladder circuit to a single equivalent resistor. To determine the input current, we use the equivalent input resistance and Ohm's law to write

$$i_S = \frac{v_S}{R_{EQ2}}$$
$$= \frac{3}{5}\frac{v_S}{R}$$

To determine the output voltage in terms of the input voltage, we apply voltage division to the series equivalent circuit in Figure 2-35b as

$$v_F = \frac{2R/3}{5R/3}\, v_S = \frac{2}{5}\, v_S$$

The reduction to $2R/3$ maintained v_F since the $2R$ and R resistors were in parallel and v_F is the common voltage across them. Further reduction, as in Figure 2-35c, causes v_F to become lost in the new equivalent resistor $5R/3$. However, i_S is not affected by the last reduction. Note the pattern of analysis in this ladder network. We reduce the ladder to a single equivalent resistance at the input and then work our way back to the output using voltage division.

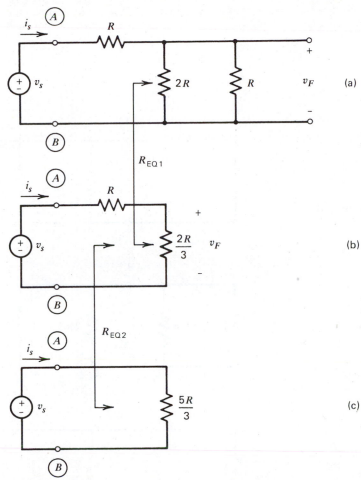

FIGURE 2-35
Analysis of a ladder circuit using equivalent resistance.

Example 2-19

Figure 2-36 shows an alternative reduction of the circuit used in Example 2-18. If we break the circuit at points Ⓐ and Ⓑ of Figure 2-36, we have a voltage source v_S in series with a resistor R. By using source transformation we can replace these two elements by a current source in parallel with the same resistor. We now have a current source v_S/R in parallel with a resistor R. We connect this equivalent circuit back to the remaining circuit at points Ⓐ and Ⓑ. The circuit now consists of three resistors in parallel that can readily be combined.

$$
\begin{aligned}
G_{EQ} &= G_1 + G_2 + G_3 \\
&= \frac{1}{R} + \frac{1}{2R} + \frac{1}{R} \\
&= \frac{5}{2R}
\end{aligned}
$$

Since all the current from our equivalent source must

go through this equivalent resistor, the output voltage is readily found using Ohm's law.

$$v_F = i_s R_{EQ}$$

$$= \frac{v_s}{R} \times \frac{2R}{S}$$

$$= \frac{2}{5} v_s$$

This is, of course, the same result as obtained in Example 2-18, but by a different sequence of circuit reductions.

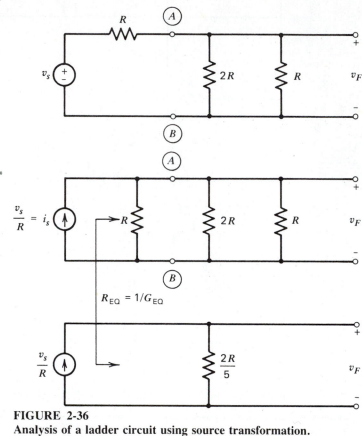

FIGURE 2-36
Analysis of a ladder circuit using source transformation.

Example 2-20

In the circuit of Figure 2-37a find v_x.

In the two previous examples the desired parameter was at one or the other extremity of the circuit. In this example the desired unknown is somewhere in the center of the network. We approach the solution by reducing the circuit from both ends toward the desired element. A source transformation at terminals Ⓐ and Ⓑ and series reduction of the two 10-

Ω resistors between terminals Ⓒ and Ⓓ yields the reduced circuit shown in Figure 2-37b. In this circuit the two pairs of 20-Ω resistors connected in parallel can be combined to produce the circuit in Figure 2-37c.

Several equally easy reduction techniques can now be applied. For example, we can do another source transformation to obtain the circuit in Figure 2-37d, and then use voltage division to find v_x. Alternately we can use current division in Figure 2-37c to find i_x and then use Ohm's law to find v_x. Both methods are shown in the following.

By voltage division in Figure 2-37d:

$$v_x = \frac{10}{10 + 10 + 10} \times 7.5 = 2.5 \text{ V}$$

By current division in Figure 2-37c:

$$i_x = \frac{10}{10 + 10 + 10} \times \frac{3}{4} = \frac{1}{4} \text{ A}$$

Hence by Ohm's law:

$$v_x = 10 \times \frac{1}{4} = 2.5 \text{ V}$$

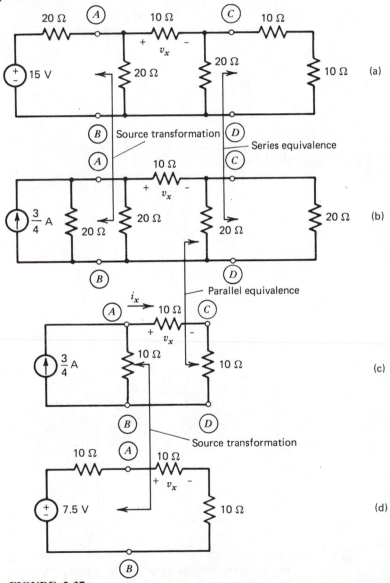

FIGURE 2-37
Analysis of a ladder circuit using various equivalent circuit techniques.

2-8   GROUND—THE REFERENCE NODE

Voltage is an across variable that is defined and measured between two nodes. It is often convenient to identify one of the nodes as the reference node, commonly called ground, and to measure (and define) the voltage at all other nodes with respect to this reference node. This concept should not be new to us. For example, the concept of elevation is similar. If one asks for the elevation of a particular mountain, one usually expects to obtain the number of feet or meters between the top of the mountain and

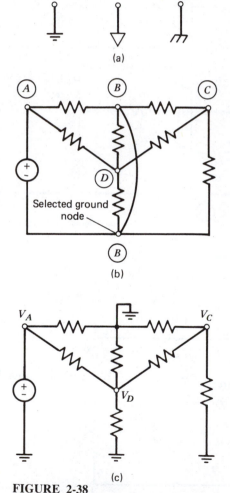

FIGURE 2-38
Ground symbols and their use in schematic diagrams. (a) Ground symbols. (b) Given schematic. (c) Redrawn schematic using ground symbol.

a reference—usually mean sea level. In electric circuits which node we select as reference is not always as obvious as mean sea level is in measuring elevations. Hence we denote the selected reference node as ground by using one of the "ground" symbols shown in Figure 2-38a. The term **node voltage** can then be used to define the voltage of any node in the circuit with respect to the selected reference node. Consider the circuit of Figure 2-38b. There are four nodes identified. If we select node Ⓑ as our reference node, then all the other nodes can have voltages defined with respect to node Ⓑ —our "sea level" for this problem. For example, the node voltage V_A shown in Figure 2-38c means the voltage of node Ⓐ with respect to the selected reference. If we had selected node Ⓐ as our reference, then we would say that the voltage at V_A was zero. It should be noted that, in general, voltages measured with respect to the reference can be positive (greater than the reference) or negative (less than the reference)—just as Death Valley has a minus elevation indicating that it is below mean sea level.

In Chapter 5 we use node voltages extensively and discuss the selection of a reference node. Once a reference node has been chosen, it can be used to advantage to reduce complex circuit drawings, called schematics, into more palatable configurations. To do this, every element connected to the reference node or ground is shown connected to the ground symbol. All elements connected to the ground symbol are assumed to be connected together at the end tied to that symbol, even though they may be located far apart on the schematic. The result is a simplified drawing with fewer interconnecting wires shown. The two circuits in Figure 2-38 are identical, except that one diagram uses the ground symbol.

SUMMARY

- Electrical quantities that are variables in the study of circuits are as follows:

	Quantity	Symbol	Unit
Basic variables {	Charge	q	Coulomb (C)
	Energy	w	Joule (J)
Signal variables {	Current	i	Ampere (A)
	Voltage	v	Volt (V)
	Power	p	Watt (W)

- Charge and energy are basic electrical quantities. Current, voltage, and power are derived **signal** variables defined as

$$i = \frac{dq}{dt} \qquad v = \frac{dw}{dq} \qquad p = \frac{dw}{dt}$$

- Current is measured as a *through* variable and voltage as an *across* variable. Power is normally determined from current and voltage by the relationship $p = vi$.

- An electric **device** is a real physical thing. A circuit **element** is a mathematical or graphical model that approximates the major features of a device.

- The associated reference mark convention for a two-terminal device is as shown in Figure 2-39. If the actual current and voltage have the same sign, then the associated power ($p = vi$) is positive and the device absorbs power. If the actual current and voltage have opposite signs, then the associated power is negative and the device delivers power.

- Two-terminal circuit elements are represented by a circuit symbol and are characterized by a single constraint on the associated current and/or voltage. The i-v constraints for the elements introduced in this chapter are reviewed in Figure 2-40.

- An electric circuit is a collection of devices interconnected at their terminals. The interconnections form nodes (points at which two or more devices are connected together) and loops (sequences of devices that form a closed path).

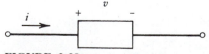

FIGURE 2-39
Reference convention.

Element	Circuit symbol	$i-v$ constraints
Resistor	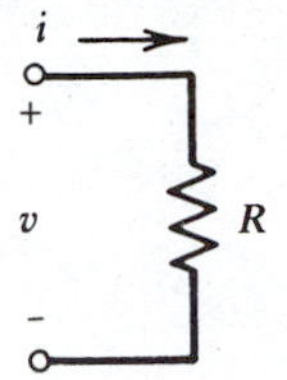	$v = Ri$ or $i = Gv$ where $G = 1/R$
Current source	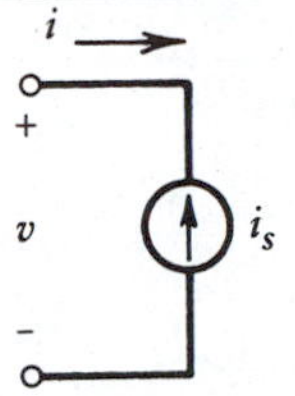	$i = -i_s$ v = depends
Voltage source	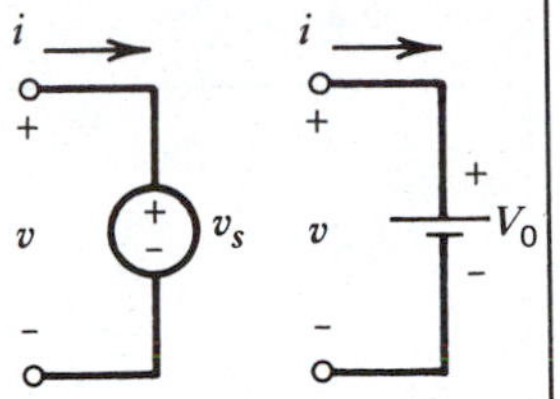	$v = v_s$ i = depends

FIGURE 2-40
Two-terminal elements.

- Device interconnections in a circuit produce two connection constraints.

Kirchhoff's Current Law (KCL)
The algebraic sum of currents at any node is zero at every instant.

Kirchhoff's Voltage Law (KVL)
The algebraic sum of voltages around any loop is zero at every instant.

- Two two-terminal devices are said to be connected in parallel if they are connected between the same pair of nodes. The same voltage appears across any two elements connected in parallel.
- Two devices are said to be connected in series if they are connected at a node to which no other devices are connected. The same current flows through any two elements connected in series.

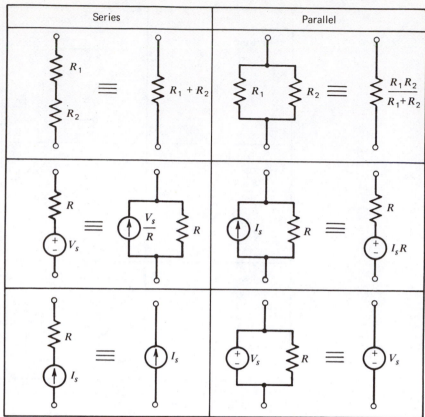

FIGURE 2-41
Series and parallel equivalent circuits.

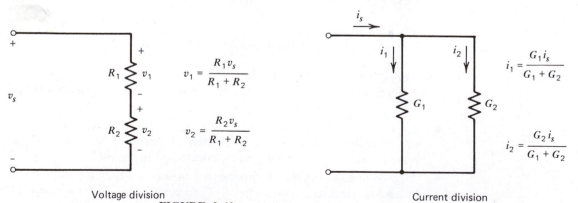

Voltage division

$$v_1 = \frac{R_1 v_s}{R_1 + R_2}$$

$$v_2 = \frac{R_2 v_s}{R_1 + R_2}$$

Current division

$$i_1 = \frac{G_1 i_s}{G_1 + G_2}$$

$$i_2 = \frac{G_2 i_s}{G_1 + G_2}$$

FIGURE 2-42
Voltage and current dividers.

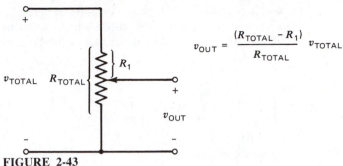

$$v_{OUT} = \frac{(R_{TOTAL} - R_1)}{R_{TOTAL}} v_{TOTAL}$$

FIGURE 2-43
The potentiometer.

- Two circuits are said to be equivalent at a specified pair of terminals if they have the same *i-v* constraints. Important series/parallel equivalent circuits rules are shown in Figure 2-41.

- Voltage and current division are very useful circuit analysis techniques and are shown in Figure 2-42.

- A potentiometer is an important application of the voltage divider, as shown in Figure 2-43.

- Source conversion permits one to change from a real-world model of a voltage source to a similar model of a current source, or conversely. The conversion is illustrated in Figure 2-44.

- Circuit reduction is a method of determining selected signal variables in series/parallel circuits. The method involves sequential application of the series/parallel equivalence rules, source conversion rules, and voltage/current division rules. The sequence used is not unique and depends on the variables to be determined and the structure of the circuit.

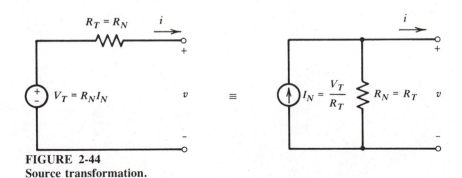

FIGURE 2-44
Source transformation.

● It is common practice in circuits to select a reference node. All other node voltages are measured with respect to it. The reference node is usually identified by using one of the ground symbols shown in Figure 2-45.

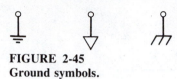

FIGURE 2-45
Ground symbols.

EN ROUTE OBJECTIVES AND RELATED EXERCISES

2-1 CIRCUIT VARIABLES (SEC. 2-1)

Given an electrical quantity described in terms of words, scientific notation, or decimal prefix notation, convert the quantity to an alternate description.

Exercises

2-1-1　Determine the number of electrons that flow through a discharge tube if all of the 2 J/C supply is dissipated in 10 ns at the cost of 3.2 W.

2-1-2　The charge flowing through a piece of wire varies as shown in the graph of Figure E2-1-2. What is the corresponding current?

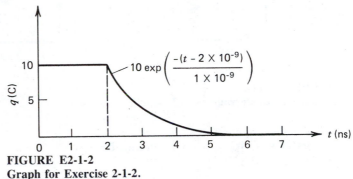

$$10 \exp\left(\frac{-(t - 2 \times 10^{-9})}{1 \times 10^{-9}}\right)$$

FIGURE E2-1-2
Graph for Exercise 2-1-2.

2-1-3　How much current is drawn from the automobile battery (12 V) of Figure E2-1-3 by a headlamp that dissipates 60 W of power? How much energy is stored in the battery if it is rated at 100 ampere-hour (Ah)?

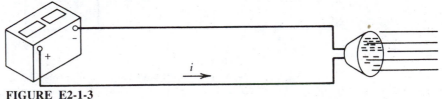

FIGURE E2-1-3
Car lamp and battery circuit.

2-1-4　A rating of 75 W on an incandescent lamp is a measure of the rate of energy use by the lamp when connected to a 120-V source. How much current flows through the lamp? Suppose the lamp is accidently left on overnight (eight hours) and electricity costs 7.2 cents/kWh. How much did it cost the owner?

2-1-5 A copper wire has a current density of 100 A/cm². If there are 5×10^{22} electrons/cm³, what is the average velocity of the electrons flowing in the wire?

2-2 ELEMENT CONSTRAINTS (SEC. 2-2)

Given a two-terminal element with one or more signal variables specified, use the element i-v constraint to determine the unspecified signal variables.

Exercises

2-2-1 Determine the unspecified signal variables associated with each of the two-terminal elements shown in Figure E2-2-1.

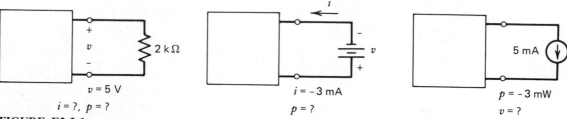

$v = 5$ V

$i = ?, \ p = ?$

$i = -3$ mA

$p = ?$

$p = -3$ mW

$v = ?$

FIGURE E2-2-1
Circuits for demonstrating signal variable relationships.

2-2-2 Compute the current through, voltage across, and power dissipated in each of the devices in Figure E2-2-2.

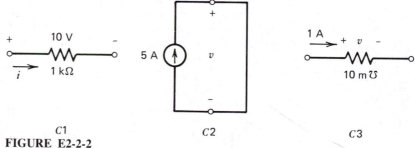

C1

C2

C3

FIGURE E2-2-2
Circuits for element constraint for Exercise 2-2-2.

2-2-3 Sketch how the *i-v* characteristics of a resistor would change as its resistance was varied from 0 Ω to ∞Ω.

2-2-4 A fuse is a device with a circuit-opening fusible link that is heated and severed (melted) by the passage of a predetermined current through it. When the fuse is intact the fusible link exhibits a fixed resistance R_F. Develop a suitable model for a fuse.

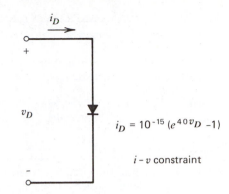

$$i_D = 10^{-15}\left(e^{40v_D} - 1\right)$$

$i\text{-}v$ constraint

Diode circuit symbol

FIGURE E2-2-5

A diode and its $i\text{-}v$ characteristics.

2-2-5 A semiconductor diode is a two-terminal circuit element with the circuit symbol and $i\text{-}v$ constraint shown in Figure E2-2-5. Find i and p for $v_D = -0.8, -0.4, -0.2, -0.1, 0, 0.1, 0.2, 0.4,$ and 0.8 V. Use these data to plot the $i\text{-}v$ characteristics of the element.

2-2-6 In this exercise linear circuit elements are tested by connecting them to circuits $C1$ and $C2$ in Figure E2-2-6 and measuring the two associated signal variables. Identify the type of circuit element under test from the measurement given.

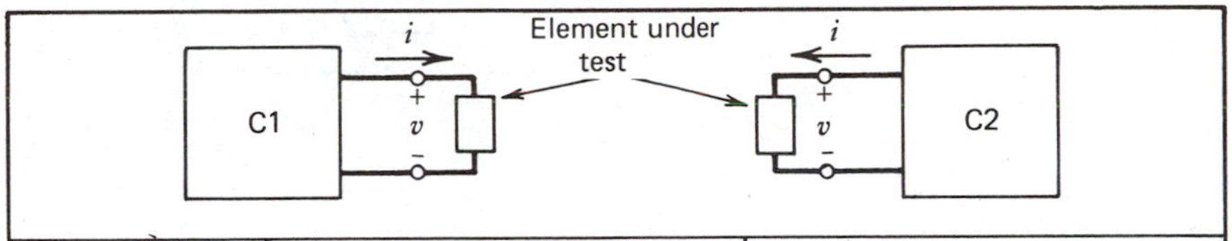

Element	Connected to C1		Connected to C2	
	$v(V)$	$i(mA)$	$v(V)$	$i(mA)$
A	5	1	-2.5	-0.5
B	5	5	5	-10
C	10	0	10	-15
D	12.5	-2.5	-2.5	-2.5

FIGURE E2-2-6

Test circuits for Exercise 2-2-6.

2-2-7 Suppose a certain device had the $i\text{-}v$ relation shown in Fig. E2-2-7. For each input given find an expression for the output.

(a) $V_{IN}(t) = 10$ V

(b) $V_{IN}(t) = 5\,t$ V

(c) $V_{IN}(t) = [5 + e^{-t}]$ V

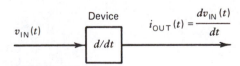

$$i_{OUT}(t) = \frac{dv_{IN}(t)}{dt}$$

FIGURE E2-2-7

A differentiator circuit.

2-3 CONNECTION CONSTRAINTS (SEC. 2-3)

Given an interconnection of two-terminal circuit elements (a circuit):

(a) Identify nodes and loops in the circuit.

(b) Identify elements connected in series and in parallel.

(c) Use the connection constraints (KCL and KVL) to determine selected signal variables.

Exercises

2-3-1 For each of the circuits in Figure E2-3-1:

(a) Identify at least three loops.

(b) Identify elements connected in series or in parallel.

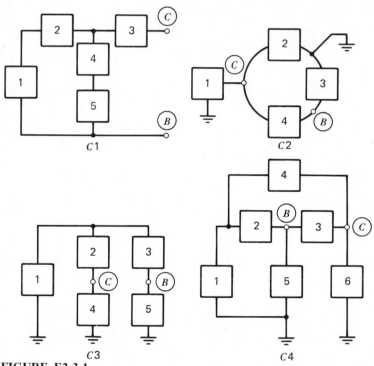

FIGURE E2-3-1
Circuits for determining series and parallel connections.

2-3-2 Repeat Exercise 2-3-1 assuming nodes ⓑ and ⓒ in each circuit are connected together.

2-3-3 Repeat Exercise 2-3-1 assuming node ⓒ is connected to ground.

2-3-4 Repeat Exercise 2-3-1 assuming element ② is removed from the circuit leaving an open circuit in its place and nodes ⓑ and ⓒ are tied together.

2-3-5 Repeat Exercise 2-3-1 assuming element ② is removed from the circuit leaving an open circuit in its place and node © is connected to ground.

2-3-6 For the circuit in Figure E2-3-2:
  (a) How many nodes are there in this circuit?
  (b) Identify elements connected in series and in parallel.

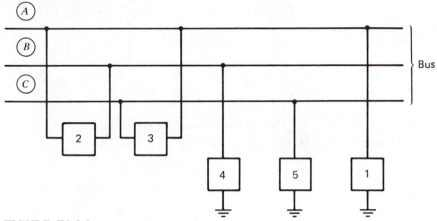

FIGURE E2-3-2
A bus circuit.

2-3-7 Show that the circuit in Figure E2-3-2 is the same as C3 in Figure E2-3-1.

2-3-8 Repeat Exercise 2-3-6 assuming line *B* is connected to ground.

2-4 COMBINED CONSTRAINTS (SEC. 2-4)

Given an interconnection of two-terminal elements (a circuit), use the combination of the connections constraints (KCL and KVL) and element constraints (i-v characteristics) to determine selected signal variables.

66   Basic Circuit Concepts

Exercises

2-4-1   Find v_x in each of the circuits of Figure E2-4-1.

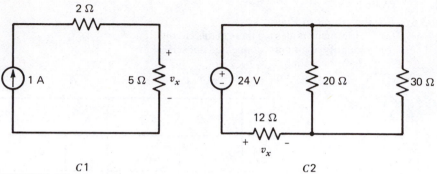

C1 C2

FIGURE E2-4-1
Circuits to determine V_x for Exercise E2-4-1.

2-4-2   Find i_x in each of the circuits of Figure E2-4-2.

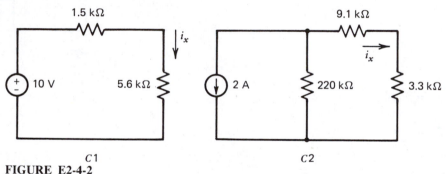

C1 C2

FIGURE E2-4-2
Circuits to determine i_x for Exercise E2-4-2.

2-4-3   In the circuit of Figure E2-4-3, use KVL to determine the voltage across each resistor. Then use Ohm's law and KCL to determine the current through every element.

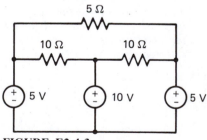

FIGURE E2-4-3
KVL circuit for Exercise 2-4-3.

2-4-4 In each of the circuits of Figure E2-4-4 determine the voltage across and current through each source.

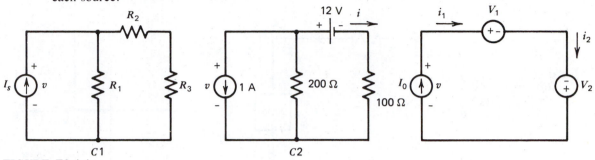

FIGURE E2-4-4
Circuits for determining unknown *i*'s or *v*'s in Exercise 2-4-4.

2-4-5 For the circuit of Figure E2-4-5 determine how much power the source must provide to the circuit by computing the total power dissipated by every element. Then verify your result by finding the power $(V_s \times I_s)$ delivered by the source.

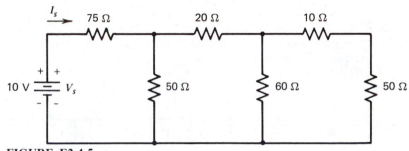

FIGURE E2-4-5
Circuit for power computation for Exercise E2-4-5.

2-4-6   Assuming the device in the box has the *i-v* characteristics shown in Figure E2-4-6, find the voltage across the 10 Ω resistor for each source.

(a) $v_s = 2$ V
(b) $v_s = \frac{1}{2}$ V
(c) $v_s = -2$ V

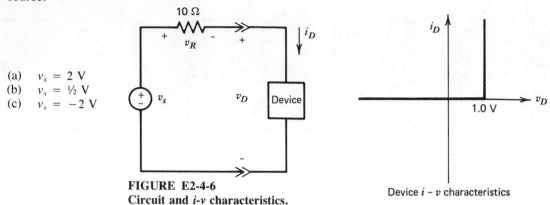

FIGURE E2-4-6
Circuit and *i-v* characteristics.

Device *i* – *v* characteristics

2-5 *EQUIVALENT CIRCUITS (SEC. 2-5)*

Given a circuit consisting of linear resistors and constant input signal sources, determine an equivalent circuit at a specified pair of terminals.

Exercises

2-5-1   Determine the equivalent resistance between terminals Ⓐ and Ⓑ of each circuit of Figure E2-5-1.

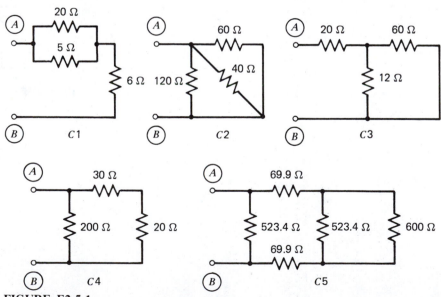

FIGURE E2-5-1
Circuits for computing equivalent resistance.

2-5-2   Determine a one-resistor, one-source, equivalent circuit between terminals Ⓐ and Ⓑ for
each of the circuits in Figure E2-5-2.

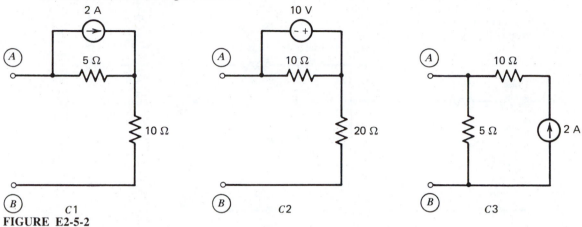

FIGURE E2-5-2
Circuits to simplify for Exercise 2-5-2.

2-5-3   The circuit of Figure E2-5-3 is an R-$2R$ resistor array package. Show how to intercon-
nect the terminals of the array to produce
equivalent resistances of $R/2$, $2R/3$, R, $8R/3$,
$5R/3$, $2R$, $3R$, and $4R$.

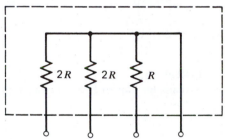

FIGURE E2-5-3
An R-$2R$ resistor array.

2-5-4   Often an engineer will make approximations
that save time and money. Considering that
the resistors you have to use have a tolerance
of ± 5 percent (they are only accurate to ±
5 percent of their stated value), simplify the
circuits of Figure E2-5-4.

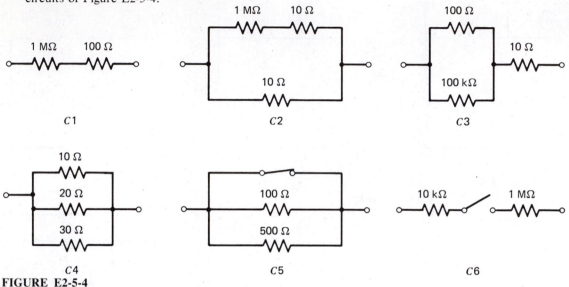

FIGURE E2-5-4
Circuits for applying engineering approximations.

2-5-5   The circuit of Figure E2-5-5 is the pin diagram of a sense amplifier resistance array (Allen-Bradley 314M130). Determine the equivalent resistance between the pins listed in the illustration.

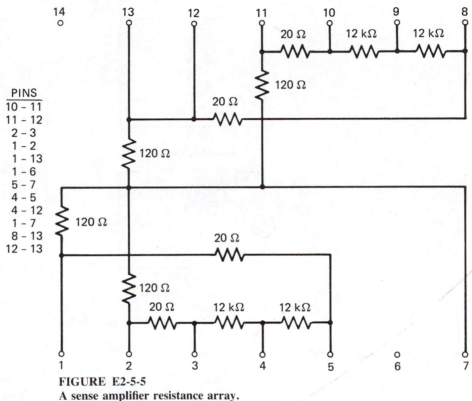

PINS
_____
10 – 11
11 – 12
2 – 3
1 – 2
1 – 13
1 – 6
5 – 7
4 – 5
4 – 12
1 – 7
8 – 13
12 – 13

FIGURE E2-5-5
A sense amplifier resistance array.

2-6 *VOLTAGE AND CURRENT DIVISION (SEC. 2-6*

Given a circuit with a number of elements connected in series or parallel, apply the rules of voltage and current division to determine specified voltages or currents.

Exercises

2-6-1   Find the desired signal variable using voltage
or current division in each of the circuits of
Figure E2-6-1.

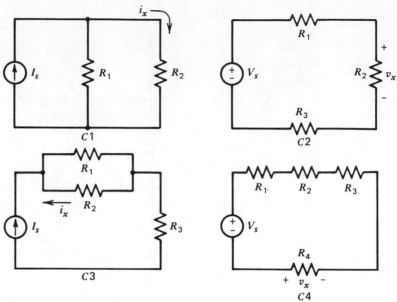

FIGURE E2-6-1
**Current and voltage divider circuits for Exercise
2-6-1.**

2-6-2   Find v_x in each of the circuits of Figure E2-
6-2.

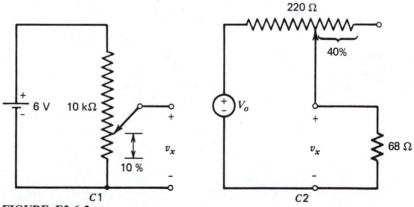

FIGURE E2-6-2
Potentiometer circuits.

2-6-3   Using voltage and current division, find the
indicated signal variable in each circuit of
Figure E2-6-3.

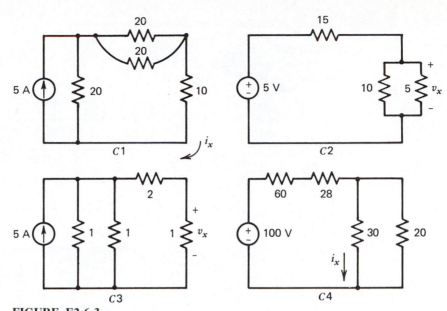

FIGURE E2-6-3
**Current and voltage divider circuits for Exercise
2-6-3.**

2-6-4 Find the indicate signal variable in each of
the circuits of Figure E2-6-4.

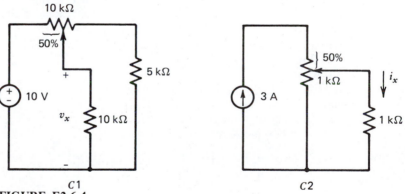

FIGURE E2-6-4
More potentiometer circuits.

2-6-5 A manufacturer's specification sheet on an electronic switch lists the switch's characteristics as shown in Figure E2-6-5. For the circuit shown in the figure, compute V_{OUT} when the switch is ON and when the switch is OFF.

G	SWITCH	R_{sw}
+5 V	ON	150 Ω
0 V	OFF	10^6 Ω

Specification

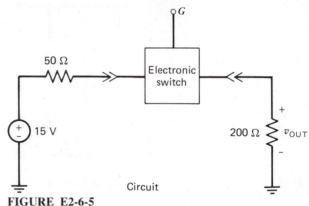

Circuit

FIGURE E2-6-5
An application of an electronic switch.

2-6-6 A D'Arsonval meter with internal resistance of 10 Ω produces full-scale deflection when 10 mA flows through it. Design an ammeter that can measure 2 A full scale.

2-6-7 Properly configured, a D'Arsonval movement can be used as a voltmeter. Design a voltmeter using the movement described in Exercise 2-6-6 to measure 10 V full scale.

2-6-8   A 2-A fuse is used to protect a small motor.
Under the conditions shown in Figure E2-6-
8 will the fuse blow? The fuse must exceed
its rating for more than 1 second to blow.

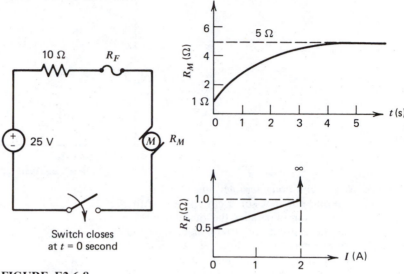

FIGURE E2-6-8
A fuse problem.

2-6-9   Design an interface that will enable the load
in Figure E2-6-9 to have $i_L = 1$ mA flowing
through it. Repeat if 10 V is desired across
the 2-kΩ load.

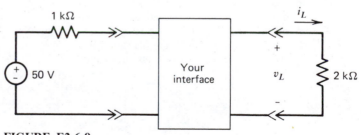

FIGURE E2-6-9
A design interface problem.

2-6-10 A 1-MΩ potentiometer is adjusted to permit 5 V across the load shown in Figure E2-6-10. At what percent of the potentiometer range should the wiper be set?

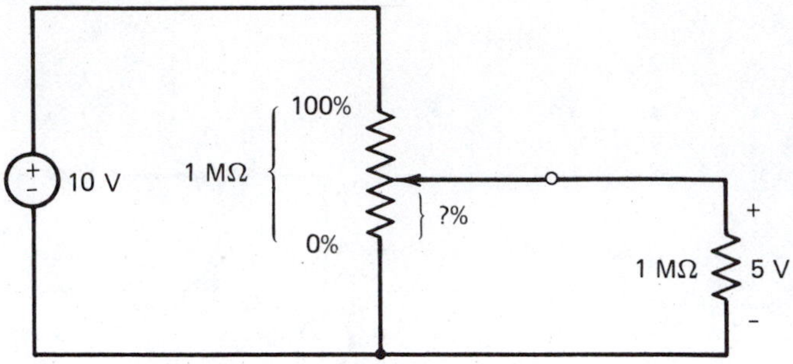

FIGURE E2-6-10
A potentiometer analysis problem.

2-6-11 We spent considerable time in Sec. 2-3 pointing out that some circuits are neither in series nor in parallel, the implication being that these circuits are irreducible. Some (but not all) may be simplified via use of a Δ-Y transformation. The transformation permits three resistors configured in a "Y" or "T" configuration to be replaced by three resistors in a "Δ" configuration. The two equivalent circuits are shown in Figure E2-6-11. The transformation between the two must be equal as far as terminals Ⓐ and Ⓑ, Ⓑ and Ⓒ, and Ⓐ and Ⓒ are concerned. For example, a circuit connected to terminals Ⓐ and Ⓑ "sees" an effective resistance of $R_1 + R_2$ for the Y or T configuration and $R_x \parallel (R_y + R_z)$ for the Δ configuration. Find

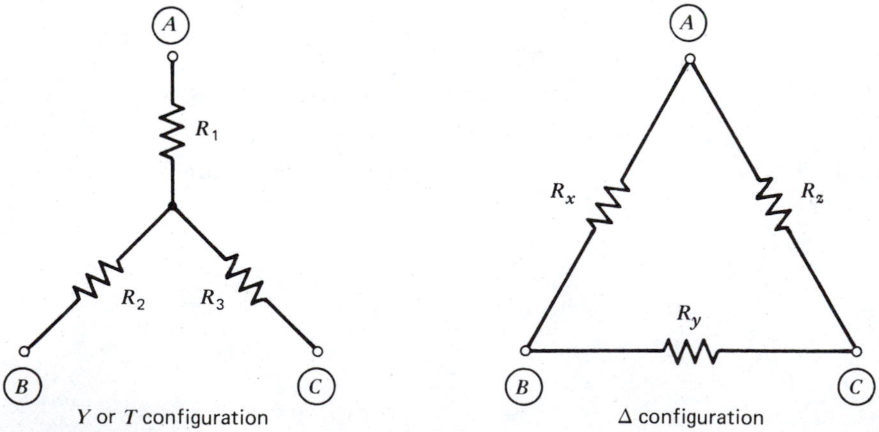

FIGURE E2-6-11
Y-Δ equivalent circuits.

the relationship between the R's, so that conversion from one configuration to the other is readily accomplished.

2-7 CIRCUIT REDUCTION (SEC. 2-7)

Given a circuit consisting of linear resistors and a constant input signal source, determine selected signal variables using successive applications or equivalence and voltage/current division.

Exercises

2-7-1 Use circuit reduction techniques to determine the indicated signal variables in each of the circuits of Figure E2-7-1.

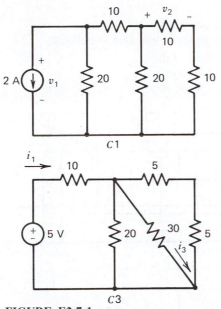

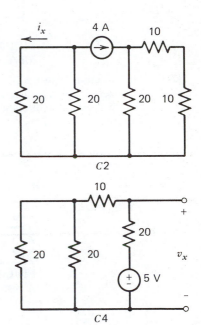

FIGURE E2-7-1
Circuits to apply reduction techniques.

2-7-2 Repeat Exercise 2-7-1 for both circuits in Figure E2-7-2.

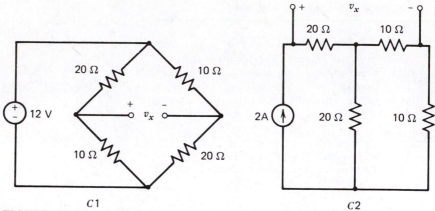

C1   C2

FIGURE E2-7-2
More circuits to apply reduction techniques.

2-7-3 The circuit of Figure E2-7-3 uses three of the R-2R resistor arrays described in Exercise 2-5-3. Use circuit reduction to determine the voltages v_1, v_2, and v_3.

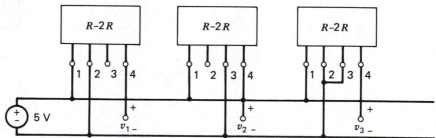

FIGURE E2-7-3
A resistor array problem.

2-7-4 The circuit shown in Figure E2-7-4 occurs extensively in electrical instrumentation. Usually, one of the resistors R_x is that of a transducer whose value varies with some external physical phenomenon—for example, temperature, pressure or light, etc. The circuit is called a *Wheatstone bridge* and works as follows: Prior to the start of any measurement one of the nontransducer resistors is adjusted until the current through R_M (usually a D'Arsonval meter movement) is zero. When the transducer is exposed to a suitable physical phenomenon, its resistance changes causing current to flow through R_M. The deflection of the meter, then, can be calibrated to the change in physical phenomenon being experienced by the transducer.

For a fixed value of R_X what should the relation be between R_1, R_2, and R_3 to ensure than $i_M = 0$?

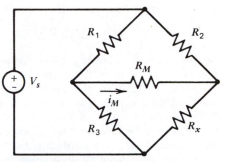

FIGURE E2-7-4
A Wheatstone bridge.

PROBLEMS

P2-1 (Analysis)
Current will flow in a semiconductor even if there is no applied voltage if there exists a concentration gradient of free charge. This current, known as diffusion current is given by

$$I(x) = AqD_n \frac{dn(x)}{dx}$$

Suppose that this current is measured to equal -1.0 mA at $x = 0$. Where A, the cross-sectional area of the semiconductor, is 10^{-8} cm² and is assumed constant for the entire length; $q = 1.6 \times 10^{-19}$ C; D_n, the diffusion constant, is 34 cm²/s; and L_n, the diffusion length, $= 0.6$ µm. How many electrons per cubic centimeter are contributing to the flow? Assume that the concentration gradient is exponential and is given by

$$n(x) = n(0) \exp[-x/L_n]$$

What current can be expected at $x = 5L_n$?

P2-2 (Design, Analysis)
The resistance of a conductive bar of length l and uniform cross-sectional area A is given by $R = \rho l/A$ where ρ is the resistivity of the material. In the manufacture of hybrid circuits thin films of conducting materials are often evaporated through a mask in a vacuum chamber to fabricate thin-film resistors. For example, the resistivity of Nichrome (a nickel-chromium alloy often used) is 150×10^{-6} Ω-cm. Since thin films are usually made uniform in thickness, these films are generally calibrated in ohms per square as the only dimensions readily available are the length and the width. The designer of a thin-film resistor then selects a certain thickness—usually between 0.05 and 5 µm. This results in a predetermined number of ohms per square. The designer then lays out a pattern of so many squares by so many squares to achieve the resistance desired. For example, suppose a certain thickness yields 100 Ω/square. A resistor of 1 unit by 2 units as shown in Figure P2-2-*a* would result in a resistance of 200 Ω between the long ends (*a* and *B*) and 50 Ω between the near sides (*C* and *D*).

(a) Using Nichrome and any thickness between 0.05 and 5 µm, design masks to make a 5-, a 150-, and a 42.5 Ω resistor.

(b) Consider the thin-film circuit of Figure P2-2*b*. What voltage would be measured between *B* and *C* if 10 V is applied at *A* and *C*?

(c) Suppose the voltage measured between *B* and *C* is too small, how can the circuit be altered

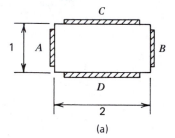

(a)

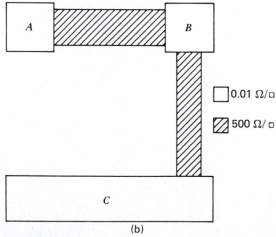

☐ 0.01 Ω/□

▨ 500 Ω/□

(b)

FIGURE P2-2
A thin-film resistor.

easily (without redesigning or reevaporating) to
achieve the desired value of V_{BC}?

P2-3   (Analysis)
The Wheatstone bridge shown in Figure P2-3 is com-
pletely balanced, that is, $R_1 = R_2 = R_3 = R_{SG} = R_B$. Suppose that R_{SG} is a strain gage, a device whose
resistance varies linearly with pressure, $R_{SG}(P)$. Now
suppose a change in pressure causes a certain change
in R_{SG}, how much will the current I_B change?
  If $V_s = 10$ V, $R = 1$ kΩ, and $\Delta R_{SG} = 1$ Ω, what is
ΔI_B?

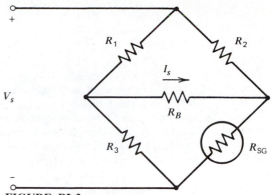

FIGURE P2-3
Strain gage problem circuit.

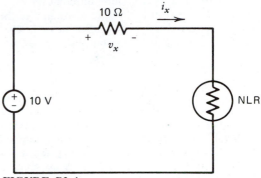

P2-4   (Analysis)
A nonlinear resistor (NLR) has an *i-v* relation given
by $1000\ i = v^3$. Suppose it is connected to the circuit
shown in Figure P2-4; find v_x and i_x.

FIGURE P2-4
A nonlinear resistor circuit.

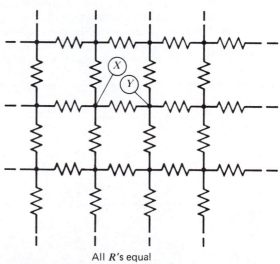

P2-5   (Analysis)
A square grid of equal resistors R, shown in Figure
P2-5, extends to infinity in all directions, at which
point they are grounded. What will be the effective
resistance between any two adjacent nodes?

All R's equal

FIGURE P2-5
An infinite resistor matrix.

P2-6 (Analysis)

Standard flashlight 1.5-V "dry" cells are made from Zn and MnO_2. As the cell is used the MnO_2 converts to Mn_2O_3, which has a much higher resistance than MnO_2. The graph of Figure P2-6 shows the internal resistance of a dry cell versus time. How much power can be delivered to a 100-Ω load when the cell is new and after 10 hours of use? How much after 100 hours of use?

P2-7 (Analysis)

An incandescent lamp is usually rated in watts. If the resistance of a 100-W bulb is measured and its rating computed using v^2/R, a quite different answer is obtained. The reason is that the resistance of the filament changes dramatically when heated. The temperature coefficient of resistance α is given by

$$\alpha = \frac{\Delta R}{\Delta T \, R_n}$$

where R_n is the resistance of a material at some known temperature. Then R at some other temperature is given by $R = R_n + \Delta R$. Assume that the resistance of the 100-W tungsten filament at room temperature ($T = 20°C$) is 12 Ω and that α for tungsten is 0.00495. If the bulb is connected to a 120-V-dc source, how hot is the filament?

P2-8 (Design)

A sealed box is known to contain only resistors. Four terminals labeled A, B, C, and D are on the box. An ohmmeter is used to measure the resistances between each pair of terminals and the results obtained are shown in Table P2-8. What is a possible circuit inside the box?

Terminal Pairs	Measured Resistance, ohms
AB	0
AC	5
AD	10
BC	5
BD	10
CD	5

Table P2-8

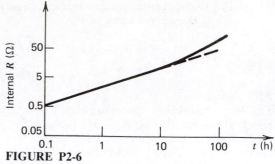

FIGURE P2-6
Battery's internal resistance versus time.

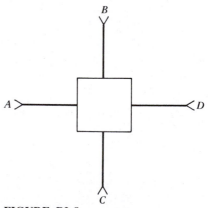

FIGURE P2-8
Unknown element problem.

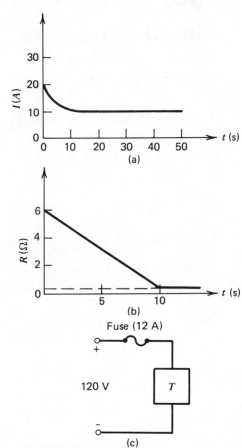

FIGURE P2-9
Slow-blow fuse problem data.

P2-9   (Analysis)

A certain 1200-W toaster is protected by a 12-A fuse (Figure P2-9c). Yet whenever the toaster was plugged in the fuse blew. After some study it was determined that the toaster draws more current when it first heats up. In fact, the manufacturer, beleagured by angry customers, realized that the current through the toaster varied versus time as shown in the graph (Figure P2-9a).

The proposed fix is to replace the 12-A fuse with a 12-A slow-blow fuse, which had the characteristics shown in Figure P2-9b. Will that solve the problem?

P2-10   (Evaluation)

A set of three identical potentiometers are cascaded as shown in Figure P2-10. The purpose is to obtain an output voltage v_o, that can be adjusted with increasing degrees of fineness. Find v_o/v_i and discuss whether the intended purpose was achieved.

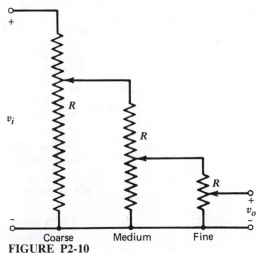

FIGURE P2-10
Cascaded potentiometers circuit.

Chapter 3
Circuit Theorems

There are in existence a very large number of concepts, mathematical relations, and design techniques relating to linear, time-invariant circuits.

Sidney Darlington

In this chapter we study a set of linear circuit principles that often simplify analysis problems. These principles are called theorems because they can be proved under very general conditions and their conclusions are relatively simple. Proportionality and superposition are extremely useful circuit principles that provide the conceptual foundation for much of the theory of linear circuits. Thévenin's and Norton's theorems are very powerful tools for dealing with circuit interfaces. The concept of an interface then leads to the question of the maximum signals that are available. In the final section we apply these principles to some simple interface circuit design problems.

3-1 PROPORTIONALITY

The subtitle of this book indicates that our interests center on the subject of **linear** circuits. The hallmark feature of such circuits is that signal outputs are linear functions of its inputs. Mathematically we say that a function is linear if it has two properties:

$$f(AX) = Af(X) \qquad \text{(homogeneity)} \tag{3-1}$$

and

$$f(X_1 + X_2) = f(X_1) + f(X_2) \qquad \text{(additive)} \tag{3-2}$$

where A is a constant. In terms of circuits the homogeneity property indicates that the output of a linear circuit is proportional to the input. The additive property means that the output due to two or more inputs can be found by adding the outputs obtained when each input is applied separately. In circuit analysis the first property is called **proportionality** and the additive property is called **superposition.** We study superposition in the next section.

If we consider the i-v characteristics of a linear resistor $v = iR$, then it is clear that if the current (input) is doubled, the voltage (output) is doubled. However, the power delivered is $p = i^2R$, hence doubling the current quadruples the power. Thus we observe that a circuit can be linear only in the current and voltage signal variables, but not power. Power is proportional to the product of current and voltage, and is inherently non-linear, even though the circuit may be linear.

The significance of linearity is that for memoryless circuits we can write input-output relationships as

$$y = Kx \tag{3-3}$$

where x is the input (current or voltage), y is an output (current or voltage), and K is a constant. We already have seen several examples of this relationship. For instance, using the voltage-divider relationship on the circuit in Figure 3-1a produces

$$V_o = \frac{R_2}{R_1 + R_2}V_s$$

where

$$x = V_8$$
$$y = V_0$$
$$K = \frac{R_2}{R_1 + R_2}$$

Similarly, applying current division to the circuit in Figure 3-1b yields

$$I_0 = \frac{G_2}{G_1 + G_2}I_s$$

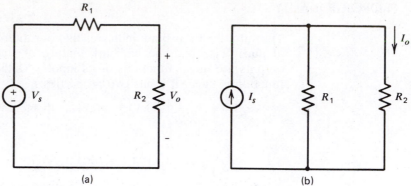

(a) (b)

FIGURE 3-1
Examples of the property of proportionality. (*a*) Voltage divider. (*b*) Current divider.

where

$$x = I_s$$
$$y = I_0$$
$$K = G_2/(G_1 + G_2)$$

The next example illustrates the determination of the input-output relationship for a more complicated circuit, and demonstrates that the proportionality constant K can be positive, negative, or even zero.

Example 3-1

To determine the input-output relationship of the bridge circuit in Figure 3-2, we observe that the circuit consists of two voltage dividers. Applying the voltage divider rule to each divider yields

$$V_A = \frac{R_3}{R_1 + R_3} V_s$$

$$V_B = \frac{R_4}{R_2 + R_4} V_s$$

But KVL allows us to write

$$V_o = V_A - V_B$$

hence

$$V_o = \frac{R_3}{R_1 + R_3} - \frac{R_4}{R_2 + R_4} V_s$$

$$= \frac{R_2 R_3 - R_1 R_4}{(R_1 + R_3)(R_2 + R_4)} V_s$$

$$= K V_s$$

Note that the output (V_o) is proportional to the input (V_s), and that the proportionality constant (K) can be positive, negative, or zero. Specifically, if

$R_2 R_3 > R_1 R_4$	then	$K > 0$
$R_2 R_3 = R_1 R_4$	then	$K = 0$
$R_2 R_3 < R_1 R_4$	then	$K < 0$

The condition $K = 0$, zero output for every input, is called bridge balance, and requires that the product of the opposite legs of the two voltage dividers be equal.

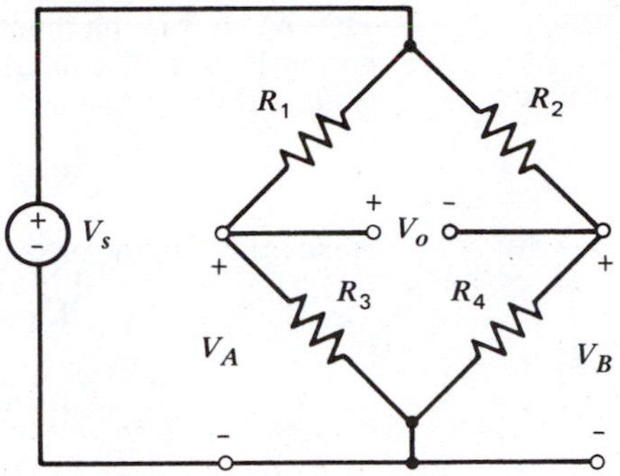

FIGURE 3-2
Bridge circuit for Example 3-1.

An analysis technique called the unit output method is directly based on the proportionality property of linear circuits. The method can be applied only to ladder circuits, and involves the following steps:

1. A unit output is assumed, that is, $V_o = 1$ V or $I_o = 1$ A.
2. The input required to produce the unit output is then found by successive application of KCL, KVL, and Ohm's law.
3. Because the circuit is linear the proportionality constant relating input and output can be found as

$$K = \frac{\text{Output}}{\text{Input}} = \frac{1}{\text{Input for unit output}}$$

4. The output for any input can then be found as

$$\text{General output} = K \text{ (General input)}$$

The unit output method works the circuit response problem backward, that is, from output to input. The next example illustrates the application of the method.

Example 3-2

To determine V_o in the circuit shown in Figure 3-3a, we first assume $V_o = 1$ as shown in Figure 3-3b. Then by Ohm's law

$$I_o = \frac{V_o}{20} = \frac{1}{20}$$

by KCL at node Ⓐ

$$I_1 = I_o = \frac{1}{20}$$

by Ohm's law

$$V_1 = 10 I_1 = \frac{1}{2}$$

by KVL around loop $L1$

$$V_2 = V_1 + V_o = \frac{1}{2} + 1 = \frac{3}{2}$$

by Ohm's law

$$I_2 = \frac{V_2}{15} = \frac{3}{30} = \frac{1}{10}$$

by KCL at node $\textcircled{B}$

$$I_3 = I_1 + I_2 = \frac{1}{20} + \frac{1}{10} = \frac{3}{20}$$

by Ohm's law

$$V_3 = 10 \, I_3 = \frac{3}{2}$$

by KVL around loop $L2$

$$V_s = V_3 + V_2 = \frac{3}{2} + \frac{3}{2} = 3$$

Thus $V_s = 3$ is the input required to produce a unit output $V_o = 1$. The input-output proportionality constant is then

$$K = \frac{V_o}{V_s} = \frac{1}{3}$$

Hence the output for the given 5-V input is

$$V_o = K \, V_s = \frac{1}{3} \times 5$$

$$= \frac{5}{3} \, V$$

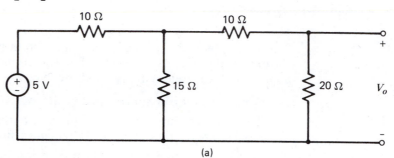

(a)

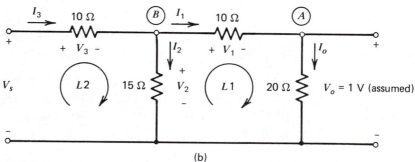

(b)

FIGURE 3-3
**Circuit for Example 3-2 used to illustrate the unit output method. (*a*) Given circuit.
(*b*) Circuit with unit output.**

3-2 SUPERPOSITION

The principle of superposition is a very useful computational and conceptual tool that finds many applications in the analysis of linear circuits with several inputs. The input-output relationships of linear circuits have

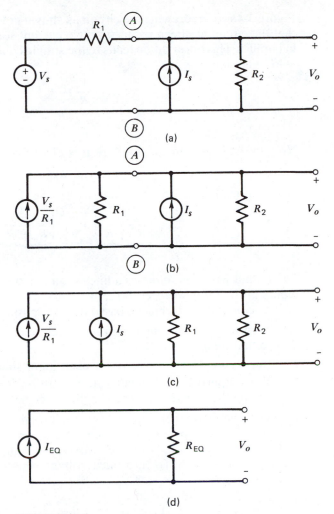

FIGURE 3-4
Illustration for superposition: linear circuit with two input signal sources. (*a*) Given circuit. (*b*) Circuit after source transformation. (*c*) Circuit redrawn. (*d*) Final equivalent circuit.

the additive property of linear functions. For linear, memoryless circuits this means that we can always write any output y as

$$y = K_1 x_1 + K_2 x_2 + K_3 x_3 \ldots \tag{3-4}$$

where $x_1, x_2, x_3, \ldots$ are circuit inputs, and $K_1, K_2, K_3, \ldots$ are constants that depend on the circuit. Briefly stated, for a linear, memoryless circuit the output is a linear combination of the various inputs.

To illustrate this principle, let us determine the output (V_o) of the two-input source circuit in Figure 3-4*a*. We first perform a source transformation to the left of terminals Ⓐ and Ⓑ to obtain the circuit shown in

Figure 3-4*b*. By redrawing the circuit as shown in Figure 3-4*c*, we observe that the circuit has been reduced to two current sources and two resistors in parallel. This observation leads to the simple equivalent circuit in Figure 3-4*d*, where

$$I_{EQ} = \frac{V_s}{R_1} + I_s$$

$$R_{EQ} = \frac{R_1 R_2}{R_1 + R_2}$$

(3-5)

Now in the final equivalent circuit it is clear that

$$V_o = I_{EQ} R_{EQ}$$

Hence in using Eq. 3-5 we obtain

$$V_o = \underbrace{\left[\frac{R_2}{R_1 + R_2}\right]}_{} V_s + \underbrace{\left[\frac{R_1 R_2}{R_1 + R_2}\right]}_{} I_s$$

$$Y = \qquad K_1 \qquad X_1 + \qquad K_2 \qquad X_2$$

(3-6)

The output has been found as a linear combination (superposition) of the inputs in the form given by Eq. 3-4.

A circuit analysis technique based on superposition proceeds as follows:

1. Turn off all input signal sources except one and find the output due to that input acting alone.

2. Repeat step 1 successively for each input signal source.

3. The output when all input sources are on is then found by simply adding the responses due to each source acting alone.

However, before we can apply this method we must discuss what it means to ''turn off'' an input signal source.

The *i-v* characteristics of the two types of input signal sources are shown in Figure 3-5. To turn off a voltage source we set its voltage to

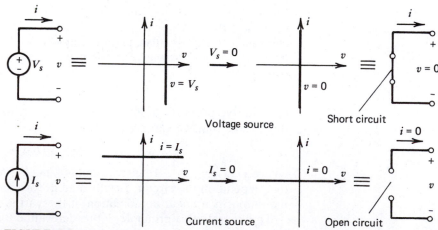

FIGURE 3-5
Turning OFF a source.

zero ($V_s = 0$). This translates the source i-v characteristic to the vertical axis, which, as we recall from Chapter 2, is the equivalent of a short circuit. Similarly, turning off a current source ($I_s = 0$) translates its i-v characteristics to the horizontal axis, which is the equivalent of an open circuit. In summary, to turn off a voltage source we replace it by a short circuit and to turn off a current source we replace it by an open circuit.

The next two examples illustrate the application of superposition in multiple-input source circuits.

Example 3-3

Let us now determine the output of the circuit in Figure 3-6a using superposition. To find the output due to the voltage source we turn off the current source (replace it by an open circuit) and obtain the circuit in Figure 3-6b.

By voltage division,

$$V_{o1} = \frac{R_2}{R_1 + R_2} I_s$$

Next we turn off the voltage source (replace it by a short circuit) and obtain the circuit in Figure 3-6c. By current division,

$$I_{o2} = \frac{R_1}{R_1 + R_2} I_s$$

But by Ohm's law,

$$V_{o2} = I_{o2}R_2$$
$$= \frac{R_1 R_2}{R_1 + R_2} I_s$$

Using superposition we then write:

$$V_o = V_{o1} + V_{o2}$$
$$= \left[\frac{R_2}{R_1 + R_2}\right] V_s + \left[\frac{R_1 R_2}{R_1 + R_2}\right] I_s$$

This is the same result we obtained in Eq. 3-6 using circuit reduction techniques on the circuit in Figure 3-4a.

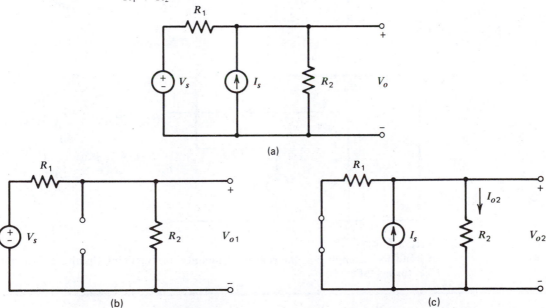

(a)

(b)

(c)

FIGURE 3-6
Circuit for Example 3-3, analysis using superposition. (*a*) **Circuit from Figure 3-4a.** (*b*) **Current source turned OFF.** (*c*) **Voltage source turned OFF.**

Example 3-4

The basic principle of superposition says that the output is a weighted sum of the inputs. Simple resistance circuits such as the one shown in Figure 3-7 are often used to implement this signal summing function. To determine the output (V_o) using superposition, we first turn off sources 2 and 3 ($V_{s2} = 0$ and $V_{s3} = 0$) to obtain the circuit in Figure 3-7b. This circuit is a simple voltage divider in which the output leg consists of two equal resistors in parallel. Hence,

$$V_{o1} = \frac{R/2}{R + R/2} V_{s1} = \frac{V_{s1}}{3}$$

But because of the symmetry of the circuit it can be seen that the same result applies to all three inputs. Hence

$$V_{o2} = \frac{V_{s2}}{3} \quad \text{and} \quad V_{o3} = \frac{V_{s3}}{3}$$

By superposition, then

$$V_o = V_{o1} + V_{o2} + V_{o3}$$
$$= \frac{1}{3} [V_{s1} + V_{s2} + V_{s3}]$$

That is, the output is proportional to the sum of the three input signals.

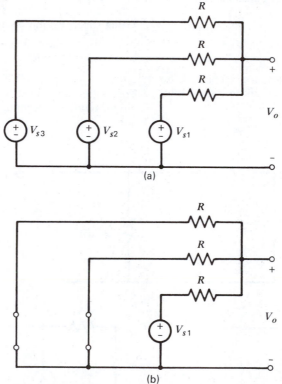

(a)

(b)

FIGURE 3-7
Circuit for Example 3-4, the resistance summer. (*a*) Summing circuit. (*b*) V_{s2} and V_{s3} turned OFF.

These two examples illustrate the use of the superposition principle to analyze circuits containing several input signal sources. The reader should not conclude that this is the primary application of the concept. In fact, unless the circuit is relatively simple it is not immediately clear that superposition actually reduces the amount of analysis effort. Presumably, if we had N sources, we would need to analyze N circuits to obtain the final result. Superposition is used primarily as a conceptual tool in the development of circuit analysis techniques. We use it repeatedly in our development.

3-3   THÉVENIN AND NORTON EQUIVALENT CIRCUITS

The subtitle of this book also suggests that we are concerned with interface circuits. An interface is a connection between two or more circuits that perform different functions. For the two-terminal interface shown in Figure 3-8a, we normally think of one circuit as the source S and the other as the load L. That is, we think of signals as being produced by the source circuit and delivered to the load. The source–load interaction at an interface is one of the central themes of this book.

A powerful tool for dealing with interfaces is the concept of the Thévenin and Norton equivalent circuits shown in Figures 3-8b and 3-8c respectively. Stated formally:

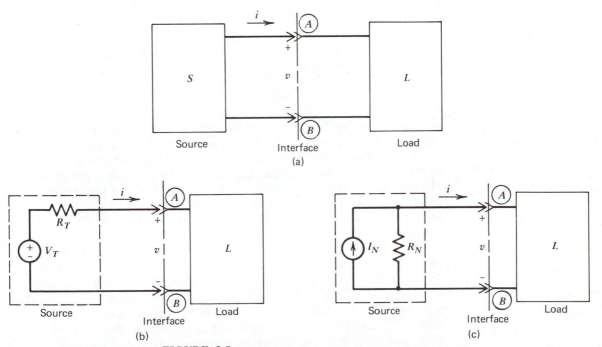

FIGURE 3-8
**Source-load equivalent circuits. (*A*) Two-terminal interface. (*b*) Thévenin equivalent.
(*c*) Norton equivalent.**

Given a two-terminal interface in which the source circuit is linear, then the same interface signals will exist if the source is replaced by its Thévenin or Norton equivalent circuit.

Note that the equivalence require that the source circuit be linear, but place no restriction on the nature of the load circuit. The load may be linear or nonlinear. Later in this section we treat a case in which the load is nonlinear. In Chapter 8 we study cases in which the load consists of elements called capacitors and inductors that are models of devices that are linear but not memoryless.

The Thévenin equivalent consists of a voltage source (V_T) in series with a resistor (R_T). The Norton equivalent is a current source (I_N) in parallel with a resistor (R_N). Since either equivalent leaves the interface signals unchanged, the two circuits must be equivalent to each other. Using KVL and Ohm's law in the Thévenin equivalent yields the *i-v* characteristics at terminals Ⓐ and Ⓑ as

$$v = V_T - iR_T \tag{3-7}$$

Applying KCL and Ohm's law to the Norton equivalent yields the *i-v* characteristics at terminals Ⓐ and Ⓑ as

$$i = I_N - v/R_N \tag{3-8}$$

or

$$v = I_N R_N - iR_N \tag{3-9}$$

Since two equivalent circuits must have identical *i-v* characteristics, if follows by comparing Eq. 3-7 and 3-9 that

$$\begin{aligned} R_N &= R_T \\ I_N R_n &= V_T \end{aligned} \tag{3-10}$$

In essence the Thévenin and Norton equivalents are related by the source transformation relationships studied in Chapter 2. This means that we do not need to find both equivalent circuits independently. Once one of them is found, the other can be determined by source transformation. Put differently, the two equivalent circuits involve four parameters (V_T, R_T, I_N, R_N), but Eqs. 3-10 provide two relations between the parameters. Hence only two conditions are required to determine both equivalent circuits.

These two conditions are easily obtained by using open-circuit and short-circuit loads. That is, if the actual load is disconnected from the source as shown in Figure 3-9a, then an open-circuit voltage v_{OC} appears between terminals Ⓐ and Ⓑ. Applying the same condition to the Thévenin equivalent reveals that $V_T = v_{OC}$. Similarly, disconnecting the load and connecting a short circuit as shown in Figure 3-9b causes a current i_{SC} to flow. Applying the same connection to the Norton equivalent indicates that $I_N = i_{SC}$. In summary, if we find the open-circuit voltage and the short-circuit current, we can determine the Thévenin and Norton equivalent circuit parameters as

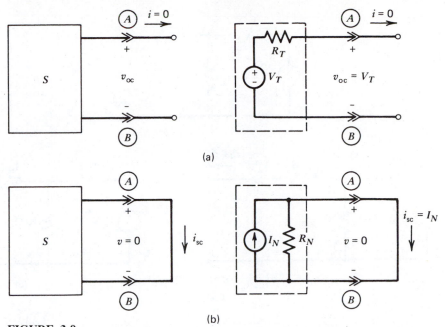

FIGURE 3-9
Finding the Thévenin and Norton equivalent circuits using the open-circuit and short-circuit connections. (*a*) Open-circuit condition. (*b*) Short-circuit condition.

$$V_T = v_{OC}$$
$$I_N = i_{SC}$$
$$R_N = R_T = v_{OC}/i_{SC}$$

(3-11)

The next two examples illustrate both finding and using Thévenin and Norton equivalent circuits.

Example 3-5

Let us determine the Thévenin and Norton equivalent circuits at terminals Ⓐ and Ⓑ in Figure 3-10*a*. For the open-circuit condition in Figure 3-10*b* no current flows through the 15-Ω resistor, hence there is no voltage across it. Thus the open-circuit voltage also appears across the 10-Ω resistor. By voltage division then

$$v_{OC} = \frac{10}{10 + 5} \times 15 = 10 \text{ V}$$

For the short-circuit condition in Figure 3-10*c*, the total current (i_x) delivered by the 15-V source can be found as

$$i_x = 15/R_{EQ}$$

where

$$R_{EQ} = 5 + \frac{10 \times 15}{10 + 15} = 11 \; \Omega$$

Hence

$$i_x = 1.36 \text{ A}$$

and by current division,

$$i_{SC} = \frac{10}{10 + 15} \times i_x$$
$$= 0.545 \text{ A}$$

and finally,

$$R_T = R_N = \frac{v_{OC}}{i_{SC}} = 18.3 \; \Omega$$

The resulting Thévenin and Norton equivalent circuts are shown in Figures 3-10*d* and 3-10*e* respectively.

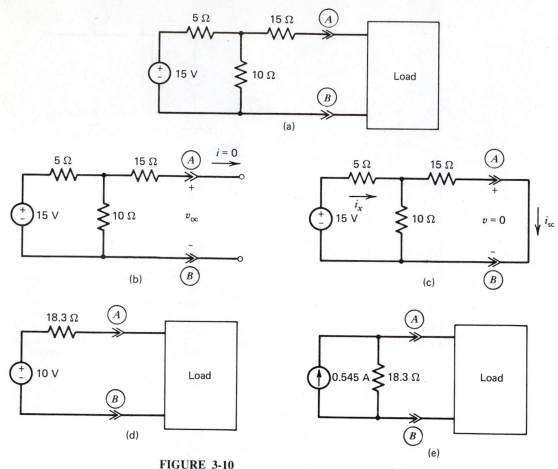

FIGURE 3-10
Computing Thévenin and Norton Equivalents for the circuits in Example 3-5. (*a*) Given circuit. (*b*) Open circuit. (*c*) Short circuit. (*d*) Thévenin equivalent. (*e*) Norton equivalent.

Example 3-6

Let us find the voltage delivered to the load resistor in Figure 3-11 using Thévenin's theorem. The source–load interface is first defined as terminals Ⓐ and Ⓑ in Figure 3-11. The open-circuit voltage that appears across these terminals was found by superposition in Example 3-4 as

$$v_{OC} = \frac{1}{3}(V_{s1} + V_{s2} + V_{s3})$$

For a short-circuit connected between terminals Ⓐ

and Ⓑ, the short-circuit current is easily found by superposition. For sources 2 and 3 off, the short-circuit current due to source 1 is seen to be

$$i_{SC1} = \frac{V_{S1}}{R}$$

Because of the symmetry of the circuit we see that this result applies to all three inputs. By superposition then:

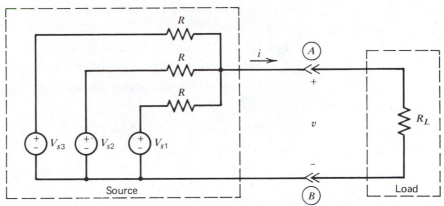

FIGURE 3-11
Summing circuit for Example 3-6.

$$i_{SC} = i_{SC1} + i_{SC2} + i_{SC3}$$

$$= \frac{V_{S1} + V_{S2} + V_{S3}}{R}$$

Hence

$$R_T = \frac{v_{OC}}{i_{SC}}$$

$$= \frac{(1/3)(V_{S1} + V_{S2} + V_{S3})}{(1/R)(V_{S1} + V_{S2} + V_{S3})}$$

$$= \frac{R}{3}$$

The Thévenin equivalent is shown in Figure 3-12. To calculate the voltage delivered to the load we use voltage division to write

$$v = \left[\frac{R_L}{R_L + R/3}\right]\left[\frac{V_{S1} + V_{S2} + V_{S3}}{3}\right]$$

$$= \left(\frac{R_L}{3R_L + R}\right)\right](V_{S1} + V_{S2} + V_{S3})$$

Note that the output remains proportional to the sum of the inputs, but the proportionality constant depends on both the source and load resistances. This source–load interaction results from interfacing two circuits at terminals Ⓐ and Ⓑ.

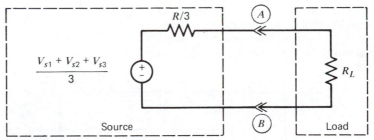

FIGURE 3-12
Thévenin equivalent circuit for Example 3-6.

An alternative, and often easier, method of determining the Thévenin resistances is outlined in Figure 3-13. If we turn off all of the signal sources, that is, replace voltage sources by short circuits and current sources by open circuits, the source circuit reduces to an interconnection of linear resistors that has some equivalent resistance (R_{EQ}) between terminals Ⓐ and Ⓑ. Applying the same condition to the Thévenin equivalent (set $V_T = 0$) indicates that the two circuits are equivalent only if

$$R_T = R_{EQ} \tag{3-12}$$

In summary, any two of the following parameters can be used to determine Thévenin and Norton equivalent circuits:

1. The open-circuit voltage at the interface.
2. The short-circuit current at the interface.
3. The equivalent source circuit resistance at the interface with all sources turned off.

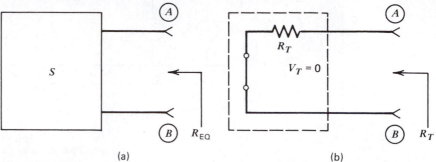

(a)   (b)

FIGURE 3-13
Alternate method of determining the Thévenin equivalent resistance. (*a*) Source circuit with all sources turned OFF. (*b*) Thévenin equivalent with source turned OFF.

Example 3-7

In Example 3-6 we found the Thévenin equivalent of the circuit in Figure 3-11 by finding the open-circuit voltage and short-circuit current. If we turn off all the voltage sources, we obtain the circuit in Figure 3-14. The equivalent resistance between terminals Ⓐ and Ⓑ is determined by three equal resistances in parallel. Hence

$$R_T = R_{EQ} = \frac{R}{3}$$

This is, of course, the same result that was obtained in Example 3-6, but by somewhat easier means.

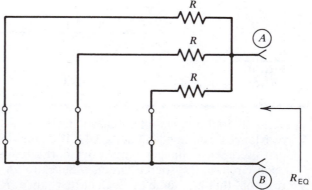

FIGURE 3-14
Finding $R_{\text{Thévenin}}$ for circuit of Figure 3-11 by the alternate method.

Notice that all three of these quantities are always determined with the actual load circuit at the interface disconnected. That is, the Thévenin and Norton equivalences characterize the source circuit and do not depend on the load.

An important use of Thévenin and Norton equivalent circuits is the graphical analysis of circuits containing a two-terminal nonlinear element. An interface is defined at the terminals of the nonlinear element and the linear part of the circuit is reduced to a Thévenin equivalent as indicated in Figure 3-15. The i-v characteristics of the Thévenin equivalent are constrained to be

$$i = \frac{1}{R_T} v + \frac{V_T}{R_T} \qquad\qquad (3\text{-}13)$$

This i-v characteristic is a straight-line in the i-v plane as shown in Figure 3-15. The straight line could logically be called the source line since it is determined by the source alone. Logic not withstanding, electrical

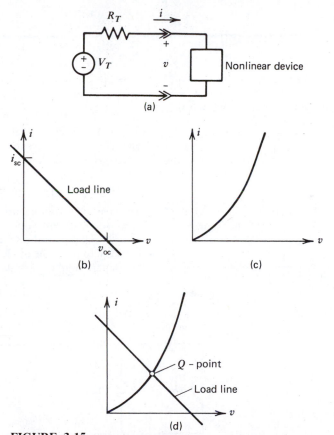

(a)

(b)

(c)

(d)

FIGURE 3-15
Using Thévenin to find current through and voltage across a nonlinear element. (*a*) Circuit. (*b*) Source i-v characteristics. (*c*) Nonlinear device i-v characteristics. (*d*) Source and device i-v characteristic showing Q-point.

engineers call this the **load line** for reasons that have become blurred by the passage of time.

As also shown in Figure 3-15, the nonlinear device has some i-v characteristic determined by its physical makeup. Mathematically this characteristic can be written as

$$i = f(v) \tag{3-14}$$

The problem is to solve Eq. 3-13 and 3-14 simultaneously. This can be done by numerical iteration if the $f(v)$ is known explicitly, but often a graphical solution is adequate.

If we graphically superimpose the source load line on the i-v characteristics of the nonlinear element, the point (or points) or intersection represent the values of i and v that satisfy both the source and load constraints. The point of intersection is called the operating point, or Q-point in the terminology of electronics. An example is used to illustrate the process.

Example 3-8

To find the power delivered to the diode shown in Figure 3-16, we first find the Thévenin equivalent at terminals Ⓐ and Ⓑ. By voltage division we see that

$$V_T = v_{OC} = \frac{100}{100 + 100} \times 5 = 2.5 \text{ V}$$

The equivalent resistance between A and B with the voltage source off is

$$R_{EQ} = 10 + \frac{100 \times 100}{100 + 100} = 60 \ \Omega$$

Therefore the i-v characteristics of the source are constrained to a load line given by

$$i = -\frac{1}{60} v + \frac{1}{60} \times 2.5$$

When this line is superimposed on the diode i-v characteristics we obtain an intersection, or Q-point, at

$$i = 15 \text{ mA} \quad \text{and} \quad v = 1.6 \text{ V}$$

This is the point (i,v) at which both the source and load constraints are satisfied. Finally, the power delivered to the diode is

$$P_D = iv$$
$$= (15 \times 10^{-3})(1.6) = 25 \text{ mW}$$

This circuit is nonlinear so it does not follow the proportionality and superposition properties of linear circuits. That is, if the source voltage were increased from 5 V to 10 V, the diode current and voltage would not simply double. Try it.

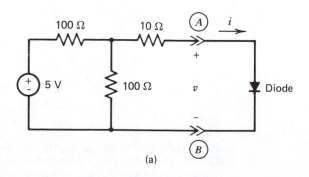

(a)

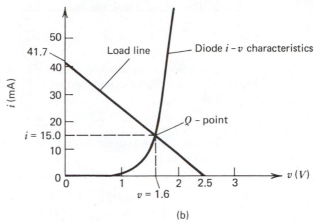

(b)

FIGURE 3-16
Q-point analysis circuit for Example 3-8. (*a*) Diode circuit. (*b*) Diode–load-line analysis.

3-4 MAXIMUM SIGNAL TRANSFER

With the advent of integrated circuits engineers spend a considerable portion of their efforts on interfacing standard, commercially available building blocks called integrated circuits (ICs) or chips. In this context interfacing means interconnecting building blocks so that they operate together in some desired fashion. One of the constraints on achieving desired circuit performance is the maximum signal levels that can be delivered at an interface. Simply put, given a **fixed** source and an **adjustable** load, what are the maximum values of the signals available at an interface?

For simplicity we will treat the case in which both the source and load circuits are linear. The source can be represented by its Thévenin equivalent and the load by an equivalent resistance R_L as shown in Figure 3-17.

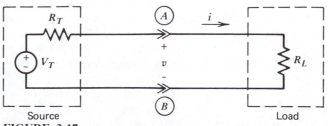

Source   Load
FIGURE 3-17
Circuit for determining maximum power transfer relationships.

By voltage division the voltage at the interface is

$$v = \frac{R_L}{R_L + R_T} V_T \tag{3-15}$$

For a fixed source and a variable load, the voltage will be a maximum if R_L is very large compared witb R_T. Ideally R_L should be infinite, in which case

$$v_{\text{MAX}} = V_T = V_{\text{OC}} \tag{3-16}$$

In sum, the maximum voltage available at the interface is the source open-circuit voltage.

The current delivered at the interface is

$$i = \frac{V_T}{R_L + R_T} \tag{3-17}$$

Again, for a fixed source and a variable load, the current will be a maximum if R_L is very small compared with R_T. Ideally R_L should be zero, in which case

$$i_{\text{MAX}} = \frac{V_T}{R_T} = I_N = i_{\text{sc}} \tag{3-18}$$

In sum, the maximum current available at the interface is the source short-circuit current.

The power delivered at the interface is the product vi. By using Eq. 3-15 and 3-17 we can write

$$\begin{aligned} p &= vi \\ &= \frac{R_L V_T^2}{(R_L + R_T)^2} \end{aligned} \tag{3-19}$$

For the constraint of a fixed source and a variable load, we find the maximum available power by differentiating Eq. 3-19 with respect to R_L.

$$\begin{aligned} \frac{\partial p}{\partial R_L} &= \frac{[\,(R_L + R_T)^2 - 2R_L(R_L + R_T)\,]\,V_T^2}{(R_L + R_T)^4} \\ &= \frac{R_T^2 - R_L^2}{(R_L + R_T)^4} \end{aligned} \tag{3-20}$$

Clearly the derivative will be zero (maximum power) if $R_T = R_L$. Thus for maximum power transfer the source and load resistances should be equal. When this occurs the source and load are generally said to be **matched.**

By substituting the matched condition ($R_T = R_L$) back into Eq. 3-19 we find the maximum power available to be

$$p_{\text{MAX}} = \frac{V_T^2}{4R_T} \tag{3-21}$$

But since $V_T = I_N R_T$, this result can also be written as

$$p_{\text{MAX}} = \frac{I_N^2 R_T}{4} \tag{3-22}$$

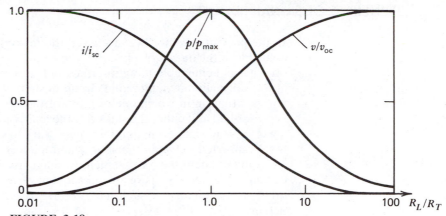

FIGURE 3-18
Normalized plots of current, voltage, and power versus R_L/R_T.

or

$$p_{MAX} = \frac{V_T I_N}{4}$$
$$= [v_{OC}/2] [i_{SC}/2] \qquad \text{(3-23)}$$

Summarizing, the maximum power available at an interface is the product of one-half the open-circuit voltage times one-half the short-circuit current.

Normalized plots of the maximum available signal levels are given in Figure 3-18 as a function of R_L/R_T. Note that the maximum power condition is not a strong function of R_L. Changing R_L by a factor of two in either direction changes the delivered power by less than 20 percent. The reader should carefully note that these maximum values were all derived for the condition of a *fixed* source and a *variable* load. The other possible source–load constraints are studied in the next section.

Example 3-9

Let us determine the maximum signals available from the source in Example 3-8. In that example we found that

$$V_T = 2.5 \text{ V} \quad \text{and} \quad R_T = 60 \text{ }\Omega$$

Hence

$$v_{MAX} = v_{OC} = 2.5 \text{ V}$$
$$i_{MAX} = \frac{V_T}{R_T} = 41.7 \text{ mA}$$
$$p_{MAX} = \frac{v_{OC} \, i_{SC}}{4} = 26.0 \text{ mW}$$

In that example we determined that the nonlinear diode actually received the following signal levels:

$$v = 1.6 \text{ V}$$
$$i = 15 \text{ mA}$$
$$P = 24 \text{ mW}$$

The signals are, of course, below the maximum levels, but note that the power delivered is rather close to the maximum available level even though the diode clearly does not "match" the source.

3-5   PRINCIPLES OF SIGNAL TRANSFER

In the previous section we discussed the maximum signal levels that are available at the output of a fixed source circuit. These results place bounds on what is achievable at an interface but do not describe all of the situations normally encountered in circuit design. Usually we are confronted with a situation in which one or more of the signal levels at the interface is prescribed and either the source or the load, or both, adjusted to achieve a workable design. Sometimes it is necessary to design an interface circuit that is inserted between the source and load as shown in Figure 3-19.

At the moment the only element we have available for interface circuit design is the linear resistor. In subsequent chapters we introduce several other useful elements, such as OP AMP (Chapter 4) and transformers (Chapter 7).

Table 3-1 lists the four canonic situations in which the interface design can arise. Situation I can be encountered when designing with discrete circuit elements. Generally we can pick both the source and the load, but the design problem may involve iteration since the source and load interact. Situation II can arise when we attempt to deliver signals to an existing load or receiver. Generally in this situation it is desirable to reduce the source resistance to a minimum, ideally to zero. There are, of course, limits on the achievement of this goal, but we will shortly see that the OP AMP can be used to approach the ideal. Situation III occurs fairly commonly. Certain types of transmission lines and laboratory signal sources have standard source resistances such as 50, 75, 300, and 600 Ω. In such situations the load must be selected to achieve the desired interface objective.

The last situation occurs all too frequently. We are often faced with the problem or interconnecting standard IC building blocks. The problem usually involves designing an interface circuit that will allow the source and load to interact in harmony. The engineer must consider a wide range of devices such as bridges, pads or potentiometers, OP AMP buffers, or tristate logic—whatever will work must be considered. However, simplicity and cost are major factors that must also influence the design decision.

Before launching into examples the student should recognize that we are now involved in a limited form of circuit design, as contrasted with circuit analysis. Although we use circuit analysis tools in design there is an important difference. A linear circuit *analysis* problem always has a

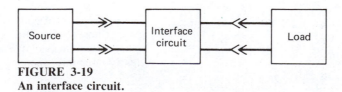

FIGURE 3-19
An interface circuit.

Situation	Source R	Load R
I	Variable	Variable
II	Variable	Fixed
III	Fixed	Variable
IV	Fixed	Fixed

Table 3-1
Interface Design Situations

unique solution. The linear circuit *design* problem may have *several* solutions. Moreover, it is possible to state design problems that have no solutions. Thus the maximum available signal levels determined in the preceding section serve as benchmarks to be used as tests for the existence of a design solution.

Example 3-10

For the circuit in Figure 3-20, select the load resistance such that $v = 10$ V. Repeat for the condition $i = 10$ mA. Since we are given a fixed source the maximum signal levels available at the interface are

$$v_{\text{MAX}} = V_T = 10 \text{ V}$$
$$i_{\text{MAX}} = V_T/R_T = 100 \text{ mA}$$

The first design requirement ($v = 10$ V) demands that the interface voltage equal the available open-circuit voltage. Hence the only solution is $R_L = \infty$. The second design condition requires a current that is less than the maximum available. The current delivered to the load is

$$i = \frac{10}{100 + R_L} = 0.01 \text{ A}$$

Hence

$$1000 = 100 + R_L$$

or

$$R_L = 900 \ \Omega$$

Note that we cannot achieve both design conditions simultaneously. For this circuit a design requirement of $v = 10$ V *and* $i = 10$ mA would be impossible to achieve.

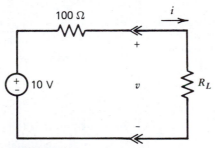

FIGURE 3-20
Circuit used to demonstrate signal transfer in Example 3-10.

Example 3-11

For the circuit in Figure 3-21, select the source resistance such that maximum power is delivered to the load. At first glance it is tempting to let $R_S = R_L = 100\ \Omega$. However, this so-called matched condition assumes that the source is fixed and the load is adjustable. Here we have situation II in Table 3-1, that is, a fixed load and an adjustable source. To determine the power delivered, we first calculate the interface current as

$$i = \frac{10}{100 + R_S}$$

and the voltage by voltage division,

$$v = \frac{10 \times 100}{100 + R_S}$$

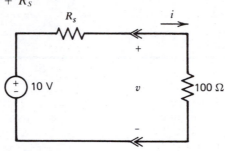

Hence the power delivered across the interface is

$$p = vi$$
$$= \frac{10^4}{(100 + R_S)^2}$$

from which it is clear that maximum power occurs for $R_S = 0$. Examination of these current and voltage expressions reveals that $R_S = 0$ produces the maximum signal levels as well. The moral of this example is that when the source resistance is adjustable, maximum current, voltage, and power are obtained by making R_S small—zero, if possible.

FIGURE 3-21
Circuit used to obtain maximum power transfer when source resistance is variable.

Example 3-12

For the circuit in Figure 3-22, select the load resistance such that exactly 100 mW is delivered across the interface. Here we have a fixed source (situation III) so that the maximum available power is

$$p_{MAX} = \frac{I_N^2 R_N}{4}$$
$$= \frac{(0.1)^2\ 100}{4}$$
$$= 250\ mW$$

Since the design requirement of 100 mW is less than maximum available power, we at least know that design solutions should exist. The interface current can be determined by current division as

$$i = \frac{100}{100 + R_L} \times 0.1$$

Hence the power is

$$p = i^2 R_L$$
$$= \frac{100\ R_L}{(100 + R_L)^2}$$

Since the power is to be 100 mW we can write

$$\frac{100\ R_L}{(100 + R_L)^2} = 0.1$$

which reduces to the quadratic equation,

$$R_L^2 - 800\ R_L + 10^4 = 0$$

which has two roots,

$$R_L = 12.7 \ \Omega, \ 787 \ \Omega$$

There are two values of load resistance that satisfy the design condition, which illustrates that design problems do not necessarily have unique solutions. If one of the roots had been negative, we would have rejected it since all resistors with which we deal must be positive.

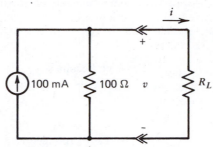

FIGURE 3-22
Circuit used to demonstrate signal transfer in Example 3-12.

Example 3-13

In the circuit in Figure 3-23 an unknown number (N) of identical 600-Ω loads are connected in parallel across the terminals of a fixed source. The design problem is to determine the number of identical load resistors that can be connected and have the interface voltage not fall below 2.5 V. Since the load resistors are identical and connected in parallel, the total load seen at the interface is

$$R_L = \frac{600}{N}$$

By voltage division

$$v = \frac{600/N}{50 + 600/N} \times 5$$

The design requirement is $v \geq 2.5$, hence

$$\frac{600 \times 5}{50 N + 600} \geq 2.5$$

or

$$1200 \geq 50 N + 600$$

and finally,

$$12 \geq N$$

In engineering terms we say that the source can *drive* up to twelve 600-Ω loads and maintain an interface voltage of at least 2.5 V.

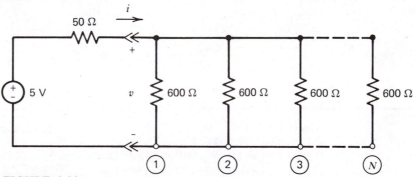

FIGURE 3-23
Use of Thévenin equivalent circuit to determine the effects of adding multiple loads in parallel to a source circuit.

Example 3-14

For the circuit in Figure 3-24, design an interface circuit so that the current delivered to the load is exactly 2 A. This is a fixed source and fixed load problem (situation IV in Table 3-1). It is clear that an interface circuit is needed because, if the source were connected directly to the load, the source current would divide equally between the two resistors, thus producing 5 A into the load. Two possible design solutions are also shown in Figure 3-24. For the parallel resistor we can use current division to obtain

$$i = \frac{1/50}{1/50 \, + \, 1/50 \, + \, G_P} \times 10$$

For the design condition $i = 2$ A, this becomes

$$\frac{10}{2 \, + \, 50 \, G_P} = 2$$

or

$$R_P = \frac{50}{3} = 16.7 \; \Omega$$

For the series resistor circuit we have performed a source conversion to simplify the analysis. For the series case,

$$i = \frac{500}{50 \, + \, 50 \, + \, R_S} = 2 \text{ A}$$

which becomes

$$250 = 100 + R_S$$

or

$$R_S = 150 \; \Omega$$

We have produced two alternative designs that both meet the stated requirement. In real life the engineer must select the best design on the basis of other considerations. The interested reader may wish to complete the design process by calculating the power dissipated in the interface circuit in each case, and by considering the standard resistor sizes that are commercially available (see Appendix C).

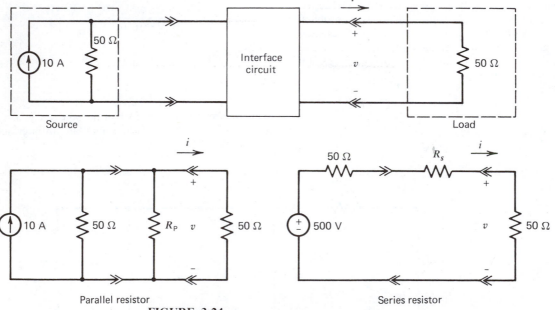

Source   Interface circuit   Load

Parallel resistor   Series resistor

FIGURE 3-24
Circuit for demonstrating interface design.

Example 3-15

A classical interface circuit that is matched at both the source and load is shown in Figure 3-25. What we want is a circuit that has an equivalent resistance of 50 Ω between terminals $\copyright$ and $\textcircled{D}$ when 300 Ω is connected between $\textcircled{A}$ and $\textcircled{B}$, and simultaneously has an equivalent resistance of 300 Ω between $\textcircled{A}$ and $\textcircled{B}$ when 50 Ω is connected between $\copyright$ and $\textcircled{D}$. This rather lengthy design statement can be written mathematically as

Terminals $\copyright$–$\textcircled{D}$

$$50 = \frac{(R_1 + 300)\, R_2}{R_1 + 300 + R_2}$$

Terminals $\textcircled{A}$–$\textcircled{B}$

$$300 = R_1 + \frac{50\, R_2}{R_2 + 50}$$

We have two equations in two unknowns. What could be simpler? But solving these equations proves to be a bit of a chore.

A heuristic approach based on the circuits in Figure 3-25 is as follows: If we let $R_2 = 50\ \Omega$, then the requirement at terminals $\copyright$ and $\textcircled{D}$ will be met, at least approximately. Similiarly, if $R_1 + R_2 = 300\ \Omega$, the $\textcircled{A}$ and $\textcircled{B}$ requirement will be approximately satisfied. In other words, try $R_1 = 250$ and $R_2 = 50$. When these values are used, the equivalent resistances turn out to be 50$\|$550 $= 45.8\ \Omega$ and 250 $+$ 50$\|$50 $= 275\ \Omega$. These are not the exact values specified, but are within 10 percent, and many electrical components are accurate to around 10 percent. For many applications this design would be adequate.

It is important to realize that in this example an exact solution was not sought, but rather a simple approximation was used to solve the problem. Finding simple, practical solutions is the real job of an engineer.

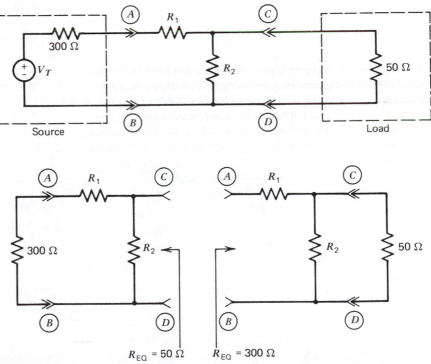

FIGURE 3-25
Circuit for demonstrating heuristic approach to design.

SUMMARY

- A function is linear if it has two properties: homogeneity and additive. In circuit analysis the first is called proportionality and the second is called superposition.

- Unit output method is an analysis technique applicable to ladder-like circuits that uses the proportionality property. In this method a unit output is assumed and the input necessary to produce that output found. The output for the actual input applied is then found by using the proportionality property.

- Superposition method is a technique for analyzing linear circuits containing multiple sources. In this method all sources except one are set to zero (replace voltage sources with short circuits, current sources with open circuits) and the response due to that single source is found. Repeat for all remaining sources. The response due to all sources applied, then, is the sum or algebraic addition of all individual responses.

- Thévenin's and Norton's theorems are two powerful analysis tools that simplify interfacing two circuits. Simply stated:

  Given a two-terminal interface in which the source circuit is linear, then the same interface signals (v and i) will exist if the source is replaced by its Thévenin or Norton equivalents.

- The Thévenin equivalent circuit is shown in Figure 3-26*a*; a Norton equivalent circuit is shown in Figure 3-26*b*. The relationships between Thévenin and Norton equivalent circuits are $R_T = R_N$ and $V_T = I_N R_T$.

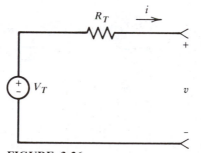 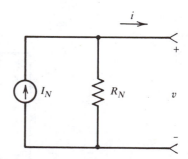

FIGURE 3-26
Thévenin and Norton equivalent circuits.

- For a fixed source and variable load interface, the maximum signal transfer conditions are:

Signal	Condition	Maximum
Voltage	$R_L = \infty$	$v_{MAX} = v_{OC}$
Current	$R_L = 0$	$i_{MAX} = i_{SC}$
Power	$R_L = R_T$	$p_{MAX} = \dfrac{v_{OC}i_{SC}}{4}$

TABLE 3-2

● The condition for maximum power transfer is called matching.

● Signal transfer conditions at an interface are specified in terms of the voltage, current, and power to be delivered to the load. Different design constraints apply depending on the signal conditions specified and the circuit parameters that are adjustable. Some conditions may require the design of an interface circuit to go between the source and the load. Based on the interface required, there may be one, many, or no solutions depending on the constraints placed on the problem.

EN ROUTE OBJECTIVES
AND RELATED EXERCISES

3-1 LINEARITY (SEC. 3-1)

Given a circuit consisting of linear resistors and one input signal source, use the response determined at one input signal level to determine the response at other input signal levels.

Exercises

3-1-1 Find the constant of proportionality K for each of the circuits of Figure E3-1-1.

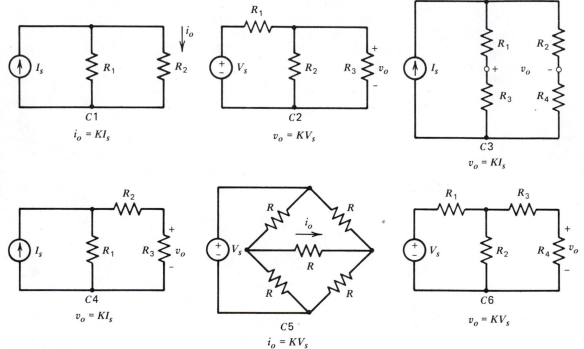

FIGURE E3-1-1
Circuits for computing constants of proportionality.

3-1-2 Certain circuit elements can be described by the *i-v* properties given below. Determine which have the additive property of linearity.

(a) $i = Gv$

(b) $v = L(di/dt)$

(c) $v = (1/C)\int_{-\infty}^{t} i\, dt$

(d) $i = I_o\,(e^{v/V_T} - 1)$

(e) $i = Kv^2$

3-1-3   Use the unit output method to determine the designated output of each of the circuits of Figure E3-1-3.

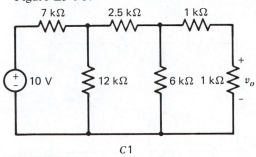

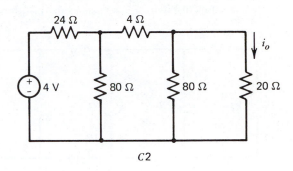

FIGURE E3-1-3
Circuits for determining a designated output using the unit output method.

3-1-4   What should the voltage V_s be so as to produce 6 V at the output of the circuit of Figure E3-1-4?

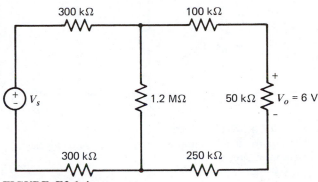

FIGURE E3-1-4
Signal transfer circuit for Exercise 3-1-4.

3-2 SUPERPOSITION (SEC. 3-2)

Given a circuit consisting of linear resistors and two or more input signal sources, determine selected signal variables using the superposition principle.

Exercises

3-2-1 Determine the output voltage in each of the circuits of Figure E3-2-1 using superposition.

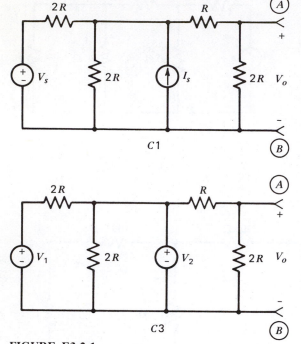

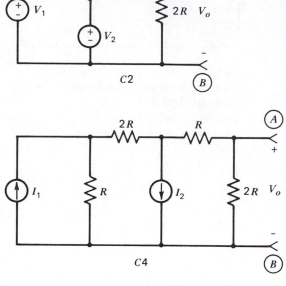

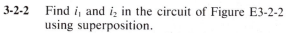

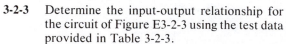

FIGURE E3-2-1
Circuits for determining the output voltage using the method of superposition.

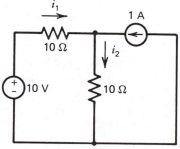

3-2-2 Find i_1 and i_2 in the circuit of Figure E3-2-2 using superposition.

FIGURE E3-2-2
Circuit for superposition Exercise 3-2-2.

3-2-3 Determine the input-output relationship for the circuit of Figure E3-2-3 using the test data provided in Table 3-2-3.

Test	V_3	V_2	V_1	V_o
1	5	0	0	1
2	5	5	0	4
3	5	5	5	6

TABLE 3-2-3

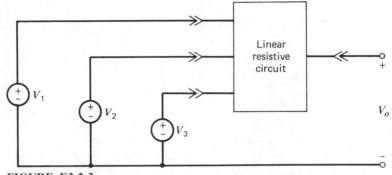

FIGURE E3-2-3
Circuit and test results for Exercise 3-2-3.

3-2-4 Show that the R-$2R$ ladder circuit of Figure E3-2-4 has the input-output relationship given below. Then determine the output for each input combination given in Table 3-2-4.
$$V_o = (V_1/8) + (V_2/4) + (V_3/2)$$

V_3	V_2	V_1	V_o
0	0	0	
0	0	5	
0	5	0	
0	5	5	
5	0	0	
5	0	5	
5	5	0	
5	5	5	

TABLE 3-2-4

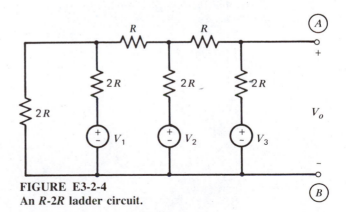

FIGURE E3-2-4
An R-$2R$ ladder circuit.

3-2-5   A resistive load in Figure E3-2-5 is driven by three different sources. If the load dissipates 0.6 W, what is the value of the load resistor?

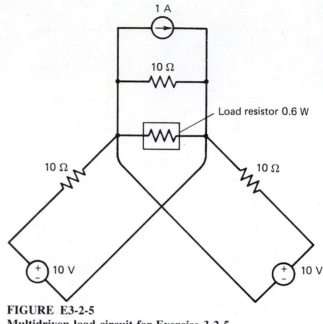

FIGURE E3-2-5
Multidriven load circuit for Exercise 3-2-5.

3-3 *THÉVENIN'S AND NORTON'S THEOREMS (SEC. 3-3)*

Given a circuit consisting of linear resistors, constant input signal sources, and not more than one nonlinear element:

(a)   Determine the Thévenin or Norton equivalent at a specified pair of terminals.

(b)   Use the Thévenin or Norton equivalent to determine the interface signals for specified loads.

Exercises

3-3-1 For each of the circuits of Figure E3-3-1, find the Thévenin and Norton equivalent circuit at terminals Ⓐ and Ⓑ.

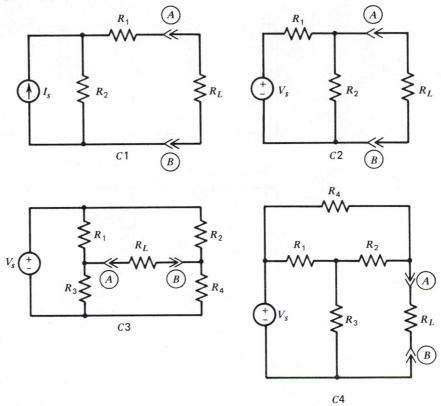

FIGURE E3-3-1
Circuits for finding Thévenin and Norton equivalents.

3-3-2 For each of the circuits of Figure E3-3-2, find the Thévenin equivalent circuit at terminals Ⓐ and Ⓑ with the switch open and with the switch closed.

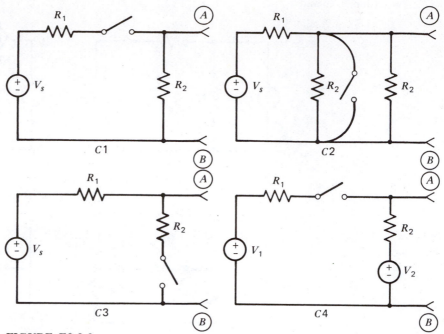

FIGURE E3-3-2
Circuits for finding Thévenin and Norton equivalents.

3-3-3 For each of the circuits of Figure E3-2-1, find the Thévenin equivalent circuit at terminals Ⓐ and Ⓑ.

3-3-4 The open-circuit voltage of a source is 10 V. When a 2-kΩ resistor is connected across the output, the voltage drops to 8 V. Determine the Thévenin equivalent of the source and predict the current, voltage, and power delivered to loads of 250 Ω, 500 Ω, and 1000 Ω.

3-3-5 For the R-$2R$ ladder of Figure E3-2-4, find the Thévenin equivalent circuit at terminals Ⓐ and Ⓑ. (*Hint:* Use successive source transformations.)

3-3-6 Find the Thévenin and Norton equivalent circuit for the network of Figure E3-3-6 at terminals Ⓐ and Ⓑ.

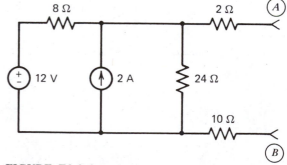

FIGURE E3-3-6
Circuit for finding Thévenin and Norton equivalents.

3-3-7 Suppose that a diode with the *i-v* characteristics shown in Figure E3-3-7*b* is connected to the circuit of Figure E3-3-7*a*. Find the power dissipated in the diode. Repeat assuming the source voltage was increased to 2.0 V.

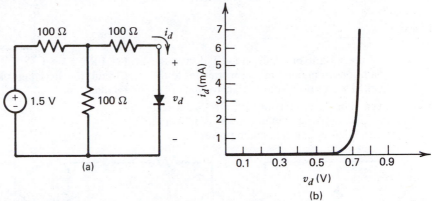

(a)

(b)

FIGURE E3-3-7
Analysis of a circuit containing a nonlinear element.

3-3-8 A nonlinear resistor with the characteristics shown in Figure E3-3-8*b* is connected to a circuit that has the Thévenin equivalent shown in Figure E3-3-8*a*.

With a voltmeter we measure the voltage *v* with and without the nonlinear resistor connected and obtain the following values: $v_{NLR} = 30$ V, $v_{OC} = 40$ V. Find the Thévenin circuit of the source.

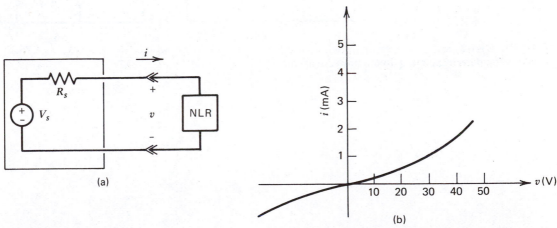

(a)

(b)

FIGURE E3-3-8
Nonlinear resistor circuit for Exercise 3-3-8. (*a*) **Circuit.** (*b*) **Nonlinear resistor's *i-v* characteristics.**

3-4 MAXIMUM POWER TRANSFER (SEC. 3-4)

Given a circuit consisting of linear resistors and constant input signal sources, determine the maximum power available at a specified pair of terminals and specify the load required to obtain the maximum power.

Exercises

3-4-1 For the circuit of Figure E3-4-1 derive the condition for maximum power transfer to the load. (Do not do a source transformation.)

3-4-2 In each of the circuits of Figure E3-4-2 select the load so that maximum power is delivered to it. What is that maximum power?

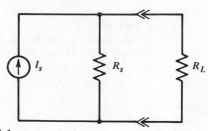

FIGURE E3-4-1
Circuit for deriving the maximum power transfer relationship.

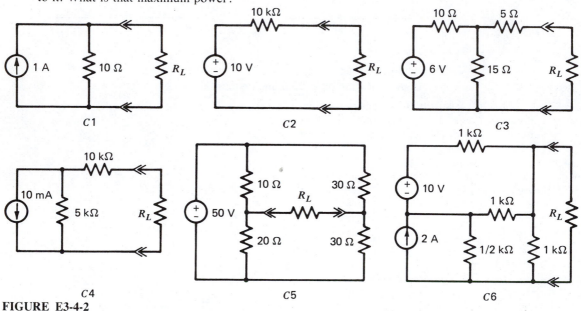

FIGURE E3-4-2
Circuits for determining maximum power transfer.

3-4-3 Suppose we could model a certain stereo speaker as a linear resistor of 8 Ω. Show how you could connect one or more speakers to each of the stereo amplifiers represented by their Thévenin or Norton equivalent circuits in Figure E3-4-3 so that maximum power is always delivered to the speaker(s).

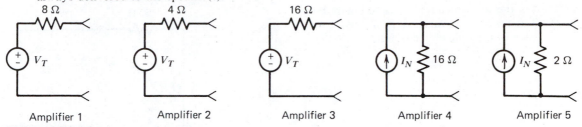

FIGURE E3-4-3
Stereo speaker connections.

3-4-4 A load R_L is connected to an interface that is connected to a source. For each of the various sources shown in Figure E3-4-4, select R_L so that maximum power is delivered.

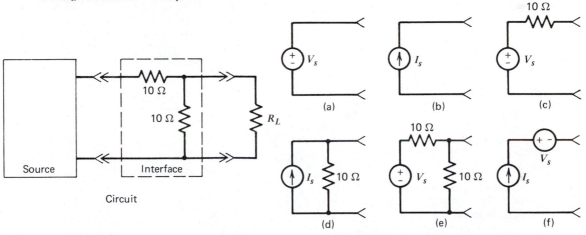

FIGURE E3-4-4
Circuits with various sources for obtaining maximum power transfer.

3-5 SIGNAL TRANSFER (SEC. 3-5)

Given a source-load interface with specified constraints, adjust the circuit parameters or design a suitable interface to achieve given signal transfer objectives.

Exercises

3-5-1   For the source-load interface of Figure E3-5-1, determine the values of R_L needed to meet each of the following constraints:

   (a) $v = 1$ V
   (b) $v = 2.5$ V
   (c) $v = 5$ V
   (d) $v = 6$ V
   (e) $i = 100$ mA
   (f) $i = 0$ A
   (g) $p = 100$ mW
   (h) $p = 125$ mW
   (i) $p = 200$ mW

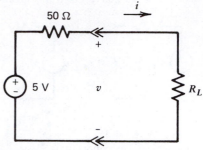

FIGURE E3-5-1
Circuit for varying R_L to meet various constraints.

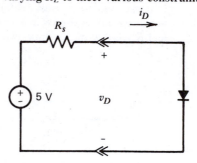

$$i_D = 10^{-15}\,(e^{40 v_D} - 1)$$

FIGURE E3-5-2
Diode circuit for exercise 3-5-2.

3-5-2   For the source-load interface of Figure E3-5-2, determine the values of R_S needed to achieve the following:

   (a) $v_D = 0.7$ V
   (b) $i_D = 0.5$ mA

3-5-3   A novel device is shown in Figure E3-5-3 that permits the operator to select a variety of outputs. Indicate to which positions you would set the two switches to obtain each of the outputs requested.

   (a) For maximum power if $R_L = 100\ \Omega$.
   (b) For 6 V if $R_L = 100\ \Omega$.
   (c) For maximum voltage if $R_L = 1$ kΩ.
   (d) For maximum current if $R_L = 50\ \Omega$.
   (e) For 40 mA if $R_L = 100\ \Omega$.
   (f) For 60 mA if $R_L = 100\ \Omega$.
   (g) For 4 V if $R_L = 50\ \Omega$.

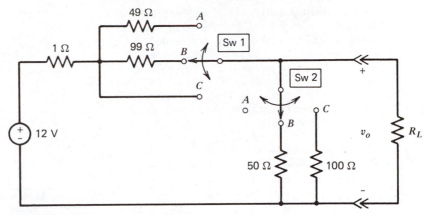

FIGURE E3-5-3
Switch circuit for Exercise 3-5-3.

3-5-4   Design an interface that will allow 2 V to be delivered to the load of Figure E3-5-4. Repeat to permit a current of 50 mA to flow through the load.

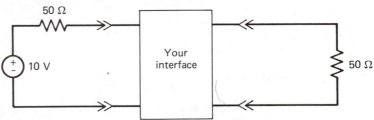

FIGURE E3-5-4
Interface design problem.

3-5-5   The source circuit of Figure E3-5-5 must always "see" an equivalent resistance of 50 Ω to avoid damaging it. Design an interface using only resistors that will permit the electronic switch to "switch" yet always permit the source to "see" ≈50 Ω and still permit at least 1 percent of the signal to be transferred to the load.

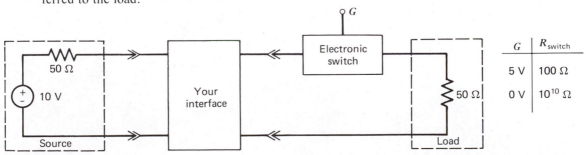

G	R_{switch}
5 V	100 Ω
0 V	10^{10} Ω

FIGURE E3-5-5
Matching design problem.

3-5-6 The 1 kΩ resistor is driven by a 110 V source shown in Figure E3-5-6. In case of power failure, a backup 12 V source is provided. Design an interface that will permit essentially uninterrupted power to the load. You should use the electronic switches provided. Note the two different kinds. The gate G draws no current but you must show how to connect it to the main source.

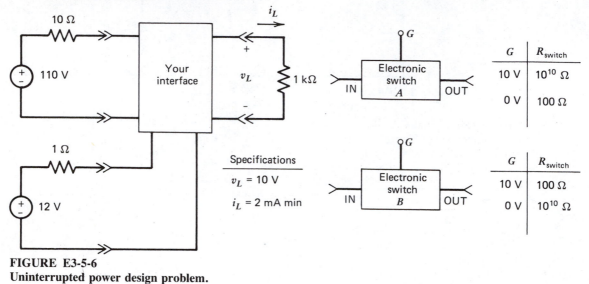

FIGURE E3-5-6
Uninterrupted power design problem.

PROBLEMS

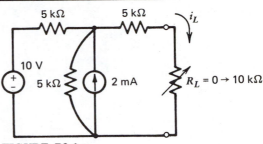

FIGURE P3-1
A signal transfer problem.

P3-1 (Analysis)
If R_L in Figure P3-1 varies from 0 to 10 kΩ, how does i_L vary?

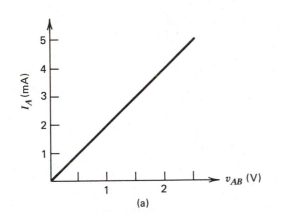

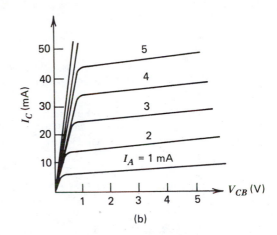

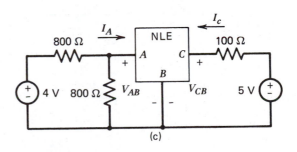

P3-2 (Analysis)
A nonlinear element (NLE) has the characteristics shown in Figure P3-2a,b. If the element is connected as shown in the circuit of Figure P3-2c, what are V_{CB} and I_C?

FIGURE P3-2
Nonlinear element characteristics and circuit. (*a*) Input characteristics. (*b*) Output characteristics. (*c*) Circuit.

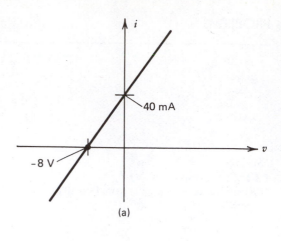

(a)

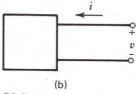

(b)

FIGURE P3-3
The *i-v* relations for unknown Thévenin circuit.

P3-3 (Analysis)
A certain circuit has the *i-v* characteristics shown in Figure P3-3*a*. What is its Thévenin equivalent?

P3-4 (Analysis)
Using the typical values provided in Table P3-4, find the input and output resistance of the hybrid-π model for a BJT (bipolar junction transistor) shown in Figure P3-4.

Typical Values

$r_{bb'}$	100 Ω
$r_{b'c}$	2 kΩ
$r_{b'c}$	20 MΩ
r_{ce}	200 kΩ
g_m	50 mA/V

TABLE P3-4

P3-5 (Evaluation)
A certain student is asked to match a 100-Ω load to a 50-Ω source for maximum power. After some thought the student adds a 100-Ω resistor in parallel to the load. Has the task been accomplished? Explain.

P3-6 (Design)
The two separate systems shown in Figure P3-6 must be connected to a 200-Ω source so that *both* have $V_T/8$ V across them simultaneously. Design a suitable interface.

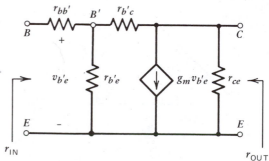

FIGURE P3-4
BJT hybrid-π model.

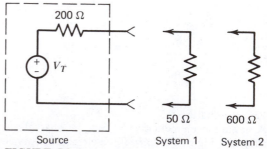

FIGURE P3-6
A signal transfer problem.

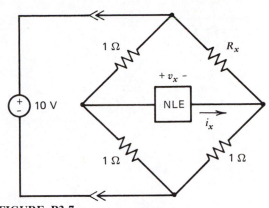

FIGURE P3-7
A nonlinear element—Wheatstone bridge problem.

P3-7 (Analysis)
The bridge circuit shown in Figure P3-7 contains a nonlinear element (NLE) with the characteristic $i = v^2$. Find an expression for i_x and v_x. Show that for $R_x = 1 \, \Omega$, $i_x = v_x = 0$.

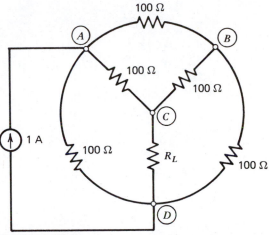

FIGURE P3-8
A Y-Δ network problem (see Problem E2-6-11).

P3-8 (Analysis)
A Y-Δ network is constructed as shown in Figure P3-8. Select R_L so that maximum power is delivered to it. How much power is delivered under those conditions, and how much power must the source supply?

Chapter 4
Active Circuits

Regeneration or feedback is of considerable importance in many applications of vacuum tubes.

Harry Nyquist

An active circuit is one capable of delivering more power to a load than it receives from the signal source. This chapter examines the amplifier, the workhorse of electronics. The transistor is the fundamental building block of current electronic amplifiers. The operational amplifier, or OP AMP, represents a second-generation improvement over the transistor as the basic building block of analog circuits. Both the transistor and the OP AMP, as well as many other active devices, can be modeled and analyzed using a new circuit element—the controlled source.

4-1 LINEAR DEPENDENT SOURCES

One of the most important signal-processing functions available to electrical engineers is signal **amplification.** This function cannot be achieved using any of the elements we have studied thus far. Circuits consisting only of linear resistors cannot produce output signal voltages, currents, or, most important, powers that are larger than their signal inputs. To obtain signal amplification we need active devices such as vacuum tubes, transistors, and operational amplifiers (OP AMPS). Simple models of these active devices require a new set of circuit elements called **dependent** or **controlled** sources.

Active devices are usually very nonlinear. However, those of interest to us in this text can be made to operate in a linear mode. When active devices are operating in a linear mode we can model their characteristics by using one or more of the four dependent source elements shown in Figure 4-1. The dominant feature of a dependent source is that the strength or magnitude of the source is proportional to—that is, controlled by—a voltage or current appearing elsewhere in the circuit. This feature should be contrasted with the characteristics of the independent signal sources we studied earlier. For example, the current delivered by an independent signal current source does not depend on the circuit to which it is connected. Likewise the voltage produced by an independent signal voltage source is not affected by the circuit. To emphasize this difference, dependent sources are represented by the diamond symbol shown in Figure 4-1, in contrast to the circle symbol used for independent sources.

Several matters of notation and symbols should be mentioned. Each of the controlled sources is characterized by a single parameter, either

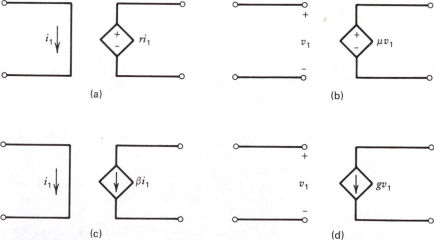

FIGURE 4-1
Dependent sources. (*a*) Current-controlled voltage source. (*b*) Voltage-controlled voltage source. (*c*) Current-controlled current source. (*d*) Voltage-controlled current source.

μ, β, r, or g. These parameters are often somewhat loosely called the *gain* of the controlled source. Strictly speaking, the parameters μ and β are dimensionless quantities called the open-circuit voltage gain and the short-circuit current gain respectively. The parameter r has the dimensions of ohms, and is called the transresistance, a contraction of transfer resistance. The parameter g is called transconductance and has the dimensions of mhos.

Dependent sources are linear elements that are used in circuit analysis. But in some respects they are conceptually different than the other elements we have used. The resistance element is an ideal model of a set of devices called resistors. The ideal switch is a model of a component called a switch. But you will not find controlled sources in parts lists and catalogs. Controlled sources in combination with other elements are used to model real devices such as transistors and OP AMPs. In this sense they are more abstract, so "one must learn by doing."

A word of caution: In this text we use the diamond symbol to indicate a linear dependent source. Many other texts do not employ this symbol, but instead use the circle symbol as for the independent sources. The reader should realize that it is the gain parameter that really identifies a source as dependent.

Example 4-1

In Chapter 3 we learned that when a voltage source is turned OFF it acts as a short circuit. Likewise when a current source is turned OFF it behaves as an open circuit. The same results apply to controlled sources, with one important difference. Controlled sources cannot be turned OFF independently because they depend on excitation supplied by independent sources. The consequences of this are illustrated in Figure 4-2. When the independent source is ON, it forces the condition $i_1 = i_S$. Through controlled source action the dependent source is ON and

$$v_o = ri_1$$
$$= ri_S$$

When the independent source is OFF ($i_S = 0$), it acts as an open circuit forcing the condition $i_1 = 0$; hence by controlled source action the dependent source is OFF and

$$v_o = ri_1$$
$$= 0$$

It then acts as a short circuit. In sum, the independent source takes the controlled source with it. Among other things this means that in applying the principle of superposition we cannot turn dependent sources ON or OFF. Their state will depend on the excitation supplied by signal sources.

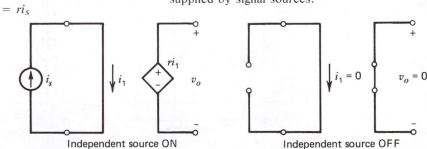

Independent source ON   Independent source OFF

FIGURE 4-2
A dependent source circuit for Example 4-1.

4-2 ANALYSIS OF CIRCUITS WITH DEPENDENT SOURCES

The analysis of circuits containing controlled sources is fundamentally the same as the passive circuits we have previously studied. Kirchhoff's laws still apply and so the connection constraints are the same. The element constraints are still independent of the connections. But we have a new set of elements so that the combined constraints yield circuit behavior that is significantly different than passive circuits. For this reason our analysis examples are chosen to highlight these differences.

A simple example of a circuit using a voltage-controlled source is shown in Figure 4-3a. Applying Kirchhoff's voltage law around loop I yields

$$v_S = R_S i + v_1$$

But $i = 0$ since there is an open circuit, and hence

$$v_S = v_1$$

The controlled voltage source output therefore is equal to μv_S. Since v_o is in parallel with the controlled source we have

$$v_o = \mu v_S \tag{4-1}$$

The output is directly proportional to the input. If μ is greater than 1, then we say that the circuit **amplified** the input and we call the circuit an **amplifier.** If μ is less than 1, we say that the circuit **attenuated** the input and we call the circuit an **attenuator.**

The real value of linear active circuits can be demonstrated by performing a similar analysis of the passive circuit in Figure 4-3b. Application of the ever-so-familiar voltage division relationship yields

$$v_o = \frac{R_L v_S}{R_S + R_L} \tag{4-2}$$

If we compare the results of our analysis of both circuits, Eqs. 4-1 and 4-2, we can make some important observations.

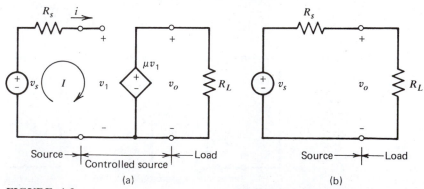

(a)   (b)

FIGURE 4-3
Comparison of active and passive circuits. (a) Active circuit. (b) Passive circuit.

First of all, the output of the circuit with the dependent source does not depend on either the source or load resistance; rather, it depends only on the gain μ. This means that the output voltage is not limited by the maximum signal transfer conditions derived in Chapter 3. In fact, if the gain μ is greater than 1, the output voltage can be greater than the input voltage. In effect the controlled source provides unilateral signal transfer that isolates the source and load. This eliminates the source–load interaction that led to the maximum signal transfer conditions.

To illustrate this further consider the power delivered to the load in the controlled source circuit. Using Eq. (4-1),

$$P_L = (V_o)^2/R_L = (\mu V_S)^2/R_L \tag{4-3}$$

The load power is not dependent on the source resistance, and hence is not limited by the maximum power transfer condition ($R_S = R_L$). In fact, the power delivered by the source is zero since it is connected to an open circuit and i_S is necessarily zero. The circuit produces an output power with zero input power. This apparent contradiction can be resolved by realizing that a dependent source is a model of devices that require an external energy source to operate. In general, we do not show the external power supply that allows the device to work. Further we assume that the external supply can indeed provide whatever power the dependent source requires. In real devices this is not always the case, and the design engineer must ensure that the limits of the external supply are not exceeded. In our discussion of OP AMPs we will consider at least some limitations of the external supply.

The next two examples illustrate the application of analysis techniques treated in Chapters 2 and 3 to active circuits.

Example 4-2

Find the output-input relationship (v_o/i_S) of the circuit in Figure 4-4.

We observe that the dependent source requires us to find i_1. Hence by current division in the input circuit,

$$i_1 = \frac{R_S i_S}{R_S + R_1}$$

To find v_o we must know i_L. We find i_L by applying current division in the output circuit.

$$i_L = \frac{R_P i_2}{R_P + R_L}$$

But by KCL $i_Z = -\beta i_1$, and then substituting for i_1, we find that

$$i_L = \frac{-R_P \beta i_1}{(R_P + R_L)} = \frac{-R_P \beta}{(R_P + R_L)} \cdot \frac{R_S i_S}{(R_S + R_1)}$$

Finally, we can find v_o by applying Ohm's law:

$$v_o = i_L R_L = \frac{-R_L R_P \beta}{(R_P + R_L)} \cdot \frac{R_S i_S}{(R_S + R_1)}$$

and the output-input relationship is

$$\frac{v_o}{i_S} = \frac{-R_L R_P \beta R_S}{(R_P + R_L)(R_S + R_1)}$$

The negative sign occurred in this example because of the orientation of the dependent source in the problem. In many active device models the dependent source is inverted, causing a negative output for a positive input, and vice versa. This special property of active devices is called **inversion.** It is important that during analysis of such circuits one keep track of these inversions.

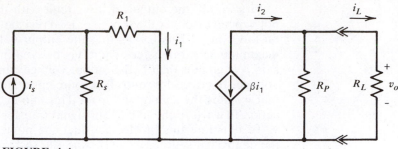

FIGURE 4-4
A dependent source circuit for Example 4-2.

Example 4-3

It often is useful to represent linear active circuits by their Thévenin or Norton equivalent. To illustrate we will find the Thévenin equivalent at the output interface of the circuit of Figure 4-5a.

The Thévenin equivalent circuit seen by the load can be found by removing the load resistor and computing the open-circuit voltage v_{OC}. This analysis proceeds in this way:

$$i_1 = \frac{v_S}{R_S + R_P}$$

because of the open circuit at the output interface,

$$v_{OC} = ri_1 = \frac{rv_S}{R_S + R_P} = V_T$$

To find the Thévenin resistance we must first find I_{SC}. We replace the open circuit with a short circuit and calculate the current flowing through the short circuit:

$$i_{SC} = \frac{ri_1}{R_o} = \frac{rv_S}{R_o(R_S + R_P)}$$

The Thévenin resistance is then found as

$$R_T = \frac{v_{OC}}{i_{SC}} = R_o$$

The Thévenin circuit seen by the load R_L is then as shown in Figure 4-5b.

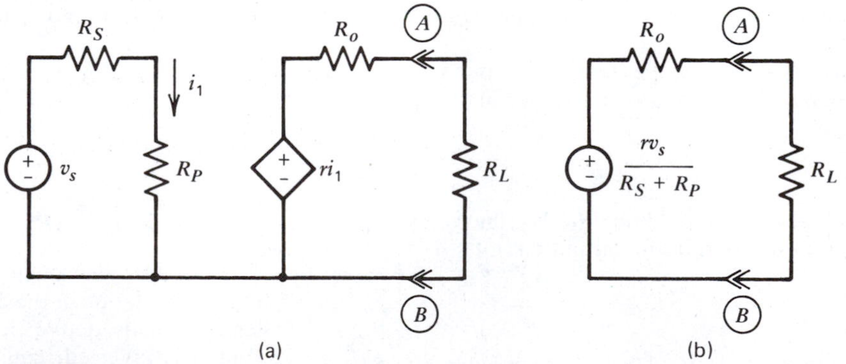

(a)

(b)

FIGURE 4-5
Thévenin analysis of circuit with a dependent source. (*a*) Active circuit. (*b*) Thévenin equivalent.

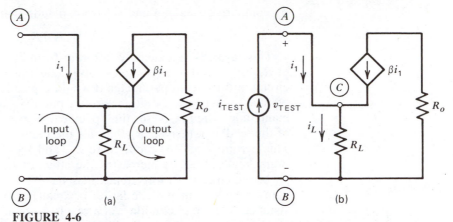

FIGURE 4-6
Analysis of dependent source circuit with feedback. (*a*) Active circuit with feedback.
(*b*) Circuit driven by test source.

The results of Example 4-3 may suggest that finding the Thévenin or Norton equivalent of an active circuit is a rather straightforward process. This is not quite the case if there is **feedback** in the circuit. Feedback occurs when there is a path from the dependent source back to its input control circuit.

To illustrate the effect we will determine the equivalent resistance R_{EQ} between terminals Ⓐ and Ⓑ in Figure 4-6*a*. The circuit in Figure 4-6*a* has no excitation from independent sources, and hence the dependent current source is off and acts as an open circuit. As a result it intuitively appears that the equivalent resistance between Ⓐ and Ⓑ should be simply R_L.

However, the effect of the dependent source has not been accounted for since the circuit has no excitation. To provide excitation we connect the test source shown in Figure 4-6*b* and analyze the circuit response. By KCL at node Ⓒ:

$$i_L = i_1 + \beta i_1$$
$$= (1 + \beta)i_1$$

By KCL at node Ⓐ $i_1 = i_{TEST}$, and hence

$$i_L = (1 + \beta)i_{TEST}$$

But by Ohm's law,

$$v_{TEST} = R_L i_L$$

Hence

$$v_{TEST} = (1 + \beta)R_L i_{TEST} \qquad (4\text{-}4)$$

Now the principle of equivalence says that two circuits are equivalent if they have the same *i-v* characteristics. Since equivalent resistance means

$$v_{TEST} = R_{EQ} i_{TEST} \qquad (4\text{-}5)$$

by comparing Eqs. 4-4 and 4-5 we conclude that

$$R_{EQ} = (1 + \beta)R_L \tag{4-6}$$

This result is significantly different from our initial intuitive estimate of $R_{EQ} = R_L$. Moreover the circuit in question is a model of a transistor circuit and the gain parameter β would typically be of the order of 100. So not only is our intuition wrong, but it is wrong by two orders of magnitude! The reason for this spectacular result is the unique placement of the feedback resistor R_L that ties the input and output loops together. This permits both the input current i_1 and the output current βi_1 to add dramatically, thereby increasing the effect of R_L on the circuit.

Active circuits with feedback may have Thévenin resistances that are many orders of magnitude larger or smaller than the apparent input resistance with no excitation. As a result, to find the Thévenin equivalent of an active circuit we should always find the open-circuit voltage and short-circuit current since both of these calculations require that the circuit be driven by independent sources. The next example shows that this approach is necessary.

Example 4-4

Determine the Thévenin equivalent at the output interface of the circuit in Figure 4-7.

If we turn the input signal source off, the control source becomes a short circuit since there is no excitation. Disconnecting the load and looking back into the circuit then suggests that the Thévenin resistances should be R_o. That this is drastically wrong is shown by the following analysis. First we determine the open-circuit voltage. Applying KVL around the perimeter of the circuit yields

$$v_1 = v_S - v_{OC}$$

This equation points out that there is feedback since the controlling signal (v_1) depends on the output and the input. With the output open no current flows through R_o; hence

$$v_{OC} = \mu v_1$$
$$= \mu (v_S - v_{OC})$$

thus

$$v_{OC} = \frac{\mu}{1 + \mu} v_S$$

We now connect a short circuit across the output to find i_{SC}. With a short circuit connected, KVL around the circuit perimeter indicates that

$$v_1 = v_S$$

hence

$$i_{SC} = \frac{\mu v_1}{R_o}$$
$$= \frac{\mu v_S}{R_o}$$

So we find

$$R_T = \frac{v_{OC}}{i_{SC}}$$
$$= \frac{R_o}{1 + \mu}$$

The circuit is a model of an OP AMP circuit called a voltage follower. For an OP AMP R_o is of the order of 100 Ω, while the gain is at least 10^5. Thus looking back into the circuit with no excitation would lead us to think that the Thévenin resistance is 100 Ω, when in fact it is less than a milliohm. As we will see, the use of high gain and feedback to produce very low output resistances is an important feature of OP AMP circuits.

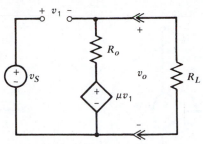

FIGURE 4-7
A dependent source circuit for Example
4-4.

4-3   TRANSISTOR MODELS

Transistors are active three-terminal semiconductor devices with very nonlinear *i-v* characteristics. Depending on how the transistor is operated, one or more different models can be used to analyze its behavior. There are many transistor models: the Ebers–Moll, hybrid-π, hybrid parameter, incremental, and large signal, to mention just a few. Because the transistor is such an important electronic device, and because it will help us in our understanding of the OP AMP in the next section, we will devote some time to studying the transistor using one of its simpler models—the large-signal model.

The large-signal model of the transistor is shown in Figure 4-8. This model is often called piecewise linear because it simplifies the very non-linear transistor characteristics into three linear regions or modes, called CUTOFF, ACTIVE, and SATURATED. Each mode has its own special model. The analysis of the large-signal transistor model requires that the operating mode be determined. Once the mode is determined, the appropriate model can be substituted for the circuit symbol. A new circuit is drawn with the appropriate substitution and then it is treated using standard analysis techniques.

It should be noted that while this is a simple model, it is applicable to the analysis of many low-speed digital switching circuits where the transistor is switched between the OFF and the SATURATED modes, or in amplifier applications ACTIVE mode where the nonlinearitites can be ignored.

The criteria for determining which mode to use are based on a knowledge of the physics that make up the transistor and are beyond the scope of this text. However, certain simple assumptions will enable one to use the models without recourse to the physics.

CRITERION 1   If the base-to-emitter voltage V_{BE} is equal to or greater than the turn-on voltage V_T (usually taken to be 0.7 V), the transistor is said to be ON; otherwise it is OFF.

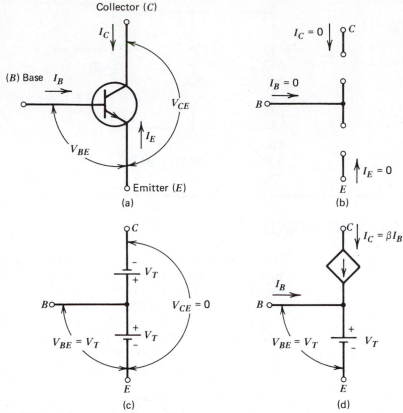

FIGURE 4-8
Large-signal transistor model. (*a*) Circuit symbol (NPN). (*b*) CUTOFF mode model
(OFF). (*c*) SATURATED mode model (ON). (*d*) ACTIVE mode model (ON).

CRITERION 2   If the collector-to-emitter voltage V_{CE} is equal to zero, the transistor is said to be SATURATED. (In reality a saturated transistor has a V_{CE} of about 0.2 V, not zero.)

CRITERION 3   If one assumes that the transistor is SATURATED, that is, V_{CE} is set to zero, and the collector current I_C is computed, this current $I_{C.SAT}$ represents the maximum current that can flow in the collector loop. If we compute βI_B, then this current represents the value of collector current that would exist if the transistor were operating in the ACTIVE mode. Now if βI_B is greater than $I_{C.SAT}$, then the transistor is SATURATED and I_C is equal to $I_{C.SAT}$. If, on the other hand, βI_B is less than $I_{C.SAT}$, the transistor is ACTIVE and the collector current simply equals βI_B.

The following example illustrates how to analyze a simple transistor circuit using these criteria.

Example 4-5

Find I_C and V_{CE} in the transistor circuit of Figure 4-9.

Applying KVL around the input loop yields

$$5 = R_B I_B + V_{BE}$$

If the transistor were CUTOFF, i_B would equal zero. With $I_B = 0$, our input loop KVL equation would have V_{BE} equal to 5 V. But if V_{BE} equals 5 V, the transistor cannot be CUTOFF since by criterion 1 $V_{BE} > V_T$. Therefore the transistor is ON and $V_{BE} = 0.7$ V.

Since we now know that the transistor is ON we can find I_B.

$$I_B = \frac{5 - 0.7}{100K} = 43\ \mu A$$

We know that the transistor is ON, but we do not know whether it is SATURATED or ACTIVE. To determine this, we assume the transistor is SATURATED by making $V_{CE} = 0$ V. Applying KVL around the output loop yields

$$5 = R_C I_C + V_{CE}$$

If we assume the transistor is SATURATED, that is, $V_{CE} = 0$, we can find $I_{C.SAT}$:

$$I_{C.SAT} = \frac{5}{R_C} = 5\ mA$$

$I_{C.SAT}$ represents the maximum collector current that can flow in this problem. Now we compute βI_B. If βI_B is greater than $I_{C.SAT}$, there is sufficient base current I_B to saturate the transistor and $I_C = I_{C.SAT}$. If βI_B is less than $I_{C.SAT}$, then the transistor is in the linear ACTIVE mode since there is insufficient I_B to saturate the transistor.

$$\beta I_B = 100 \times (43\ \mu A) = 4.3\ mA$$

4.3 mA is less than 5 mA $I_{C.SAT}$. By criterion 3, the transistor is operating as a linear amplifier in the ACTIVE mode.

In digital switching circuits the transistor switches between CUTOFF and SATURATED, passing through the ACTIVE mode very quickly. If this transistor were to be used in such a circuit, either R_B would have to be decreased (increasing I_B and ensuring saturation) or R_C increased (decreasing I_C and ensuring saturation). Increasing β would work, but β is a function of the transistor and cannot be changed without selecting a different transistor.

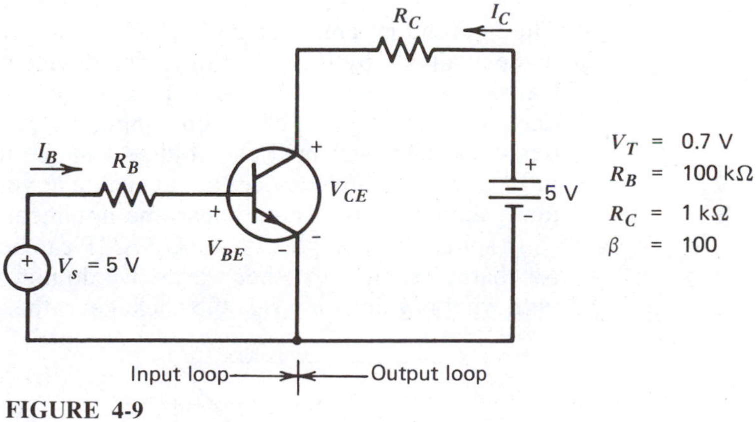

FIGURE 4-9
Large-signal model analysis of a transistor circuit.

Example 4-6

Consider the circuit of Figure 4-9. Suppose that the input signal v_S increased as a function of time as $v_S(t) = t$ V. Draw a sketch of v_{CE} versus t.

Referring to the analysis already accomplished in Example 4-5, we note that until the input is equal to or greater than 0.7 V the transistor is OFF. If the transistor is OFF, V_{CE} must equal 5 V. As the voltage increases with time beyond 0.7 V, the transistor is ON. When $v_S(t) = 0.7$ V, the transistor is just turned on and current I_C starts to flow. The greater the input voltage, the more "on" the transistor becomes and the more collector current I_C flows. The more collec-

tor current flows, the more voltage is dropped across R_C and the less across V_{CE}. When $i_C R_C$ equals 5 V, v_{CE} must equal zero. Beyond this no more collector current can flow; the transistor is said to be SATU-RATED. Table 4-1 was used to develop the sketch shown in Figure 4-10.

t (second)	$v_S(t)$ (volts)	I_B (amperes)	I_C (ampere)	V_{CE} (volts)
0.00	0.00	0.00	0.00	5.00
0.69	0.69	0.00	0.00	5.00
0.71	0.71	0.1μ	10.0μ	4.99
1.00	1.00	3.0μ	0.3m	4.70
2.00	2.00	13.0μ	1.3m	3.70
3.00	3.00	23.0μ	2.3m	2.70
4.00	4.00	33.0μ	3.3m	1.70
5.00	5.00	43.0μ	4.3m	0.70
5.70	5.70	50.0μ	5.0m	0.00
6.00	6.00	53.0μ	5.0m	0.00

TABLE 4-1

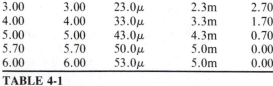

FIGURE 4-10
Response of circuit in Figure 4-9 to $V_S = t$ input.

4-4   THE OPERATIONAL AMPLIFIER

The operational amplifier (OP AMP) is the premier linear active device made available by IC technology. The device itself is a complex array of transistors, resistors, diodes, and capacitors, all fabricated and interconnected on a tiny piece of silicon commonly called a chip. The completed device is then packaged and sold as a single unit, such as those shown in Figure 4-11. Such devices are available at very low unit costs and have many applications in both linear and nonlinear circuits.

In spite of its complexity, the OP AMP can be modeled by rather simple i-v characteristics. In other words, we do not need to concern ourselves with what is going on inside the package; rather we can treat the OP AMP

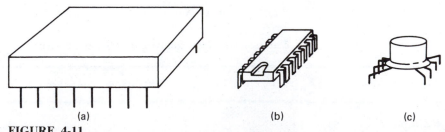

(a)   (b)   (c)

FIGURE 4-11
OP AMP packages. (*a*) Encapsulated hybrid. (*b*) DIP package (eight or fourteen pins). (*b*) TO can.

Historical Note

The term *operational amplifier* apparently was first used in the literature in a 1947 paper by John R. Ragazzini and his colleagues reporting work carried out for the National Defense Research Council during World War II. The paper described the use of high-gain dc amplifiers in the development of circuits that performed various mathematical operations (addition, subtraction, multiplication, division, integration, etc.); thus the name "operational" amplifier was coined. For more than a decade the most important applications were general and special-purpose analog computers. By 1960 vacuum tube amplifiers for such computers were available at unit costs of the order of $100. In the early 1960s discrete-transistor, general-purpose, operational amplifiers became available commercially, most notably from the Burr-Brown Corporation. In those years the Burr-Brown *Handbook of Operational Amplifier Applications* was a primary source of information on the use of this device in signal processing. In the middle 1960s the first commercial IC OP AMPs became available. Because of IC fabrication processes, these devices were differential amplifiers with both "plus" and "minus" input terminals.

The earlier vacuum tube amplifiers had only a single "inverting" input, so it was natural that the additional input came to be called the "noninverting" input. The initial IC OP AMPs sold for unit prices of the order of $10, but by 1970 the price had dropped to less than $1. In the transition from vacuum tubes to integrated circuits OP AMPs decreased in size, power comsumption, and cost by more than two orders of magnitude. During the 1970s the IC version became the dominant active device in almost all nondigital circuits. The price drop illustrates the powerful influence of the economy of scale offered by IC technology. At the time IC production was economically feasible only if a large number of nominally identical units could be produced. Conversely a large number of units could be sold only if there were many potential applications. The operational amplifier was ideally suited to this situation. It both was amenable to IC fabrication and had a wide variety of applications. Thus IC OP AMPs are not useful *because* they are inexpensive; they are inexpensive *because* they are useful.

as a circuit element with certain terminal characteristics. In what follows we learn how to analyze and design OP AMP circuits using what is called its **ideal** model.

Before we develop this model there are certain matters of notation and nomenclature that must be discussed. The OP AMP to be presented is a five-terminal device as shown in Figure 4-12. The " + " and " − " symbols identify the input terminals and are a shorthand notation for the noninverting and inverting input terminals respectively. These symbols simply identify the two input terminals and have nothing to do with the polarity of the voltages applied. The other terminals are the output and the plus and minus supply voltage, usually labeled $\pm V_{CC}$. While some OP AMPs have more terminals than these, these five are always present and are the only ones we will use in this text.

It is especially important to note the two terminals marked $\pm V_{CC}$.

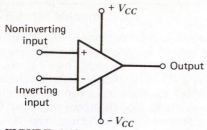

FIGURE 4-12
OP AMP circuit symbol.

These two terminals usually are not included in OP AMP circuits so as to keep the diagrams uncluttered; however, they are always there. It is through these terminals that the power for the operation of the OP AMP must flow. The supply voltages, as $\pm V_{CC}$ is often called, provide the power for amplification and determine the upper and lower voltage limits on the OP AMP output. In general these voltages mark the boundary between linear and nonlinear operation of the OP AMP.

Figure 4-13 shows a complete set of voltage and current variables for the OP AMP, and also shows the abbreviated set of signal variables we will use. All voltages are defined with respect to a common reference node, usually ground. Figure 4-13*b* shows the shorthand notation we use to describe the signal voltages. The voltage symbol is written beside the terminal and we describe these voltages as the output voltage, or the voltage at the inverting input, etc. By such terminology we really mean the voltage between the output terminal and the reference node, or the voltage between the inverting input terminal and ground.

The reference directions for the currents are the traditional ones (in at

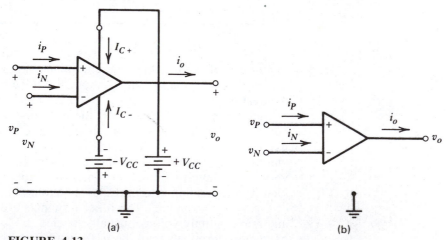

FIGURE 4-13
OP AMP currents and voltages. (*a*) Complete set of variables. (*b*) Signal variables only.

inputs and out at outputs), but the use of the abbreviated set of signal variables can cause conceptual problems. If we were to write a global KCL equation for the complete set of variable in Figure 4-13a, it would read

$$i_o = I_{C+} + I_{C-} + i_P + i_N \quad \text{(correct)} \tag{4-7}$$

A similar equation for the signal variables in Figure 4-13b reads

$$i_o = i_P + i_N \quad \text{(incorrect)} \tag{4-8}$$

This latter equation is not correct as it does not include all of the currents. More important, it implies that the output current comes from the inputs. In fact this is wrong. The input currents are very small, ideally zero. Thus the output current comes from the supply voltages as Eq. 4-7 points out, even though these terminals are not shown on the diagram.

The dominant feature of the OP AMP is its transfer characteristics shown in Figure 4-14. These characteristics provide the relationships between the two input voltages (v_P, the noninverting input, and v_N, the inverting input) and the output voltage (v_o). The transfer characteristic is divided into linear and saturation ranges. In the linear range the output is proportional to the difference between the two inputs, and consequently the OP AMP is called a **differential** amplifier. The slope of the line in the linear region is called the OP AMP gain or open-loop gain, and is denoted as μ. Thus in the linear range the input-output relation can be written as

$$v_o = \mu \, (v_P - v_N) \tag{4-9}$$

The open-loop gain of an OP AMP is very large, usually greater than 10^5. As long as the net input ($v_P - v_N$) is small, the output will be proportional to the input. However, when μ times the net input lies outside the range defined by the supply voltages $\pm V_{CC}$, the output is limited by the supply

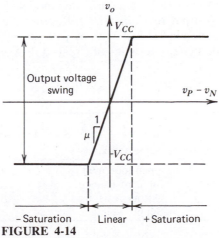

FIGURE 4-14
OP AMP transfer characteristics.

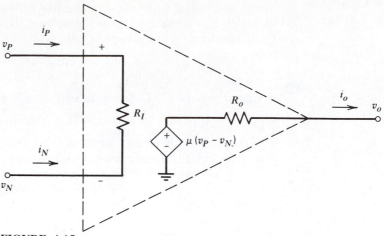

FIGURE 4-15
Controlled source model of linear OP AMP.

voltages (less some small internal losses). When this occurs, the OP AMP is said to be saturated, and the output is no longer proportional to the input but is determined by the supply voltages.

There are applications such as the **comparator** or **Schmidt trigger** where the OP AMP is intentionally driven into the saturation regions. Such nonlinear uses of the OP AMP find applications in many electronic circuits. (We briefly discuss the comparator in Section 4-8.) However, in our treatment for now, we shall restrict the use of the OP AMP to its linear region. In general we wish to analyze and design circuits that do not saturate for the given input(s).

A controlled source model of the OP AMP is shown in Figure 4-15. In addition to the open-loop gain μ, this model includes an input resistance (R_I) and an output resistance (R_o). Representative values for these parameters are given in Table 4-2, along with the values for the ideal OP AMP. More detailed characteristics of "real" OP AMPS are given in Appendix B.

Parameter	Name	Typical Values	Ideal OP AMP Values
μ	Open-loop gain	10^5–10^7	∞
R_I	Input resistance	10^6–10^{13}	∞
R_o	Output resistance	10–100	0
$\pm V_{CC}$	Supply voltages	± 15 V	± 15 V

TABLE 4-2
OP AMP Parameters

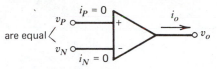

FIGURE 4-16
Ideal OP AMP characteristics.

The controlled source model can be used to develop the *i-v* character-istics of the ideal model. We have restricted our treatment to the linear range of operation. This means that

$$- V_{CC} \leq v_o \leq + V_{CC}$$

By using Eq. 4-9, we can write this bound as

$$- \frac{V_{CC}}{\mu} \leq (v_P - v_N) \leq + \frac{V_{CC}}{\mu}$$

The supply voltage $\pm V_{CC}$ is typically ± 15 V while μ is a very large number, usually 10^5 or greater. Consequently $v_P \simeq v_N$ for linear operation. For the ideal OP AMP the open-loop gain is infinite ($\mu = \infty$) and this fuzzy equality becomes an exactitude. Moreover the input resistance of the ideal OP AMP model is also infinite so no current is drawn at either input. In sum, the *i-v* characteristics of the ideal OP AMP are

$$\begin{aligned} v_P &= v_N \\ i_P &= i_N = 0 \end{aligned}$$
(4-10)

These characteristics are illustrated on the OP AMP circuit symbol in Figure 4-16.

At first glance the element constraints of the ideal OP AMP appear to be fairly useless. They actually look more like connection constraints and are totally silent about the output quantities (v_o and i_o), which are usually the signals of interest. The answer to this dilemma is that in linear ap-plications feedback is always present. That is, for the OP AMP to operate in a linear mode it is necessary that there be some feedback paths from the output to one or both of the inputs. These feedback paths allow us to analyze OP AMP circuits using the ideal OP AMP constraints.

To illustrate this process let us determine the input-output character-istics of the circuit in Figure 4-17. This circuit has a feedback path from the output to the inverting input via a voltage divider. Since the ideal OP AMP draws no current at either input ($i_P = i_N = 0$), we can use the voltage division rule to determine the voltage at the inverting input:

$$v_N = \frac{R_2}{R_1 + R_2} v_o$$
(4-11)

The input source connection at the noninverting input requires that

$$v_P = v_S$$
(4-12)

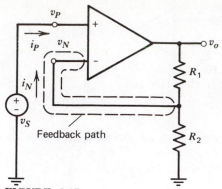

FIGURE 4-17
The noninverting OP AMP circuit.

But the ideal OP AMP constraint demands that $v_P = v_N$; hence we can equate the right sides of Eqs. 4-11 and 4-12 to obtain the input-output relationship of the circuit as

$$v_o = \frac{R_1 + R_2}{R_2} v_S \qquad \text{(4-13)}$$

The analysis strategy is to use the input signal source constraint together with the feedback path to determine the two OP AMP input voltages. The ideal OP AMP constraint requires that these voltages be equal and this equality is used to determine the overall circuit input-output relationship.

The circuit in Figure 4-17 is called the **noninverting** amplifier. The input-output relationship is of the form $v_o = Kv_S$, which reminds us that the circuit is linear. The constant

$$K = \frac{R_1 + R_2}{R_2} \qquad \text{(4-14)}$$

is called the closed-loop gain since it includes the effect of the feedback path. In discussing OP AMP circuits it is necessary to distinguish between two types of gains. The first is the **open-loop** gain provided by the OP AMP. This gain is a very large number and ideally it is infinite. Then there is the **closed-loop** gain of the OP AMP circuit, which includes a feedback path. This gain must be much smaller and is determined by the elements in the feedback path. For example, the closed-loop gain in Eq. 4-14 is really the voltage division rule upside down. Thus the feedback converts the very high but imprecisely known open-loop gain into a much smaller but precisely known closed-loop gain.

Example 4-7

Design an amplifier with a closed-loop gain $K = 10$. To obtain this gain, we will use the noninverting OP AMP amplifier circuit. The design problem is to select the values of the resistors in the feedback path. From Eq. 4-14 the design constraint is

$$10 = \frac{R_1 + R_2}{R_2}$$

We have one constraint with two unknowns. Hence if we select $R_2 = 10$ kΩ, we find $R_1 = 90$ kΩ. These resistors would normally have high precision to produce a precise closed-loop gain.

We should pause here and reflect for a moment on the rather dramatic properties of the ideal OP AMP. The model has an infinite open-loop gain, which is converted into a finite closed-loop gain through feedback. Real OP AMPS have very large, but finite open-loop gains. The question we now want to address is whether the difference between an infinite gain (the ideal model) and a large but finite gain (the real thing) is really significant. To accomplish this we will analyze the noninverting amplifier using a controlled source model of the OP AMP.

The circuit in Figure 4-18 shows the model we will use. We have taken the input resistance R_I to be infinite. The actual values of OP AMP input resistance range from 10^6 to 10^{12} Ω so no important effect is left out by

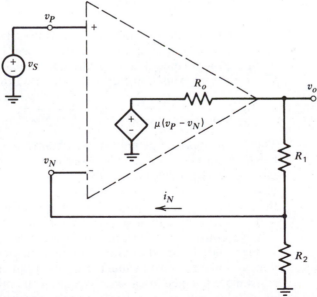

FIGURE 4-18
The noninverting OP AMP circuit with a controlled source model.

ignoring this resistance. In examining the circuit we see that the noninverting input voltage is determined by the input signal source and the inverting input can be found by voltage division since the current i_N is zero. In other words, Eqs. 4-11 and 4-12 apply to this circuit as well. We next determine the output voltage in terms of the controlled source voltage using voltage division on the three resistors R_o, R_1, and R_2 connected in series.

$$v_o = \frac{R_1 + R_2}{R_o + R_1 + R_2} \mu (v_P - v_N)$$

Substituting v_P and v_N from Eqs. 4-11 and 4-12 yields

$$v_o = \frac{R_1 + R_2}{R_o + R_1 + R_2} \mu \left[v_S - \frac{R_2}{R_1 + R_2} v_o \right] \tag{4-15}$$

This intermediate result reminds us that there is feedback because the output depends on the input (v_S) and itself. Solving for v_o yields

$$v_o = \frac{\mu(R_1 + R_2) v_S}{R_o + R_1 + R_2(1 + \mu)} \tag{4-16}$$

In this result we see the first indication of the effect of high open-loop gain. That is, if μ is a very large (but finite) number, Eq. 4-16 can be written as

$$v_o \simeq \frac{R_1 + R_2}{R_2} v_S = K v_S$$

where K is the closed-loop gain we previously found using the ideal OP AMP model that has an infinite gain. To see this more clearly we ignore R_o in Eq. 4-16 since it is generally quite small compared with $R_1 + R_2$. With this approximation Eq. 4-16 can be written as

$$v_o = \frac{K}{1 + K/\mu} v_S \tag{4-17}$$

Thus as μ approaches infinity we obtain the closed-loop gain predicted by the ideal OP AMP model. Moreover, we see that the ideal OP AMP model yields good results as long as $K \ll \mu$. In other words, the ideal element is a valid model as long as the closed-loop gain is much less than the actual open-loop gain of the real OP AMP.

The feedback path also affects the apparent output resistance. To see this we construct a Thévenin equivalent circuit using the open-circuit voltage and the short-circuit current. Equation 4-17 is the open-circuit voltage and we need only find the short-circuit current to find the Thévenin resistance. Connecting a short circuit at the output in Figure 4-18 forces $v_N = 0$ but leaves $v_P = v_S$. Hence the short-circuit current is

$$i_{sc} = \mu v_S / R_o$$

and as a result

$$R_T = v_{OC}/i_{SC}$$
$$= \frac{K/\mu}{1 + K/\mu} R_o$$

If the closed-loop gain is much smaller than the open-loop gain, this reduces to

$$R_T = \frac{K}{\mu} R_o \simeq 0$$

Thus the OP AMP circuit with feedback has an output resistance that is very much smaller than the output resistance of the OP AMP itself. In fact it is essentially zero.

At this point we should summarize our discussion. We have introduced the OP AMP as a five-terminal device, including two supply terminals that are not normally shown on circuit diagrams. We then developed an ideal model of this device and used that model to analyze and design circuits that have feedback. Such feedback is always present and is necessary for the device to operate in a linear mode. The most dramatic feature of the ideal model is the assumption of infinite gain. We then used the controlled source model of the OP AMP to explore the consequences of the large but finite gain of real OP AMPs. We found that the ideal model predicts the feedback amplifiers response very well as long as the closed-loop gain is much less than the actual open-loop gain of the OP AMP. We also discovered that the output resistance of the feedback amplifier is essentially zero.

In summary, the *i-v* constraints (Eqs. 4-10) of the ideal OP AMP model can be used to analyze the types of circuits of interest in this text. Such circuits have essentially zero output resistance, which means that any reasonable load connected to the output does not affect the output voltage. Such loads must be within the power-handling capability of the actual device of course. But the same restrictions apply to all of the models we use, including the linear resistance model of the load itself.

Hereinafter when we use the term OP AMP we mean the ideal model thereof. That is, the term OP AMP will mean the ideal OP AMP element and all that is implied by that model.

Example 4-8

The circuit in Figure 4-19 has a direct feedback path from the output to the noninverting input. As a result of this connection $v_N = v_o$. The input signal source is directly connected to the noninverting input, and hence $v_P = v_S$. The ideal OP AMP model requires $v_P = v_N$; hence we conclude that $v_o = v_S$. The closed-loop gain is $K = 1$, which means that the output follows the input, and the name voltage follower is derived. The voltage follower finds applications in interface circuits as discussed later in this chapter. Notice that the input-output relationship can be obtained without regard to the load R_L. This points out that the circuit has zero output resistance and can therefore supply any reasonable load. We can calculate the output current actually delivered to the load

$$i_o = v_o/R_L$$

But since $v_o = v_s$ in this circuit, the output current is

$$i_o = v_s/R_L$$

Thus the input voltage is transferred directly to the load. If we now write a KCL equation at the reference node, we discover an apparent dilemma:

$$i_P = i_o$$

Now $i_P = 0$ for our model but i_o is not zero unless v_s is zero. Thus it appears that KCL is violated. The dilemma is resolved by noting that the circuit diagram does not include the supply terminals. The output current actually comes from the power supply and not from the input. This dilemma only arises at the reference node. In OP AMP circuits KCL is satisfied at all nodes including the reference node providing we recall that the supply terminals are not shown.

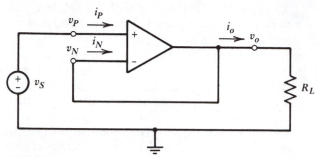

FIGURE 4-19
The voltage follower.

4-5 BASIC LINEAR OP AMP CIRCUITS

Analog signal-processing systems are often constructed using interconnections of relatively simple OP AMP circuits in a building block fashion. In this section we introduce a basic set of circuits that can be used as these building blocks. We have already introduced one of these circuits—the noninverting amplifier discussed in the preceding section. The other circuits are the **inverting** amplifier, the **summer,** and the **subtractor.** The key to using the building block approach is to recognize the feedback pattern and to isolate the basic circuit as a building block. Our first example illustrates this process.

Example 4-9

Determine the input-output relationship of the circuit in Figure 4-20.

When the circuit is partitioned as shown we should recognize the noninverting amplifier. Since this circuit has zero output resistance the load has no effect on the output voltage. Hence we can write, by inspection,

$$v_o = \frac{R_3 + R_4}{R_4} v_P$$

since R_3 and R_4 form the feedback path. Because the input current at the noninverting terminal is zero, we can determine v_P using voltage division.

$$v_P = \frac{R_2}{R_1 + R_2} v_S$$

Thus the overall circuit input-output relationship is

$$v_o = \frac{R_3 + R_4}{R_4} \times \frac{R_2}{R_1 + R_2} v_S$$

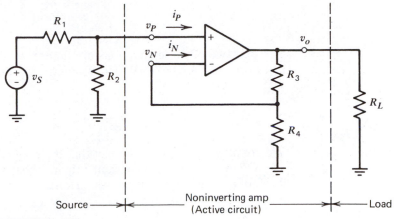

FIGURE 4-20
OP AMP circuit for Example 4-9.

Our next basic OP AMP circuit is the inverting configuration shown in Figure 4-21. The distinguishing feature of this circuit is that the input signal and the feedback are both applied at the inverting input. The noninverting input is grounded, which means that $v_P = 0$. We make use of this important observation shortly. The inverting circuit is usually analyzed by applying KCL at node Ⓐ.

$$i_1 + i_2 = i_N \tag{4-18}$$

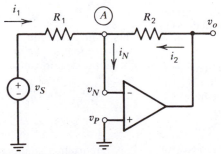

FIGURE 4-21
The inverting OP AMP circuit.

Since this is a sum of currents, node Ⓐ is often called the summing point. The element constraints for this circuit can then be written as

$$i_1 = \frac{(v_S - v_N)}{R_1}$$

$$i_2 = \frac{(v_o - v_N)}{R_2} \tag{4-19}$$

$$i_N = 0$$

The first two of these equations are an expression of Ohm's law and the last is one of the OP AMP $i\text{-}v$ characteristics. Substituting Eqs. 4-19 into 4-18 yields

$$\frac{v_S - v_N}{R_1} + \frac{v_o - v_N}{R_2} = 0 \tag{4-20}$$

We are now ready to use the fact that the noninverting input is grounded. The OP AMP voltage constraint requires that $v_P = v_N$, but since $v_P = 0$, it follows that $v_N = 0$ as well. Hence we can solve Eq. 4-20 for the input-output relationship as

$$v_o = -\frac{R_2}{R_1} v_S \tag{4-21}$$

This result is of the form $v_o = K v_S$, where K is the closed-loop gain. In this case K is negative, indicating a signal inversion between input and output, and hence the name **inverting** amplifier. Again we observe that the closed-loop gain depends only on the resistors forming the feedback path. Our next example shows that more complex circuits can often be reduced to the inverting amplifier configuration.

Example 4-10

Determine the input-output relationship of the circuit in Figure 4-22a. The circuit to the right of node Ⓑ is an inverting amplifier. The load resistor has no effect on its transfer characteristics since the circuit has zero output resistance. If we construct a Thevenin equivalent of the circuit to the left of node Ⓑ, we obtain the circuit shown in Figure 4-22b, where

$$v_T = \frac{R_2}{R_1 + R_2} v_S \qquad R_T = \frac{R_1 R_2}{R_1 + R_2}$$

The Thévenin resistance is connected in series with the input resistor R_3, and using series equivalence the circuit can be reduced to that shown in Figure 4-22c, where

$$R_{EQ} = R_3 + R_T$$

This reduced circuit is in the form of the inverting configuration, so we can write the overall circuit input-output relationship as

$$v_o = -\frac{R_4}{R_{EQ}} v_T$$

$$= -\frac{R_1 R_4 + R_2 R_4}{R_1 R_2 + R_1 R_3 + R_2 R_3} v_S$$

It is important to note that the overall gain is not simply the ratio R_4/R_3 but depends on the equivalent resistance at the input—a fact that must be taken into account when designing with inverters.

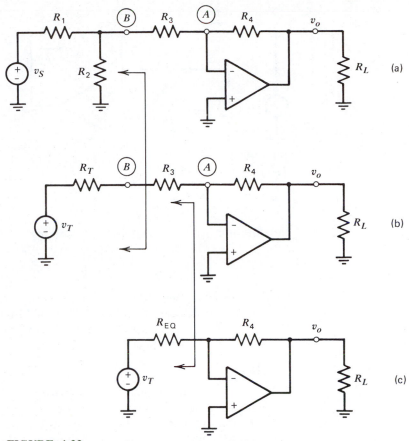

FIGURE 4-22
Circuit and analysis for Example 4-10. (*a*) Given circuit. (*b*) Thévenin's theorem applied at Ⓑ. (*c*) Equivalent inverting circuit.

Our next building block is the **summer** or **adder** circuit shown in Figure 4-23. This circuit has two inputs connected at the summing point (node Ⓐ) and the noninverting input is again grounded; hence $v_P = 0$. This configuration is quite similar to the inverting amplifier and so we proceed by applying KCL at the summing point.

$$i_1 + i_2 + i_F = i_N \qquad (4\text{-}22)$$

Making use of the fact that $v_P = v_N$ due to the OP AMP constraint, and hence $v_N = 0$, we write the element constraints as

$$i_1 = \frac{v_1}{R_1} \qquad i_2 = \frac{v_2}{R_2}$$

$$i_F = \frac{v_o}{R_F} \qquad i_N = 0 \qquad (4\text{-}23)$$

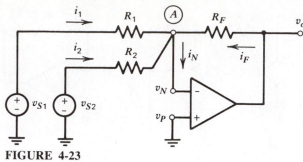

FIGURE 4-23
The OP AMP summer.

Substituting Eqs. 4-23 into 4-22 and solving for the output yields

$$v_o = -\frac{R_F}{R_1}v_1 - \frac{R_F}{R_2}v_2 \qquad (4\text{-}24)$$

The output is a weighted sum of the two inputs. The weighting factors, or gains as we will call them, are determined by the ratio of the feedback resistor R_F to the input resistor for each input. In the special case $R_F = R_1 = R_2$, Eq. 4-24 reduces to

$$v_o = -(v_1 + v_2)$$

Hence we have the name summer, or more precisely, inverting summer.

There are at least two noteworthy items regarding this circuit. First of all it is a multiple-input circuit, and as such we cannot refer to the circuit gain; rather there are two gains to contend with, one for each input. Second, if one were to put a finger over the input v_2 and its input resistor R_2, the circuit would appear to be identical to the noninverting amplifier. It should then be a simple step for the reader to expand our two-input summer into one with any number of inputs using superposition.

Example 4-11

Design an inverting summer to implement the input-output relationship

$$v_o = -(5v_1 + 13v_2)$$

The design problem is to select input and feedback resistors such that

$$\frac{R_F}{R_1} = 5 \quad \text{and} \quad \frac{R_F}{R_2} = 13$$

One solution is to select $R_F = 65\ \text{k}\Omega$, which requires $R_1 = 13\ \text{k}\Omega$ and $R_2 = 5\ \text{k}\Omega$. This design is shown in Figure 4-24. There is nothing particularly wrong with this design, except that the resistance values are not the standard values available in precision resistors. An alternative is to select $R_F = 100\ \text{k}\Omega$, which leads to $R_1 = 20\ \text{k}\Omega$ and $R_2 = 7.69\ \text{k}\Omega$. The first two values are standard sizes and only R_2 would require special consideration.

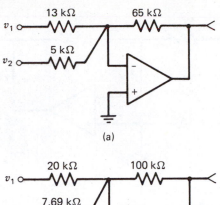

(a)

(b)

FIGURE 4-24
Possible design solutions for Example 4-
11. (*a*) First design. (*b*) Alternative design.

The final circuit in our collection of basic building blocks is the **differential** amplifier or **subtractor** shown in Figure 4-25. This circuit also has two inputs, one applied at the inverting and one at the noninverting input. The input-output relationship can be obtained by applying superposition and recognizing certain features of the circuit. First we replace the input v_2 by a short circuit, in which case there is no excitation at the noninverting input, and hence $v_P = 0$. In other words, the noninverting input is effectively grounded and the circuit acts as an inverting amplifier with the result that

$$v_{o1} = -\frac{R_2}{R_1} v_1 \qquad (4\text{-}25)$$

Now, turning v_2 back ON and replacing v_1 by a short circuit, we see that the circuit looks like a noninverting amplifier with a voltage divider connected at its input. This case was treated in Example 4-9, so we can write

$$v_{o2} = \frac{R_4}{R_3 + R_4} \times \frac{R_1 + R_2}{R_1} v_2 \qquad (4\text{-}26)$$

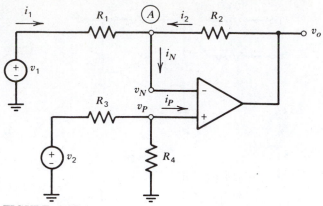

FIGURE 4-25
The OP AMP subtractor.

Using superposition we can then combine Eqs. 4-25 and 4-26 to obtain

$$v_o = v_{o1} + v_{o2}$$
$$= \frac{R_4}{R_3 + R_4} \frac{R_1 + R_2}{R_1} v_2 - \frac{R_2}{R_1} v_1 \qquad \textbf{(4-27)}$$

In the special case of $R_1 = R_2 = R_3 = R_4$ this expression reduces to

$$v_o = v_2 - v_1$$

and hence the name subtractor.

We have now introduced four basic OP AMP circuit building blocks. It is often convenient to represent these circuits in block diagram form. The reason is that much of analog system analysis and design take place at the **functional** or **block diagram** level. A major advantage of OP AMP circuits is that there is a near one-to-one correspondence between the block diagrams and the circuits that actually implement the system function. For linear systems there are only the two block diagram symbols shown in Figure 4-26. The amplifier or gain block simply says that the output is obtained by multiplying the input by the block gain, that is, $v_o = Kv_S$. The summing symbol is almost self-explanatory. It indicates that the output is simply $v_S = v_1 + v_2$. The amplifier block and the summing symbol can then be combined to produce a weighted summer as shown in Figure 4-26(c).

The block diagram representation of our four basic circuits is shown in Figure 4-27. The noninverting and inverting amplifiers are represented as gain blocks. The inverting summer and subtractor require both gain blocks and the summing symbol. Considerable care must be used in translating from the block diagram to the circuit since some of the block may involve negative gains, that is, inversions. For example, the gain of the

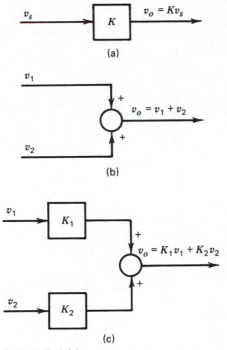

FIGURE 4-26
Block diagram symbols. (*a*) Amplifier block. (*b*) Summer. (*c*) Combined funcitons.

inverting amplifier is negative, as are the K_1 and K_2 gains of the summer and the K_1 gain of the subtractor. This minus sign is sometimes moved to the summing symbol and the gain within the block changed to a positive number. Since there is no standard convention for doing this, it is important to keep track of the signs associated with gain blocks.

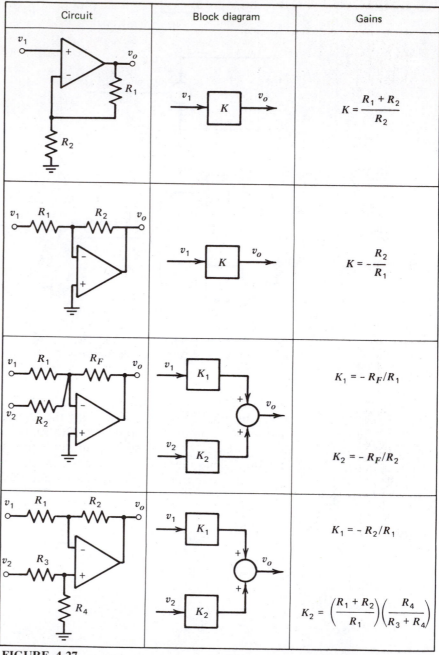

FIGURE 4-27
Block diagram representations of some basic linear OP AMP circuits.

Example 4-12

Construct a block diagram for the circuit in Figure 4-28a.

We recognize the circuit as a subtractor. The two gains are

$$K_1 = -\frac{40}{10} = -4$$

$$K_2 = \frac{40 + 10}{10} \times \frac{10}{10 + 10} = 2.5$$

This leads to the block diagram shown in Figure 4-

28b, and we can write the input-output relationship by inspection as

$$v_o = 2.5\, v_2 - 4\, v_1$$

The illustration also shows an alternative block diagram, Figure 4-28c, in which the negative gain is replaced by a positive gain and the minus sign moved to the summing symbol.

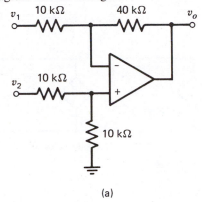

(a)

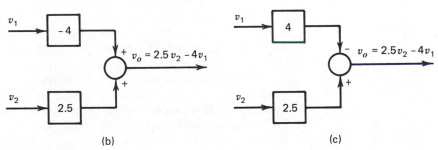

(b)   (c)

FIGURE 4-28
Circuit and block diagrams of a subtractor. (a) Circuit. (b) Block Diagram.
(c) Alternate block diagram.

4-6   CASCADED OP AMP CIRCUITS

One of the important advantages of OP AMP circuits is that we can connect them in cascade without worrying about the loading effects of one circuit on the other, provided we take a few simple precautions. There are several reasons for connecting circuits in cascade; the main ones are

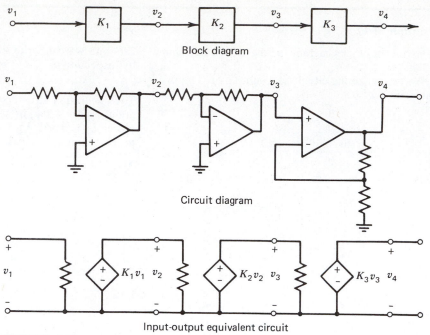

FIGURE 4-29
Cascade connection.

to achieve more complex input-output relationships and to obtain higher gains without sacrificing bandwidth (a trade-off we discuss in a later chapter). Both of these motivations are illustrated shortly.

An example of a cascade connection is shown in Figure 4-29. From a block diagram point of view we would say that the overall gain of the circuit is simply K_1 times K_2 times K_3, where these individual circuit gains were calculated with no load at the output. The input-output equivalent circuit of the cascade reminds us that the OP AMP circuits have essentially zero output resistance, so that generally it is legitimate to consider the individual circuits independently. However, this is strictly true only if the load presented by the succeeding stage is reasonable, that is, within the power-handling capability of the OP AMP. For this and other reasons, it is important to know the input resistance of an OP AMP circuit.

Figure 4-30 shows the circuit diagram for determining the input resistance for the two basic OP AMP circuit configurations. To obtain input resistance we conceptually apply a test signal and determine R_{IN} from equivalence.

$$R_{\text{IN}} = \frac{v_{\text{TEST}}}{i_{\text{TEST}}}$$

For the noninverting circuit the test current is necessarily zero of the ideal OP AMP model. Therefore apparently

$$R_{\text{IN}} = \infty \quad \text{(noninverting amplifier)}$$

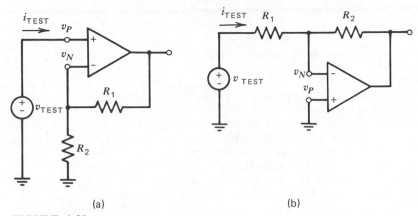

FIGURE 4-30
Input resistance test for the basic OP AMP circuits. (*a*) Non-Inverting. (*b*) Inverting.

If we were to use the detailed controlled source model of the OP AMP studied in Sec. 4-4, we would find that the actual input resistance is approximately $\mu R_1/K$, where μ is the open-loop gain, K is the closed-loop gain, and R_1 is the OP AMP input resistance. For any reasonable closed-loop gain this is a resistance of the order of 10^{10} to 10^{15} Ω. Thus for all practical purposes the noninverting amplifier has an infinite input resistance.

For the noninverting amplifier the input current is

$$i_{TEST} = \frac{v_{TEST} - v_N}{R_1}$$

For the ideal OP AMP model $v_P = v_N$, and since the noninverting input is grounded $v_N = 0$. Thus the input resistance is

$$R_{IN} = R_1 \qquad \text{(inverting amplifier)}$$

Again if we were to use the complete controlled source model of the OP AMP this result would be modified very slightly by a parallel resistance of the order of $\mu R_1/K$, a resistance whose value is of the order of 10^{10} to 10^{15} Ω. Thus for all practical purposes the input resistance of the inverting amplifier is R_1.

The impact of this is that in designing inverting amplifiers we must pick R_1 with two goals in mind. The first is to achieve the specified gain and the second is to minimize the load presented in the preceding stage. This is particularly true when the preceding stage is *not* the output of an OP AMP circuit. We saw this in Example 4-10, where we found that the gain and effective input resistance were influenced by the driving circuit. It is particularly important to control the input resistance of an inverting amplifier when interfacing with signal transducers at the input to an amplifier cascade. The important subject of transducer interfacing is discussed in a later section.

Example 4-13

Determine the input-output relationship of the circuit in Figure 4-31a.

This circuit is a cascade of three OP AMP circuits. The first stage is an inverting amplifier with $K = -0.33$, the second stage is a unity gain inverting summer, and the final stage is another inverting amplifier of gain $-5/9$. Armed with this information we can construct the block diagram shown in the figure. In tracing the signal through the various blocks we begin with the 9.7-V input of the first stage and find its output as $(-0.33) \times (9.7) = -3.2$. This input is added to the variable input v_F in the second stage to produce $(-1)(-3.2) + (-1)(v_F) = 3.2 - v_F$. This result is then multiplied by the gain of the final stage $(-5/9)$ to obtain the circuit output as

$$v_C = \frac{5}{9}(v_F - 3.2)$$

The significance of this relationship is that if $v_F = \theta_F/10$, where θ_F is a temperature in degrees Farenheit, then $v_C = \theta_C/10$, where θ_C is the equivalent temperature in degrees centigrade. Each stage is an inverting amplifier in which the input resistance is about 10 kΩ. Since $\pm V_{CC} = \pm 15$ V, this means that the maximum current drawn at the input of any stage is about $15/10^4 = 1.5$ mA. This is well within the power capabilities of most OP AMPs. The reader is now invited to show that the output of any stage will not exceed ± 15 V if the input temperature is in the range $-118°$ and $+182°$ Farenheit (F).

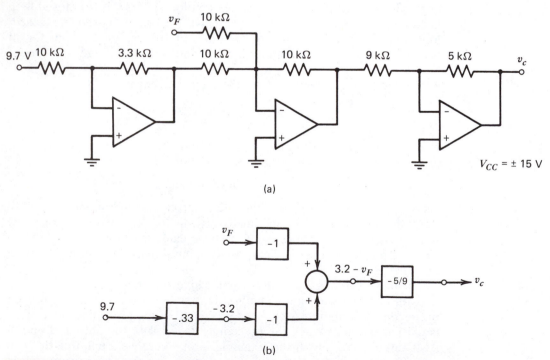

(a)

(b)

FIGURE 4-31
Circuit and block diagram for Example 4-13. (a) Circuit diagram. (b) Block diagram.

Example 4-14

Design an amplifier with a gain of $+1000$.

As with many design problems there are, of course, many possible solutions. Several are shown in Figure 4-32. The single-stage design used only one OP AMP but requires a rather large spread in resistor values, essentially 1 kΩ to 1 MΩ. It also may be pushing the closed-loop gain a bit close to the OP AMP open-loop gain. The second design involves a cascade of two inverting amplifiers. The first stage has a gain of -50 and the second -20 so that the overall gain is $(-50)(-20) = +1000$ as required. The circuit has an input resistance of 10 kΩ, which may or may not be important, depending on what is driving the circuit. Incidentally, the order in which the stages are connected is not important since $K_1 K_2 = K_2 K_1$. The

three-stage design is a cascade of three identical non-inverting amplifiers, each with $K = 10$, so that overall the gain is $(10)(10)(10) = +1000$. This design is simply a replication of the single-stage design from Example 4-7. The three-stage design has a very high input resistance, as does the single-stage design, because the first stage is a noninverting amplifier. Of course its power consumption would be greater since three OP AMPS are required. There are also many other alternatives and the final design selection would be based on such factors as element spread, input resistance, power consumption, gain bandwidth, and other constraints that may be unique to the design situation.

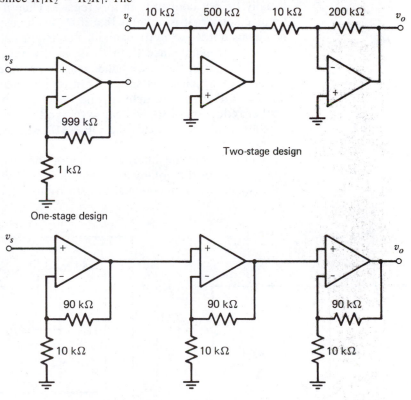

FIGURE 4-32
Possible design alternatives to achieve a gain of 1000.

4-7   INTERFACING WITH OP AMPS

Designing circuits to interface with each other is greatly simplified by using operational amplifiers. This simplification is primarily attributable to the isolation properties of the OP AMPs, which enable an engineer to design one circuit without worrying about the interaction of connecting circuits. A drawback of OP AMP interfacing is that supply voltages must be provided to operate the amplifier. Often, the low cost of OP AMPs and the ease of designing using OP AMPs will outweigh any problems associated with providing supply voltages. Furthermore, in many applications, the supply voltages are already available to drive other parts of the system.

To understand the utility of OP AMPs in interfacing, consider the circuit of Figure 3-33a. The problem is to provide a constant voltage to a load that can vary over several orders of magnitude.

What we really want is an ideal voltage source, but signal sources usually do not cooperate in this regard. Thus if we are limited to a real voltage source, it is obvious that if R_L is changed, so too must R_S or V_S be changed if we want to maintain a constant voltage. This could become very inconvenient since V_S or R_S is usually not accessible. The circuit of Figure 4-33b, on the other hand, readily provides a constant voltage regardless of which R_L is selected. In reality there is a maximum current that the OP AMP can deliver, and once this maximum is reached, any further reduction of R_L will cause the voltage to drop. However, for any reasonable range of load, the OP AMP will look like an ideal voltage source.

A most important interfacing tool is shown in Figure 4-34. This is a special case of the noninverting amplifier first introduced in Figure 4-17. In this case R_1 is made equal to zero by using a short circuit in place of

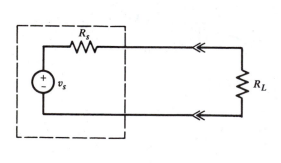

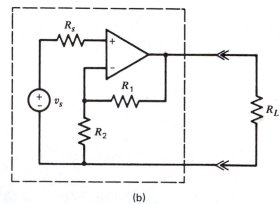

(a)   (b)

FIGURE 4-33
Source–load interface with an OP AMP circuit. (*a*) Real voltage source. (*b*) Near ideal voltage source.

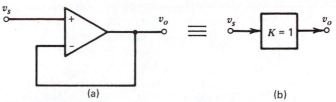

FIGURE 4-34
The OP AMP voltage follower. (*a*) Circuit representation. (*b*) Block diagram.

R_1. If R_1 is zero, the gain is one, regardless of what R_2 equals. Therefore R_2 can be removed from the circuit without effect.

The result is that

$$\frac{v_o}{v_s} = 1 \tag{4-28}$$

Would this circuit be of any value? After all, a piece of wire has the same transfer characteristics! The key to the answer lies in the fact that $i_P = 0$ and $R_{IN} = \infty$.

To illustrate an application for a voltage follower, consider the problem of measuring the voltage across a 1-MΩ resistor with a typical voltmeter of 1-MΩ input resistance. This problem is shown in Figure 4-35. In circuit

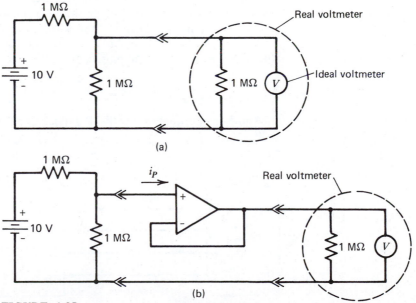

FIGURE 4-35
A voltage follower application. (*a*) Voltmeter connected directly. (*b*) Follower added.

4-35a, the 1-MΩ resistor in the voltmeter is in parallel with the 1 MΩ that is being measured. The result is a 500-kΩ resistor in series with the rest of the circuit. By voltage division the voltage measured by the voltmeter is

$$v_{\text{METER}} = \frac{500 \text{ k}}{500 \text{ k} + 1 \text{ M}} \, 10 = 3.3 \text{ V}$$

This 3.3 V is not the 5.0 V expected. By inserting a voltage follower between the meter and the voltage measured, the current flowing in the circuit will no longer split as i_P is zero. Hence we measure the true voltage of 5 V across the 1-MΩ resistor. Another way of looking at this effect is that the voltage follower places a resistance of infinity in parallel with the resistor that is being measured. The parallel combination of an infinite resistor and any other resistor is the other resistor.

Another very important application of OP AMPs is in interfacing digital systems with analog systems.

Consider the diagram of Figure 4-36. The parallel 4-bit output is a digital representation of a signal. Each bit is weighted so that v_1 is worth eight times v_4, v_2 is worth four times v_4, and v_3 is worth two times v_4. We call v_4 the least significant bit (LSB) and v_1 the most significant bit (MSB). Yet each bit can only have two values, a high value, or "1," and a low value, or "0." To convert this digital signal to an analog signal, we must weight the various bits so that the analog output v_o will be

$$v_o = 8v_1 + 4v_2 + 2v_3 + v_4 \qquad \textbf{(4-29)}$$

A simple way to implement this function is to use a summer with properly weighted resistors. Figure 4-37a shows a **binary-weighted resistance** D/A converter. Figure 4-37b shows the same function in block diagram form.

The output of the OP AMP will be the binary-weighted sum of the digital input, scaled or amplified by $-R_F/R$. That is,

$$v_o = \frac{-R_F}{R} (8v_1 + 4v_2 + 2v_3 + v_4) \qquad \textbf{(4-30)}$$

Another approach is the **R-2R ladder** D/A converter shown in Figure 4-38.

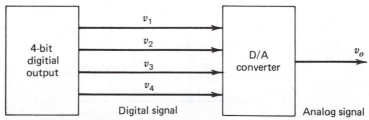

FIGURE 4-36
A digital-to-analog (D/A) converter.

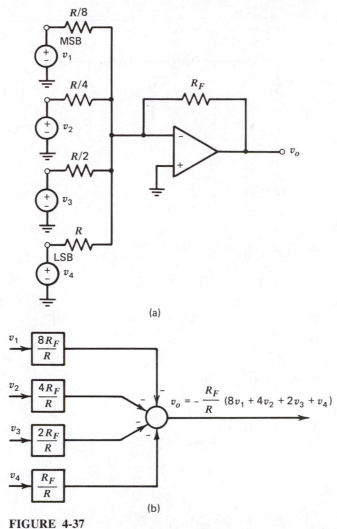

FIGURE 4-37
A binary-weighted resistance D/A converter. (*a*) Circuit. (*b*) Block diagram.

If one were to find the Thévenin equivalent circuit of R-2R network (between node Ⓐ and ground), the circuit of Figure 4-39 would be obtained. The output is readily found using the inverting amplifier relation

$$v_o = \frac{-R_F}{R_T} v_T$$

$$v_o = \frac{-R_F}{R} \left(\frac{v_1}{2} + \frac{v_2}{4} + \frac{v_3}{8} + \frac{v_4}{16} \right) \tag{4-31}$$

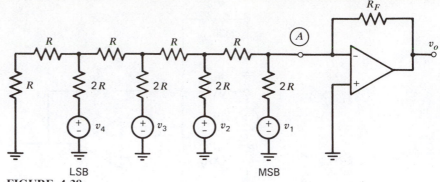

FIGURE 4-38
An *R*-2*R* ladder D/A converter.

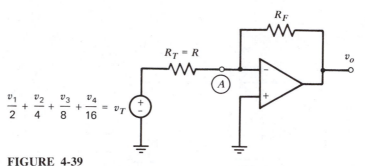

FIGURE 4-39
Thévenin equivalent of the *R*-2*R* ladder D/A converter.

If we pick R_F to equal 16 R, we get

$$v_o = -(8v_1 + 4v_2 + 2v_3 + v_4) \tag{4-32}$$

which is the desired result. It may appear that this technique is not really any different than the earlier method. Indeed, in theory they are equivalent; however, in practice the latter method is far superior. First of all, the first method requires many precision resistors of different values. This is difficult to achieve, especially for 8- or 16-bit converters. Second, the input resistance that each binary source sees will vary with each bit and could cause a loading problem. The *R*-2*R* ladder, on the other hand, presents the same input resistance to each binary input. Furthermore, *R*-2*R* ladders are easily manufactured to high precision by thin- or thick-film technology, resulting in readily available low-cost networks.

Example 4-15

A thermocouple is a transducer that converts heat energy to electric energy through a phenomenon known as the Seebeck effect. It consists of a welded junction of two dissimilar metals, for example, copper and an alloy called constantan, or chromel (chromium-nickel alloy) and alumel (aluminum-nickel alloy). The output from this junction is a weak voltage that varies proportionally to the temperature of the junction. For example, when measuring as wide a temperature range as 75K to 2200K, the voltage output will vary only from 5 to 50 μV/K. To use this sensor, it must be connected to a voltmeter capable of measuring such a small change of voltage. Furthermore neither side of the thermocouple can be grounded since a noise voltage can easily override the small thermocouple signal. Common voltmeter movements can only detect voltages of 1 mV or larger. The problem then is twofold. First, provide isolation from ground until the signal can be amplified; second, amplify the signal so it can be read adequately on a readily available voltmeter.

A common solution is to use an OP AMP as a difference amplifier (a subtractor). We can model our thermocouple as a nonideal variable-voltage source and our voltmeter as shown in Figure 4-40. Both the series resistance and the output voltage are functions of temperature, but $R(T)$ never exceeds about 100 Ω. We want to use a voltmeter with a standard range of 0 to 1 V. Since the signals will range from $\approx$375 μV to less than 100 mV, we will need an amplifier with a variable-voltage gain of 10 to 1000. The OP AMP circuit in Figure 4-40 is a subtractor; hence

$$v_o = \left[\frac{R_2}{R_1 + R_2} \times \frac{R_F + R_S + R(T)}{R_S + R(T)} \right] v_2 - \frac{R_F}{R_S + R(T)} v_1$$

if we pick $R_F = R_S = R_1 = R_2 \gg R(T)$, we get

$$v_o = v_2 - v_1$$

which is precisely $- V(T)$. We have isolated the thermocouple from ground. While selecting specific values of R_S, R_F, R_1, and R_2 will enable one to obtain any gain desired, a more obvious (if not less expensive) way is to feed the output of the difference amplifier to an inverting amplifier with a selectable feedback resistor. This complete circuit is shown in Figure 4-41. A more sophisticated meter could be calibrated in degrees Kelvin or Centigrade instead of in volts.

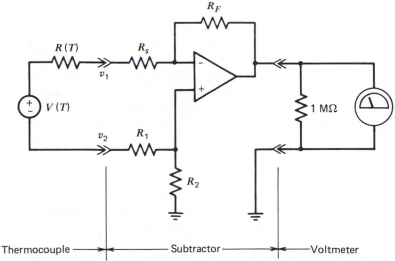

FIGURE 4-40
Application of OP AMP interface design.

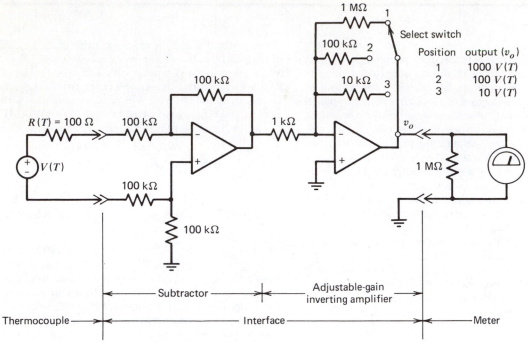

FIGURE 4-41
Enhanced application of OP AMP interface design.

The problem of interfacing linear transducers as demonstrated in Example 4-15 is a very common one. A generalized block diagram approach to this interfacing problem is shown in block diagram form in Figure 4-42. The transducer converts some physical parameter to a voltage that is proportional to the physical parameter being measured. The output of the transducer is usually very weak and has to be greatly amplified. To achieve a desired output, the amplified transducer output often has to be offset, a process known as adding **bias.** A final example will demonstrate the ease of designing an interface for a linear transducer using OP AMPs..

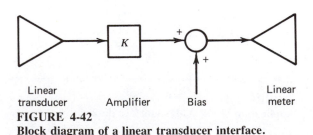

FIGURE 4-42
Block diagram of a linear transducer interface.

Example 4-16

In a laboratory experiment it is desired to measure the amount of light incident on a certain photocell between 5 and 20 lumens. The output is to be displayed on a 0- to 10-V voltmeter. Design the interface if the photocell output is as shown in the graph of Figure 4-43a. Furthermore it is desired that 5 lumens indicate 0 V and 20 lumens indicate 10 V.

The output of the transducer will change (linearly) $(0.5 - 0.2)$mV = 0.3 mV for a light intensity change between 5 and 20 lumens. This 0.3-mV change must produce a $(10 - 0)$-V change in the meter. Thus our signal needs to be amplified:

$$K = \frac{\text{Desired range}}{\text{Available range}}$$

$$= \frac{(10 - 0)\ \text{V}}{(0.5 - 0.2) \times 10^{-3}\ \text{V}} = 3.3 \times 10^4$$

If we multiply the transducer's output voltage range by the gain, we obtain an output range of 6.67 to 16.67 V instead of the desired 0 to 10-V spread. However our meter is only good from 0 to 10 V. To satisfy this interface requirement we must subtract 6.67 V from the amplifier's output. The 6.67 V is the bias. The block diagram solution is shown in Figure 4-43b. To realize the solution using OP AMPs, we design each piece of the interface independently and then cascade them. Figure 4-44a shows a suitable scalar multiplier of gain 3.3×10^4. Figure 4-44b shows the subtractor or bias operation. Be aware that the designer did not actually use a subtractor but a summer, taking advantage of the inverting properties of the scalar multiplier and summer used. The total circuit is shown in Figure 4-44c.

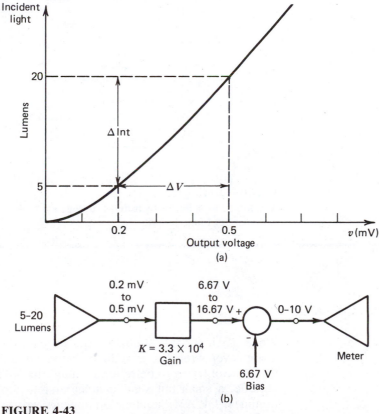

FIGURE 4-43
A linear transducer OP AMP interface example. (a) Transducer characteristics. (b) Block diagram.

Notice that to achieve the 33,000 gain would require a relatively small input resistor and a very large, untypical resistor in the feedback. In the final design the gain of the multiplier was reduced by 100 to permit use of better resistor values. The missing gain was then made up in the summer circuit.

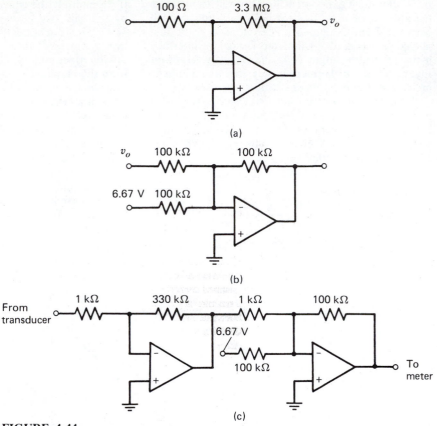

(a)

(b)

(c)

FIGURE 4-44
A solution to a linear transducer design example. (*a*) Scalar multiplier. (*b*) Bias summer. (*c*) Final design.

4-8   THE COMPARATOR

Since this text deals almost exclusively with linear circuits, it seems unfitting, even wrong, to introduce a nonlinear application. However, under the guise of trying to better understand the *interface* aspects of our text's subtitle, a small but significant incursion into one simple yet powerful nonlinear OP AMP application appears worthy of seeking forgiveness.

The **comparator** is perhaps the most rudimentary application of the OP AMP. An OP AMP with no feedback applied is capable of functioning as a comparator. For example, consider our newly discovered device with

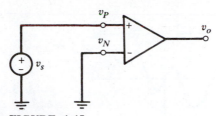

FIGURE 4-45
A simple comparator circuit.

one terminal v_N, connected to ground and the other terminal v_P, connected to a variable source v_S, as shown in Figure 4-45.

Since the OP AMP is connected in its open-loop configuration, Eq. 4-9 applies

$$v_o = \mu(v_P - v_N)$$

In our example we made $v_P = v_S$ and $v_N = 0$ (ground), so we can write

$$v_o = \mu v_S \qquad \textbf{(4-33)}$$

Recalling that μ is a very large number, it takes a very small v_S to drive v_o to saturation. For example, from Appendix B we see that μ for a general-purpose OP AMP is approximately 10^5. Assuming a $+V_{CC}$ of 15 V, an input of only $+0.00015$ V will force the output to saturate at a value near but less than $+V_{CC}$, which we call V_H, while an input of -0.00015 V would saturate the OP AMP at a value near but greater than $-V_{CC}$, which we call V_L. In sum, then, the output of our comparator has only two values, V_H and V_L, if we ignore the range of values between V_H and V_L that occur when the input is in transition between $+0.00015$ V and -0.00015 V and the OP AMP is temporarily in its linear range.

To demonstrate more vividly the power of the simple circuit of Figure 4-45, consider v_S to be as shown in Figure 4-46a. The output of the comparator is easily found to be as shown in Figure 4-46b. Every time v_S exceeded zero (actually exceeded zero by 0.00015 V) the output was V_H; otherwise it was V_L. The circuit has detected very time v_S crossed zero! Our circuit of Figure 4-45 is aptly called a "zero crossing detector."

The zero crossing detector does not illustrate the full versatility of the comparator. Consider the circuit shown in Figure 4-47a. In this circuit a constant 2-V source is connected between the inverting input and ground, and the same signal applied to the noninverting input. How will the output change?

Essentially the input voltage is now compared with 2 V instead of zero. Our Eq. 4-9 can be written as

$$v_o = 10^5(v_S - 2) \qquad \textbf{(4-34)}$$

Values of $v_S > 2$ V will force the output to saturate at V_H and $v_S < 2$ V to saturate at V_L. The result is shown in Figure 4-47c. Further, we can see that changing the value of the fixed source will cause the input signal

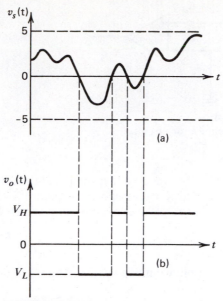

FIGURE 4-46
Input and output of a zero-crossing detector.

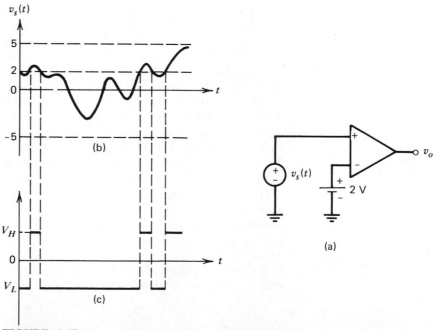

FIGURE 4-47
Comparator circuit showing results of comparing V_s with 2 V.

to be compared with different values. If we made the fixed source, say, 10 V, the output of our comparator would always be V_L since v_S would never exceed V_L. If we made the fixed source -10 V, the output would always be V_H for an analogous reason.

Finally, there is no reason not to generalize this discussion further: v_P and v_N can be connected to either fixed or varying signals. The resulting output can then be summarized as

$$\text{If } v_P > v_N, \text{ then } v_o = V_H$$
$$\text{If } v_P < v_N, \text{ then } v_o = V_L \tag{4-35}$$

The equal sign is omitted intentionally since it represents an unstable or unpredictable state, that is, we cannot say with any degree of certainty what the output will be if $v_P = v_N$.

Example 4-17

Consider as a practical application two sensors that are detecting two temperatures in, say, a rocket engine. The temperature T_1 of transducer 1 must always be less than T_2 that of transducer 2 for safe operation of the engine. If T_1 is ever greater than T_2, an alarm should sound. Design a suitable alarm system.

The problem is shown in Figure 4-48a. It is assumed that two calibrated temperature transducers are expertly interfaced to produce two voltages that are proportional to the temperature they measure. Each voltage, a function of the respective temperature it is measuring and time, is connected to a comparator as shown in Figure 4-48b. The output of the comparator will be 5 V (V_H) whenever $v_1 > v_2$, and 0 V (V_L) whenever $v_2 > v_1$. This output is connected to another circuit, usually digital, that sounds an alarm whenever it senses 5 V at its input. Figures 4-48c and 4-48d demonstrate a possible scenario.

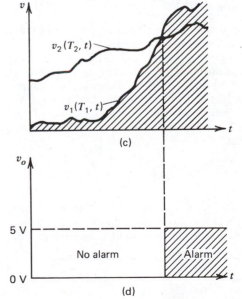

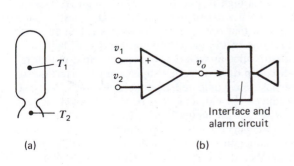

FIGURE 4-48
A comparator application circuit problem.

As a final word, Example 4-17 used a comparator with a V_H of 5 V and a V_L of 0 V. This brings us to a discussion of real comparators. While any OP AMP operating in its open-loop configuration will operate as a comparator, commercially available devices are optimized to function as such. OP AMPs often require some time to recover after becoming saturated—a property not desirable in a comparator. Furthermore OP AMPs saturate at values near $\pm V_{CC}$, the supply voltages. For example, the general-purpose OP AMP usually has a $+V_{CC}$ of 15 V and a $-V_{CC}$ of -15 V. These signal values are not compatible with the digital systems with which comparators usually interface. Hence commercial comparators usually will be single sided, driven by a single supply voltage compatible with the kind of digital logic with which it will interface. Two typical types are V_{CC}'s of 5 V and 0 V for TTL (transistor-transistor logic) and 15 V and 0 V for CMOS (complementary metal-oxide semiconductor) logic.

Example 4-18

Figure 4-49 shows a digital application of the comparator. Determine its function.

In this circuit the inputs V_1 and V_2 are digital signals, which means that they have only two possible values: high (5 V) and low (0 V). Likewise the comparator output high level is 5 V and the low level is 0 V. The input circuit is a linear resistance summer of the type studied in Chapter 3. By superposition it is easy to see that

$$v_P = \frac{1}{3} V_1 + \frac{1}{3} V_2$$

Likewise v_N is determined by a simple voltage divider, and it is easily seen that $v_N = 2.5$ V. We can now write all possible input-output results for this circuit.

V_1	V_2	v_P	v_N	V_o
0	0	0	2.5	0
0	5	1.67	2.5	0
5	0	1.67	2.5	0
5	5	3.33	2.5	5

Notice the pattern in the first two and last columns. Only when both inputs are high is the output high. The circuit implements the logical statement that the output is high if, and only if, V_1 is high AND V_2 is high. This is called the logical AND operation, and a circuit that operates in this way is called an AND gate, one of the basic building blocks of digital circuits. Notice as well that the circuit input-output relationship does not obey superposition. That is, superposition would tell us that if $V_1 = 5$, $V_2 = 0$ produces $V_o = 0$; and $V_1 = 0$, $V_2 = 5$ also produces $v_o = 0$; then the inputs $V_1 = 5$, $V_2 = 5$ should produce $V_0 = 0$. This does not happen in this circuit because the comparator is a nonlinear device. The reader should remember that only linear circuits obey superposition, and that nonlinear circuits can "disobey" superposition in very useful ways.

It is left to the student to show that an OR function could be obtained by simply changing v_N to lie comfortably between 0 and 1.67 V—say, 1.0 V, a NAND operation by switching v_P and v_N, and a NOR operation if v_P and v_N are switched *and* v_P is changed from 2.5 to 1.0 V.

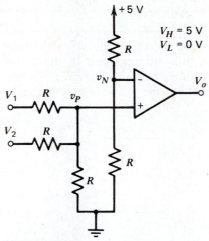

FIGURE 4-49
A digital application of the comparator.

SUMMARY

● A dependent source is one whose output is controlled by a signal appearing in a different part of the circuit. Dependent sources are linear circuit elements that are used to model active devices and are represented by one of the diamond-shaped sources shown in Figure 4-50.

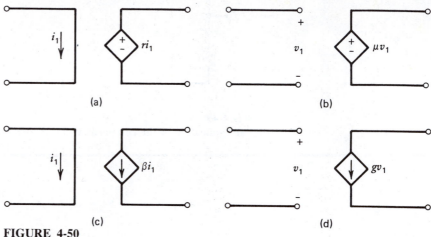

FIGURE 4-50
Dependent sources.

● Dependent sources cannot be set to zero when applying superposition to circuits containing dependent and independent sources. The Thévenin resistance of a circuit containing dependent sources cannot be found by the "look back" resistance method. It must be found by determining the open-circuit voltage and short-circuit current, or by applying a test source at the terminals in question.

● An active circuit is capable of delivering more power to its load than it receives from the input signal source. The additional power comes from supply sources that often are not shown in the circuit diagram.

● Active devices are usually nonlinear but can be modeled using dependent sources and other linear elements by restricting their operation to a linear region of their nonlinear characteristics.

● The transistor is a three-terminal, nonlinear, active, semiconductor device. The large-signal or piecewise linear model of the transistor can be used for many dc or low-speed applications. In this model the transistor's nonlinear characteristics are divided into three regions or modes. Each mode is represented by the linear models shown in Figure 4-51.

● The OP AMP is a nonlinear, active, semiconductor device with at least five terminals: inverting input, noninverting input, output, and

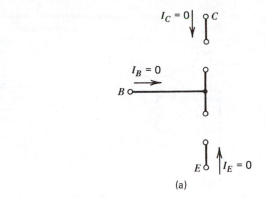

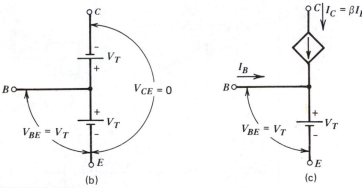

FIGURE 4-51
Large-signal transistor models. (*a*) CUTOFF mode. (*b*) SATURATED mode.
(*c*) ACTIVE mode.

two power supply terminals. The power supply terminals usually are not shown in circuit diagrams. In IC form the OP AMP is a differential amplifier with a very high open-loop gain (μ).

- The OP AMP can be made to operate in a linear mode by providing a feedback circuit from the output to the inverting input and by restricting its output voltage to the range.

$$-V_{CC} \leq v_o \leq +V_{CC}$$

where $\pm V_{CC}$ are the supply voltages.

- The OP AMP can be modeled using a finite-gain dependent source, but the ideal OP AMP model is adequate for many purposes. In the ideal model the gain and input resistance are infinite, and the output resistance is zero. The *i-v* characteristics of the ideal OP AMP model are

$$i_P = i_N = 0 \qquad v_P = v_N$$

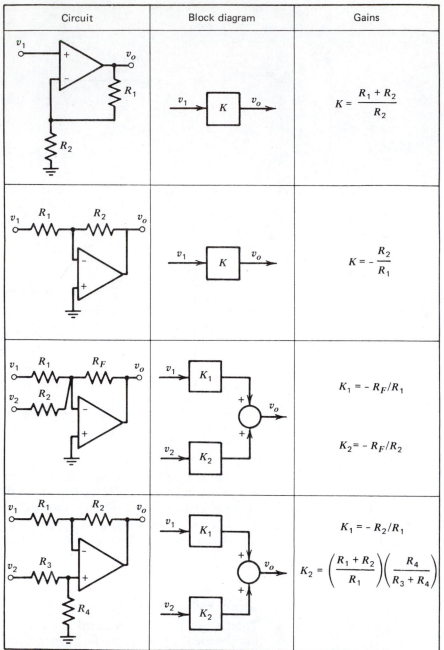

FIGURE 4-52
Some linear OP AMP circuits.

- OP AMP circuits can perform many different signal-processing functions. Four of the most common functions and their block diagram representations are shown in Figure 4-52.

- OP AMP circuits can be cascaded to obtain more complicated signal-processing functions. The individual stages in the cascade can be designed separately provided that some consideration is given to the input resistance of the following stage.

- The comparator is a nonlinear signal-processing device that can be obtained by operating an OP AMP without feedback. The comparator has two analog inputs and its output is a digital signal. Its input-output relationships are

$$\text{If } v_P > v_N, \text{ then } v_o = V_H$$
$$\text{If } v_P < v_N, \text{ then } v_o = V_L$$

where v_P and v_N are the analog inputs, and V_H and V_L are the high (logical one) and low (logical zero) digital signal levels at the output, respectively.

182 Active Circuits

EN ROUTE OBJECTIVES
AND RELATED EXERCISES

4-1 CONTROLLED SOURCES (SECS. 4-1 and 4-2)

Given a circuit consisting of linear resistors, controlled sources, and input signal sources, determine selected output signal variables or input-output relationships or equivalent circuits at specified terminals.

Exercises

4-1-1 For each circuit of Figure E4-1-1:
  (a) Determine the voltage gain v_2/v_s.
  (b) Show that the two gains are equal if $\mu = gR_o$.

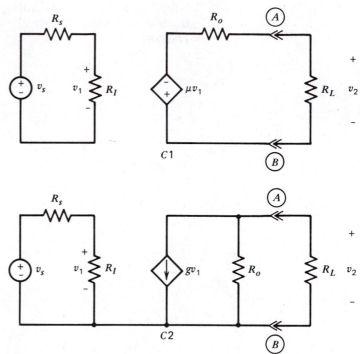

FIGURE E4-1-1
Dependent source problems.

4-1-2 For each circuit of Figure E4-1-2:
  (a) Determine the current gain i_2/i_s.
  (b) Show that the two gains are equal if $\beta = gR_1$.

4-1-3 Determine the voltage gain v_2/v_s for both circuits of Figure E4-1-3.

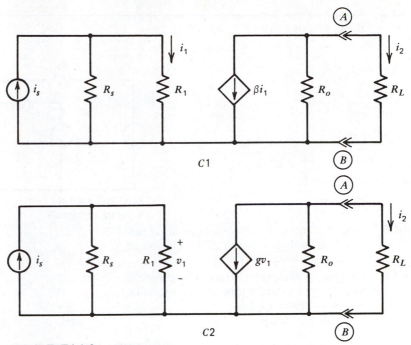

FIGURE E4-1-2
Dependent source problems.

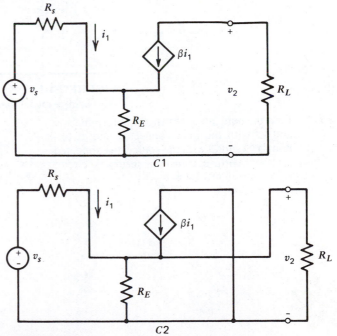

FIGURE E4-1-3
Dependent source problems with feedback.

4-1-4 Show that the input-output relationship for the circuit of Figure E4-1-4 is of the form:

$$v_2 = \beta(k_1 v_{S1} + k_2 v_{S2})$$

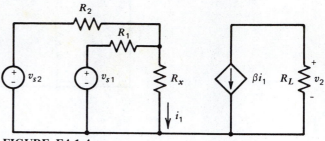

FIGURE E4-1-4
A summing circuit.

4-1-5 Find the Thévenin equivalent circuit at terminals Ⓐ and Ⓑ for each of the circuits of Exercises 4-1-1 and 4-1-2.

4-1-6 Find the equivalent resistance at terminals Ⓐ and Ⓑ for the hybrid-π transistor circuit model shown in Figure E4-1-6.

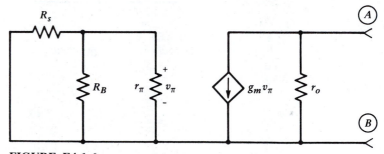

FIGURE E4-1-6
A hybrid-π model dependent source problem.

4-1-7 Find the equivalent resistance at terminals Ⓐ and Ⓑ with the load connected and at terminals Ⓒ and Ⓓ with the source disconnected for the hybrid-π transistor feedback circuit shown in Figure E4-1-7.

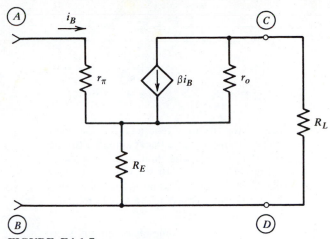

FIGURE E4-1-7
A hybrid-π model problem with feedback.

4-1-8 Find the Norton equivalent seen by the load resistor for each of the circuits in Exercise 4-1-3.

4-2 TRANSISTOR CIRCUITS (SEC. 4-3)

Given a circuit consisting of linear resistors, constant-input signal sources, and one transistor, use the large signal model to determine the operating state of the transistor or adjust circuit parameters to obtain a specified operating state.

Exercises

4-2-1 Select R_B in the circuit of Figure E4-2-1 to ensure that the transistor will be in the CUT-OFF mode when $v_S = 0$ and in the SATURATED mode when $v_S = 5$. Assume that $\beta = 50$.

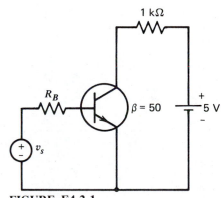

FIGURE E4-2-1
Transistor circuit for Exercise 4-2-1.

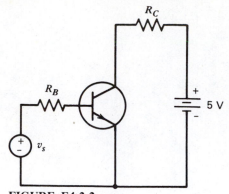

FIGURE E4-2-2
Transistor circuit for Exercise 4-2-2.

4-2-2 Refer to the circuit of Figure E4-2-2. We want to use a transistor as a linear amplifier (in the ACTIVE mode). The input signal we know will never exceed + 10 V and never be less than + 1 V. Select R_B and R_C to ensure that the transistor is always in the ACTIVE mode. Assume that $\beta = 100$.

4-2-3 Often the output of some digital circuit is required to turn on a small incandescent lamp. Usually the digital circuit's output is insufficient to provide the power necessary to light the lamp. To overcome this limitation engineers employ a discrete transistor as in the circuit of Figure E4-2-3.

Select R_B so that exactly 2 W is dissipated in the lamp when it is ON.

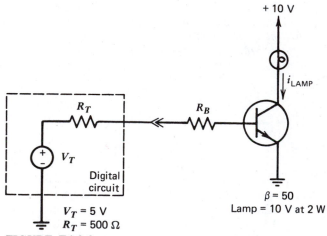

FIGURE E4-2-3
An incandescent lamp driver circuit.

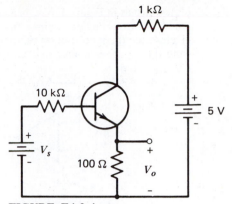

4-2-4 For the circuit of Figure E4-2-4 find V_o if V_S = 10 V. Repeat assuming V_S = 5 V. Repeat assuming V_S = 1 V.

FIGURE E4-2-4
Transistor circuit with feedback.

4-3 OP AMP CIRCUIT ANALYSIS (SECS. 4-4 to 4-7)

Given a circuit consisting of linear resistors, OP AMPs, and constant-input signal sources, determine selected output signals or input-output relationships in equation or block diagram form.

Exercises

4-3-1 For the circuits in Figure E4-3-1:
 (a) Determine the current voltage and power delivered to the 1-kΩ output resistor.
 (b) Identify the source of the delivered output power.

FIGURE E4-3-1
OP AMP circuits for Exercise 4-3-1.

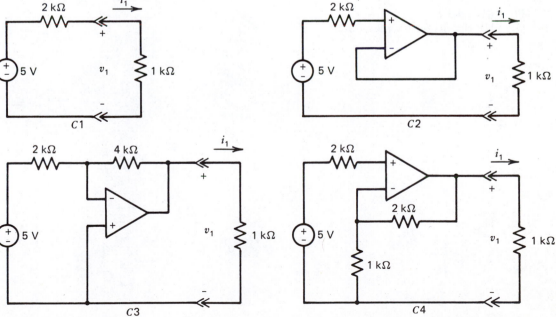

4-3-2   Each circuit in Figure E4-3-2 is a variation of
the basic inverting or noninverting amplifier.
Determine the input-output relationship v_o/v_s
and the input resistance in each circuit.

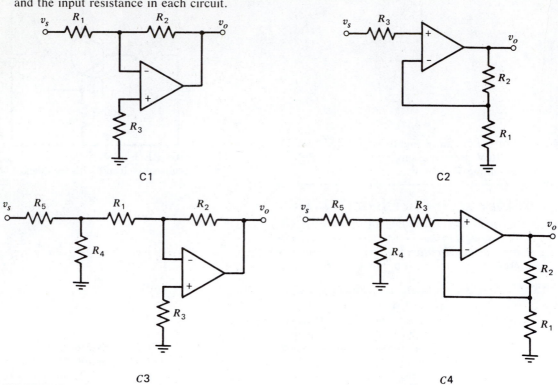

FIGURE E4-3-2
OP AMP circuits for Exercise 4-3-2.

4-3-3 Show that the OP AMP circuits in Figure E4-3-3 have the same input-output relationship. Draw a block diagram of this relationship. Discuss the advantages and disadvantages of each circuit.

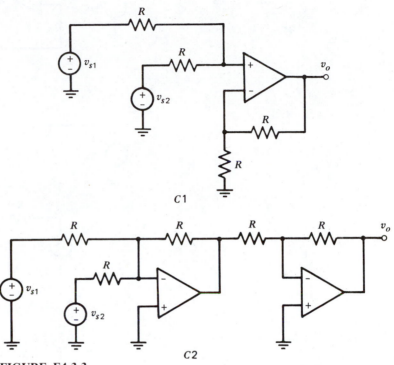

FIGURE E4-3-3
Summing circuits.

4-3-4   Repeat Exercise 4-3-3 for the circuits in Figure E4-3-4.

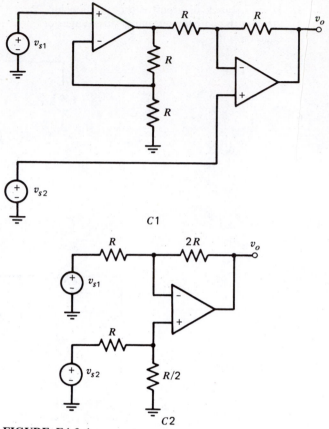

FIGURE E4-3-4
OP AMP circuits for Exercise 4-3-4.

4-3-5   Find the input-output relationship for the circuit in Figure E4-3-5 and draw a block diagram of the relationship.

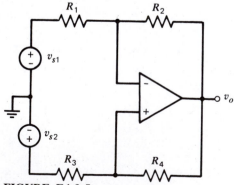

FIGURE E4-3-5
A difference amplifier circuit.

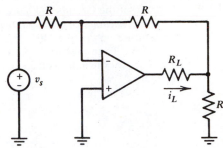

FIGURE E4-3-6
Near-ideal constant-current source circuit.

4-3-6 For the circuit in Figure E4-3-6, show that the current $i_L = -2\,v_S/R$ regardless of the value of R_L.

4-3-7 For the OP AMP circuit in Figure E4-3-7, draw an equivalent block diagram of the circuit, and then use the diagram to determine the input-output relationship.

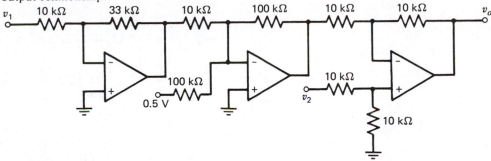

FIGURE E4-3-7
A cascade circuit.

4-3-8 For each circuit in Figure E4-3-8, determine the relationship between i_S and v_o. Then determine the input resistance seen by the current source.

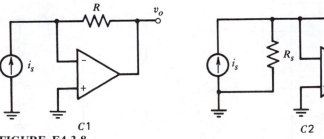

FIGURE E4-3-8
OP AMP circuits for Exercise 4-3-8.

4-3-9 For each of the circuits in Figure E4-3-9, determine the range of input voltage that will not saturate any of the OP AMPs for $\pm V_{CC}$ $= \pm 15$ V. Then plot the input-output characteristics for the input voltage over the range of ± 15 V.

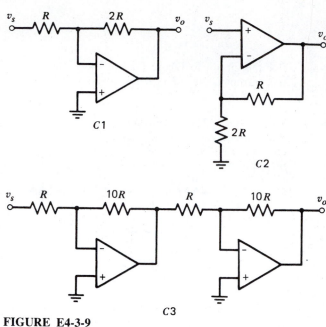

FIGURE E4-3-9
OP AMP circuits for Exercise 4-3-9.

4-3-10 For the two cases listed, plot the output of the 3-bit D/A converter shown in Figure E4-3-10 for the given inputs.

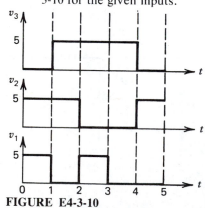

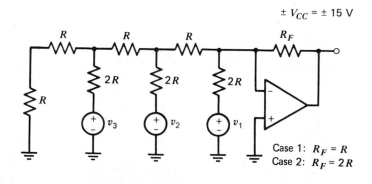

FIGURE E4-3-10
An R-$2R$ D/A circuit problem.

4-3-11 Using the controlled source model of the OP AMP shown in Figure E4-3-11, show that the voltage follower gain, input resistance, and output resistance are

$$K \simeq \frac{\mu}{1 + \mu}$$

$$R_{\text{IN}} \simeq \mu R_I \qquad R_{\text{OUT}} \simeq \frac{R_o}{\mu}$$

if μ is large and R_I much greater than R_o.

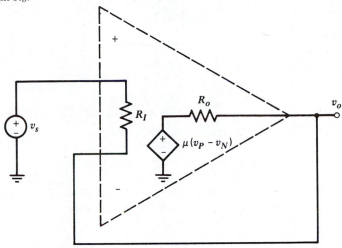

FIGURE E4-3-11
OP AMP follower analysis circuit.

4-4 OP AMP CIRCUIT DESIGN (SECS. 4-4 to 4-7)

Given an input-output relationship, devise a circuit consisting of linear resistors and OP AMPs that implements that relationship.

Exercises

4-4-1 Design a circuit using resistors and OP AMPs to implement the following input-output relationships:
(a) $V_o = 3V_1 - 3V_2$
(b) $V_o = 2V_1 - V_2$
(c) $V_o = 2V_1 + 4V_2$

4-4-2 Show how to interconnect a single OP AMP and the *R*-2*R* resistor array shown in Figure E4-4-2 to obtain voltage gains of ±3, ±2, ±1, and ±½.

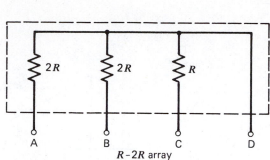

FIGURE E4-4-2
***R*-2*R* Array and OP AMP design problem.**

4-4-3 Design an interface using an OP AMP so that a 300-Ω source will see 300-Ω input with a 50-Ω load connected and the load will see 50 Ω with the source connected.

4-4-4 For the circuit of Figure E4-4-4, design an OP AMP network that will output the difference of ¼ V_A and ½ V_B into a load resistor and produce 10 mW of power.

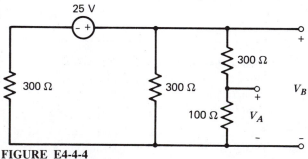

FIGURE E4-4-4
OP AMP design problem.

4-4-5 A strain gage is a device that, when properly mounted, has a resistance that varies in proportion to the applied pressure. The resistance of the strain gage for this problem is equal to 1000 Ω at $P = 10$ pounds per square inch (psi) and 10 kΩ at $P = 100$ psi. Assume that the relation is linear between pressure and resistance between the two calibration points, $P = 10$ psi and $P = 100$ psi. Design a circuit that will provide a linear output such that 10 V corresponds to 100 psi and 1 V to 10 psi.

4-4-6 Design an interface that will permit the transducer whose characteristics are shown in Figure E4-4-6 to be used with a 0-V to 1-V voltmeter. The meter should indicate $-20°C$ at 0 V and $+120°C$ at 1 V.

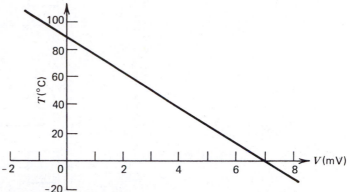

FIGURE E4-4-6
Transducer characteristics.

4-4-7 A diode has the following *i-v* characteristics: $i_D = 10^{-14}\,v_{D}/0.026$. Design an OP AMP circuit such that the output voltage is proportional to the natural logarithm of the input voltage $v_o = K \ln (v_S)$.

4-4-8 Design a circuit that can be used to convert from $-50°C$ to $+150°C$ to an equivalent degrees Fahrenheit. The OP AMPs that you must use have the following specifications:

$$\mu = 10^{10}$$
$$R_I = 5.0 \times 10^9 \ \Omega$$
$$R_o = 2 \ \Omega$$
$$\pm V_{CC} = \pm 15 \text{ V}$$

4-4-9 A 5-bit parallel digital signal must be converted to an equivalent analog signal. The digital signal has a logic "1" equal to 0 V and a logic "0" equal to 5 V. When all 5 bits are logic ones—that is, 11111—the converted analog output must be 23.25 V, and when all are logic zero—that is, 00000—the output must be zero. The OP AMPs you are to use have $\pm V_{CC}$ of ± 25 V, and the resistors you are to use are ± 1 percent tolerance. After you have completed your design check it by inserting a 00001-, 00010-, 00100-, etc., bit input and noting the output. (We have written the digital signals with the most significant bit first.) Finally, comment on the ± 1 percent tolerance resistors. Are they adequate?

4-4-10 Design a circuit using OP AMPs and resistors that will implement each of the block diagrams in Figure E4-4-10.

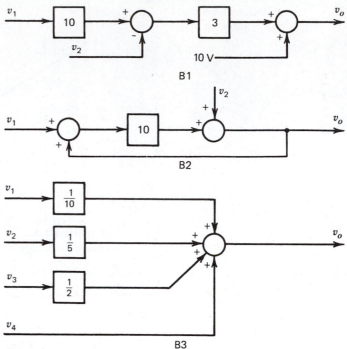

FIGURE E4-4-10
Block diagrams for Exercise 4-4-10.

4-4-11 The circuit in Figure E4-4-11 has been designed to implement a certain input-output relationship. Determine the input-output relationship and develop an alternative design that uses only one OP AMP.

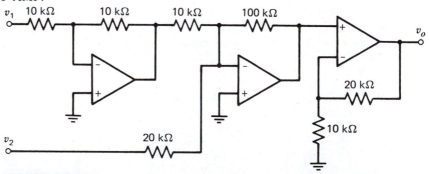

FIGURE E4-4-11
OP AMP circuit for Exercise 4-4-11.

4-5 COMPARATOR CIRCUITS (SEC. 4-8)

(a)   *Design: Given a need to detect which of two signals is greater, design a suitable interface circuit using a comparator.*

(b)   *Analysis: Given a circuit with one or more comparators, find the output for given input(s).*

Exercises

4-5-1   For each of the circuits in Figure E4-5-1 determine the condition for the input to produce a V_H output. Repeat for a V_L output.

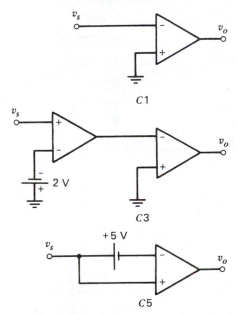

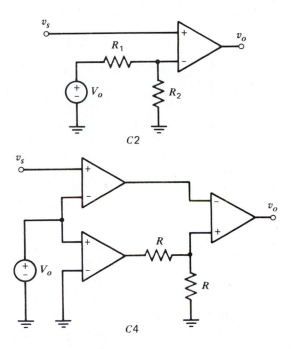

FIGURE E4-5-1
Comparator circuits for Exercise 4-5-1.

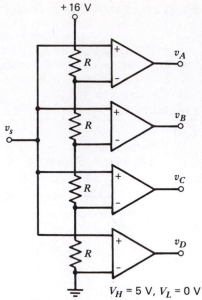

FIGURE E4-5-2
A flash A/D converter.

4-5-2   A simple yet very fast A/D converter is shown in Figure E4-5-2. For $v_S = t$, sketch the outputs v_A, v_B, v_C, and v_D versus t for $0 \le t \le 16$.

4-5-3   Determine the output of the circuit of Figure E4-5-3a if the input is as shown in Figure E4-5-3b.

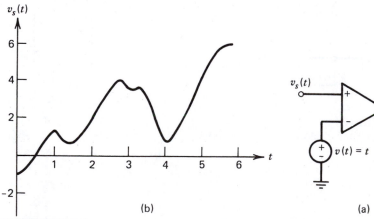

(b)

(a)

FIGURE E4-5-3
A comparator circuit with time-varying signals.
(a) Circuit. (b) Input signal.

4-5-4   A certain signal varies from -10 to $+10$ V. Design a circuit that will light a yellow light whenever the signal is outside the range ± 7.5 V.

4-5-5   A signal varies as $v = 2t - 12$. Design a circuit that will detect when the signal crosses 0 V. At what time does this occur?

4-5-6 A particular transducer has the transfer characteristics shown in Figure E4-5-6. Design a circuit that lights a green light whenever the temperature is less than 120°C and a red light whenever it is greater than 150°C. Both lights must be lit between 120 and 150°C.

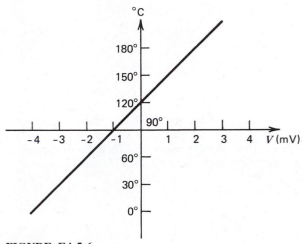

FIGURE E4-5-6
Transducer characteristics for comparator design problem.

PROBLEMS

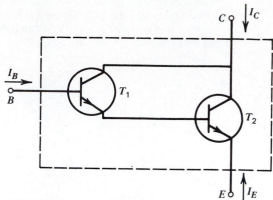

P4-1 (Analysis)

A Darlington pair consists of two transistors connected as shown in Figure P4-1 acting together as one device. Find an expression for I_C/I_B if the transistors are operating in the active region. Assume that $\beta_1 = \beta_2$.

FIGURE P4-1
Darlington pair problem.

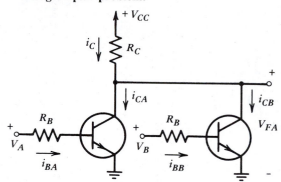

Circuit A

P4-2 (Analysis)

Before the advent of the integrated circuit, logic operations were designed around discrete devices. Transistors and resistors were configured to perform certain logic operations (resistor-transistor logic or RTL). Two such circuits are shown in Figure P4-2. Assume that the transistors are biased to switch from cutoff to saturated by the application of a logic input ("0" = 0 V, "1" = 5 V). Complete the table provided (Table P4-2) and determine what type of logic function each performs.

R_B = 10 kΩ
R_C = 1 kΩ
V_{CC} = 5 V
β = 100

V_A	V_B	Circuit A						Circuit B
		V_{FA}	i_{BA}	i_{BB}	i_C	i_{CA}	i_{CB}	V_{FB}
0	0							
0	5							
5	0							
5	5							

TABLE P4-2

Circuit B

FIGURE P4-2
RTL logic circuits.

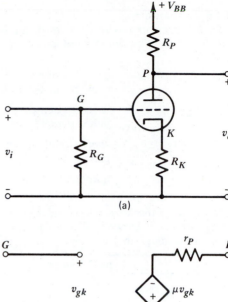

P4-3 (Analysis)

In days not too long gone, vacuum tubes ruled electronic circuits. Figure P4-3a shows such a circuit using a vacuum triode. Use the dependent source model of Figure P4-3b and compute the circuit's incremental gain: $(\Delta v_o/\Delta v_i)$ and v_o if $v_i = 3$ V. Assume the circuit is operating in the linear region (i.e., the model holds for all values of input) and that $\mu = 50$, $r_P = 30$ kΩ, $V_{BB} = 300$ V, $R_K = 1$ kΩ, $R_P = 20$ kΩ, $R_G = 5$ MΩ.

FIGURE P4-3
Vacuum triode circuit and model.

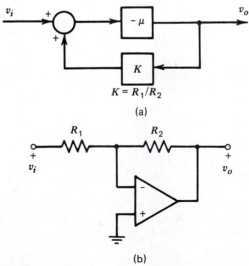

P4-4 (Analysis)

Show that the block diagram in Figure P4-4a provides the same output as the OP AMP circuit of Figure P4-4b if $\mu \to \infty$.

FIGURE P4-4
OP AMP inverter and its block diagram equivalent.

P4-5 (Analysis)

The *h*-parameter model of a transistor is given in Figure P4-5a. The definition of each parameter is as follows:

$$h_{ie} = \left. \frac{v_b}{i_b} \right|_{v_c = 0} = \text{input resistance with output short-circuited}$$

$$h_{re} = \left. \frac{v_b}{v_C} \right|_{i_b = 0} = \text{reverse open-circuit voltage amplification}$$

$$h_{fe} = \left. \frac{i_C}{i_b} \right|_{v_C = 0} = \text{short-circuit current gain or } \beta$$

$$h_{oe} = \left. \frac{i_e}{v_C} \right|_{i_b = 0} = \text{output conductance with input open-circuited}$$

Find each of the *h*-parameter values for the typical hybrid-π model shown in Figure P4-5b. Use the typical values found in Table P4-5.

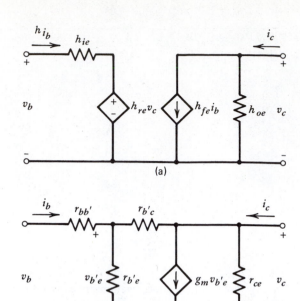

(a)

(b)

FIGURE P4-5
H-parameter and hybrid-π models of a BJT. (*a*) *H*-parameter model. (*b*) Hybrid-π model.

Typical Values		
$r_{bb'}$	$\simeq$	100 Ω
$r_{b'e}$	$\simeq$	2 kΩ
$r_{b'c}$	$\simeq$	20 MΩ
r_{ce}	$\simeq$	200 kΩ
g_m	$\simeq$	50 m$\mho$

TABLE P4-5

P4-6 (Analysis)

A transistor difference amplifier can be modeled as shown in Figure P4-6. Find v_o as a function of v_1 and v_2.

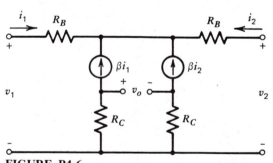

FIGURE P4-6
A transistor difference amplifier.

P4-7 (Design)

A thermistor is to be used to measure accurately the temperature of a motor from lowest expected room temperature ($-10°C$) to the highest motor operating temperature ($+150°C$). Data on the available thermistor are given in Table P4-7. The temperature is to be displayed on a 0- to 5-V voltmeter. Design a suitable system. Assume you have general-purpose OP AMPs available with $\pm V_{CC} = \pm 15$ V.

Temp (°C)	R (ohms)
−10	12.5K
0	7.4K
10	4.5K
20	2.8K
30	1.8K
40	1.2K
50	860
60	560
70	400
80	280
90	210
100	150
110	115
120	90
130	65
140	50
150	40

TABLE P4-7

Chapter 5
General Circuit Analysis

Watch your step when you immediately know the one way to do any-thing. Nine times out of ten, there are several better ways.

W. B. Given, Jr.

The methods of circuit analysis that we have studied so far require the recognition of certain connection patterns that allow simplification of the circuit. These methods help us develop an intuition about circuits and are useful because in using them we work directly with the circuit model. This intuitive approach cannot always be used, consequently there is a need for a more formal approach. General methods of circuit analysis can be divided into two phases. In the first phase we select a set of circuit variables and formulate a set of linear equations that describe the circuit. In the solution phase we manipulate this system of equations to determine the signal variables of interest. The solution phase uses standard mathematical techniques and does not require reference to the circuit model. We may lose thereby some insight into circuit behavior, but such is the price of generality.

5-1   DEVICE AND CONNECTION ANALYSIS

To develop general methods of circuit analysis we must first understand the foundations on which all methods of circuit analysis are based. The behavior of a circuit, an interconnection of devices, is fundamentally rooted in constraints of two types. First are the connection constraints that are embodied in Kirchhoff's laws. These constraints depend only on the manner in which the circuit is interconnected and not on the nature of the individual devices. Device individuality is captured by its i-v characteristics, which, in turn, do not depend on the nature of the circuit connections. In sum then, equilibrium in a circuit is the result of the balancing of two independent types of constraints: (1) connection constraints (Kirchhoff's laws), and (2) device constraints (i-v characteristics).

This observation provides a way to formulate a system of independent, linear equations that completely describe the circuit behavior. We first identify a current and voltage variable with every element in the circuit. We then write the connection constraints in terms of these variables using Kirchhoff's laws. The device constraints also can be expressed in terms of these variables using the element i-v characteristics. Collectively these steps provide a complete description of the circuit. More formally, for a circuit with N nodes and E elements we need to complete four steps:

1. Identify a current and a voltage variable with every element in the circuit.
2. Write KCL connection constraints in terms of the element currents at $N - 1$ nodes.
3. Write KVL connection constraints in terms of the element voltages around $E - N + 1$ loops.
4. Write device constraints in terms of the element currents and voltages using the element i-v characteristics.

These steps lead to $N - 1$ KCL connection equations, $E - N + 1$ KVL connection equations, and E element equations. Collectively this leads to a total of $(N - 1) + (E - N + 1) + E = 2E$ independent equations, which is precisely the number needed to solve for all of the variables identified in step 1. The next two examples illustrate the process.

Example 5-1

In Figure 5-1 we have identified a current and voltage with every element (step 1), and labeled three nodes (Ⓐ, Ⓑ, Ⓒ) and two loops (①, ②). To complete step 2 we write $N - 1 = 2$ KCL connection constraints.

Node Ⓐ $I_S - I_1 - I_2 = 0$
Node Ⓑ $I_2 - I_3 = 0$

Step 3 requires the application of KVL around $E - N + 1 = 4 - 3 + 1 = 2$ loops,

Loop ① $-V_o + V_1 = 0$
Loop ② $-V_1 + V_2 + V_3 = 0$

In step 4 we write the four device constraints as

I_S = known input
$V_1 = I_1R_1$
$V_2 = I_2R_2$
$V_3 = I_3R_3$

We have eight equations in eight variables. But since I_S is a known input we end up with seven linear equations in seven unknowns. In principle this allows us to solve for any or every voltage or current in the circuit.

The perceptive reader may wonder why we did not write a KCL connection constraint at node ©.

Node © $-I_S + I_1 + I_3 = 0$

This would have added a connection equation and perhaps eliminated the need for one of the KVL equations. But the node © equation can be obtained by adding the node Ⓐ equation and the node Ⓑ equation. In other words, the node © equation is not independent of the two KCL equations we have already written. This illustrates the general principle that there are exactly $N - 1$ independent KCL constraints and $E - N + 1$ independent KVL constraints.

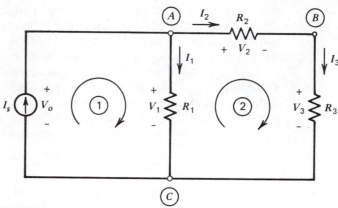

FIGURE 5-1
Circuit for demonstrating device and connection analysis in Example 5-1.

Example 5-2

In Figure 5-2 we have identified five element currents and five element voltages (step 1), and labeled four nodes and two loops. To complete step 2 we write $N - 1 = 3$ KCL equations:

Node Ⓐ $I_o - I_1 = 0$
Node Ⓑ $I_1 - I_2 - I_3 = 0$
Node © $I_2 + I_4 = 0$

To complete step 3 we write $E - N + 1 = 2$ KVL equations:

Loop ① $-V_{S1} + V_1 + V_3 = 0$
Loop ② $-V_3 + V_2 + V_{S2} = 0$

Finally we write the element constraints (step 4):

V_{S1} = known input
V_{S2} = known input
$V_1 = I_1R_1$
$V_2 = I_2R_2$
$V_3 = I_3R_3$

We have written 10 equations in 10 variables. But because there are two known inputs we end up with eight equations in eight unknowns. In principle we have the capability to solve for every current and voltage in the circuit.

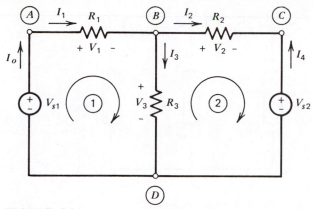

FIGURE 5-2
Circuit for demonstrating device and connection analysis in Example 5-2.

These examples illustrate that we can formulate a system of linear equations that describe circuit behavior by applying the device and connection constraints. As a practical matter, however, this approach leads to a large number of equations that must be manipulated simultaneously to obtain solutions. For Example 5-2, we would need to treat eight equations in eight unknowns, a prospect that is not of sudden and irresistible appeal. Therefore we will develop other methods such as node analysis and mesh analysis, which can greatly reduce the number of equations that must be dealt with simultaneously. Nonetheless device and connection analysis is conceptually important because it provides the foundation upon which all other methods of circuit analysis are based. This method is sometimes used in computer-aided circuit analysis since the equations are easy to formulate and the drudgery of solving many linear equations simultaneously can be handled by the computer.

5-2   BASIC NODE ANALYSIS

For most electronic circuits, particularly those containing OP AMPs, the most convenient solution variables are the node voltages.[1] To define a set of node voltages we first select a reference node. The node voltages are then defined as the voltages at the remaining $N - 1$ nodes with respect to the selected reference node. Figure 5-3 shows the notation and inter-

[1]In most circuits it is easier to make voltage measurements than current measurements. Recall that to make a voltage measurement one need only measure the voltage between two points. The circuit is generally not disturbed. In current measurements the circuit must be dissected or cut to divert the desired current through the measuring ammeter. It is primarily for this reason that many schematics often list the voltages expected at various nodes—all with respect to the referenced ground node.

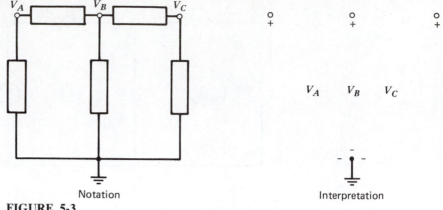

FIGURE 5-3
Definition of node voltages.

pretation of this definition. The reference node is indicated by the ground symbol, and the node voltage is identified by a voltage symbol written adjacent to the remaining nodes. This notation means the positive reference mark is located at the node in question while the negative mark is at the reference node.

To use these variables to formulate equilibrium equations for a circuit we proceed as in device and connection analysis, except that the KVL connection equations are not explicitly written down. Instead the KVL constraints are used to express the device constraints in terms of the node voltages. More formally, to develop a set of node voltage equations we need four steps:

1. Select a reference node and identify a node voltage at each of the remaining $N - 1$ nodes. Identify a current with every element in the circuit.

2. Write KCL connection constraints in terms of the element currents at the $N - 1$ nonreference nodes.

3. Use KVL and the element i-v characteristics to express the element currents in terms of the node voltages.

4. Substitute the device constraints from step 3 into the KCL connection constraints from step 2 and arrange the resulting $N - 1$ equations in a standard form.

This process leads to $N - 1$ linear equations in $N - 1$ unknown node voltages. Thus the node voltage formulation leads to a significant reduction in the number of linear equations that must be manipulated simultaneously, particularly if the circuit contains a large number of devices connected in parallel. The next three examples illustrate the node voltage analysis process.

Example 5-3

The circuit in Figure 5-4 was used in Example 5-1 to illustrate the formulation of device and connection equations. To apply node analysis we have identified a reference node (indicated by the ground symbol), two node voltages (V_A and V_B), and four element currents (I_S, I_1, I_2, I_3). This completes step 1 of the algorithm. Step 2 requires writing KCL constraints at two nonreference nodes.

Node Ⓐ $\qquad I_S - I_1 - I_2 = 0$

Node Ⓑ $\qquad I_2 - I_3 = 0$

To complete step 3 we write the device constraints as

$$I_S = \text{known input}$$
$$I_1 = G_1 V_A$$
$$I_2 = G_2(V_A - V_B)$$
$$I_3 = G_3 V_B$$

The right sides of device equations involve only known signals or unknown node voltages. Substituting these equations into the connection constraints yields

Node Ⓐ $\quad I_S - G_1 V_A - G_2 (V_A - V_B) = 0$

Node Ⓑ $\quad G_2 (V_A - V_B) - G_3 V_B = 0$

which can be arranged in the form:

Node Ⓐ $\quad (G_1 + G_2) V_A - G_2 V_B = I_S$

Node Ⓑ $\quad - G_2 V_A + (G_2 + G_3) V_B = 0$

We have reduced the formulation to two linear equations in the two unknown node voltages. The coefficients of the equations ($G_1 + G_2$, G_2, G_2, $G_2 + G_3$) depend only on circuit parameters, while the right side contains only known input signals (forcing functions or sources).

In formulating these node voltage equations it does not appear that we have used KVL. In fact we have. In writing the third device equation we have implicitly used KVL to write

$$\text{Voltage across } R_2 = V_A - V_B$$

Thus although KVL equations are not explicitly written in node analysis, they are always there. Put differently, any method of developing equilibrium equations must satisfy the KCL, KVL, and device constraints.

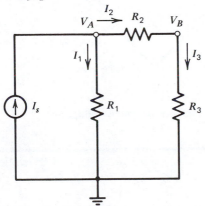

FIGURE 5-4
Circuit for demonstrating node voltage analysis.

Example 5-4

To formulate node voltage equations for the circuit in Figure 5-5 we first identify a reference node. Then we apply KCL at the three nonreference nodes:

Node Ⓐ $I_{S1} - I_1 - I_2 = 0$

Node Ⓑ $I_1 - I_3 + I_{S2} = 0$

Node Ⓒ $I_2 - I_4 - I_{S2} = 0$

This completes step 2. To complete step 3 we write

$$I_{S1} = \text{known input}$$
$$I_{S2} = \text{known input}$$
$$I_1 = G_1 (V_A - V_B)$$
$$I_2 = G_2 (V_A - V_C)$$
$$I_3 = G_3 V_B$$
$$I_4 = G_4 V_C$$

When these device equations are substituted into the KCL constraints and the result arranged in standard form, the result is three linear equations in the three unknown node voltages:

Node Ⓐ
$$(G_1 + G_2)V_A - G_1 V_B - G_2 V_C = I_{S1}$$

Node Ⓑ
$$-G_1 V_A + (G_1 + G_3)V_B = I_{S2}$$

Node Ⓒ
$$-G_2 V_A + (G_2 + G_4)V_C = -I_{S2}$$

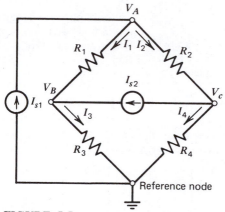

FIGURE 5-5
Bridge circuit example for node voltage analysis.

Some readers may have observed that the node equations derived in Example 5-4 have a certain symmetrical pattern. In the node Ⓐ equation the coefficient of V_A is the sum of the conductances connected to node Ⓐ. The coefficients of V_B and V_C are the negatives of the conductances connected between node Ⓐ and node Ⓑ and node Ⓐ and node Ⓒ respectively. The right side of the equation is the input signal source current into node Ⓐ. This pattern is repeated in the node Ⓑ and node Ⓒ equations.

This symmetrical pattern will always occur for circuits containing only resistors and independent current sources. To see this, consider a general conductance G that is connected to, say, node Ⓐ. There are then only two possibilities. Either the other terminal of G is connected to the reference node, in which case the current *leaving* node Ⓐ is

$$I = G (V_A - 0) = GV_A$$

or else it is connected to another nonreference node, say node Ⓑ, in which case the current *leaving* node Ⓐ is

$$I = G (V_A - V_B)$$

The pattern is easy to see. The sum of the currents leaving node Ⓐ via conductances is

1. V_A times the sum of conductances connected to node Ⓐ.

2. Minus V_B times the sum of conductances connected between nodes Ⓐ and Ⓑ.

3. Minus similar terms for all other nodes.

This sum must equal the sum of currents *injected* into node Ⓐ by independent current sources.

The process outlined allows us to write node voltage equations by inspection without going through the intermediate steps involving the KCL constraints and the device equations. We have developed this process assuming that the circuit contains only resistors and independent current sources. The method can be extended to circuits containing dependent sources by temporarily treating the dependent source as an independent source. The next example illustrates this extension.

Example 5-5

The circuit in Figure 5-6 contains an independent current source and voltage-controlled dependent current source. We will treat the dependent source temporarily as if it were an independent source, and write node voltage equations by inspection. Starting with node Ⓐ, the sum of conductances connected is $G_1 + G_2$. The conductance between Ⓐ and Ⓑ is G_2, and between Ⓐ and Ⓒ there is none. Further, I_S is injected into node Ⓐ and hence it would appear as a positive current on the right-hand side of the equation. Consequently the node Ⓐ equation is

Node Ⓐ $(G_1 + G_2)V_A - G_2V_B - 0V_C = I_S$

Similarly at node Ⓑ the sum of conductances is $G_2 + G_3$. Between Ⓑ and Ⓐ the conductance is again G_2, while there is no conductance between nodes Ⓑ and Ⓒ. Now treating the dependent source as if it were independent:

Node Ⓑ $(G_2 + G_3)V_B - G_2V_A - 0V_C = gV_x$

By inspection at node Ⓒ:

Node Ⓒ $G_4V_C - 0V_A - 0V_C = -gV_x$

We have derived a set of symmetrical node equations by treating the dependent source as independent. We are now ready for the *coup de grace*, since by KVL we can write

$$V_x = V_A - V_B$$

When this result is substituted into the node equations and all unknowns moved to the left side, there results:

Node Ⓐ
$$(G_1 + G_2)V_A - G_2V_B = I_S$$

Node Ⓑ
$$-(G_2 + g)V_A + (G_2 + G_3 + g)V_B = 0$$

Node Ⓒ
$$gV_A - gV_B + G_4V_C = 0$$

We now have three equations in three unknowns, including the effect of the dependent source.

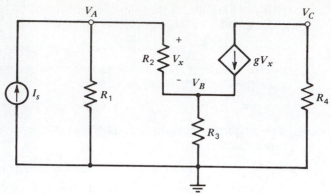

FIGURE 5-6
Dependent source circuit for Example 5-5.

The controlled source in Example 5-5 upset the symmetry of the node equations. Note particularly that V_C does not appear in the first two equations. This occurs because the controlled source is a unilateral element. That is, the voltage V_x causes a current gV_x to flow out of node ©, but injecting current into node © does not affect V_x. This unilateral signal coupling is an important property of active circuits that shows up as a lack of symmetry in the node equations.

5-3 SOLUTION TECHNIQUES

In the introduction to this chapter we indicated that general circuit analysis can be divided into a formulation phase and a solution phase. For the linear, memoryless circuits considered so far we have shown that the formulation phase leads to sets of linear algebraic equations. We now turn to the solution of these equations.

The solution phase is essentially a mathematical process of manipulating linear equations. We thereby lose some contact with the circuit, but such is the price of a general approach. Actually our treatment need not be extremely formal since in this book we never need to solve more than three linear equations simultaneously. Thus the largest set of equations we will need to deal with can be written as

$$a_{11}X_1 + a_{12}X_2 + a_{13}X_3 = b_1$$
$$a_{21}X_1 + a_{22}X_2 + a_{23}X_3 = b_2 \qquad \text{(5-1)}$$
$$a_{31}X_1 + a_{32}X_2 + a_{33}X_3 = b_3$$

In this form the X's represent unknown voltages or currents, the a's are determined by circuit parameters (conductances and the like), and the b's are known input signal source terms that drive the circuit. The objective is, of course, to solve for the X's in terms of the a's and b's.

Now there are a number of ways of solving Eq. 5-1, including simply floundering around algebraically until the desired result finally bubbles to the surface. However, Cramer's rule is one systematic method that allows us to write down the solution process in an especially compact form. Briefly Cramer's rule states that the solutions of Eqs. 5-1 are

$$X_1 = \frac{D_1}{D} \qquad X_2 = \frac{D_2}{D} \qquad X_3 = \frac{D_3}{D} \tag{5-2}$$

where D, D_1, D_2, and D_3 are three by three determinants defined as

$$D = \begin{vmatrix} a_{11} & a_{12} & a_{13} \\ a_{21} & a_{22} & a_{23} \\ a_{31} & a_{32} & a_{33} \end{vmatrix} \qquad D_1 = \begin{vmatrix} b_1 & a_{12} & a_{13} \\ b_2 & a_{22} & a_{23} \\ b_3 & a_{32} & a_{33} \end{vmatrix}$$

$$D_2 = \begin{vmatrix} a_{11} & b_1 & a_{13} \\ a_{21} & b_2 & a_{23} \\ a_{31} & b_3 & a_{33} \end{vmatrix} \qquad D_3 = \begin{vmatrix} a_{11} & a_{12} & b_1 \\ a_{21} & a_{22} & b_2 \\ a_{31} & a_{32} & b_3 \end{vmatrix} \tag{5-3}$$

The determinant D is called the circuit determinant since it depends only on the a's in Eqs. 5-1, that is, it depends only on the circuit parameters. The other determinants are constructed by replacing a column in the circuit determinant by the corresponding b's, that is, the corresponding source terms. For example, D_1 is found by replacing the first column in D by the b's. D_2 and D_3 are constructed in an analogous manner.

Thus Cramer's rule reduces the solution of linear equations to a matter of evaluating determinants. To evaluate a 2×2 determinant:

$$D = \begin{vmatrix} a_{11} & a_{12} \\ a_{21} & a_{22} \end{vmatrix} = a_{11}a_{22} - a_{21}a_{12} \tag{5-4}$$

We take the difference of the products of the elements down the diagonal to the right and the product of the elements down the diagonal to the left. This process also works for a 3×3, except there are more diagonals:

$$= (a_{11}a_{22}a_{33} + a_{12}a_{23}a_{31} + a_{13}a_{32}a_{21})$$
diagonals to the right
$$-(a_{13}a_{22}a_{31} + a_{12}a_{21}a_{33} + a_{11}a_{32}a_{23})$$
diagonals to the left

The method of diagonals works well for 2×2 and 3×3 determinants, which are the largest we will encounter in this book. For larger determinants the method of minors is required. However, for larger systems of linear equations Cramer's rule is not usually used anyway. Computer-based circuit analysis programs use some form of Gauss reduction, which is a numerical algorithm that is far more efficient than Cramer's rule for large systems of equations.

For our purposes Cramer's rule will suffice. Its application in circuit analysis is illustrated in the next two examples.

Example 5-6

In Example 5-3 we formulated node voltage equations for a circuit repeated in Figure 5-7. The node equations derived in Example 5-3 are

$$(G_1 + G_2)V_A \qquad\qquad - G_2V_B = I_S$$
$$- G_2V_A + (G_2 + G_3)V_B = 0$$

Applying Cramer's rule we write

$$V_A = \frac{D_A}{D} = \frac{\begin{vmatrix} I_S & -G_2 \\ 0 & G_2 + G_3 \end{vmatrix}}{\begin{vmatrix} G_1 + G_2 & -G_2 \\ -G_2 & G_2 + G_3 \end{vmatrix}}$$

$$= \frac{(G_2 + G_3)I_S}{G_1G_2 + G_2G_3 + G_1G_3}$$

$$V_B = \frac{D_B}{D} = \frac{\begin{vmatrix} G_1 + G_2 & I_S \\ -G_2 & 0 \end{vmatrix}}{D} = \frac{G_2I_S}{G_1G_2 + G_2G_3 + G_1G_3}$$

Having found the node voltages we can now determine every voltage and every current in the circuit by trivial calculations. For example, the element currents shown in Figure 5-7 are

$$I_1 = G_1V_A \qquad I_2 = G_2(V_A - V_B) \qquad I_3 = G_3V_B$$

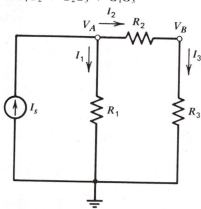

FIGURE 5-7
Circuit used in Examples 5-3 and 5-6 to illustrator analysis technique.

Example 5-7

In Example 5-5 we formulated node voltage equations for the circuit repeated in Figure 5-8 as

$$(G_1 + G_2)V_A \qquad\qquad - G_2V_B \qquad = I_S$$
$$-(G_2 + g)V_A + (G_2 + G_3 + g)V_B \qquad = 0$$
$$gV_A \qquad\qquad - gV_B + G_4V_C = 0$$

Normally we are not required to find every voltage and current in a circuit, but only selected ones. As shown in Figure 5-8, in this case we want to determine the voltage and current delivered to the output resistor (R_4). The output voltage corresponds to the node voltage V_C, hence by Cramer's rule we can write

$$V_o = V_C = \frac{D_C}{D} = \frac{\begin{vmatrix} G_1 + G_2 & -G_2 & I_S \\ -G_2 - g & G_2 + G_3 + g & 0 \\ g & -g & 0 \end{vmatrix}}{\begin{vmatrix} G_1 + G_2 & -G_2 & 0 \\ -G_2 - g & G_2 + G_3 + g & 0 \\ g & -g & G_4 \end{vmatrix}}$$

$$= \frac{-g\,G_3 I_S}{G_4(G_1 G_2 + G_1 G_3 + G_2 G_3 + gG_1)}$$

The output current is

$$I_o = G_4 V_o$$

$$= \frac{-g\,G_3 I_S}{G_1 G_2 + G_1 G_3 + G_2 G_3 + gG_1}$$

Notice the minus associated with the outputs. This means that they have the opposite polarity to that of the input signal I_S. The inversion results from the reference direction of the controlled source. Note as well that the output current does not depend on the output resistor since the controlled source drives the load directly and hence there is *no* current division in the output circuit.

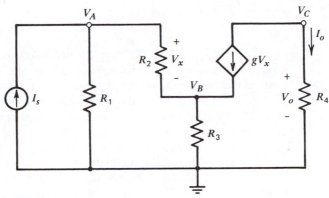

FIGURE 5-8
Circuit used in Examples 5-5 and 5-7 to illustrate analysis technique.

5-4   MODIFIED NODE EQUATIONS

In developing the formulation of node equations we have tacitly assumed that the circuit contains only independent or dependent *current* sources. It might appear that inclusion of *voltage* sources (dependent or independent) would complicate node analysis since the current through a voltage source is a dependent variable. In fact, the opposite is often the case. The presence of voltage sources often simplifies node analysis.

There are basically three ways to deal with voltage sources, as shown in Figure 5-9. If there is a resistor in series with the voltage source, then method 1 applies. The series combination of voltage source and resistor can be replaced at terminals Ⓐ and Ⓑ by an equivalent current source using source conversion. We can then develop node equations in the usual way. It is evident that this method eliminates node Ⓒ from the circuit, which simplifies the analysis.

If there is no resistor in series with the source, then we can use method 2 in Figure 5-9. Here we have selected node Ⓑ as the reference node. As a result the node voltage V_A is no longer an unknown since it must be

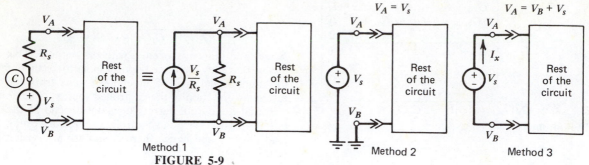

FIGURE 5-9
Methods of handling voltage sources in the formulation of node equations.

equal to the input V_S. We now do not need to develop a node voltage equation at node Ⓐ since its voltage is known. We then write node equations at the remaining nodes in the usual way, but as a final step we move all terms involving V_A to the right side since it is a known input and not an unknown response. This reduces the number of node equations as one is not needed at node Ⓐ.

If the circuit contains two or more voltage sources that do not share a common node, and if none are connected in series with a resistor, then we must use method 3 shown in Figure 5-9. In this case we identify an unknown current I_x with the voltage source. This adds one additional unknown to our problem. However, we can also write a KVL constraint

$$V_A - V_B = V_S \qquad \text{(5-6)}$$

which adds an additional equation.

We call the resulting equations modified node equations since we either modify the circuit (method 1), do not write node equations at all nodes (method 2), or introduce an unknown current to the list of variables (method 3). It should be noted that the three methods are not mutually exclusive. In a complicated circuit we might appeal to all three methods. All three methods will be illustrated by examples.

Example 5-8

In Example 5-2 we developed device and connection equations for the circuit repeated in Figure 5-10. The given circuit contains two voltage sources, but both are connected in series with resistors. Therefore method 1 applies, and using source conversion we obtain the modified circuit shown. The modified circuit has only two nodes. For the indicated selection of a reference node, we can then write *one* node equation by inspection:

$$(G_1 + G_2 + G_3)V_A = G_1 V_{S1} + G_2 V_{S2}$$

We do not need to appeal to Cramer's rule here since

$$V_A = \frac{G_1}{G_1 + G_2 + G_3} V_{S1} + \frac{G_2}{G_1 + G_2 + G_3} V_{S2}$$

Note the example of superposition here. The voltage V_A is expressed as a linear combination of the two inputs. Knowing V_A, we now can solve for every voltage and current in the circuit by trivial calcula-

tions. In this case the voltage sources greatly simplified the analysis, especially compared with the device and connection equation formulation in Example 5-2, which involved eight equations in eight unknowns.

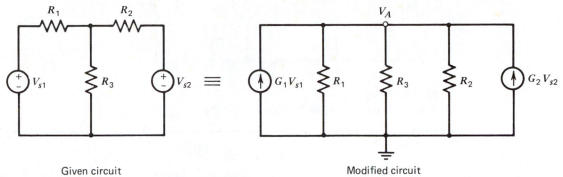

Given circuit   Modified circuit

FIGURE 5-10
Circuit used in Examples 5-2 and 5-8 demonstrating source conversion simplification.

Example 5-9

The circuit in Figure 5-11 contains two voltage sources, one independent input source V_S, and a voltage-controlled dependent voltage source. Since these two sources share a common node at the indicated reference node, we can apply method 2 to this circuit. We can immediately write

$$V_A = V_S$$

which eliminates one node voltage as an unknown. We also observe that the control voltage V_x is none other than V_B, and hence we can write

$$V_C = -\mu V_B$$

The minus sign is due to the polarity indicated by the reference marks on the controlled source. In any case we see that there is only one independent unknown node voltage, that is, V_B. V_A is known and V_C is proportional to V_B. Therefore we only need to write a node equation at Node Ⓑ. By inspection,

$$(G_1 + G_2 + G_3)V_B - G_1V_A - G_3V_C = 0$$

But using the two constraint equations above this becomes

$$(G_1 + G_2 + G_3 + \mu G_3)V_B = G_1V_S$$

Hence

$$V_B = \frac{G_1V_S}{G_1 + G_2 + G_3(1 + \mu)}$$

To solve for the output current and voltage indicated in Figure 5-11 we note that

$$V_o = V_C = -\mu V_B$$

Hence

$$V_o = \frac{-\mu G_1 V_S}{G_1 + G_2 + G_3(1 + \mu)}$$

and

$$I_o = G_4 V_o$$
$$= \frac{-\mu G_1 G_4 V_S}{G_1 + G_2 + G_3(1 + \mu)}$$

Notice that the signal is inverted. This inversion between input and output is due to the polarity of the controlled source. Note as well that the output voltage does not depend on the output resistor R_4, since the output is taken across a perfect, though dependent, voltage source.

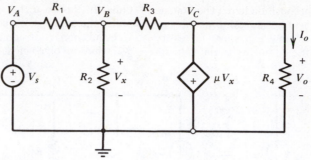

FIGURE 5-11
Dependent source circuit for Example 5-9.

Example 5-10

The circuit in Figure 5-12 contains two voltage sources that do not share a common node, and neither is connected in series with a resistor. This means that neither method 1 nor method 2 will work here, and we must apply method 3. Toward this end we have identified a current I_x associated with source V_{S2}. This current must be one of our solution variables. For the selected reference node we can immediately write

$$V_B = V_{S1}$$

Hence we need write node equations only at nodes Ⓐ and Ⓒ. By temporarily treating I_x as a known input, we can write these equations by inspection:

Node Ⓐ $\quad (G_1 + G_2)V_A - G_2V_B = I_x$

Node Ⓒ $\quad (G_3 + G_4)V_C - G_3V_B = -I_x$

But V_B is known and I_x is unknown, and to clarify this we rearrange the equations as

Node Ⓐ $\quad (G_1 + G_2)V_A - I_x = G_2V_{S1}$

Node Ⓒ $\quad (G_3 + G_4)V_C + I_x = G_3V_{S1}$

We now have two equations in three unknowns, which hardly seems like progress. But we have not yet used the constraint implied by V_{S2}. If we apply KVL around the perimeter of the circuit, we find that

$$V_A - V_C = V_{S2}$$

That is, V_{S2} fixes the difference between the voltages at nodes Ⓐ and Ⓒ. This constraint supplies the missing third equation.

Turning now to the solution phase we could simply apply Cramer's rule to these three equations. After a moment's reflection we observe that by adding the two node equations we eliminate I_x, and hence

$$(G_1 + G_2)V_A + (G_3 + G_4)V_C = (G_2 + G_3)V_{S1}$$
$$V_A - \qquad V_C = V_{S2}$$

are two equations in two unknowns. This result will always occur since the current I_x must always leave one node of the ungrounded voltage source and enter the other. Now using Cramer's rule,

$$V_A = \frac{D_A}{D} = \frac{\begin{vmatrix} (G_2 + G_3)V_{S1} & G_3 + G_4 \\ V_{S2} & -1 \end{vmatrix}}{\begin{vmatrix} G_1 + G_2 & G_3 + G_4 \\ 1 & -1 \end{vmatrix}}$$

$$= \frac{(G_2 + G_3)V_{S1} + (G_3 + G_4)V_{S2}}{G_1 + G_2 + G_3 + G_4}$$

and

$$V_C = \frac{D_C}{D} = \frac{\begin{vmatrix} G_1 + G_2 & (G_2 + G_3)V_{S1} \\ 1 & V_{S2} \end{vmatrix}}{D}$$

$$= \frac{(G_2 + G_3)V_{S1} - (G_1 + G_2)V_{S2}}{G_1 + G_2 + G_3 + G_4}$$

We now have found all three node voltages (V_B included) and so are in a position to determine every voltage and current in the circuit.

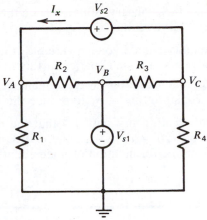

FIGURE 5-12
Circuit with source connecting two un-
grounded nodes for Example 5-10.

5-5 OP AMP CIRCUIT ANALYSIS

Modified node analysis also can be easily applied to OP AMP circuits. The general situation is shown in Figure 5-13. We are usually interested in determining the OP AMP output v_o, relative to ground, alias the reference node in the present context. Hence we must assign a node voltage variable to the output. However, as we saw in Chapter 4, the output of an ideal OP AMP acts as a controlled-voltage source connected between the output terminal and ground. According to method 2 of modified node analysis we do not need to formulate a node equation at such a node.

If we then formulate node equations at the remaining nodes, it would appear that we would have one more variable than the total number of equations available. However, the ideal OP AMP forces its net input to be zero; hence $v_P = v_N$ in Figure 5-13. This forces these two nodes to have identical voltages, thereby eliminating one unknown.

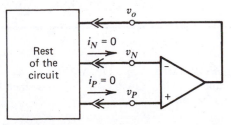

FIGURE 5-13
A general OP AMP circuit.

To summarize these observations we state the following rules for node analysis of OP AMP circuits:

1. Always identify a node voltage variable at the OP AMP's output but do *not* formulate a node equation at this node.

2. Formulate node equations at the remaining nodes and use the ideal OP AMP constraint $v_P = v_N$ to reduce the number of unknowns.

In addition, we recall from Chapter 4 that the ideal OP AMP draws no current at its inputs; $i_P = i_N = 0$ in Figure 5-13, and hence these currents can be ignored in the formulation of node equations.

Example 5-11

The circuit in Figure 5-14 is the familiar inverting amplifier configuration. As drawn the circuit has three nonreference nodes. We do not need to formulate a node equation at node Ⓐ because it is connected to ground via the input voltage source. As discussed, we also do not need to write an equation at node Ⓒ, the output. Only one node equation then is required at node Ⓑ. By inspection,

Node Ⓑ $(G_1 + G_2)V_B - G_1V_A - G_2V_C = 0$

Now invoking the OP AMP input constraint ($v_P = v_N$), we see that V_B must be zero since the noninverting input is connected to the reference node (grounded). Hence

$$V_C = -\frac{R_2}{R_1}V_A \quad \text{or} \quad V_o = -\frac{R_2}{R_1}V_S$$

which is the same result as found in Chapter 4.

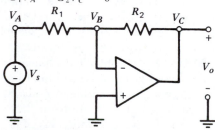

FIGURE 5-14
OP AMP circuit for Example 5-11.

Example 5-12

In the circuit shown in Figure 5-15 neither OP AMP input is grounded. We need only formulate node equations at nodes Ⓑ and Ⓓ since nodes Ⓐ and Ⓒ are constrained. By inspection:

Node Ⓑ $(G_1 + G_2)V_B - G_1V_A - G_2V_C = 0$
Node Ⓓ $(G_3 + G_4)V_D - G_3V_C = 0$

The OP AMP input constraint ($v_P = v_N$) demands that $V_B = V_D$, and since $V_A = V_S$, we have

$$(G_1 + G_2)V_B - G_2V_C = G_1V_S$$
$$(G_3 + G_4)V_B - G_3V_C = 0$$

or two equations in two unknowns. Solving for the output:

$$V_o = V_C = \frac{D_C}{D} = \frac{\begin{vmatrix} G_1 + G_2 & G_1 V_S \\ G_3 + G_4 & 0 \end{vmatrix}}{\begin{vmatrix} G_1 + G_2 & -G_2 \\ G_3 + G_4 & -G_3 \end{vmatrix}}$$

$$= \frac{G_1 G_4 + G_1 G_3}{G_1 G_3 - G_2 G_4} V_S$$

which, after a bit of algebraic manipulation, yields

$$V_o = \frac{R_2 R_3 + R_2 R_4}{R_2 R_4 - R_1 R_3} V_S$$

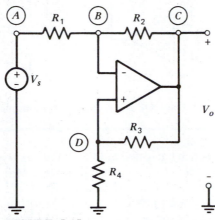

FIGURE 5-15
Analysis of an OP AMP subtractor circuit
for Example 5-12.

Example 5-13

The circuit in Figure 5-16 has two OP AMPs and a total of five nonreference nodes. However, nodes Ⓒ and Ⓔ are connected to OP AMP outputs, and node Ⓐ is connected to the grounded input source. Hence we need only two modified node equations. By inspection,

Node Ⓑ
$(G_1 + G_2 + G_5)V_B - G_1 V_A - G_2 V_C - G_5 V_E = 0$
Node Ⓓ
$(G_3 + G_4 + G_6)V_D - G_4 V_A - G_3 V_C - G_6 V_E = 0$
But both noninverting inputs are grounded and the $v_P = v_N$ contraints then mean that $V_B = 0$ and $V_D = 0$. Since $V_A = V_S$, these equations reduce to

$$G_2 V_C + G_5 V_E = -G_1 V_S$$
$$G_3 V_C + G_6 V_E = -G_4 V_S$$

and we end up with two equations in two unknowns (the two OP AMP outputs). These equations are not symmetrical, which points out that the OP AMP is a unilateral active device.

Solving for the circuit output:

$$V_o = V_E = \frac{D_E}{D} = \frac{\begin{vmatrix} G_2 & -G_1 V_S \\ G_3 & -G_4 V_S \end{vmatrix}}{\begin{vmatrix} G_2 & G_5 \\ G_3 & G_6 \end{vmatrix}}$$

$$= \frac{(G_1 G_3 - G_2 G_4)}{(G_2 G_6 - G_3 G_5)} V_S$$

Happily we see that this rather formidable appearing OP AMP circuit quickly reduces to two straightforward equations using modified node analysis.

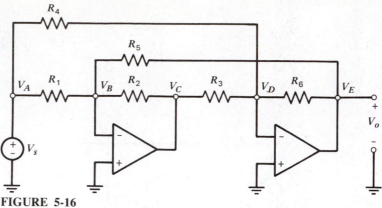

FIGURE 5-16
Feed forward– feedback OP AMP circuit for Example 5-13.

We have seen in basic and modified node analysis that the node voltages are a very useful set of analysis variables. Many large computer circuit analysis programs such as ECAP and SPICE are based on node equations since the formulation process is very algorithmic. But node voltages are also very useful in the laboratory. To evaluate a circuit, we connect one terminal of a voltmeter to ground (the reference node) and probe the remaining nodes with the other voltmeter terminal, thereby measuring V_A, V_B, V_C, and so on. Thus node voltages are useful for both analysis and experimentation.

5-6 MESH ANALYSIS

Mesh currents are an alternative set of solution variables that are useful for the analysis of circuits containing many elements connected in series, and hence there are many nodes. Stated formally, a mesh is a loop that contains no elements within it. For example, the circuit in Figure 5-17 has three loops, but only two meshes ($L1$ and $L2$). We restrict our development of mesh analysis to planar circuits, that is, circuits that can be drawn on a flat surface without crossovers. Such circuits can always be drawn in the window-pane form illustrated in Figure 5-18. Actually the restriction to planar circuits is not significant for the complexity of circuits we will consider, but ultimately does limit the applicability of mesh analysis.

Mesh currents are depicted as flowing through the elements around the perimeter of the mesh. A mesh current is then defined by labeling the meshes (I_1, I_2, I_3, etc.) and assigning a reference direction to each current. Mesh currents are all assumed to flow in a clockwise sense, but there is no fundamental reason for this, except perhaps tradition. We then think of these mesh currents as flowing around their respective meshes, as indicated in Figure 5-18.

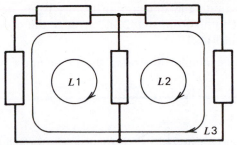

FIGURE 5-17
Circuit with three loops and two meshes
(L1 and L2).

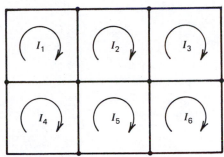

FIGURE 5-18
Meshes in planar circuits.

We should emphasize that we are not describing the physical process that takes place in a circuit. The electrons do not somehow get assigned to I_1 or I_2, and so on. We are here defining a set of variables that can be used to determine the voltage across and current through the circuit elements. In other words, mesh currents are variables of convenience that are used in analysis, but are only somewhat abstractly related to the physical process.

To use mesh currents to formulate circuit equations, we proceed as in device and connection analysis, except that the KCL connection constraints are not explicitly written down. Instead the KCL constraints are used to express the device constraints in terms of the mesh currents. Stated formally, there are four steps in mesh analysis:

1. Identify a mesh current with every mesh and identify a voltage with every circuit element.

2. Write KVL connection constraints in terms of the element voltages around every mesh.

3. Use KCL and the element *i-v* characteristics to express the element voltages in terms of the mesh currents.

4. Substitute the device constraints from step 3 into the connection constraints from step 2 and arrange the resulting equations in a standard form.

The next example illustrates the application of this basic formulation process.

Example 5-14

The circuit in Figure 5-19 was previously analyzed using node analysis (see Example 5-8). In the figure we have defined two mesh currents and all element voltages (step 1). We now write KVL constraints around each mesh (step 2).

Mesh ① $\qquad -V_{S1} + V_1 + V_3 = 0$

Mesh ② $\qquad -V_3 + V_2 + V_{S2} = 0$

We next write the device constraints in terms of the mesh currents (step 3).

$$V_1 = R_1 I_1$$
$$V_2 = R_2 I_2$$
$$V_3 = R_3(I_1 - I_2)$$
$$V_{S1} = \text{known input}$$
$$V_{S2} = \text{known input}$$

Note that we have in effect used KCL at node ④ in Figure 5-19 to write the third element equation. Finally we substitute the element equation into the connection equations and arrange the result in standard form (step 4).

$$(R_1 + R_3)I_1 \qquad - R_3 I_2 = \quad V_{S1}$$
$$- R_3 I_1 + (R_2 + R_3) I_2 = - V_{S2}$$

We have completed the formulation process with two equations in two unknown mesh currents.

Turning now to the solution phase we use Cramer's rule.

$$I_1 = \frac{D_1}{D} = \frac{\begin{vmatrix} V_{S1} & -R_3 \\ -V_{S2} & (R_2 + R_3) \end{vmatrix}}{\begin{vmatrix} (R_1 + R_3) & -R_3 \\ -R_3 & (R_2 + R_3) \end{vmatrix}}$$

$$= \frac{(R_2 + R_3)V_{S1} - R_3 V_{S2}}{R_1 R_2 + R_1 R_3 + R_2 R_3}$$

and

$$I_2 = \frac{D_2}{D} = \frac{\begin{vmatrix} R_1 + R_3 & V_{S1} \\ -R_3 & V_{S2} \end{vmatrix}}{D}$$

$$= \frac{R_3 V_{S1} - (R_1 + R_3)V_{S2}}{R_1 R_2 + R_1 R_3 + R_2 R_3}$$

These mesh currents can now be substituted into the device constraints to determine every voltage in the circuit. The reader is invited to do this and to demonstrate that the same results are obtained as in Example 5-8 where node analysis was applied to the same circuit.

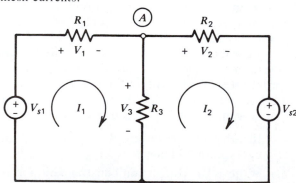

FIGURE 5-19
Circuit for demonstrating mesh analysis.

The mesh equations derived in Example 5-14 have a symmetrical pattern that is similar to the symmetry observed in node equations. The coefficient of I_2 in the first equation is the sum of the resistances in the first mesh. The coefficient of I_2 is the negative of the resistance common to both mesh ① and ②. The right side of the equation is the input source voltage in mesh ①. This pattern is repeated in mesh ②.

This symmetrical pattern will always occur for circuits containing resistors and independent voltage sources. To see this, consider a general resistance R that is contained in, say, mesh ①. There are only two possibilities. Either R is not contained in any other mesh, in which case the voltage across it is

$$V = R(I_1 - 0) = RI_1$$

or else it is also contained in only one adjacent mesh, say, mesh ②, in which case,

$$V = R(I_1 - I_2)$$

The pattern is now easy to see. The sum of the voltages around a mesh is

1. I_1 times the sum of the resistances in mesh ①.
2. Minus I_2 times the sum of resistances common to mesh ① and mesh ②.
3. Minus similar terms for any other mesh.

This sum equals the sum of the input voltage around mesh ① due to independent voltage sources.

The process outlined allows us to write mesh current equations by inspection without going through the time-consuming intermediate steps involving the KVL connection constraints and the device constraints.

Example 5-15

In this example we will write mesh current equations by inspection for the circuit in Figure 5-20.

Mesh ① $(R_1 + R_2)I_1 - R_2I_3 = -V_{S1}$
Mesh ② $(R_3 + R_4)I_2 - R_3I_3 = +V_{S1}$
Mesh ③ $(R_2 + R_3)I_3 - R_2I_1 - R_3I_2 = -V_{S2}$

or in standard form,

$(R_1 + R_2)I_1 \qquad\qquad - R_2I_3 = -V_{S1}$
$\qquad + (R_3 + R_4)I_2 \quad R_3I_3 = +V_{S1}$
$-R_2I_1 \quad - R_3I_2 + (R_2 + R_3)I_3 = -V_{S2}$

The use of symmetry greatly simplifies the formulation of mesh equation.

Turning now to the solution phase, Figure 5-20 indicates that we are asked to find the voltage V_o across R_4. In terms of our solution variables this means that we must solve for I_2. Using Cramer's rule to solve first for I_2 as

$$I_2 = \frac{D_2}{D} = \frac{\begin{vmatrix} (R_1 + R_2) & -V_{S1} & -R_2 \\ 0 & V_{S1} & -R_3 \\ -R_2 & -V_{S2} & (R_2 + R_3) \end{vmatrix}}{\begin{vmatrix} (R_1 + R_2) & 0 & -R_2 \\ 0 & (R_3 + R_4) & -R_3 \\ -R_2 & -R_3 & (R_2 + R_3) \end{vmatrix}}$$

$$I_2 = \frac{R_1(R_2 + R_3)V_{S1} - R_3(R_1 + R_2)V_{S2}}{R_1R_2R_3 + R_1R_2R_4 + R_1R_3R_4 + R_2R_3R_4}$$

Then by inspection we can write

$$V_o = I_2R_4$$

Replacing I_2 in this expression by the solution found above yields,

$V_o =$

$$\frac{(R_1R_2R_4 + R_1R_3R_4)V_{S1} - (R_1R_3R_4 + R_2R_3R_4)V_{S2}}{R_1R_2R_3 + R_1R_2R_4 + R_1R_3R_4 + R_2R_3R_4}$$

This result can be compared with the node voltage solution of the same circuit found in Example 5-10 to show that either approach produces the same answer. Which do you think was easier to obtain?

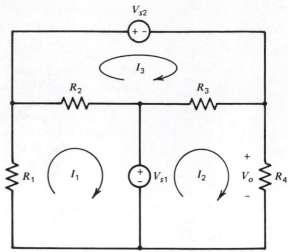

FIGURE 5-20
Three-mesh circuit for Example 5-15.

In developing mesh analysis we have tacitly assumed that the circuit contains only voltage sources. If there are current sources (independent or dependent), then in a dual fashion to what we called modified node equations, we can use modified mesh analysis. Specifically there are three possibilities.

1. If the current source is connected in parallel with a resistor, then it can be converted to an equivalent voltage source by source conversion.

2. If only one mesh current flows through a current source, then that mesh current is no longer an unknown but is determined by the source current. We can write mesh equations around the remaining mesh in the usual way and move the known mesh current to the source side of the equations in the final step.

3. If two mesh currents flow through a current source, then their difference is fixed and we must then introduce an unknown voltage across the current source as one of our solution variables.

These approaches are not mutually exclusive (we might apply all three in a very complicated circuit), and are the duals of the three methods used in developing modified node equations. The reader should attempt to develop a figure analogous to Figure 5-9 for the three cases noted. Our last example applies the first approach to a dependent source.

Example 5-16

The circuit in Figure 5-21 is a small-signal model of a transistor circuit called an *emitter follower*. The given circuit contains a voltage-controlled current source that can be converted to a voltage-controlled voltage source using source conversion, as shown in the modified circuit. We now temporarily treat the dependent voltage source as an independent source and write two mesh equations by inspection.

Mesh ① $(R_1 + R_2 + R_3)I_1 - R_3I_2 = V_S - gR_3V_x$

Mesh ② $(R_3 + R_4)I_2 - R_3I_1 \quad = gR_3V_x$

We observe that the control voltage V_x can be written in terms of mesh currents as

$$V_x = R_2I_1$$

When this result is substituted into mesh equations and all unknowns shifted to the left side of the resulting equations, we get

$$(R_1 + R_2 + R_3 + gR_2R_3)I_1 - R_3I_2 = V_S$$
$$- (R_3 + gR_2R_3)I_1 + (R_3 + R_4)I_2 = 0$$

The controlled source is a unilateral element and therefore the resulting mesh equations are *not* symmetrical.

Turning to the solution phase we solve first for the indicated output current.

$$I_o = I_2 = \frac{D_2}{D} = \frac{\begin{vmatrix} R_1 + R_2 + R_3 + gR_2R_3 & V_S \\ -(R_3 + gR_2R_3) & 0 \end{vmatrix}}{\begin{vmatrix} R_1 + R_2 + R_3 + gR_3R_2 & -R_3 \\ -(R_3 + gR_2R_3) & R_3 + R_4 \end{vmatrix}}$$

$$= \frac{(R_3 + gR_2R_3)V_S}{R_1R_3 + R_1R_4 + R_2R_3 + R_2R_4 + R_3R_4 + gR_2R_3R_4}$$

And finally,

$$V_o = I_oR_4$$

$$= \frac{(R_3R_4 + gR_2R_3R_4)V_S}{R_1R_3 + R_1R_4 + R_2R_3 + R_2R_4 + R_3R_4 + gR_2R_3R_4}$$

If g is a very large number, then the last terms in the numerator and denominator dominate and

$$V_o \simeq \frac{gR_2R_3R_4}{gR_2R_3R_4} V_S = V_S$$

Hence the name voltage follower is often used for the emitter follower circuit since the output is the same as (follows) the input.

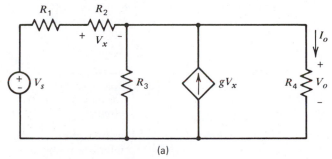

(a)

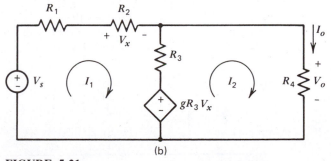

(b)

FIGURE 5-21
Dependent source circuit for applying mesh analysis. (*a*) Given circuit. (*b*) Modified circuit.

5-7   SUMMARY OF ANALYSIS METHODS

At this point we have completed our study of memoryless circuits, and so it is wise to sit back and reflect on what we have learned. We have studied a number of methods of analyzing linear circuits, including circuit reduction, unit output, superposition, Thevenin's and Norton's theorems, device and connection analysis, node analysis, and finally mesh analysis. A summary of the major advantages and disadvantages of these methods is given in Table 5-1.

Are all of these methods indeed necessary? Why do we not learn one, and learn it well? Certainly if we were to put all our eggs in one basket, then node analysis would probably provide the best basket. But more important, to an engineer circuit analysis is not the end product, but a means to an end. Put differently, an engineer is almost never required to analyze a circuit outside of the context of circuit design or evaluation. In

Technique	Advantages	Disadvantages
Circuit reduction	Involves working directly with the circuit model	Not general
Unit output	Straightforward application of KCL, KVL, and Ohm's law	Ladder circuits only
Superposition	Powerful conceptual tool for multisource circuits	Requires repeated analysis
Thévenin/Norton	Describes the interface problem well	Difficult to apply to active circuits
Device and connection analysis	Provides the foundations for all general analysis methods	Difficult because of many variables
Node analysis	Easily generalized and applied to active circuits	Difficult to apply to transformers (Chapter 7)
Mesh analysis	Easily applied to planar circuits	OP AMPs not easily handled

TABLE 5-1
Summary of Analysis Methods

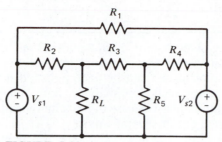

FIGURE 5-22
An example circuit.

this context analysis efficiency is a primary concern, and the most efficient method depends on what needs to be determined.

To illustrate the point consider the circuit in Figure 5-22. If we were to perform a general analysis, then node analysis would be more efficient than mesh analysis because there are only two unknown nodes but four meshes. However, if the purpose is to determine R_L for maximum power transfer, then neither node nor mesh analysis would provide the answer readily. In this case either Thévenin's or Norton's theorem would be a better choice. Likewise, if the effect of V_{S1} on the circuit is needed, then superposition is appropriate. In other circuits the unit output method, or source conversion, or circuit reduction might be more efficient than all of these. In sum, an engineer should know how to use many different tools, and also know how to select the one best suited to the task at hand.

A few general clues about analysis are as follows:

1. Circuits with many elements connected in parallel are probably best treated using node analysis.

2. Conversely circuits with many elements in series are probably best treated using mesh analysis.

3. Single-input source ladder circuits are easily treated using successive source conversions, circuit reduction, or the unit output method.

4. The circuit should be simplified by combining elements in series or parallel whenever possible.

5. OP AMP circuits are best treated using node analysis.

6. If the effect of individual inputs to a circuit is desired, then the application of superposition is suggested.

7. Thévenin's or Norton's theorem is useful in interface situations where maximum signal transfer is desired, or when a nonlinear device is involved.

Unfortunately no single technique always fits the task. In most circuit problems several different techniques will be needed. Only practice and experience will provide the insight needed to select the best set of tools in each new situation.

SUMMARY

- All general methods of circuit analysis can be divided into two phases: formulation of linear equations and solution of these equations. The formulation phase can be accomplished in several ways, including device and connection analysis, node analysis, and mesh analysis. The solution phase is essentially a mathematical process that can also be carried out in several ways.

- Device and connection analysis forms the basis of all general methods of formulating circuit equilibrium equations. This method produces $N - 1$ independent KCL connection equations, $E - N + 1$ independent KVL connection equations, and E element equations. Collectively this produces $2E$ equations that must be treated in the solution phase.

- Node analysis is probably the most useful general analysis technique. It involves identifying a reference node and the voltages at the remaining $N - 1$ nonreference nodes. The KCL connection constraints at the $N - 1$ nonreference nodes are combined with the device constraints written in terms of the node voltages to produce $N - 1$ linear equations in the unknown node voltages.

- Mesh analysis is the dual of node analysis. It involves identifying a set of somewhat artificial mesh currents that flow around the perimeter of each mesh. The KVL connection constraints are written around each of the $E - N + 1$ meshes and combined with the device constraints written in terms of the mesh currents to produce $E - N + 1$ linear equations in the unknown mesh currents.

- Modified forms of both node and mesh analysis often simplify the analysis of certain types of circuits. The possible modifications include source transformations, unknown variable elimination by appropriate selection of solution variables, and the introduction of mixed sets of solution variables.

- OP AMP circuits can be easily treated using modified node analysis. A node voltage must be identified at each OP AMP output, but a node equation need not be written at these nodes. Node equations are then written at the remaining nodes and the ideal OP AMP input voltage constraint ($v_P = v_N$) used to reduce the number of unknowns. In formulating these equations the input currents i_P and i_N can be ignored since they are zero for an ideal OP AMP.

- Cramer's rule is a systematic method for solving systems of linear equations obtained in the formulation phase of analysis. This rule reduces the solution process to a matter of evaluating ratios of determinants. In principle it can be applied to any number of equations, but in this text we will treat no more than three equations.

● Many different circuit analysis techniques have been studied in the first five chapters of this book. All of these techniques find applications in engineering circuit analysis problems. Table 5-1 summarizes the major advantages and limitations of these techniques. Some general guidelines exist, but only experience and practice lead to the ability to select the best technique in a given situation.

EN ROUTE OBJECTIVES
AND RELATED EXERCISES

5-1 *GENERAL CIRCUIT ANALYSIS (SECS. 5-1 to 5-6)*

Given a circuit consisting of linear resistors, controlled sources, OP AMPs, and input signal sources:

(a)   Formulate a complete set of node-voltage or mesh-current equations that describe the circuit equilibrium.

(b)   Solve these equations for selected signal variables or input-output relationships.

Exercises

5-1-1   For each of the circuits of Figure E5-1-1 find the indicated variable(s) by:

(a)   Device and connection analysis

(b)   Node-voltage   analysis   (modify   if applicable)

(c)   Mesh-current   analysis   (modify   if applicable)

Then compare the different techniques for each circuit.

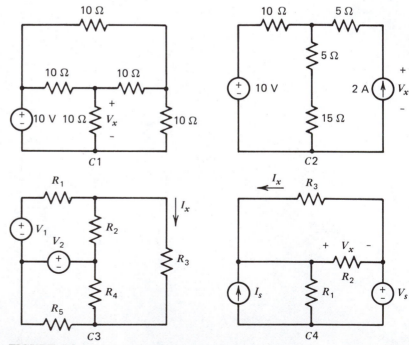

FIGURE E5-1-1
Circuits for applying various circuit analysis techniques.

5-1-2   Formulate node-voltage equations for each
circuit of Figure E5-1-2.

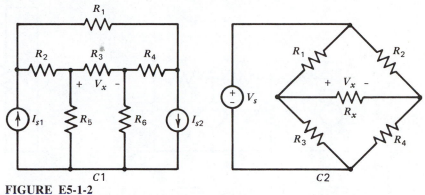

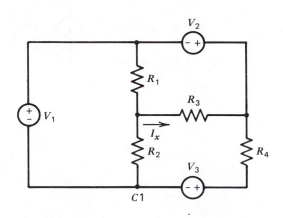

FIGURE E5-1-2
Circuits for solution via node-voltage analysis.

5-1-3   Formulate mesh-current equations for each
circuit of Figure E5-1-3.

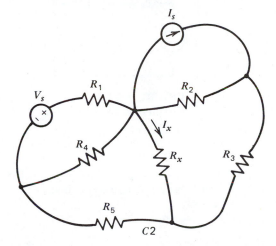

FIGURE E5-1-3
Circuits for solution via mesh-current analysis.

5-1-4 Solve for the designated variable (V_o, V_E, or I_o) in each of the circuits of Figure E5-1-4.

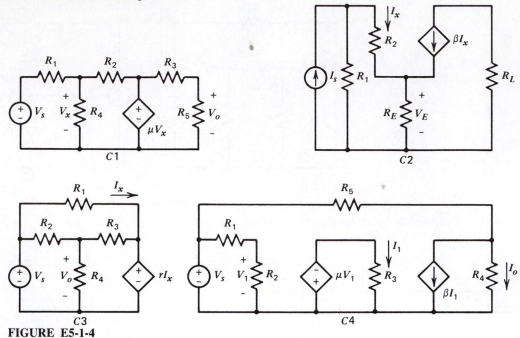

FIGURE E5-1-4
Circuits for Exercise 5-1-4.

5-1-5 Solve for the unknown variables in Exercise 5-1-2 if all R's are 1 Ω, V's are 10 V, and I's are 1 A.

5-1-6 For each OP AMP circuit of Figure E5-1-6, find V_o as a function of the input.

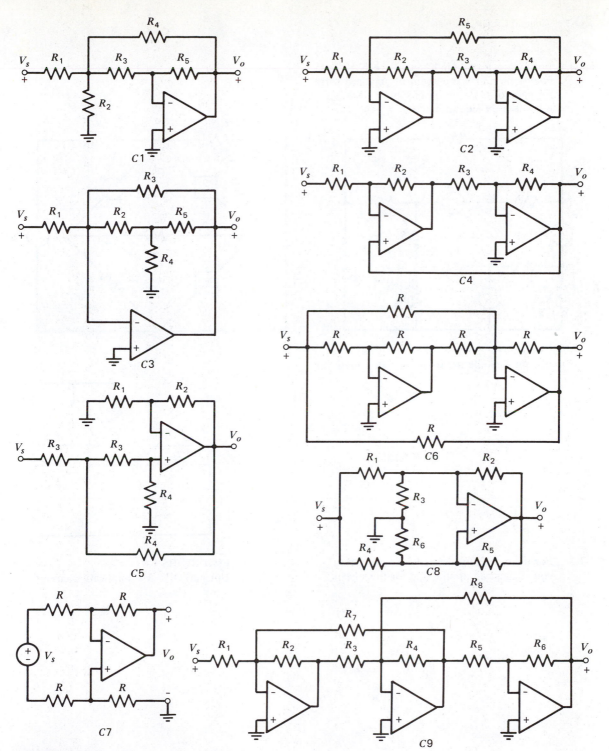

FIGURE E5-1-6
OP AMP circuits for Exercise 5-1-6

235

5-1-7 Formulate a complete set of mesh or node equations and write them in standard form for each of the circuits of Figure E5-1-7. Do not solve the equations.

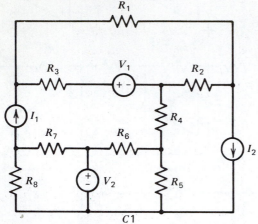

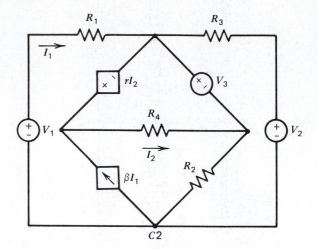

FIGURE E5-1-7
Circuits for formulating mesh or node equations.

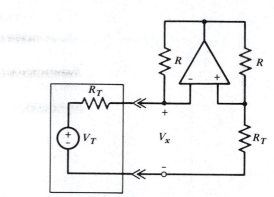

5-1-8 For the circuit in Figure E5-1-8, find V_x. Can you suggest an application for this circuit?

FIGURE E5-1-8
Unusual OP AMP application circuit.

5-1-9 A cube has a series resistor—voltage source on each edge as shown in Figure E5-1-9. If all sources produce 1 V and each resistor is 1 Ω, what should R_L be for maximum power transfer *and* what is the maximum power transferred?

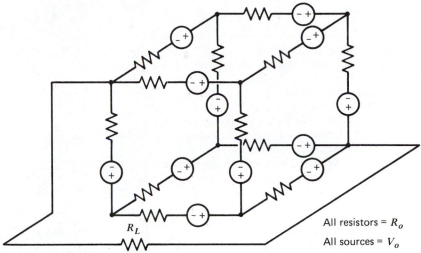

All resistors = R_o

All sources = V_o

FIGURE E5-1-9 Cube circuit.

5-1-10 The circuit shown in Figure E5-1-10 is a model for the midband operation of a MOS (metal-oxide semiconductor) two-transistor amplifier. If V_1 is the signal to be amplified, what is the output V_o?

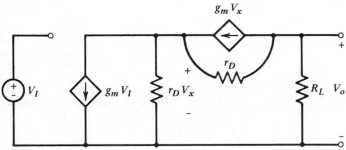

FIGURE E5-1-10
MOS two-transistor *cascode* amplifier.

5-2 COMPARISON OF ANALYSIS TECHNIQUES (SEC. 5-7)

Given a circuit consisting of linear resistors, controlled sources, OP AMPs, and input signal sources, identify and compare alternative analysis techniques for determining specified circuit variables or parameters.

Exercises

5-2-1 Consider the circuit in Figure E5-2-1. Discuss what analysis technique you would use to solve for the following.
 (a) Current through R_2.
 (b) Voltage across R_5.
 (c) Selecting R_4 for maximum power.
 (d) Determining the contribution of I_S to the current in R_2.
 (e) Adjusting V_A so to obtain a certain voltage across R_1.

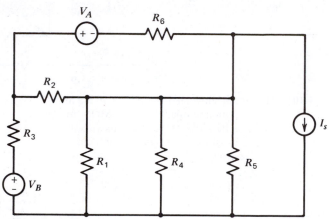

FIGURE E5-2-1
Circuit for Exercise 5-2-1.

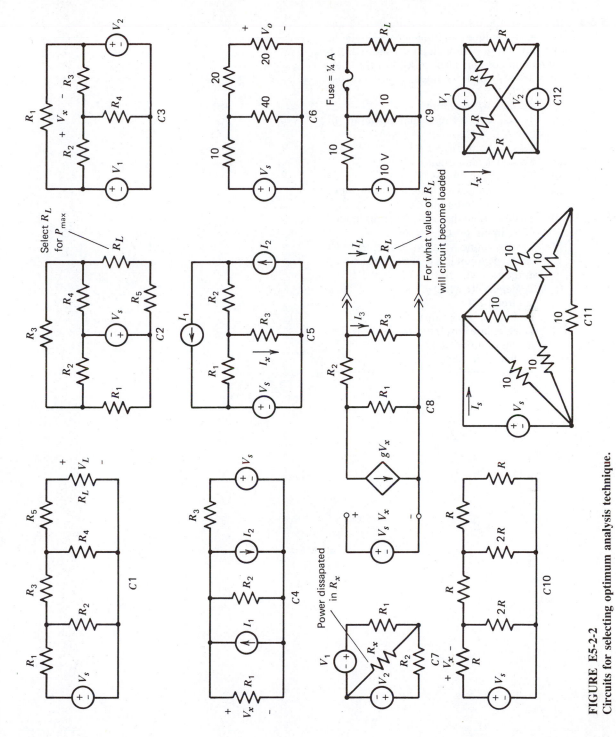

FIGURE E5-2-2
Circuits for selecting optimum analysis technique.

239

5-2-2   In each of the circuits of Figure E5-2-2 one analysis technique is optimum for solving the question asked (although one or more different techniques may also work). For the questions asked, which technique would you use and why?

(a)   To find V_L in circuit $C1$.

(b)   To select R_L for maximum power in circuit $C2$.

(c)   To find V_x in circuit $C3$.

(d)   To find V_x in circuit $C4$.

(e)   To find I_x in circuit $C5$.

(f)   To determine what value of V_S to achieve 10 V at V_o in circuit $C6$.

(g)   To determine the total power dissipated in R_x in circuit $C7$.

(h)   To determine what values of R_L will load the circuit in $C8$. We define loading in the problem to mean $I_L \geq 0.1\ I_3$.

(i)   To determine whether the fuse will blow in circuit $C9$.

(j)   To find V_x in circuit $C10$.

(k)   To calculate how much power V_S is supplying to circuit $C11$.

(l)   To find I_x in circuit $C12$.

PROBLEMS

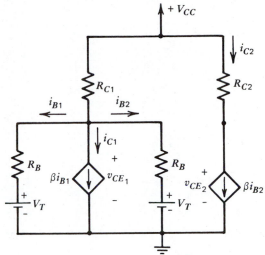

FIGURE P5-1
Two-transistor small-signal amplifier problem.

P5-1 A certain two-transistor small-signal amplifier is shown in Figure P5-1. Find the quiescent point (i_C, v_{CE}) of each transistor. Assume β is very large.

P5-2 A **gyrator** is a nonreciprocal network with unique transfer relationships. In terms of the two-port network shown in Figure P5-2a, these characteristics are $I_1 = -V_2/R_o$ and $I_2 = V_1/R_o$, where R_o is a positive real quantity called the **gyration resistance.** Show that the OP AMP circuit of Figure P5-2b realizes the gyrator transfer characteristics.

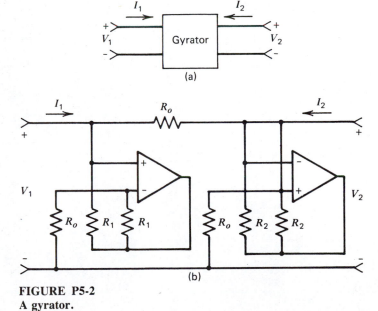

FIGURE P5-2
A gyrator.

BLOCK II
CIRCUITS WITH MEMORY

BLOCK OBJECTIVES

ANALYSIS

Given a linear circuit containing not more than two memory elements with an input signal waveform that is a step function, an impulse, or a sinusoid, determine prescribed output signals or input-output relationships using either time domain or s-domain techniques.

DESIGN

Devise an active or passive circuit containing not more than two memory elements or modify an existing circuit to obtain a specified output signal for a given input signal or to implement a given input-output relationship.

EVALUATION

Given two or more circuits that perform the same signal-processing function, select the best circuit based on given criteria such as performance, cost, parts count, power dissipation, and simplicity.

Chapter 6
Signal Waveforms

Introduction
The Step Waveform
The Exponential Waveform
The Sinusoidal Waveform
Composite Waveforms
Partial Waveform Descriptors

Under the sea, under the sea
Mark how the telegraph motions to me
Under the sea, under the sea
Signals are coming along

James Clerk Maxwell

To analyze circuits it is necessary to develop mathematical models for both electrical components and electric signals. In this chapter we begin the study of signals and the means used to describe and quantify their characteristics. Signals are the physical manifestation of data and energy in a circuit, and so the study of the characteristics is clearly an important matter. Signals are represented by mathematical models that are approximations or idealizations of the actual physical processes involved. Yet to be useful these models must capture the essential features of the situation. This balancing of complexity of reality against simplicity of models is one of the central issues in engineering. To achieve this balance we often use partial signal descriptors that do not provide an exhaustive account of signal properties, but are measurable quantities that highlight essential signal characteristics.

6-1 INTRODUCTION

Electrical engineers normally consider a signal to be an electric current or voltage. Such a signal could be the voltage produced by an input signal source or the current at the output of a circuit. In any case, a signal is generally a time-varying quantity. The pattern of time variation is called a **waveform.** More formally:

A waveform is an equation or graph that describes the signal characteristics as a function of time.

Some examples of signal waveforms are shown in Figure 6-1.

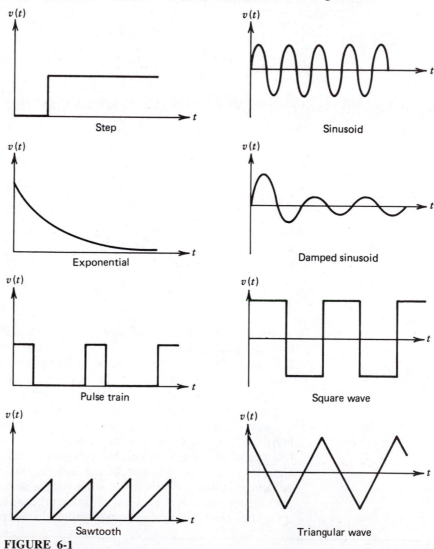

FIGURE 6-1
Some example waveforms.

It might seem that we are faced with the uninviting task of compiling a lengthy catalog of waveforms. However, it turns out that we do not need a long list, and in fact we can develop most of the waveforms of interest using just three basic signal models: the step function, the exponential function, and the sinusoidal function. This situation illustrates the great value that engineers find in the use of mathematical models. The real, physical world of signals is indeed complex, but its study is greatly simplified by the use of a few relatively simple models that approximate that reality.

There are two matters of notation and convention that must be discussed before we continue. First, quantities that are constant (non-time-varying) are usually represented by uppercase letters (V_A, I, T_o, A_1) or lowercase letters in the early part of the alphabet (a, b, c, d_o). Time-varying quantities are represented by lowercase letters that are not in the early part of the alphabet. This time variation is expressly indicated when we write a term such as $v_1(t)$, $i_A(t)$, or $u(t)$. The time variation is implied when such terms are written as v_1, i_A, or u.

Second, the signal variables in a circuit are normally provided with reference marks as shown in Figure 6-2. It is important to remember that these reference marks do not indicate the polarity of a voltage or the direction of flow of a current. As Figure 6-2 shows, the signal waveforms in a circuit can assume both positive and negative values. The purpose of these marks is to provide a reference against which the actual value

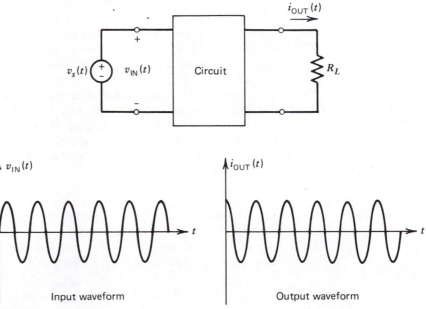

FIGURE 6-2
Example input and output waveforms.

of the signal waveform can be compared. When the actual polarity of a voltage or direction of current flow coincides with the reference direction, then we say that the signal is positive. When the opposite occurs, then the signal is considered negative.

Finally, in this chapter we generally use the symbol $v(t)$ to represent a signal waveform. The reader should remember, however, that a signal can be either a current $i(t)$, a voltage $v(t)$, or power $p(t)$. In this chapter the distinction is not important but in subsequent chapters it will again be important to remember the difference between the various signal variables.

6-2   THE STEP WAVEFORM

To develop the general step function waveform, let us first consider the unit step function defined by the relationship:

$$u(t) = \begin{cases} 0 & \text{for } t < 0 \\ 1 & \text{for } t \geq 0 \end{cases} \tag{6-1}$$

The unit step function is zero when its argument (t) is negative, and it is unity when its argument is zero or positive. Mathematically the unit step function $u(t)$ has a jump or discontinuity at $t = 0$.

It is not possible to generate a true step function since no physical variable can undergo a jump change in zero time. Practically speaking it is possible to generate very good approximations to the step function. What is required is that the finite switching time be very short compared with the time the variable remains in its new state. Actually the generation of approximate step functions is an everyday occurrence, since people are forever turning out lights and slamming doors (usually in that order). Once the light is out and the door slammed shut, they normally remain in that state for a period of time that is long compared with the time required to attain those states.

On the surface, it might appear that the step function is not a very exciting waveform or, at best, only a source of temporary excitement. However, we shall see that the step waveform is a versatile signal and is used to construct models of a wide range of interesting waveforms. First of all, we can multiply $u(t)$ by a real constant A to obtain

$$A\, u(t) = \begin{cases} 0 & \text{for } t < 0 \\ A & \text{for } t \geq 0 \end{cases} \tag{6-2}$$

This constant A is called the **amplitude** of the waveform. In addition, we can shift the time at which the step occurs by replacing (t) by $(t - T_S)$. Since the definition of the step function indicates that the jump occurs when the argument is zero, the function $u(t - T_S)$ takes on the values

$$u(t - T_S) = \begin{cases} 0 & \text{for } t - T_S < 0 \text{ or } t < T_S \\ 1 & \text{for } t - T_S \geq 0 \text{ or } t \geq T_S \end{cases} \tag{6-3}$$

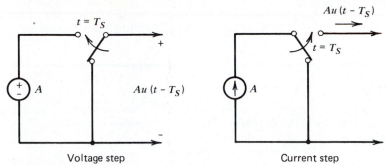

Voltage step   Current step

FIGURE 6-3
Circuits that generate step waveforms.

Thus we can delay or advance the time at which the step occurs by changing the argument. The constant T_S is called the **step time**.

The general step waveform is then written as $A\,u(t - T_S)$, and we see that it takes two constants to define the signal. The constant A determines the amplitude of the step, and normally carries the units of volts, amperes, or watts in an electrical context. The constant T_S determines the time at which the step occurs, and carries the units of time, usually seconds. Either A or T_S, or both, can be negative. Figure 6-3 shows how ideal sources and switches can be used to generate a step waveform

Figure 6-4 shows the effects on the step waveform of changing the amplitude A and the step time T_S. By changing A and T_S and adding

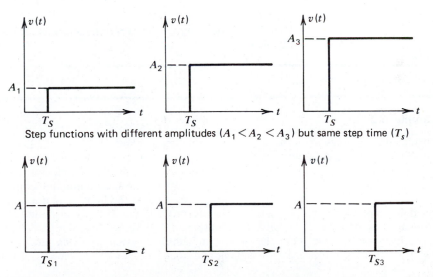

Step functions with different amplitudes ($A_1 < A_2 < A_3$) but same step time (T_s)

Step functions with the same amplitude (A) but different step times ($T_{S1} < T_{S2} < T_{S3}$)

FIGURE 6-4
Effects of amplitude and step time on the general step waveform, $Au(t - T_s)$.

several step functions, we can represent a number of important waveforms. Some of the possibilities are illustrated in the following examples.

Example 6-1

The rectangular pulse in Figure 6-5a can be expressed as a sum of step functions. The amplitude of the pulse jumps to a value of 3 V at $t = 1$ s; therefore $3\,u(t - 1)$ is part of the expression for the waveform. The pulse goes to zero at $t = 3$ s, and so a step of amplitude -3 V at $t = 3$ s must be added. Putting these together, we express the rectangular pulse as

$$v(t) = 3\,u(t - 1) - 3\,u(t - 3)$$

Figure 6-5b shows how the two step functions com-bine to produce the pulse. From this we see that a signal of the form

$$v(t) = A\,u(t - T_1) - A\,u(t - T_2)$$

is a rectangular pulse of amplitude A, which turns on at $t = T_1$ and off at $t = T_2$. By adding a number of such pulses we can represent the long sequences of pulses commonly used in digital systems.

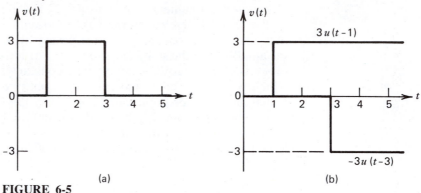

FIGURE 6-5
Use of step functions to realize a rectangular pulse. (*a*) A rectangular pulse. (*b*) Step functions.

Example 6-2

Consider the rectangular pulse labeled v_1 in Figure 6-6a. Such a pulse can be written as

$$v_1(t) = \frac{1}{T}\,[u(t) - u(t - T)]$$

This pulse has an amplitude that is inversely proportional to its duration. If we halve the duration and double the amplitude, we obtain the pulse v_2 in Figure 6-6a. By repeating the process of halving and dou-bling we obtain v_3. For all three pulses we observe that if we integrate with respect to time (find the area under the waveform), we obtain unity. Now consider the special function obtained by carrying this process to the limit. The duration approaches zero, the amplitude becomes infinite, but the area remains unity. Such a function finds wide application in circuit and system analysis, and goes by the name **impulse** or Dirac delta function.[1] Its mathematical symbol is $\delta(t)$

[1]The Dirac delta function is named in honor of the English mathematician and physicist Paul Andrien Maurice Dirac, who, along with Fermi, pioneered a novel distribution function that is applicable to semiconductors and is known as Fermi-Dirac statistics.

and its graphical representation is shown in Figure 6-6b. More formally, the unit impulse is defined as

$$\delta(t) = 0 \qquad \text{for } t \neq 0$$

$$\int_{-\infty}^{t} \delta(x) \, dx = u(t)$$

This latter expression suggests that the impulse is the derivative of a step function.

$$\frac{d \, u(t)}{dt} = \delta(t)$$

This cannot be justified using elementary mathematics since the function $u(t)$ has a discontinuity at $t = 0$, and therefore its derivative does not exist at that point in the usual sense. However, the concept can be justified using the theory of distributions as shown in advanced texts on circuits and systems. The general impulse is written as $C \, \delta(t - T_S)$ and its graphical representation is shown in Figure 6-6c. Notice that the magnitude of the impulse is represented by its area and not its amplitude, which is infinite.

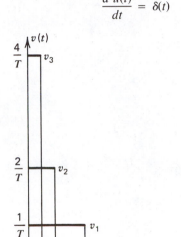

(a)

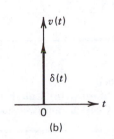

(b)

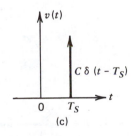

(c)

FIGURE 6-6
Impulse waveforms. (a) Pulse waveforms. (b) Unit impulse. (c) General impulse.

Example 6-3

The **unit ramp waveform** can be derived from the step function by integration.

$$r(t) = \int_{-\infty}^{t} u(x) dx$$

or

$$r(t) = t \, u(t)$$

For negative times the ramp waveform is zero and for positive time it is simply equal to t. The unit ramp waveform is shown in Figure 6-7. Notice that its slope is unity. The general ramp waveform can then be written as $B \, r(t - T_S)$ and is shown graphically in Figure 6-7. The general ramp is zero until $t = T_S$, and then increases with a constant slope of B. Thus the magnitude of a ramp waveform is measured by its slope. The ramp is useful in representing triangular and sawtooth waveforms used as timing or sweep signals in both digital and analog circuits.

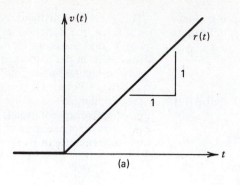

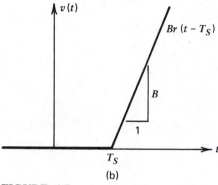

FIGURE 6-7
**Ramp waveforms. (*a*) Unit ramp wave-
forms. (*b*) General ramp waveforms.**

The impulse, step, and ramp form a set of signals that are commonly
referred to as singularity functions. The three waveforms can be related
by integration as

$$u(t) = \int_{-\infty}^{t} \delta(x)\, dx$$

$$r(t) = \int_{-\infty}^{t} u(x)\, dx$$

$$(6\text{-}4)$$

or by differentiation as

$$\delta(t) = \frac{du(t)}{dt}$$

$$u(t) = \frac{dr(t)}{dt}$$

$$(6\text{-}5)$$

These waveforms are often used as standard forcing functions or inputs
in the study of circuits and systems.

6-3   THE EXPONENTIAL WAVEFORM

The waveform defined by the relationship

$$v(t) = A\ u(t)\ e^{-t/T_C} \tag{6-6}$$

is an exponentially decreasing function. A plot of this waveform is shown in Figure 6-8. Because of the step function in the definition, the waveform is zero for negative time and jumps to an amplitude of A at $t = 0$. Thereafter the waveform monotonically decays toward zero as time approaches infinity. The two parameters required to define the waveform are the **amplitude** (A) and the **time constant** (T_C). The parameter A has the same units as the signal quantity (usually volts, amperes, or watts) and represents the initial ($t = 0$) amplitude of the waveform. The time constant carries the units of time, and determines the rate at which the waveform decays. Figure 6-9 shows the effects on the waveform of changing A and T_C.

The time constant is of special significance since it determines the decay rate. For $t = T_C$, $v(T_C) = Ae^{-1}$, which is approximately 0.368 A. Therefore an exponential waveform decays to about 37 percent of its initial amplitude in a time span of one time constant. At $t = 5T_C$, the value of the waveform is Ae^{-5}, which is approximately 0.00674 A. Thus an exponential signal decays to less than 1 percent of its amplitude in a time span of five time constants. While theoretically an exponential waveform endures forever, practically speaking the amplitude becomes negligible after about $5T_C$. For this reason we will define the **signal duration** as $5T_C$. This definition is somewhat arbitrary and is based on practical considerations.

The exponential waveform has two important properties regarding the rate at which the signal decays.

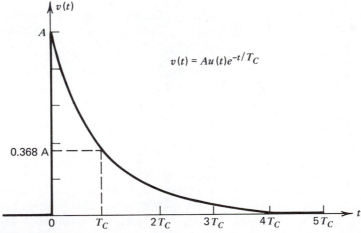

FIGURE 6-8
The exponential waveform.

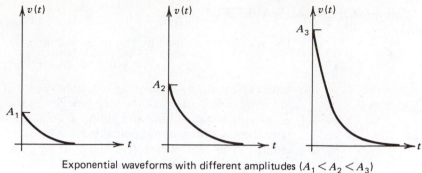

Exponential waveforms with different amplitudes $(A_1 < A_2 < A_3)$
but the same time constant (T_C)

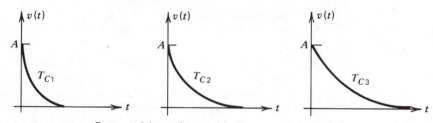

Exponential waveforms with the same amplitude (A)
but different time constants $(T_{C1} < T_{C2} < T_{C3})$

FIGURE 6-9
Effect of amplitude and time constant on the exponential waveform.

Decrement Property

The value of the waveform for $t > 0$ is given by

$$v(t) = A \, e^{-t/T_C} \qquad \text{(6-7)}$$

Note that the step function $u(t)$ has been omitted since its value is unity
for $t > 0$. After an additional time period Δt has gone by the amplitude
is

$$v(t + \Delta t) = A \, e^{-(t + \Delta t)/T_C} \qquad \text{(6-8)}$$

The ratio of these two amplitudes is

$$\frac{v(t + \Delta t)}{v(t)} = \frac{A \, e^{-(t + \Delta t)/T_C}}{A \, e^{-t/T_C}} = e^{-\Delta t/T_C} \qquad \text{(6-9)}$$

This ratio is independent of the starting time and the amplitude A. In
other words, in any fixed time period Δt, the fractional decrease in the
amplitude of an exponential waveform depends only on the time constant.
The decrement property can be expressed as equal percent change in
equal time intervals. For example, if an exponential waveform decreases
to one-half of its initial amplitude in, say, 15 ms, then it will be reduced
by the same 50 percent factor in every subsequent 15-ms interval.

Slope Property

The slope of the exponential waveform (for $t > 0$) is found by taking the derivative of Eq. 6-7 with respect to time:

$$\frac{dv(t)}{dt} = \frac{-A}{T_C}\, e^{-t/T_C} = \frac{-1}{T_C}\, v(t) \tag{6-10}$$

The slope of the exponential waveform is inversely proportional to the time constant. This means that small time constants lead to large slopes or rapid decays, while large time constants describe signals with shallow slopes and long decay times. All of this is summed up in the fact that the derivative of an exponential signal is itself an exponential with the same time constant but a different amplitude.

Example 6-4

To plot the waveform

$$v(t) = 17\, u(t)\, e^{-100t}$$

we must first recognize that $A = 17\text{V}$ and $T_C = 1/100$ or 10 ms. The maximum value of the waveform is 17 V and its duration is about $5T_C$ or 50 ms. These observations allow us to select appropriate scales for plotting the waveform. We can then calculate the values of $v(t)$ as

$v(0.00) = 17e^{-0} = 17.0 \text{ V}$
$v(0.01) = 17e^{-1} = 6.25 \text{ V}$
$v(0.02) = 17e^{-2} = 2.30 \text{ V}$
$v(0.03) = 17e^{-3} = 0.846 \text{ V}$
$v(0.04) = 17e^{-4} = 0.311 \text{ V}$
$v(0.05) = 17e^{-5} = 0.114 \text{ V}$

A plot of these data is shown in Figure 6-10.

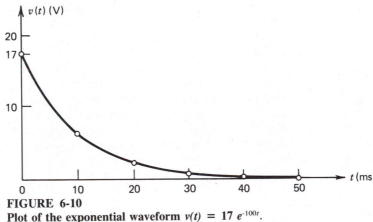

FIGURE 6-10
Plot of the exponential waveform $v(t) = 17\, e^{-100t}$.

The exponential waveform is a good representation of a number of natural phenomena occurring in the physical and biological sciences. The decay in the pressure of a punctured tire and the transfer of nutrients through a cell membrane are but two examples. The term *half-life* is often used to describe the exponential decay of physical phenomena. The next example relates the half-life of a waveform to its time constant.

Example 6-5

Half-life is defined as the time required for the signal to decay to one-half of its initial amplitude. Denoting this time as T_H, for an exponential waveform we have

$$A\,e^{-T_H/T_C} = A/2$$

or

$$e^{T_H/T_C} = 2$$

Finally,

$$T_H = \ln(2) \times T_C = 0.693T_C$$

In words, the exponential waveform decreases by 50 percent in a time interval of about 69.3 percent of its time constant. The half-life of a radioactive mass can be used to identify different elements. For example, the half-life of beryllium-8 is 3×10^{-16} s, while the half-life of carbon-10 is about 19 s.

6-4   THE SINUSOIDAL WAVEFORM

The cosine and sine functions arise frequently in science and engineering, and the corresponding time-varying signal is especially hallowed by electrical engineers. This sacred waveform is shown in Figure 6-11.

In contrast to the step and exponential waveforms, the sinusoid extends indefinitely in time in both the positive and negative directions. It has neither a beginning nor an end! While it may seem unrealistic to invent a signal model that will last from before creation to beyond judgment, nonetheless electrical engineers cleave unto the eternal sine wave with a remarkable conviction. We admit that a real signal must have been turned on at some time in the past. We acknowledge that in all likelihood it will be turned off at some time in the future. While all real signals are surely finite in duration, the eternal sine wave turns out to be both a very convenient artifice and a good approximation.

Since this model of the sinusoid is infinite in extent, it turns out to be an endless repetition of the same old thing—a periodic oscillation between positive and negative values. Since it is unendingly repetitive, we need not examine the entire signal but only a representative segment called a cycle. Some cycles of sinusoidal waveforms are shown in Figure 6-12. Since the signal does not "start" somewhere, the location of the time origin ($t = 0$) is somewhat uncertain, or at least arbitrary.

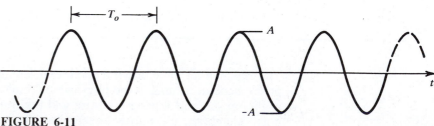

FIGURE 6-11
The eternal sinusoid.

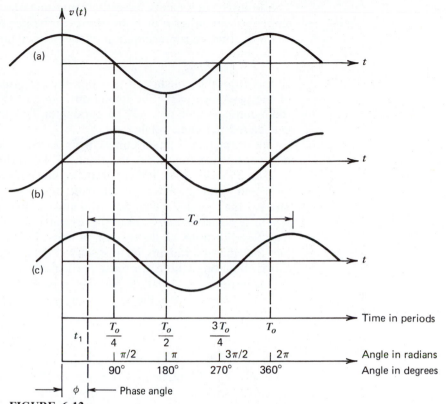

FIGURE 6-12
Representative cycles of a sinusoidal waveform.

If the time origin is located as in Figure 6-12a, then the most natural way to describe the sinusoid is

$$v(t) = a \cos (2\pi t/T_o) \tag{6-11}$$

On the other hand, if the time origin is located as in Figure 6-12b, then the natural expression for the waveform is

$$v(t) = b \sin (2\pi t/T_o) \tag{6-12}$$

In either case, the **period** T_o is the interval of time over which the waveform repeats itself and carries the units of time.

The general case is shown in Figure 6-12c. This waveform can be expressed as a linear combination of the cosine and sine functions.

$$v(t) = a \cos (2\pi t/T_o) + b \sin (2\pi t/T_o) \tag{6-13}$$

The constants a and b are called the **Fourier coefficients** of the sinusoid and can be defined as

$$a = v(0) \qquad b = v(T_o/4) \tag{6-14}$$

The Fourier coefficients have the same units as the waveform (volts or amperes) and either a or b, or both, can be negative.

An alternative representation of the general sinusoid is

$$v(t) = A \cos (2\pi t/T_o - \phi) \tag{6-15}$$

In this form the constant A is called the **amplitude** and carries the units of the waveform $v(t)$. The maximum value of the waveform is $+A$ and the minimum value is $-A$. The parameter T_o is the period as before, and ϕ is called the **phase angle.**

The term phase angle comes from the circular interpretation of the cosine and sine functions. We think of the period as being divided into $360°$ or 2π radians, as indicated at the bottom of Figure 6-12. From this viewpoint, and since we are referring our angle to the cosine, the phase angle is the "angle" between $t = 0$ and the first positive peak (the place where a cosine begins) after the origin (t_1).[2] The phase angle can be expressed in units of degrees or radians, but note that the term $2\pi t/T_o$ has the units of radians in this sense. Therefore care must be used in those situations where we must perform the addition operation implied by the expression $(2\pi t/T_o - \phi)$ to ensure that both terms have the same units. We can express the relation between ϕ and t_1 as

$$\phi = \frac{t_1}{T_o} (360°) = \frac{t_1}{T_o} (2\pi) \qquad \text{(radians)} \tag{6-16}$$

Since we have two representations of the general sinusoid, we are naturally led to inquire into the relationship between them. One way to derive these relationships is to apply the definitions in Eqs. 6-14 to the representation of the sinusoid in Eq. 6-15. Substituting $t = 0$ into Eq. 6-15 yields

$$a = v(0) = A \cos (-\phi)$$

Hence

$$a = A \cos (\phi) \tag{6-17}$$

Likewise, substituting $t = T_o/4$ into Eq. 6-15 yields

$$b = v(t_o/4) = A \cos (\pi/2 - \phi)$$

or

$$b = A \sin (\phi) \tag{6-18}$$

The converse problem also arises—that is, given the Fourier coefficients a and b, find the constants A and ϕ. These relationships can be derived by squaring Eqs. 6-17 and 6-18, and adding the result to obtain

$$A = \sqrt{a^2 + b^2} \tag{6-19}$$

---

[2]If we wished to refer our signal to the sine instead of the cosine, the angle θ would be measured from $t = 0$ to the first positive going zero crossing—between $3T_o/4$ and T_o in Figure 6-12c. We would then write

$$v(t) = A \sin (2\pi t/T_o - \theta)$$

Likewise, dividing Eqs. 6-18 by 6-17 and solving for ϕ yields

$$\phi = \tan^{-1}(b/a) \qquad \text{(6-20)}$$

It is also customary to describe the time variation of the sinusoid in terms of frequency. Frequency is defined as the number of periods per unit time, or cycles per unit time. Clearly the period T_o is the number of seconds per cycle; hence the number of cycles per second is

$$f_o = \frac{1}{T_o} \qquad \text{(6-21)}$$

where f_o represents the **cyclic frequency** or, as it is more commonly called, simply the frequency. The unit of frequency (cycles per second) is called a hertz (Hz). The frequency can also be expressed as an angular quantity in radians per second. This method of representing frequency is called **angular frequency** ω_o, where

$$\omega_o = 2\pi f_o = 2\pi/T_o \qquad \text{(6-22)}$$

since there are 2π radians per cycle. Thus we have two ways to express frequency: cyclic frequency (f_o, hertz) and angular frequency (ω_o, radians per second). The two frequencies differ by a factor of 2π, as shown by Eq. 6-22.

In working with signals, engineers prefer to express frequency in cyclic fashion— for example, we tune our radios to 690 kHz (AM) or 101 MHz (FM). However, when working with systems, engineers often design and analyze using radian frequency for reasons that will become clear in later chapters.

In summary then there are several equivalent ways to describe the general sinusoid:

$$v(t) = a\cos(2\pi t/T_o) + b\sin(2\pi t/T_o) = A\cos(2\pi t/\text{T}_o - \phi)$$
$$v(t) = a\cos(2\pi f_o t) + b\sin(2\pi f_o t) = A\cos(2\pi f_o t - \phi) \qquad \text{(6-23)}$$
$$v(t) = a\cos(\omega_o t) + b\sin(\omega_o t) = A\cos(\omega_o t - \phi)$$

In any case we need three constants to characterize a sinusoid—either the Fourier coefficients a and b or the amplitude A and phase angle ϕ, together with one of the possible time–frequency parameters, T_o, f_o, or ω_o. Throughout this text we will use different forms of the sinusoid so it is important that the reader thoroughly understand the relationships between the various parameters as they are given in Eqs. 6-17 through 6-22.

Historical Note

The hertz is named in honor of the German theorist and experimenter Heinrich Rudolf Hertz. It was Hertz's experiments with electromagnetic waves that made possible much of our modern radio, television, and radar communications. His most famous experiments dealt with the transmission and detection of electromagnetic waves using an induction spark coil and a spark-gap detector. He published his efforts in 1887.

Example 6-6

An oscilloscope is a laboratory instrument used to display signal waveforms. It is nothing more than a voltmeter that visibly displays amplitude versus time. Figure 6-13 shows an oscilloscope display of a sinusoid. The horizontal axis is calibrated in units of time, 0.1 ms per division for this example. The vertical axis is calibrated in volts, 5 V per division in this case. To determine the parameters of the sinusoid we first note that there are four divisions between successive zero crossings. Hence

$$T_o = 2(4 \text{ div}) (0.0001 \text{ s/div}) = 0.0008 \text{ s}$$

and thus,

$$f_o = 1/T_o = 1250 \text{ Hz}$$

If the time origin ($t = 0$) is taken at the left edge of the display, then the amplitude of the waveform at $t = 0$ is -1.5 divisions. Using the first of Eqs. 6-14 we find:

$$a = v(0) = (-1.5 \text{ div}) (5 \text{ V/div}) = -7.5 \text{ V}$$

The $t = T_o/4$ point must occur 2 divisions from the left edge, at which point the amplitude is -4 divisions. Using the second of Eqs. 6-14 we have

$$b = v(T_o/4) = (-4 \text{ div}) (5 \text{ V/div}) = -20$$

We can now compute the amplitude and phase angle of the waveform as

$$A = \sqrt{(-7.5)^2 + (-20)^2} = 21.4 \text{ V}$$
$$= \tan(-20/-7.5) = 4.35 \text{ radians} = 249°$$

Thus the displayed sinusoid can be written in either of the following forms:

$$v(t) = -7.5 \cos(2500\pi t) - 20 \sin(2500\pi t)$$
$$= 21.4 \cos(2500\pi t - 249°)$$

This example illustrates the two equivalent forms of the sinusoid and the fact that either or both of the Fourier coefficients can be negative.

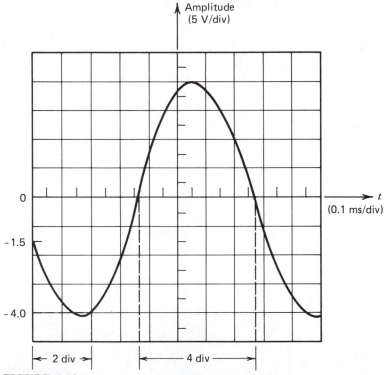

FIGURE 6-13
Sinusoid displayed on an oscilloscope screen.

Example 6-7

The three sinusoidal voltages described by

$$v_A(t) = A \cos (\omega t)$$
$$v_B(t) = A \cos (\omega t - 120°)$$
$$v_C(t) = A \cos (\omega t - 240°)$$

are a set of balanced three-phase voltages. If the amplitudes were not exactly equal and the phase displacements exactly 120° and 240°, then the set would be said to be unbalanced. For this set we say that $v_B(t)$ lags $v_A(t)$ by 120°, and that $v_C(t)$ lags $v_A(t)$ by 240°. Balanced three-phase voltages are produced by commercial power generators and are distributed by the three wires commonly seen on high-voltage transmission lines. Because of the physical makeup of the generator the three voltages are constrained to have exactly the same frequency.

Three important properties of the sinusoid are discussed in the following.

Periodic Property

Waveforms that are endlessly repetitious are called **periodic.** The definition of a periodic waveform can be formalized in the statement:

A waveform is said to be periodic if

$$v(t + T_o) = v(t)$$

for all values of t.

The constant T_o is called the period of the waveform if it is the smallest nonzero interval for which this equation is true. The sinusoid is the premier example of a periodic waveform since

$$v(t + T_o) = A \cos [2\pi(t + T_o)/T_o - \phi]$$
$$= A \cos (2\pi t/T_o - \phi + 2\pi)$$

But $\cos (x + 2\pi) = \cos (x) \cos (2\pi) - \sin (x) \sin (2\pi) = \cos (x)$, and hence

$$v(t + T_o) = A \cos (2\pi t/T_o - \phi) = v(t) \tag{6-24}$$

for all t. The sawtooth, square, and triangular waves in Figure 6-1 are also examples of periodic waveforms. Signals that are not periodic are called **aperiodic.**

Additive Property

If two sinusoids with the same frequency are added, we get a sinusoid with different amplitude parameters but the same frequency. To illustrate this consider the sinusoids

$$v_1(t) = a_1 \cos (\omega t) + b_1 \sin (\omega t)$$
$$v_2(t) = a_2 \cos (\omega t) + b_2 \sin (\omega t) \tag{6-25}$$

The waveform $v_3 = v_1 + v_2$ can be written as

$$v_3(t) = (a_1 + a_2) \cos (\omega t) + (b_1 + b_2) \sin (\omega t) \tag{6-26}$$

and we obtain $a_3 = a_1 + a_2$, and $b_3 = b_1 + b_2$. In sum, we can get the Fourier coefficients of the sum of two sinusoids of the same frequency by simply adding the coefficients. *A word of caution:* The addition must take place in Fourier coefficient form. We cannot find the sum of two sinusoids by adding their amplitudes and phase angles.

Derivative and Integral Properties

The sinusoid maintains its waveshape when it is differentiated or integrated.

$$\frac{d(A \cos \omega t)}{dt} = -\omega A \sin \omega t = \omega A \cos (\omega t + 90°)$$

$$\int A \cos (\omega t) \, dt = \frac{A}{\omega} \sin (\omega t) = \frac{A}{\omega} \cos (\omega t - 90°)$$

(6-27)

These operations change the *amplitude* and *phase angle* of the sinusoid but do not change the *frequency*. The fact that the frequency is unchanged under differentiation and integration is a unique property of the sinusoid. No other periodic waveform has this shape-preserving property.

Example 6-8

In this example we will find the parameters of the sinusoid that is the sum of the two sinusoids:

$$v_1 = 17 \cos (200t - 30°)$$
$$v_2 = 12 \cos (200t + 30°)$$

We can use the additive property since the two sinusoids have the same frequency. However, beyond this the actual value of the frequency plays no further role in the calculation. The two given sinusoids are in amplitude–phase angle form and must be converted to the Fourier coefficient form using Eqs. 6-16 and 6-17.

$$a_1 = 17 \cos (30°) = 14.7 \text{ V}$$
$$b_1 = 17 \sin (30°) = 8.50 \text{ V}$$
$$a_2 = 12 \cos (-30°) = 10.4 \text{ V}$$
$$b_2 = 12 \sin (-30°) = -6.00 \text{ V}$$

The Fourier coefficients of the signal $v_3 = v_1 + v_2$ are found as

$$a_3 = a_1 + a_2 = 25.1 \text{ V}$$
$$b_3 = b_1 + b_2 = 2.50 \text{ V}$$

The amplitude and phase angle parameters of v_3 are then

$$A = \sqrt{a_3^2 + b_3^2} = 25.2 \text{ V}$$
$$= \tan^{-1} (2.5/25.1) = 5.69°$$

Hence the two equivalent representations of the sinusoid v_3 are

$$v_3(t) = 25.1 \cos (200t) + 2.5 \sin (200t)$$

and

$$v_3(t) = 25.2 \cos (200t - 5.69°)$$

6-5   COMPOSITE WAVEFORMS

So far, three basic waveforms have been introduced: the **step** function, the **exponential** function and the **sinusoidal** function. These three signals are the mainstays—the constituents of almost all of the other signals we will need. They form the basis for most of the signals used in electrical engineering. By adding and multiplying these basic waveforms we can generate almost all of the signals covered in this text. The subject of this section is the construction of some of the composite waveforms. We illustrate the process by a series of examples.

Example 6-9

Our first example is obtained by taking the difference between a step function and an exponential. This gives the waveform shown in Figure 6-14 and is called an **exponential rise**. Mathematically the waveform is obtained by writing

$$v_1(t) = A\, u(t)$$
$$v_2(t) = A\, u(t)\, e^{-t/T_C}$$

and forming the exponential rise as

$$v_3(t) = v_1(t) - v_2(t) = A\, u(t)\, (1 - e^{-t/T_C})$$

As time becomes large the waveform approaches a final value of A. Practically speaking the amplitude is within less than 1 percent of this final value at $t = 5T_C$. At $t = T_C$, $v_3(T_C) = A(1 - 0.368) = 0.632\,A$. Thus the waveform rises to about 63 percent of its final value in a time span of one time constant. This waveform is also often called a "charging exponential" and represents the behavior of signals that occur in the charging of circuits that contain memory elements. We study this behavior in Chapters 7, 8 and 11.

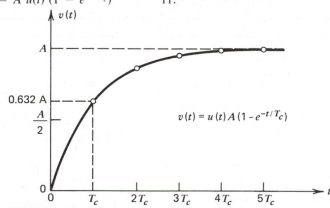

FIGURE 6-14
The exponential rise.

Example 6-10

The next composite waveform is obtained by multiplying a ramp and an exponential. Let

$$v_1(t) = r(t)$$
$$v_2(t) = A\, u(t)\, e^{-t/T_C}$$

then

$$v_3(t) = A\, r(t)\, e^{-t/T_C}$$

For $t > 0$ this waveform can be written as

$$v_3(t) = A\, t\, e^{-t/T_C}$$

As time becomes large the ramp term increases without bound while the exponential decays to zero. Since

the composite waveform is the product of these terms it is important to determine which effect predominates. A single application of L'Hospital's rule will convince the reader that the exponential wins the race. That is, the exponential decay is stronger than the unbounded behavior of the ramp. For this reason the waveform has the shape shown in Figure 6-15. We call this waveform the damped ramp because of the powerful influence of the exponential decay on the ramp waveform.

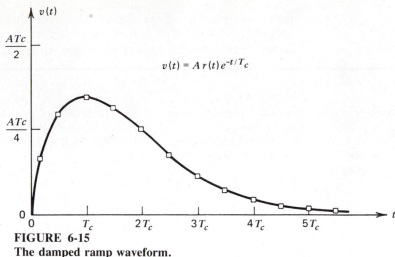

FIGURE 6-15
The damped ramp waveform.

Example 6-11

Our next example is realized by multiplying an exponential and a sinusoid. Let

$$v_1(t) = A\ u(t)\ e^{-t/T_C}$$
$$v_2(t) = \cos(\omega t - \phi)$$

Then define

$$v_3(t) = v_1(t) \times v_2(t)$$
$$= A\ u(t)\ e^{-t/T_C} \cos(\omega t - \phi)$$

The resulting composite waveform is called a damped sinusoid for reasons that are apparent in its graph shown in Figure 6-16. We can view this signal as a sinusoid whose amplitude is not constant but decays with time. The decay is provided by the exponential, and although theoretically the waveform does not reach zero in finite time, practically the signal duration is about $5T_C$. This waveform occurs in the pulse response of amplifiers and is given the descriptive name *ringing*, which is a dirty word among audio purists.

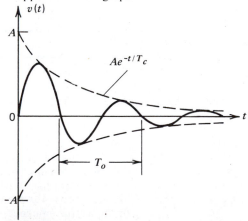

FIGURE 6-16
The damped sinusoidal waveform.

Example 6-12

Our final example of a composite waveform is the double exponential obtained by subtracting two exponential signals. Let

$$v_1(t) = A\ u(t)\ e^{-t/T_1}$$
$$v_2(t) = A\ u(t)\ e^{-t/T_2}$$

Then define

$$v_3(t) = v_1(t) - v_2(t) = A\ u(t)\ (e^{-t/T_1} - e^{-t/T_2})$$

For $T_1 > T_2$ the waveform is illustrated in Figure 6-17.

At $t = 0$ the waveform is zero since $v_3(0) = A(1 - 1)$. Similarly, as time becomes large the waveform is again zero since both exponentials decay to zero. Note that for $t > 5T_2$ the second exponential is essentially zero so that v_3 is essentially equal to v_1.

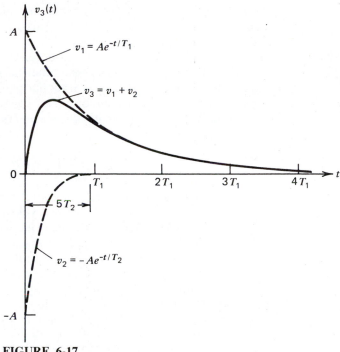

FIGURE 6-17
The double exponential waveform.

6-6   PARTIAL WAVEFORM DESCRIPTORS

A waveform is an equation or graph that gives a complete description of the signal. However, we often work with quantities such as peak-to-peak value and root-mean-square (rms) value that are only partial descriptions of the waveform that highlight certain characteristics but do not completely specify the signal. For example, the rms value is a measure of the average power carried by the signal, while peak-to-peak value is an indication of the total excursion of the waveform. In other words, partial descriptors bring out important attributes such as the amount of energy or data that can be carried by a waveform.

Generally a waveform $v(t)$ varies between two extreme values, which we will denote as V_{MAX} and V_{MIN}. The **peak-to-peak** (V_{PP}) describes the total excursion and is defined as

$$V_{\text{PP}} = V_{\text{MAX}} - V_{\text{MIN}} \tag{6-28}$$

This definition means that V_{PP} is always positive even if V_{MAX} and V_{MIN} are both negative. The **peak value** (V_{P}) is the maximum absolute value of the waveform. That is, V_{P} is $|V_{\text{MAX}}|$ or $|V_{\text{MIN}}|$, whichever is larger. The peak value is always a positive number that indicates the maximum excursion of the waveform. The waveforms in Figure 6-18 illustrate the definitions of these two descriptors.

The peak and peak-to-peak values describe the hills and valleys of a waveform. The average value, on the other hand, involves using integration to level the hills and valleys into a flat plane. Basically, the average value is the area under the waveform over some period of time T, divided by that time period. Mathematically we define average value over the time T as

$$V_{\text{AVE}} = \frac{1}{T} \int_0^T v(t) \, dt \tag{6-29}$$

For periodic signals the averaging interval is taken to be one period (T_o). Technically this would be called the one-cycle average. But for a periodic waveform it is simply called the average value without any qualifiers about the averaging interval, which is understood to be one period.

For many periodic waveforms the integration in Eq. 6-29 can be simplified by observing that the net area under the waveform is equal to the area above the time axis minus the area below the time axis. For example, the sinusoid in Figure 6-19a obviously has zero average value since the area above the axis exactly equals the area below. The sawtooth in Figure

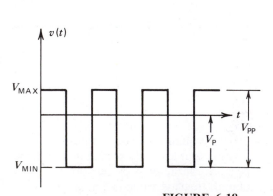

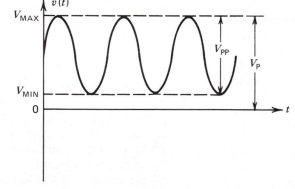

FIGURE 6-18
Peak and peak-to-peak values.

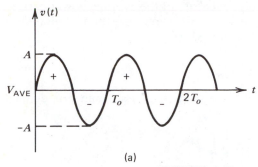 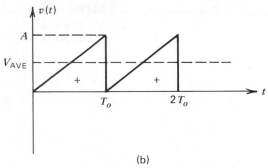

FIGURE 6-19
Average value of some periodic waveforms. (*a*) Sinusoid. (*b*) Saw tooth.

6-19*b* clearly has a positive average value. For this waveform the net area is $AT_o/2$, and hence its average value is simply $A/2$.

The average value measures the offset or asymmetry of the waveform with respect to $v = 0$; that is, it indicates how completely the hills fill in the valleys in the leveling process. It is also called the direct-current or dc component of the waveform.

Example 6-13

Figure 6-20 shows the input and output waveforms of a signal processor. The input is a sinusoid and its amplitude descriptors are

$$V_{PP} = 2A \qquad V_P = A \qquad V_{AVE} = 0$$

The output is obtained by clipping off the negative half-cycle of the input sinusoid. The amplitude descriptors of the output waveform are

$$V_{PP} = V_P = A$$

The output has a nonzero average value since there is a net positive area under the waveform. In using Eq. 6-29, the upper limit can be taken as $T_o/2$ since

the waveform is zero from $T_o/2$ to T_o.

$$V_{AVE} = \frac{1}{T_o} \int_0^{T_o/2} A \sin (2\pi t/T_o) \, dt$$
$$= \frac{A}{2} \left[-\cos (2\pi t/T_o) \right]_0^{T_o/2}$$
$$= A/\pi$$

This mysterious "signal processor" is called a **half-wave rectifier** and finds widespread application in electronic circuits where ac signals must be changed to dc.

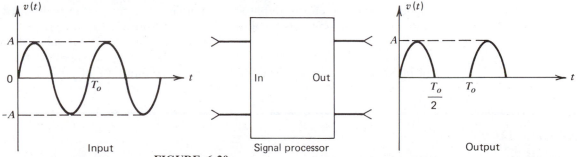

FIGURE 6-20
Input and output waveforms of a half-wave rectifier.

The **rms value** of a waveform is a measure of the average power carried by the signal. To derive this measure we first note that the instantaneous power delivered to a resistor by a voltage $v(t)$ is

$$p(t) = \frac{1}{R}[v(t)]^2 \qquad (6\text{-}30)$$

The average power delivered to the resistor in a time span T is defined as

$$P_{AVE} = \frac{1}{T}\int_0^T p(t)\,dt \qquad 6\text{-}31$$

Combining Eqs. 6-30 and 6-31 then yields

$$P_{AVE} = \frac{1}{R}\left[\frac{1}{T}\int_0^T [v(t)]^2 dt\right]$$

The quantity inside the large brackets in this equation is the average value of the square of the waveform over the time interval T. This term is a partial descriptor that is related to the average power carried by the waveform. The units of this term are volts squared. It is customary to use the square root of this term to define the descriptor.

$$V_{RMS} = \sqrt{\frac{1}{T}\int_0^T [v(t)]^2 dt} \qquad (6\text{-}32)$$

The resulting partial descriptor is called the rms value since it is obtained by taking the square *root* of the average (*mean*) of the *square* of the waveform. For periodic signals the averaging interval is taken as one cycle since such a waveform duplicates itself every T_o seconds. Thus V_{RMS} for a periodic waveform can be obtained from Eq. 6-32, except that T is replaced by T_o.

We can now write the average power in terms of V_{RMS} as

$$P_{AVE} = \frac{1}{R}V_{RMS}^2 \qquad (6\text{-}33)$$

The average power is proportional to the square of the rms value of the signal. If the waveform amplitude is doubled, the rms value is doubled, and the average power is quadrupled. If the purpose is to transfer power, then the signal level should be as high as possible. Commercial electric power systems often use transmission voltages on powerlines in the range of hundreds of kilovolts (rms) to transfer large amounts of power.

Example 6-14

Let us determine the rms value of the sawtooth and the sinusoid in Figure 6-19. For the sawtooth we can write

$$V_{RMS}^2 = \frac{1}{T_o}\int_0^{T_o}(At/T_o)^2 dt = \frac{A^2}{T_o^3}\left[\frac{t^3}{3}\right]_0^{T_o} = \frac{A^2}{3}$$

Thus the rms value of the sawtooth is $A/\sqrt{3}$. The rms value of the sinusoid is derived as

$$V_{RMS}^2 = \frac{A^2}{T_o} \int_0^{T_o} \sin^2 (2\pi t/T_o)\, dt$$

$$= \frac{A^2}{T_o} \left[\frac{t}{2} - \frac{\sin (4\pi t/T_o)}{8\pi/T_o} \right]_0^{T_o}$$

$$= A^2/2$$

Hence the rms value of the sinusoid is $A/\sqrt{2}$.

We often speak of 110-V "power" being delivered to our homes. The 110 V is the rms value of a sinusoidal signal that has an amplitude of $110\sqrt{2}$ or 155.6 V peak or 311 V peak to peak.

SUMMARY

- A waveform is an equation or graph of a time-varying voltage or current.

- There are a large number of waveforms of potential interest in electrical engineering. Most signals of interest can be constructed from three basic waveforms: the step, the exponential, and the sinusoid.

- The general step waveform can be written as

$$v(t) = A\, u(t - T_S)$$

where A is the amplitude (volts, amperes, or watts) and T_S is the step time (seconds).

- The general exponential waveform can be written as

$$v(t) = Au(t)\, e^{-t/T_C}$$

where A is the amplitude (volts, amperes, or watts) and T_C is the time constant (seconds). An exponential waveform decays to 0.368 A at $t = T_C$. For practical purposes the exponential waveform duration is about $5T_C$.

- The general sinusoidal waveform can be written in two equivalent forms:

Fourier coefficient form

$$v(t) = a \cos (\omega t) + b \sin (\omega t)$$

Amplitude–phase angle form

$$v(t) = A \cos (\omega t - \phi)$$

where a and b are the Fourier coefficients (volts, amperes, or watts); A is the amplitude (volts, amperes, or watts); ϕ is the phase angle (degrees or radians); $\omega = 2\pi f_o = 2\pi/T_o$; f_o is the frequency in hertz; T_o is the period (seconds).

- The connection between the two forms of the sinusoid are the right-triangle relationships

$$a = A \cos (\phi) \qquad A^2 = \sqrt{a^2 + b^2}$$
$$b = A \sin (\phi) \qquad \phi = \tan^{-1} (b/a)$$

- The sum of two sinusoids with the same frequency and the derivative or integral of a sinusoid produce another sinusoid with the frequency unchanged but with a different amplitude and phase angle.

- A large number of composite waveforms can be derived using the three basic waveforms. Some important examples are the impulse, ramp, damped ramp, damped sinusoid, double exponential, and exponential rise.

- A periodic waveform repeats itself at regular intervals and satisfies the relation

$$v(t + T_o) = v(t)$$

for all value of t. The constant T_o is called the period. Waveforms that are not periodic are said to be aperiodic.

- A complete description of a signal provides the value of its waveform at every instant of time. A partial descriptor is some important feature of the signal that does not completely specify the waveform.

- Partial descriptors of the amplitude of a waveform include:

V_P—peak value or the maximum absolute value of the waveform.

V_{PP}—peak-to-peak value or the difference between the maximum and the minimum value of the waveform.

V_{AVE}—average value or the net area under the waveform over a specified time interval divided by the time interval.

V_{RMS}—rms value or the square root of the average value of the square of the waveform.

- For periodic signals the averaging interval is taken to be one period, and hence

$$V_{AVE} = \frac{1}{T_o} \int_0^{T_o} v(t)dt$$

$$V_{RMS} = \sqrt{\frac{1}{T_o} \int_0^{T_o} [v(t)]^2 dt}$$

- The average power carried by a signal is proportional to the square of its rms value

$$P_{AVE} = \frac{1}{R} V_{RMS}^2$$

EN ROUTE OBJECTIVES
AND RELATED EXERCISES

6-1 BASIC WAVEFORMS (SECS. 6-2 TO 6-4)

Given a complete description (equation, graph, or word description) of a basic signal waveform (step, exponential, sinusoid):

(a) *Construct an alternative description of the waveform.*

(b) *Determine the waveform obtained by a prescribed processing of the waveform.*

Exercises

6-1-1 Construct a graph of the following step function waveforms.

(a) $v_1(t) = 5\,u(t)$

(b) $v_2(t) = -5\,u(t)$

(c) $v_3(t) = 5\,u(t - 1)$

(d) $v_4(t) = -10\,u(t - 1)$

6-1-2 Construct a graph of the waveforms obtained by adding the step functions defined in Exercise 6-1-1.

(a) $v_A(t) = v_1(t) + v_2(t)$

(b) $v_B(t) = v_1(t) + v_3(t)$

(c) $v_C(t) = v_1(t) + v_4(t)$

(d) $v_D(t) = v_1(t) - v_4(t)$

6-1-3 Construct a graph of the waveforms obtained by integrating and differentiating the step functions defined in Exercise 6-1-1.

6-1-4 Sketch a graph of an exponential waveform that starts at $t = 0$, and has an amplitude of 10 V and a time constant of 20 ms. Write a mathematical expression for the waveform.

6-1-5 Construct a graph of the following exponential waveforms and determine their time constants.

(a) $v_1(t) = 10u(t)\,e^{-2t}$

(b) $v_2(t) = 10u(t)\,e^{-t/2}$

(c) $v_3(t) = -10u(t)\,e^{-20t}$

(d) $v_4(t) = -10u(t)\,e^{-t/20}$

6-1-6 Construct a graph of the waveforms obtained by adding the exponential waveforms defined in Exercise 6-1-5.

(a) $v_A(t) = v_1(t) + v_2(t)$

(b) $v_B(t) = v_1(t) + V_3(t)$

(c) $v_C(t) = v_1(t) + v_4(t)$

(d) $v_D(t) = v_2(t) + v_4(t)$

6-1-7 Construct a graph of the waveform obtained by integrating and differentiating the exponential waveforms defined in Exercise 6-1-5.

6-1-8   Show that the exponential waveform is a solution of the differential equation

$$\frac{dv}{dt} + Kv = 0$$

provided that $T_C = 1/K$.

6-1-9   Sketch the waveform of a sinusoid with an amplitude of 20 V, a period of 10 ms, and its first positive peak at $t = 5$ ms. Write a mathematical expression for the waveform.

6-1-10   A sinusoid has a frequency of 50 Hz, a value of 10 V at $t = 0$, and reaches its first positive peak at $t = 2.5$ ms. Determine its amplitude, phase angle, and Fourier coefficients.

6-1-11   Determine the period, frequency, amplitude, and phase angle of each of the following sinusoids. Sketch the waveform.
(a) $v_1(t) = 10 \cos (2000\pi t) + 10 \sin (2000\pi t)$
(b) $v_2(t) = -30 \cos (2000\pi t) - 20 \sin (2000\pi t)$
(c) $v_3(t) = 10 \cos (2\pi t/10) - 10 \sin (2\pi t/10)$
(d) $v_4(t) = -20 \cos (800\pi t) + 30 \sin (800\pi t)$

6-1-12   Determine the frequency, period, amplitude, and phase angle of the sum of the first two sinusoids in Exercise 6-1-11.

6-1-13   Construct a graph of the waveforms obtained by integrating and differentiating the sinusoidal waveforms defined in Exercise 6-1-11.

6-1-14   Determine the frequency, period, and Fourier coefficients for each of the following sinusoids. Sketch the waveforms.
(a) $v_1(t) = 20 \cos (4000\pi t - 180°)$
(b) $v_2(t) = 20 \cos (4000\pi t - 90°)$
(c) $v_3(t) = 30 \cos (2\pi t/400 - 45°)$
(d) $v_4(t) = 60 \sin (2000\pi t + 45°)$

6-1-15   Determine the frequency, period, phase angle, and amplitude of the sum of the first two sinusoids in Exercise 6-1-14.

6-1-16   Construct a graph of the waveforms obtained by integrating and differentiating the sinusoids defined in Exercise 6-1-14.

6-1-17   Show that the sinusoid

$$v(t) = a \cos (\omega t) + b \sin (\omega t)$$

is a solution of the differential equation

$$\frac{d^2v}{dt^2} + K^2v = 0$$

provided that $K = \omega$.

6-2 COMPOSITE WAVEFORMS (SEC. 6-5)

Given a complete description (equation, graph, or word description) of a composite waveform:
(a) *Construct an alternative description of the waveform.*
(b) *Determine the waveform obtained by a prescribed signal processing of the waveform.*

Exercises

6-2-1 Prepare an accurate sketch of each of the following composite waveforms.
(a) $v_1(t) = 20\, u(t)\,(1 - e^{-100t})$
(b) $v_2(t) = 20\, u(t) \cos(10\pi t)$
(c) $v_3(t) = 20\, u(t)e^{-t} \cos(10\pi t)$
(d) $v_4(t) = 20\,[u(t) - u(t - 1)] \cos(10\pi t)$

6-2-2 Write a mathematical expression for each of the waveforms in Figure E6-2-2.

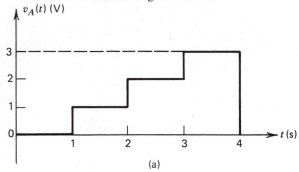

(a)

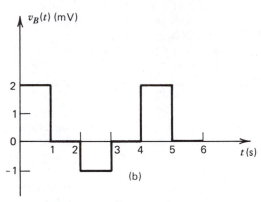

(b)

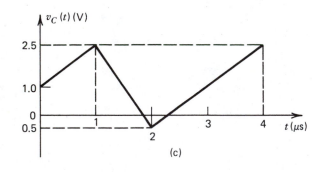

(c)

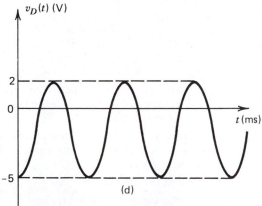

(d)

FIGURE E6-2-2
Waveforms for Exercise 6-2-2.

6-2-3 Construct a graph of the waveforms obtained by integrating and differentiating the waveforms given in Figure E6-2-2.

6-2-4 Sketch the waveform

$$v(t) = A \left[u(t) - u(t - T) \right] e^{-t/T_C}$$

for $T >> T_C$. Repeat for $T = T_C$ and $T << T_C$.

6-2-5 Sketch the waveform

$$v(t) = Au(t)e^{-t/T_C} \cos (2\pi t/T_o)$$

for $T_C >> T_o$. Repeat for $T_C = T_o$ and $T_C << T_o$.

6-2-6 Prepare a sketch of each of the following ramp waveforms. Then sketch and write a mathematical expression for the derivative of each waveform.
(a) $v_1(t) = r(t - 1)$
(b) $v_2(t) = -2 \, r(t - 1)$
(c) $v_3(t) = r(t) - 2 \, r(t - 2)$
(d) $v_4(t) = r(t) - 2 \, r(t - 1) + r(t - 2)$

6-2-7 Sketch the double exponential waveform

$$v(t) = A \left(e^{-t/T_1} - e^{-t/T_2} \right)$$

for $T_1 > T_2$. Then show that the peak value of the waveform occurs at

$$T_{\text{peak}} = \frac{T_1 T_2}{T_1 - T_2} \ln (T_1/T_2)$$

6-2-8 Show that the peak value of the damped ramp waveform

$$v(t) = Ar(t)e^{-t/T_C}$$

occurs when $t = T_C$.

6-2-9 Show that the waveform

$$v(t) = u(t)t^n \, e^{-t/T_C}$$

approaches zero as t approaches infinity provided that $T_C > 0$ and n is a finite, positive integer.

6-2-10 Write a mathematical expression for the waveforms in Figure E6-2-10.

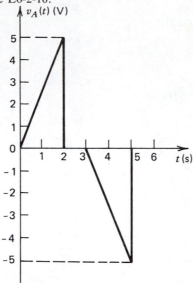
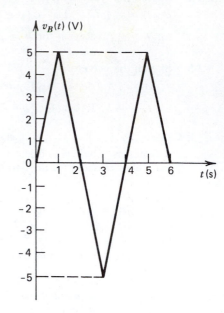

FIGURE E 6-2-10
Waveforms for Exercise 6-2-10.

6-2-11 Construct a graph of the waveforms obtained by integrating and differentiating the waveforms shown in Figure E6-2-10.

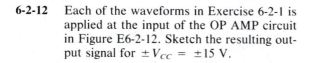

6-2-12 Each of the waveforms in Exercise 6-2-1 is applied at the input of the OP AMP circuit in Figure E6-2-12. Sketch the resulting output signal for $\pm V_{CC} = \pm15$ V.

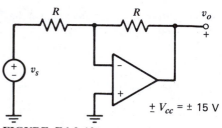

FIGURE E6-2-12
OP AMP circuit for Exercise 6-2-12.

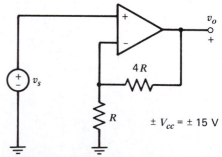

6-2-13 Each of the waveforms in Exercise 6-2-2 is applied at the input of the OP AMP circuit shown in Figure E6-2-13. Sketch the resulting output signals for $\pm V_{CC} = \pm15$ V.

6-2-14 Repeat Exercise 6-2-13 for the waveforms in Figure E6-2-10.

FIGURE E6-2-13
OP AMP circuit for Exercise 6-2-13.

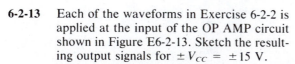

6-2-15   Show that the damped sinusoid

$$v(t) = Ae^{-at} (\cos bt + \sin bt)$$

is a solution of the differential equation

$$\frac{d^2v}{dt^2} + K_1 \frac{dv}{dt} + K_o v = 0$$

provided that $a = K_1/2$ and $K_o = a^2 + b^2$.

6-3 PARTIAL WAVEFORM DESCRIPTORS (SEC. 6-6)

Given a complete description of a basic or composite waveform, determine partial waveform descriptors.

Exercises

6-3-1   Determine V_P and V_{PP} for the following waveforms. Also determine V_{AVE} and V_{RMS} for the periodic waveforms.
(a) $v_1(t) = 10u(t) (1 - e^{-100t})$
(b) $v_2(t) = 10 \cos (1000t) - 10 \sin (1000t)$
(c) $v_3(t) = 10 [u(t) - u(t - 1)] \cos (1000t)$
(d) $v_4(t) = 100u(t)e^{-t} \cos (1000t)$

6-3-2   Repeat Exercise 6-3-1 for each of the waveforms in Exercise 6-1-6.

6-3-3   Repeat Exercise 6-3-1 for each of the waveforms in Exercise 6-1-11.

6-3-4   If the waveform in Figure E6-2-2a is periodic with a period of 5 s, determine V_P, V_{PP}, V_{AVE}, and V_{RMS}.

6-3-5   If the waveform in Figure E6-2-2b is periodic with a period of 3 s, determine V_P, V_{PP}, V_{AVE}, and V_{RMS}.

6-3-6   The waveform

$$v(t) = A_o + 10 \sin \omega t$$

is applied at the input of the OP AMP circuit in Figure E6-2-12. What range of the constant A_o will ensure that the OP AMP will not saturate for this input?

6-3-7   Show that the waveform

$$v(t) = A_o + A_1 \cos (2\pi t/T_o)$$

is periodic. Then determine V_P, V_{PP}, V_{AVE}, and V_{RMS}.

6-3-8   A loudspeaker has a power rating of 100 W and a resistance of 8 Ω. What is the maximum rms voltage that can be applied to the speaker? If the applied waveform is a sinusoid, what is the maximum peak-to-peak voltage?

6-3-9 A certain periodic waveform $v(t)$ has a zero average value and delivers 1 W to a 10-Ω resistor. What average power would be delivered by the waveform $v(t) + 5$?

6-3-10 The crest factor of a periodic waveform is defined as the ratio of its peak value to its rms value. What is the crest factor of a sinusoid? What is the crest factor of a square wave?

PROBLEMS

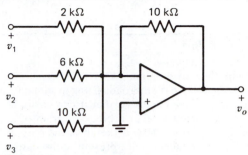

FIGURE P6-1
Fourier series square-wave demonstration problem.

P6-1 (Analysis)
Sketch the output of the circuit of Figure P6-1. Assume all signals are in phase (i.e., they all start at $t = 0$).

$$v_1 = 1.0 \sin \pi \, 100t$$
$$v_2 = 1.0 \sin \pi \, 300t$$
$$v_3 = 1.0 \sin \pi \, 500t$$

P6-2 (Analysis)
A comparator is a device that compares two inputs (see sect 4-8,) $x_1(t)$ and $x_2(t)$. If $x_1 > x_2$, it outputs a constant value y_1. If $x_2 > x_1$, then it outputs a different constant value y_2. Suppose $x_1(t) = e^t$ V, and $x_2(t) = 5 \cos 2\pi t$ V; draw a sketch of the output if $y_1 = +5$ V and $y_2 = -5$ V.

P6-3 (Analysis)
In analog to digital conversion **a sample hold** is used to capture the signal and hold it constant while the analog signal is converted to its digital equivalent. In essence the sample hold is a switch that is turned on just long enough for the hold part of the device to acquire the signal. The signal that turns on the sample portion of the device is a gating signal given by

$$v_s(t) = \sum_{n=o}^{\infty} [u(t - 0.001n)$$
$$- u(t - 0.001n - 0.00001)] \text{ V}$$

If the analog signal to be sampled is $v(t) = e^{-1000t}$, draw a sketch of the gating signal, the signal to be sampled, and the output $y(t)$ of the sample hold for the first 5 ms.

P6-4 (Analysis)
Find and sketch the instantaneous power dissipated in a 1-Ω resistor excited by $v(t) = \sin \pi t$ V.

P6-5 (Analysis)
Consider the operation of signal multiplication. In this operation two signals are multiplied together to produce a third signal. A block diagram symbol for multiplication is shown in Figure P6-5. For each case of v_1 and v_2 given in Table P6-5, find and draw a sketch of v_3.

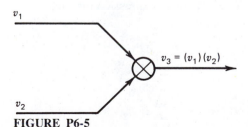

FIGURE P6-5
A function multiplier.

Case	v_1	v_2
1	5	$\sin 2\pi t$
2	e^{-t}	$\sin 2\pi t$
3	$\sin 16\pi t$	$\sin 2\pi t$

TABLE P6-5

P6-6   (Analysis)

Amplitude modulation (AM) is a signal-processing scheme used in radiobroadcasting to shift information to higher frequencies so that it can easily be transmitted. Figure P6-6 is a block diagram of a typical AM modulator. Draw a sketch of the output if $m = 1$. What frequencies are present at the output? How would the output change if $m = \frac{1}{2}$?

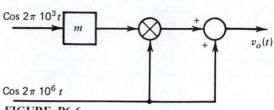

FIGURE P6-6
AM modulator block diagram.

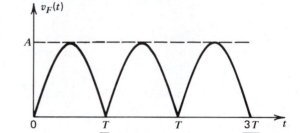

(a)

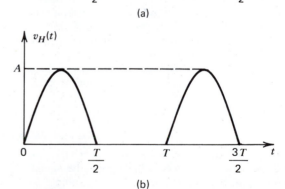

(b)

P6-7   (Analysis)

Figure P6-7a represents the ideal waveform obtained when a sine wave is fully rectified. Figure P6-7b represents the ideal waveform output of a half-wave rectifier. Write an expression for each.

FIGURE P6-7
Full- and half-wave rectified sinusoid waveforms.

Chapter 7
Memory Elements

I remember, I remember.

A poem by
Thomas Hood

People remember, and as a result they can shape their future behavior on the basis in part of their past experiences. So too circuits that remember permit outputs not possible with memoryless circuits. Capacitors and inductors remember voltage and current values. Through these new elements a whole new repertoire of circuits whose outputs depend on both present and past inputs become possible. These elements produce outputs that depend on the time rate of change of their inputs. As a result, they can perform as integrators and differentiators. With these new linear functions a whole host of signal processors such as filters and analog computers become possible.

Placing two or more inductors in close proximity permits mutual couplings and the transfer of a signal from one coil to a second without an electrical connection. This mutual coupling is utilized in the transformer. Transformers have many uses in electronics, but one of the more important is in matching sources and loads.

7-1   THE CAPACITOR

The first memory element we study is the ideal **capacitor.** A capacitor is a two-terminal device. The simplest form of this device is constructed from two parallel conducting plates separated by an insulator or dielectric. Figure 7-1 shows a parallel plate capacitor, the circuit symbol, and examples of actual devices.

From electrostatics it can be shown that a pair of equal but opposite charges, q^+ and q^-, on each plate, respectively, and separated by a distance d will give rise to an electric field $\mathscr{E}$ between the plates.[1] If the separation d is small compared with the dimension of the plates, then the electric field can be written as

$$\mathscr{E}(t) = q(t)/\epsilon A \qquad (7\text{-}1)$$

where ϵ is the permittivity of the dielectric, A is the area of the plates and $q(t)$ the magnitude of the charge on each plate. With d small compared with the dimensions of the plates, the electric field can be assumed to be uniform. We can then write

$$\mathscr{E}(t) = v_C(t)/d \qquad (7\text{-}2)$$

Substituting into Eq. 7-1 and solving for q, we find

$$q(t) = v_C(t)\,[\epsilon A/d] \qquad (7\text{-}3)$$

The quantity in the brackets of Eq. 7-3 is called the **capacitance** C of the capacitor, that is,

$$C \equiv \epsilon A/d \qquad (7\text{-}4)$$

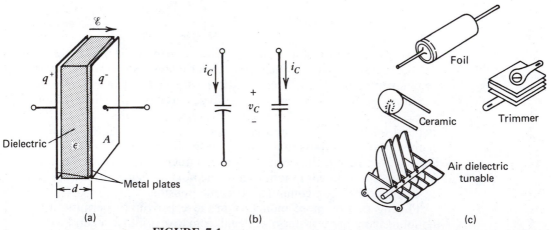

(a)   (b)   (c)

FIGURE 7-1
Capacitors. (*a*) Parallel plate capacitor. (*b*) Circuit symbols. (*c*) Actual devices.

[1]Electric field $\mathscr{E}$ is a vector quantity. In this simple example, we assume the entire field to be between the two plates and in a direction perpendicular to the two plates, that is, in the direction of the separation d.

so that we can rewrite Eq. 7-3 as

$$q(t) = Cv_C(t) \tag{7-5}$$

Capacitance is measured in **farads** (F) and depends only on the physical properties of the dielectric and the dimensions of the plates. A more thorough discussion of the capacitor is contained in Appendix C.

In practice it is difficult to measure charge. Rather it is customary to measure the time rate of change of charge, or current. If we take the time derivative of Eq. 7-5, we get

$$\frac{d}{dt}[q(t)] = \frac{d}{dt}[Cv_C(t)]$$

But since C is considered a constant, and $i = \dfrac{dq}{dt}$

$$i_C(t) = C\frac{dv_C(t)}{dt} \tag{7-6}$$

This i-v relation or element constraint for a capacitor is shown graphically in Figure 7-2.

The reader should be aware that Eq. 7-6 represents a mathematical model of an ideal element that at times might break down when working with real devices. For example, we assume that the resistance of the dielectric between the plates is sufficiently high that it can be considered infinite so that the charges q^+ and q^- on each plate cannot recombine through the dielectric.

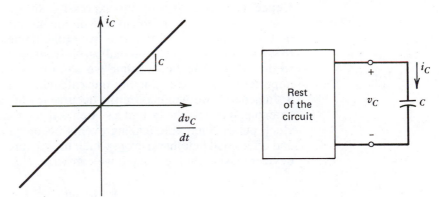

FIGURE 7-2
The i-v characteristics of a capacitor.

Historical Note

The unit of capacitance is named in honor of the British physicist Michael Faraday (1791–1867). Faraday was an assistant to the Royal Institute in London. There he found that a changing magnetic field would produce a flow of electricity. His work led to the development of the electric generator and later the telegraph.

In Eq. 7-6 we express the capacitor current in terms of the capacitor voltage. To find a representation of the capacitor voltage in terms of current, we must perform an integration.

$$\int dv_C(t) = \int \frac{i}{C} \, i_C(t) dt$$

Selecting the limits of integration requires some thought. In general, we can assume that at some time long ago, possibly when the capacitor was first built, there was no voltage across the capacitor. Hence we can reasonably assume the voltage at $t = -\infty$ to be zero. How much voltage there is at some other time will depend on how i behaved since $t = -\infty$ to the time t that we want to determine v_C. Therefore we can write

$$\int_0^{v_C} dv_C(t) = \frac{1}{C} \int_{-\infty}^{t} i_C(t) dt$$

which we can partially integrate as

$$v_C(t) = \frac{1}{C} \int_{-\infty}^{t} i_C(t) dt \qquad (7\text{-}7)$$

This expression states that the voltage at any time t is proportional to the integral of the current flowing through the capacitor from time immemorial to time t. This integral may be easy to interpret but it is difficult to evaluate since we usually have no idea of how the current behaved from time immemorial to when we were ready to use it in our circuit. Hence we need to modify the expression so we can use it.

A way around our problem is to divide our integral into two parts. The first part represents all the time from time immemorial to when we want to start our problem. The second part is from the time we start our problem to the present. The exact time we choose to divide our problem is not important but we can simplify our calculation if we let that time be $t = 0$. In general, we will choose the occurrence of some event as $t = 0$; for example, it can be the instant a switch is thrown or the start of a particular clock pulse. Such epic-making events occur quite frequently in circuits and offer good landmarks from which to measure the behavior of memory devices. Using this technique we can write Eq. 7-7 as

$$v_C(t) = \underbrace{\frac{1}{C} \int_{-\infty}^{0} i_C(t) dt}_{v_C(0)} + \frac{1}{C} \int_{0}^{t} i_C(t) dt \qquad (7\text{-}8)$$

The first term on the right of Eq. 7-8 can be evaluated and is equal to a constant $v_C(0)$. It represents the value of the voltage across the capacitor at time $t = 0$ due to the current flowing through it from time immemorial to $t = 0$. It may or may not be possible to calculate the value of $v_C(0)$, but it can always be measured. Regardless of how the current is changing, there always is only one value of $v_C(t)$ at any instant of time including $t = 0$. Hence we can rewrite Eq. 7-8 as a sum of two terms: an integral that represents the positive time variation of the voltage and a constant that represents the **initial condition** of the capacitor, that is,

$$v_C(t) = \frac{1}{C} \int_0^t i_C(t)dt + v_C(0) \qquad \text{(7-9)}$$

Let us pause and reflect on Eq. 7-9 for a moment. At $t = 0$ the left side of Eq. 7-9 clearly is equal to $v_C(0)$. At $t = 0^+$, an instant after $t = 0$, $v_C(0^+)$ is given as

$$v_C(0^+) = \frac{1}{C} \int_0^{0^+} i_C(t)dt + v_C(0)$$

For any finite value of $i_C(t)$ (any realizable signal) the integral from $t = 0$ to $t = 0^+$ is zero. This means that

$$v_C(0^+) = v_C(0) \qquad \text{(7-10)}$$

The implication of Eq. 7-10 is that *the voltage across a capacitor cannot change instantaneously*. To support our implication, suppose for a moment that it was possible to change the voltage across a capacitor instantaneously. This means that $dv_C(t)/dt$ is infinite. But from Eq. 7-6 this would mean that the current $i_C(t)$ would be infinite—a physical impossibility. Thus we can say that the voltage across a capacitor cannot change instantaneously since it would require an infinite current. This simple but important realization is called the **continuity** relation for capacitors and is usually written as

$$v_C(t^+) = v_C(t^-) \qquad \text{(7-11)}$$

Example 7-1

A ½-F capacitor has the voltage of Figure 7-3a applied across it. What is the resulting current flowing through it?

The current through a capacitor is given by Eq. 7-6. From $t = 0$ to $t = 2$ s the slope is a positive 5, and the current is $(\frac{1}{2}) \times (5) = 2.5$ A. Between $t = 2$ s and $t = 3$ s the slope is a negative 5. The current

now changes direction and is given by $(\frac{1}{2}) \times (-5) = -2.5$ A. Finally from $t = 3$ s on, the voltage is equal to a constant value of 5 V. The resulting current then is zero, $(\frac{1}{2}) \times (0) = 0$ A. The behavior of the current versus time can be depicted graphically as shown in Figure 7-3b.

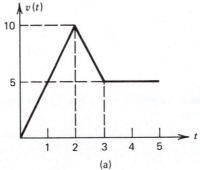

(a)

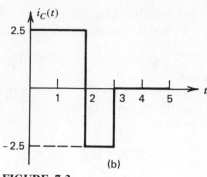

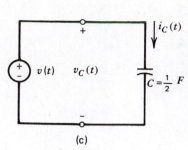

FIGURE 7-3
Circuit and signals for Example 7-1. (*a*) Input voltage $v(t)$. (*b*) Output current $i_C(t)$. (*c*) Circuit.

Example 7-2

The current through a certain capacitor is given by $i_C(t) = I_o e^{-t/T_C}$. What is the voltage across that capacitor if $v_C(0) = 0$ V?

$$v_C(t) = \frac{1}{C} \int_0^t I_o e^{-t/T_C} dt + 0$$

$$= \frac{I_o}{C} (-T_C) e^{-t/T_C} \Big|_0^t$$

$$= \frac{-I_o T_C}{C} (e^{-t/T_C} - e^{-0})$$

$$= \frac{I_o T_C}{C} (1 - e^{-t/T_C})$$

Figure 7-4a shows the capacitor current while 7-4b shows the resulting voltage.

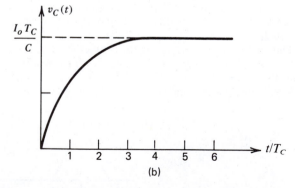

FIGURE 7-4
Waveforms demonstrating the *i-v* relationship of a capacitor (*a*) Capacitor current. (*b*) Capacitor voltage.

7-2   POWER, ENERGY, AND MEMORY

Consider the capacitor shown in Figure 7-5. The relationship between current, voltage, and power is given by (assuming our normal convention for positive power).

$$p_C(t) = i_C(t)\, v_C(t) \qquad (7\text{-}12)$$

which can be rewritten using the capacitor's *i-v* characteristics as

$$p_C(t) = C v_C(t)\, \frac{dv_C(t)}{dt} \qquad (7\text{-}13)$$

From Eq. 7-13 we can see that the power can be either positive or negative since the time rate of change of voltage can be positive or negative regardless of the absolute value of voltage. This is in stark contrast to a resistor where the power can only be positive. We should recall from Chapter 2 that positive power implies power *delivered* to an element, while negative power suggests power *supplied* by that element. The ability to supply power implies the ability to store energy. Let's focus our attention on this special ability.

The power expression of Eq. 7-13 is a perfect derivative.

$$p_C(t) = C v_C(t)\, \frac{dv_C(t)}{dt} = \frac{d}{dt}\left(\tfrac{1}{2}\, C v_C^2(t) \right) \qquad (7\text{-}14)$$

Since power is the time rate of change of energy, the quantity inside the brackets must represent the energy stored in the electric field of the capacitor, that is,

$$w_C(t) = \frac{1}{2}\, C v_C^2(t) \qquad (7\text{-}15)$$

The amount of energy stored at any instant of time is dependent only on the instantaneous value of the voltage. Because of the $v_C^2(t)$ term, the energy stored is always positive. Looking back at Eq. 7-13 we can see further justification for the fact that the voltage across a capacitor cannot change instantaneously (Eq. 7-11). An instantaneous change would mean that $dv_C(t)/dt$ is infinite, which, in turn, would mean that infinite power would be required, an impossible situation. The important idea here is that the capacitor can store energy, and that the energy is proportional to the square of the voltage across it. Voltage, then, is the **memory** variable (or **state** variable) for the capacitor. In later sections we contrast this with the inductor.

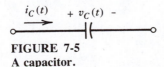

FIGURE 7-5
A capacitor.

Example 7-3

The voltage across the ½-F capacitor varies as shown in Figure 7-6a. Find the current through, the energy stored in, and the power delivered to or supplied by the capacitor.

The current through the capacitor was found in Ex-

ample 7-1. The power is simply given by the product of the voltage curve and the current curve. The energy is found by either integrating the power curve or by finding $\frac{1}{2}(Cv^2_C(t))$ point by point. The resulting graphs are shown in Figure 7-6b–d.

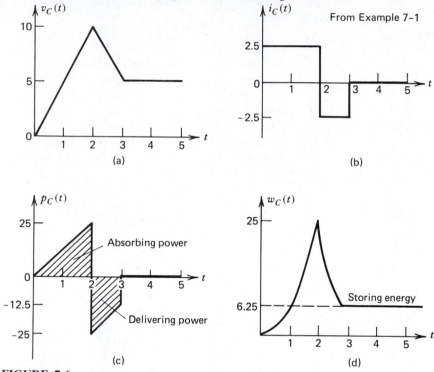

FIGURE 7-6
Waveforms showing the relationships between i, v, p, and w in a capacitor. (a) Voltage. (b) Current. (c) Power. (d) Energy.

Example 7-4

The current through a certain capacitor is as shown in Figure 7-7a. Find the voltage across, the energy stored in, and the power delivered to or supplied by the capacitor.

The voltage was found in Example 7-2. The power is found by multiplying the expressions for the current and the voltage. That is,

$$p_C(t) = i_C(t)\, v_C(t)$$

$$p_C(t) = I_o\, e^{-t/T_C} \cdot \frac{I_o T_C}{C}\,(1 - e^{-t/T_C})$$

$$p_C(t) = \frac{I_o^2 T_C}{C}\,(e^{-t/T_C} - e^{-2t/T_C})$$

This result is sketched in Figure 7-7c. The energy is found easily from

$$w_C(t) = \frac{1}{2} C v_C^2(t)$$

$$w_C(t) = \frac{1}{2} \frac{(I_o T_C)^2}{C} (1 - e^{-t/T_C})^2$$

The time variation of the energy is shown in Figure 7-7d.

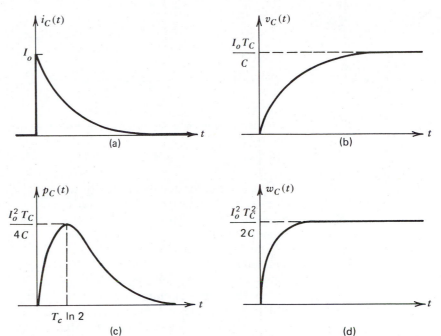

FIGURE 7-7
Graphs for Example 7-4. (*a*) Current. (*b*) Voltage (*c*) Power. (*d*) Energy.

Let us return to the concept that capacitors can remember a particular voltage. Consider the circuit of Figure 7-8*a*. The voltage across the capacitor must *track* the voltage supplied by the source since they are connected in parallel.

Suppose at some time $t = T_o$ the switch is opened. The source will continue to vary and produce a voltage $v_S(t)$. However, the capacitor is disconnected from the circuit, and lacking a path for the charges on its plates to recombine, will *hold* the voltage it had across it at the time the switch was opened, that is, $v_S(T_o)$. Another way to look at this is that at $t = T_o$, the current is forced to zero, $i_C(t) = 0$, since the switch is open. If $i_C(t) = 0$, then the voltage across the capacitor is a constant, that is,

$$i_C(t) = 0 = C \left[dv_C(t)/dt \right]$$

but

$$0 = C \left[d(\text{constant})/dt \right]$$

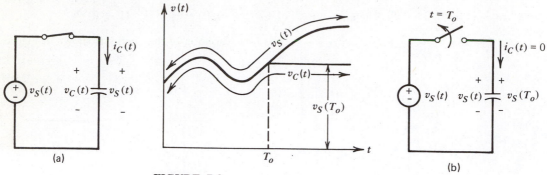

FIGURE 7-8
Demonstration of capacitor memory. (*a*) Switch closed, *track mode*. (*b*) Switch open, *hold mode*.

Our ideal capacitor theoretically will hold (remember) the voltage applied at the instant the switch was opened forever. In reality, because of parasitic effects real capacitors gradually will lose energy in a manner explained in the next chapter, and hence the remembered value gradually will decrease. For this reason, if the actual value remembered is to be retained for a long period of time, it occasionally must be refreshed. This concept of memory and energy storage is a fundamental property of many of the electronic circuits used for data computation in computers and ultimately provides a limitation to the speed with which these computations can occur. (See, for example, MOS devices in electronic texts such as Millman, *Microelectronics,* McGraw-Hill, New York, 1979.)

7-3 THE INDUCTOR

Our second memory element is the ideal **inductor.** It also is a two-terminal device. A piece of wire with a current flowing through it has a magnetic field surrounding it. This phenomenon was discovered by Oersted in the early 1800s. The magnetic field produced by this current can be related to a magnetic flux ϕ, which forms closed loops around the current flowing in the wire and is proportional to that current. If the wire is wound into a coil as shown in Figure 7-9*a*, the lines of flux become concentrated along the axis of the coil. The more turns N the coil has, the more flux.

The relationship among flux, current, and turns is given by

$$\phi(t) = kNi_L(t) \qquad (7\text{-}16)$$

where k is a constant of proportionality. This flux, in turn, will tie together or link all the turns of the inductor. The more turns, the more linkage. This total linkage is called the **flux linkage** and has the symbol λ and the units of Weber (Wb). Flux linkage is another basic variable like charge and energy and is the dual of charge. We will develop the *i-v* relations

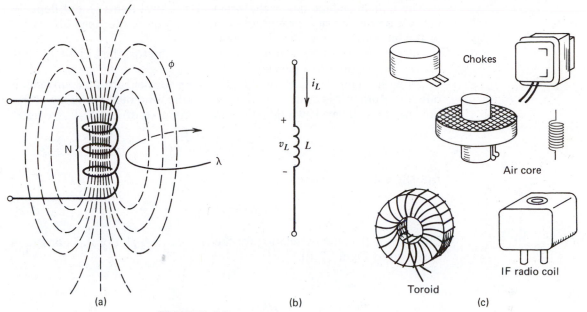

FIGURE 7-9
Inductors. (*a*) A coil. (*b*) Circuit symbol. (*c*) Actual devices.

for the inductor in a manner dual to the capacitor. Flux linkages and flux are related by

$$\lambda(t) = N\phi(t) \tag{7-17}$$

so that by substitution we get

$$\lambda(t) = [kN^2]\, i_L(t) \tag{7-18}$$

The quantity in the brackets in Eq. 7-18 is called the **inductance** L of the coil, that is,

$$L \equiv kN^2 \tag{7-19}$$

so that we can rewrite Eq. 7-18 as

$$\lambda(t) = Li_L(t) \tag{7-20}$$

Historical Note

The weber is named in honor of the German experimental physicist Wilhelm Eduard Weber, who, along with the German mathematician Carl Friedrich Gauss, established the absolute system of electrical measurement. Their joint paper in 1833 originated a system of force measurements using the metric units of time, mass, and length.

The unit of inductance is the **henry** (H) (plural henrys) and depends entirely on the physical properties of the coil. A more thorough discussion of the inductor is contained in Appendix C.

Figure 7-9 shows, in addition to a simple inductor, the circuit symbol, and examples of actual devices.

As with trying to measure charge, trying to measure flux linkages is difficult. Rather it is customary to measure the time rate of change of flux linkages or voltage. If we take the time derivative of Eq. 7-20, we get

$$\frac{d}{dt}(\lambda(t)) = \frac{d}{dt}(Li_L(t))$$

But since L is considered a constant,

$$v_L(t) = L\frac{di_L(t)}{dt} \tag{7-21}$$

Equation 7-21 is the i-v relation for the inductor. A plot of this constraint is shown in Figure 7-10.

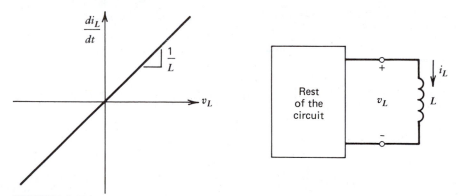

FIGURE 7-10
The i-v characteristics of an inductor.

Once again the reader is cautioned that Eq. 7-21 represents a mathematical model of a real device. Real inductors are far less close to their ideal models than capacitors or resistors. For example, we assume that inductor coils are wound with zero-resistance wire, but in practice this

Historical Note

Named for the American scientist and inventor Joseph Henry. Henry and Faraday independently discovered the law of magnetic induction almost simultaneously. The inductor is the circuit representation of magnetic induction and its unit was named to honor Henry rather than Faraday for reasons that have been lost with the passage of time. The unit of capacitance (the farad) honors Faraday, although capacitive action has absolutely nothing to do with magnetic induction.

is never the case. Figure 7-11 shows a more realistic model for a real inductor.

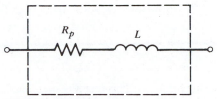

FIGURE 7-11
Model of an inductor that accounts for parasitic resistance.

In spite of this realization we will continue to assume that the inductor is lossless and use the ideal model for the remainder of this text.

In Eq. 7-21 we represent the inductor voltage in terms of the inductor current. To find a representation of the inductor current in terms of its voltage, we must integrate Eq. 7-21:

$$\int di_L(t) = \int \frac{1}{L} v_C(t)dt$$

Using analogous arguments as for the capacitor in selecting limits of integration, we can write

$$i_L(t) = \frac{1}{L} \int_{-\infty}^{t} v_L(t)dt \qquad (7\text{-}22)$$

This equation states that the current at any time t is proportional to the integral of the voltage across the inductor from time immemorial to time t. Just as with the capacitor, we divide our integral into two parts using $t = 0$ as our benchmark. The result then is

$$i_L(t) = \underbrace{\frac{1}{L} \int_{-\infty}^{0} v_L(t)dt}_{i_L(0)} + \frac{1}{L} \int_{0}^{t} v_L(t)dt \qquad (7\text{-}23)$$

The first term on the right can be evaluated and is equal to a constant, $i_L(0)$. It represent the value of the current flowing through the inductor at time $t = 0$ due to the voltage impressed across it from time immemorial to $t = 0$. Like the voltage across the capacitor, it may or may not be possible to calculate this value of current, but it can always be measured. There is only one value of current $i_L(t)$ flowing through the inductor at any instant of time including $t = 0$. Hence we can rewrite Eq. 7-23 as a sum of two terms: an integral that represents the positive time variation of the current and a constant that represents the initial condition of the inductor, that is

$$i_L(t) = \frac{1}{L} \int_{0}^{t} v_L(t)dt + i_L(0) \qquad (7\text{-}24)$$

Following a similar argument as for the capacitor we should realize that *the current through an inductor cannot change instantaneously*. That is,

$$i_L(t^+) = i_L(t^-) \tag{7-25}$$

This simple but important relation is called the **continuity** relation for inductors.

Example 7-5

A 2-mH inductor has the current shown in Figure 7-12*a* flowing through it. What voltage is produced as a result?

The equation for the current $i(t)$ in Figure 7-12*a* is

$$i(t) = 2 + 2 \sin 1000\pi t = i_L(t)$$

The voltage is found by applying the *i-v* relation (Eq. 7-21):

$$
\begin{aligned}
v_L(t) &= L \frac{di_L(t)}{dt} \\
&= 0.002 \, [0 + 2 \, (1000\pi) \cos 1000\pi t] \\
v_L(t) &= 12.57 \cos 1000\pi t \text{ V}
\end{aligned}
$$

This result is sketched in Figure 7-12*b*. An interesting observation is that the voltage waveform is exactly 90° ahead of the current waveform. It is customary to say that the current **lags** the voltage by 90° or the voltage **leads** the current by 90°. This nomenclature is used extensively in sinusoidal steady-state analysis discussed in Chapter 12.

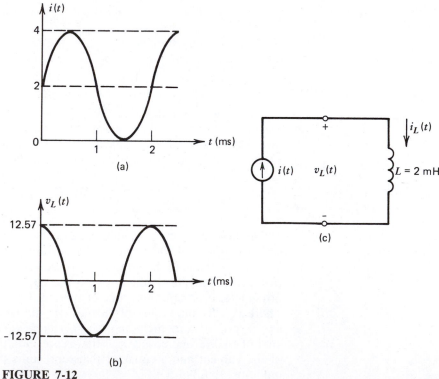

FIGURE 7-12
Circuit and signals for Example 7-5. (*a*) Current. (*b*) Inductor voltage. (*c*) Circuit.

An inductor, like the capacitor or any other element has the following relationship among voltage, current, and power:

$$p_L(t) = i_L(t)\, v_L(t) \tag{7-26}$$

which can be rewritten using the inductor's *i-v* relation as

$$p_L(t) = L\, i_L(t)\, \frac{di_L(t)}{dt} \tag{7-27}$$

As with the capacitor, the power can be either positive or negative, which means that an inductor can absorb or provide power. The power relation in Eq. 7-27 is a perfect derivative, that is,

$$p_L(t) = \frac{d}{dt}\left(\frac{1}{2}L\, i_L^2(t)\right) \tag{7-28}$$

Since power is the time rate of change or energy, the quantity inside the bracket represents the energy stored in the magnetic flux of the inductor, hence,

$$w_L(t) = \frac{1}{2}L\, i_L^2(t) \tag{7-29}$$

The amount of energy stored at any instant of time is dependent only on the instantaneous value of the current. The energy stored in always positive. Like the capacitor, an inductor can store energy, except it stores it in its magnetic rather than electric field. An inductor can remember the last current to flow through it. Current then is the memory variable (or state variable) for the inductor.

Example 7-6

A small box contains a memory element. The waveform shown in Figure 7-13a is input into the box and the output measured is shown in Figure 7-13b. What element is in the box, what is its value, and how much energy is stored in the element at $t = 1$ s?

Clearly the device performs a differentiation. The only memory element that differentiates a current to produce a voltage is an inductor, that is,

$$v_L(t) = L(di_L(t)/dt)$$

For the period between 0 and 1, di/dt is $+\,10$ A/s. The voltage during that same period is 100 mV, and hence the inductance must be 10 mH. The energy stored at $t = 1$ s is

$$w_L = \frac{1}{2}L i_L^2(t = 1) = \frac{1}{2}(0.01)(10)^2 = \frac{1}{2}\text{ J}$$

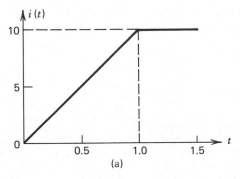

(a)

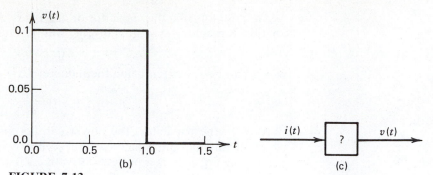

FIGURE 7-13
Waveforms for mystery element of Example 7-6. (*a*) Input current. (*b*) Output voltage. (*c*) Mystery element.

Inductors are the dual of capacitors.[2] This means that the *i-v* characteristics of one element give rise to similar *v-i* characteristics of the other. That is, inductors and capacitors give rise to similar looking equations except that the roles of current and voltage are interchanged. One may be tempted to believe that inductors and capacitors can be interchanged freely in circuits. In some ways this is true—especially in theory. In practice, however, real inductors tend to be lossy and heavy as discrete components, and are rarely produced on ICs because of severe limitations in building them on a planar surface. The end result is that capacitors are the principal memory elements used in electronics. Inductors are used in special applications where both types of memory elements are necessary, such as in resonant circuits; in applications where interaction between two or more inductors is desired, such as in transformers; and in other applications where size does not matter, such as in power applications. Some of these applications of inductors are discussed in subsequent chapters.

[2]More on duality. It should have caught the student's eye that the *i-v* relations for inductors and capacitors are remarkably similar. In fact, if one replaces *C* by *L* and *i* by *v*, the two equations become identical. This phenomenon is the result of the concept of duality and can be extremely useful in the solution of problems. The solution of a series *RL* circuit will have the same form as a parallel *GC* circuit. Table 7-1 lists some of the dual quantities studied in this course. The quantity in the first column is the dual of the quantity in the second column, and vice versa.

Charge	Flux linkage
Voltage	Current
Resistance	Conductance
Inductance	Capacitance
Open	Short
Series	Parallel
Node	Loop

TABLE 7-1
Some Dual Quantities

7-4   SIMPLE OP AMP MEMORY CIRCUITS

The *i-v* relationships of memory elements when combined with the special properties of operational amplifiers give rise to several important applications.

Consider the circuit of Figure 7-14. Except for the capacitor in the feedback loop, this circuit should remind us of the inverting OP AMP configuration studied in Chapter 4. We began our analysis by writing a KCL equation at node Ⓐ:

$$i_R(t) + i_C(t) = i_N(t)$$

and the element equations are

$$i_C(t) = C \frac{d[v_o(t) - v_N(t)]}{dt}$$

and

$$i_R(t) = \frac{1}{R}[v_S(t) - v_N(t)]$$

Finally, if we use the properties of an ideal OP AMP,

$$v_N(t) = v_P(t)$$

and

$$i_N(t) = i_P(t) = 0$$

we can substitute all these element constraints into our original KCL equation or connection constraint. In doing so we should also note that $v_P(t)$ is connected to ground and hence is zero, making $v_N(t)$ equal to zero as well.

We get

$$\frac{v_S(t)}{R} + C \frac{dv_o(t)}{dt} = 0$$

which we can rewrite to solve for $v_o(t)$ in terms of $v_S(t)$ as

$$dv_o(t) = -\frac{1}{RC} v_S(t)dt$$

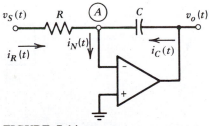

FIGURE 7-14
An OP AMP integrator.

If we integrate both sides, we arrive at our desired result:

$$v_o(t) = -\frac{1}{RC} \int_{-\infty}^{t} v_S(t)dt \tag{7-30}$$

The output of our OP AMP circuit is proportional to the integral of the input. We call this circuit an **integrator.** Note that the proportionality constant $(-1/RC)$ has the units of inverse time (second^{-1}).

Example 7-7

Analyze the circuit of Figure 7-15 and find its input-output relation.

Following the same analysis pattern as for the integrator, we start with a KCL equation at Ⓐ:

$$i_1(t) + i_R(t) = i_N(t)$$

The element costraints are

$$i_1(t) = C \frac{d[v_S(t) - v_N(t)]}{dt}$$

and

$$i_R(t) = \frac{v_o(t) - v_N(t)}{R}$$

and

$$i_N(t) = 0 \qquad v_P(t) = 0 = v_N(t)$$

so that upon substituting back into our connection constraint we find

$$C \frac{dv_S(t)}{dt} + \frac{v_o(t)}{R} = 0$$

which we can write as

$$v_o(t) = -RC \frac{dv_S(t)}{dt}$$

We readily observe that in this configuration, our OP AMP circuit performs as a **differentiator,** and the constant $(-RC)$ has the units of time (seconds).

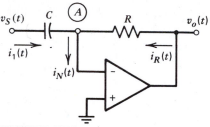

FIGURE 7-15
An OP AMP differentiator.

Example 7-8

Replace the capacitor in Example 7-7 with an inductor and repeat the analysis.

Everything remains essentially the same except that the element constraint for the inductor is used to find $i_1(t)$, that is,

$$i_i(t) = \frac{1}{L} \int_{-\infty}^{t} [v_S(t) - v_N(t)]dt$$

Proceeding with our analysis we find

$$\frac{1}{L} \int_{-\infty}^{t} v_S(t)dt + \frac{v_o(t)}{R} = 0$$

Another integrator! However, we should recall our comments from the last section that inductors are

rarely used in these applications. By the way, this constant of proportionality $(-R/L)$ also has units of inverse time (second^{-1}).

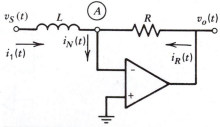

FIGURE 7-16
Another OP AMP integrator.

We have introduced two more building blocks to our collection of functional mathematical operations. Along with the scalar multipliers and summers introduced in Chapter 4 we now add the mathematical operations of integration and differentiation. The operational amplifier received its name from its ability to perform so many varied mathematical operations. These two new building blocks are collected and represented along with circuit, block diagram, and gains in Figure 7-17.

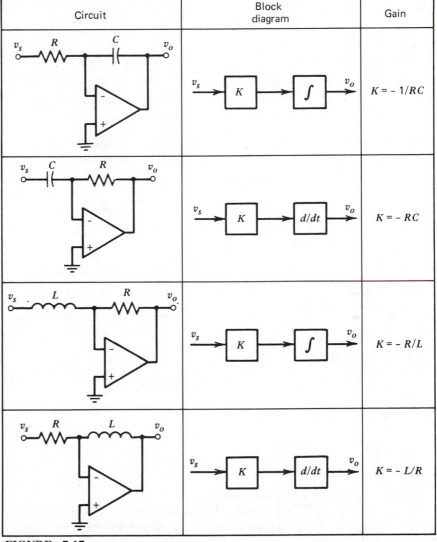

FIGURE 7-17
Block diagram representation of OP AMP integrators and differentiators.

This entire collection of OP AMP realizable mathematical operations forms the basis of the so-called analog computer that finds extensive application in simulating systems described by differential or integrodifferential equations.

Example 7-9

Using the building blocks contained in Figures 4-27 and 7-17, devise a solution to the following differential equation. Assume all initial conditions to be zero.

$$\frac{d^2y}{dx^2} + 6\frac{dy}{dx} + \frac{1}{2}y = x$$

The solution to our equation is y. Hence we begin by solving our equation for y.

$$y = 2x - 2\frac{d^2y}{dx^2} - 12\frac{dy}{dx}$$

Now we must realize that y is the output of a summer that has all the terms on the right as inputs (see Figure 7-18a). If we have y, and we differentiate it, we obtain dy/dx, and if we then multiply this output by -12,

we have one of the necessary inputs to our summer. Continuing this process, if we differentiate dy/dx, we get our second derivative, which we can scale by -2 to obtain our second input. Figure 7-18b shows the result of our block diagram analysis. From this block diagram we can then design each individual block, being careful to consider the gains and inverting properties of each block.

Practical realities make our approach to this problem unrealistic. Differentiators are rarely used as they accentuate noise, and noise is always present. An alternative design uses integrators in lieu of differentiators as shown in the figure. It is left as an exercise for the reader to show that the block diagram of Figure 7-18c will produce the same result as that of Figure 7-18b.

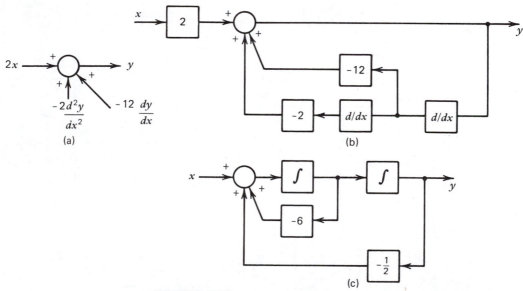

FIGURE 7-18
Example use of block diagrams to solve differential equations. (*a*) Summing operation. (*b*) Solution using differentiators. (*c*) Solution using integrators.

7-5 LINEARITY AND EQUIVALENT CIRCUITS

In Chapter 3 we studied some of the consequence of linearity. For example, the principle of superposition and the Thévenin and Norton theorems were possible because the circuits to which they were applied were linear. Capacitors and inductors, like the resistor, are also linear elements. Circuits composed of R's, L's, and C's are linear circuits and all the analysis tools studied previously apply. The analysis of linear circuits with memory elements is left for subsequent chapters, however, we would like to demonstrate that linearity applies to these two elements.

Consider the capacitor. Recall that it has the following i-v characteristics:

$$v_C(t) = \frac{1}{C} \int_{-\infty}^{t} i_C(t)dt \tag{7-31}$$

or equivalently

$$i_C(t) = Cdv_C(t)/dt \tag{7-32}$$

To show that the capacitor is linear it must have the properties of proportionality and superposition.

Let us demonstrate both. For proportionality, if we multiply the input variable by K, a constant, the output variable will be multiplied by the same constant. Suppose $i_C(t)$ is multiplied by K in Eq. 7-31:

$$v_x(t) = \frac{1}{C} \int_{-\infty}^{t} Ki_C(t)dt$$

Since K is a constant, we can pull it outside the integral:

$$v_x(t) = K\frac{1}{C} \int_{-\infty}^{t} i_C(t)dt = Kv_C(t) \tag{7-33}$$

For the derivative relation of Eq. 7-32, we find

$$i_x(t) = C\frac{d[Kv_C(t)]}{dt}$$

But we can pull K outside the derivative since it is a constant, so that

$$i_x(t) = K[Cdv_C(t)/dt] = K i_C(t) \tag{7-34}$$

The proportionality property is upheld. To prove the superposition property consider Eq. 7-31 again and the voltage produced by current $i_{C1}(t)$ and that produced by $i_{C2}(t)$. If the same voltage can be produced by applying the sum of both currents, superposition will hold.

$$v_{C1}(t) = \int_{-\infty}^{t} i_{C1}(t)dt$$

$$v_{C2}(t) = \int_{-\infty}^{t} i_{C2}(t)dt$$

$$v_{C3}(t) = \int_{-\infty}^{t} [i_{C1}(t) + i_{C2}(t)]dt$$

If $v_{C3}(t) = v_{C1}(t) + v_{C2}(t)$, then superposition holds

$$v_{C1}(t) + v_{C2}(t) = \int_{-\infty}^{t} i_{C1}(t)dt + \int_{-\infty}^{t} i_{C2}(t)dt$$

Since the limits are the same the two integrals can be combined:

$$v_{C1}(t) + v_{C2}(t) = \int_{-\infty}^{t} [i_{C1}(t) + i_{C2}(t)]dt$$

so that

$$v_{C1}(t) + v_{C2}(t) = v_{C3}(t) \tag{7-35}^3$$

Similarly for the differential relation, Eq. 7-32

$$i_{C1}(t) = C\, dv_{C1}(t)/dt$$
$$i_{C2}(t) = c\, dv_{C2}(t)/dt$$
$$i_{C3}(t) = C\, d[v_{C1}(t) + v_{C2}(t)]/dt$$

If $i_{C3}(t) = i_{C1}(t) + i_{C2}(t)$, then superposition will hold.

$$i_{C1}(t) + i_{C2}(t) = C\, dv_{C1}(t)/dt + C\, dv_{C2}(t)/dt$$

which can be combined as

$$i_{C1}(t) + i_{C2}(t) = C\, d[v_{C1}(t) + v_{C2}(t)]/dt \tag{7-36}$$

so that

$$i_{C1}(t) + i_{C2}(t) = i_{C3}(t)$$

This completes our demonstration that the capacitor fulfills all the requirements to be a linear element. Since the inductor is the dual of the capacitor, a similar analysis will show it also to be a linear element.

One of the first and simplest applications of the linearity property is in establishing a rule for equivalent circuits. There are many occasions where two or more capacitors are connected in series or parallel. Analysis of such circuits would be simplified if those capacitors could be combined and replaced by a simpler equivalent. Consider the capacitors in Figure 7-19a. They are obviously connected in parallel.

A KCL equation at node Ⓐ yields

$$i(t) = i_1(t) + i_2(t) + \cdots + i_N(t)$$

But $i_1(t) = C_1 dv_1(t)/dt$, $i_2(t) = C_2 dv_2(t)/dt$, and so forth. Since the capacitors are connected in parallel,

$$v_1(t) = v_2(t) = v_N(t) = v(t)$$

so that

$$i(t) = C_1 \frac{dv(t)}{dt} + C_2 \frac{dv(t)}{dt} + \cdots + C_N \frac{dv(t)}{dt}$$

---

[3]A word of caution. When the integral relations were multiplied by K in Eq. 7-31, the initial conditions were also multiplied by K. In our demonstration we skirted the issue by considering the limits to be from $-\infty$ to t rather than 0 to t and including the initial condition.

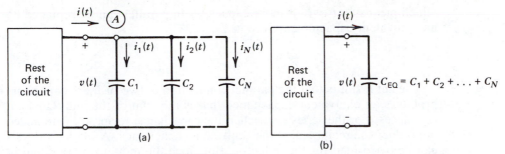

FIGURE 7-19
Capacitors in parallel. (*a*) Original circuit. (*b*) Equivalent.

But we can factor out $dv(t)/dt$ to obtain

$$i(t) = (C_1 + C_2 + \cdots + C_N)\, dv(t)/dt$$

which means that two or more capacitors in parallel can be replaced by an equivalent capacitor equal to the sum of the individual capacitances, that is,

$$C_{EQ} = C_1 + C_2 + \cdots + C_N \qquad (7\text{-}37)$$

This result is shown in Figure 7-19*b*.

Consider now the capacitors in series shown in Figure 7-20*a*.

Applying Kirchoff's voltage law around the loop produces the following equation:

$$v(t) = \frac{1}{C_1} \int_{-\infty}^{t} i_1(t)dt + \frac{1}{C_2} \int_{-\infty}^{t} i_2(t)dt + \cdots + \frac{1}{C_N} \int_{-\infty}^{t} i_N(t)dt$$

But since the capacitors are in series, the same current flows through each of them. We can then factor the integral out of the terms on the right to obtain the following:

$$v(t) = \left(\frac{1}{C_1} + \frac{1}{C_2} + \cdots + \frac{1}{C_N} \right) \int_{-\infty}^{t} i(t)dt$$

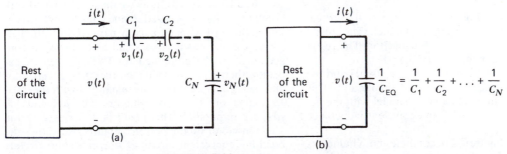

FIGURE 7-20
Capacitors in series. (*a*) Original circuit. (*b*) Equivalent.

which means that two or more capacitors in parallel can be replaced by an equivalent capacitance given in Eq. 7-38.

$$\frac{1}{C_{EQ}} = \frac{1}{C_1} + \frac{1}{C_2} + \cdots + \frac{1}{C_N} \tag{7-38}$$

The net result is that the parallel connection of two or more capacitors produces an equivalent capacitance that is the sum of the individual capacitances. For the series connection the capacitance reciprocals are added to produce the reciprocal of the equivalent capacitance. A similar analysis can be performed for the inductor. But since the inductor is the dual of the capacitor the equivalence rules are interchanged. That is, for the series connection of inductors the equivalent inductance is found by adding the individual inductances, while for the parallel connection the reciprocals are added. These observations on combining capacitors or inductors in series or parallel are summarized in Table 7-2. A few examples will demonstrate the usefulness of equivalent circuits.

Element	Series	Parallel
Capacitors	$\frac{1}{C_{EQ}} = \frac{1}{C_1} + \frac{1}{C_2} + \cdots + \frac{1}{C_N}$	$C_{EQ} = C_1 + C_2 + \cdots + C_N$
Inductors	$L_{EQ} = L_1 + L_2 + \cdots + L_N$	$\frac{1}{L_{EQ}} = \frac{1}{L_1} + \frac{1}{L_2} + \cdots + \frac{1}{L_N}$

TABLE 7-2
Equivalent Circuits

Example 7-10

Find the equivalent circuit for each of the circuits in Figure 7-21.

For the circuit of Figure 7-21a, the two 0.5-μF capacitors add since they are in parallel. This result then is in series with the 1-μF capacitor. The result is an equivalent capacitance of 0.5 μF.

For the circuit of Figure 7-21b, the 10-mH and the 30-mH inductors are in series and simply add. This new equivalent inductance is in parallel with the 80-mH inductor. The result is an equivalent inductance of 26.67 mH.

The circuit of Figure 7-21c combines both inductors and capacitors. In subsequent chapters we will learn under what conditions we may combine all of these together. However, for now we must be satisfied with combining inductors together and capacitors together only. The 5-pF capacitor is in parallel with the 0.1-μF capacitor, but it is so small as to have no consequence to the combined total. The $\approx$0.1-μF capacitor then is in series with another 0.1-μF capacitor and their equivalent result is in parallel with the 0.05-μF capacitor. The two inductors, 700 μH and 300 μH, are in parallel. This result is in series with the 1-mH inductor on the bottom. Figure 7-22 shows the resulting equivalent circuits for each of the circuits of Figure 7-21.

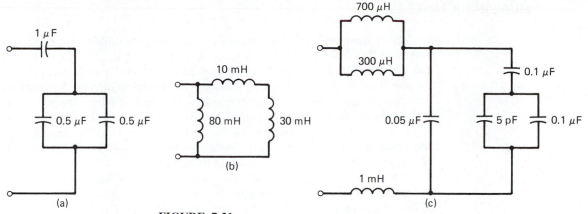

FIGURE 7-21
Examples of circuit that can be reduced by combining similar elements in series or parallel.

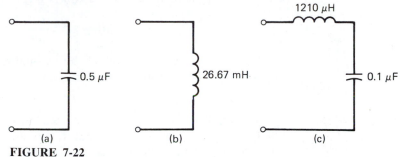

FIGURE 7-22
Results of combining elements in Figure 7-21.

Example 7-11

You have just completed a calculation for a capacitor in a circuit you designed. The desired value is 0.534 μF. Refer to Appendix C for a table of standard values. The circled values cost $0.50 while the uncircled values cost $1.50. Assume you will buy all tubular capacitors with ± 5 percent tolerance. What capacitor(s) would you buy and how would you connect them together.?

Since the best tolerance available is ± 5 percent, your 0.534-μF capacitor really cannot be guaranteed. Any capacitor with a value between 0.51 and 0.56 μF

will probably work. Clearly either the 0.51-μF or the 0.56-μF capacitor might satisfy our design but either would cost $1.50 and could be out of range—the 0.51-μF capacitor could be 5 percent to the low side and the 0.56-μF capacitor 5 percent to the high side. A better solution would be to connect a 0.22-μF and a 0.33-μF capacitor in parallel. This would result in a 0.55-μF capacitor, which is fairly close to the desired value and would cost only $1.00. Other results are possible, but considering cost, this probably is the best.

7-6 MUTUAL INDUCTANCE

We saw in Section 7-3 that a current flowing through a coiled wire produces magnetic flux lines that form closed loops about the conductor. We also saw that the number of turns in the coil enhances that flux. Furthermore we saw that if the flux lines or linkages were changing with respect to time, a voltage would be generated across the coil of wire. Suppose that a second coil of wire was brought close to the first coil. It is reasonable to assume that the flux linkages of the first coil would cut the second coil and generate a voltage across it. In this section we will look at the effect of one coil on another, called **mutual** inductance.

Consider the coils of wire in close proximity as shown in Figure 7-23.

Suppose a current $i_1(t)$ flows in the first coil of turns N_1. The amount of flux linkages produced by this current is given by Eq. 7-18:

$$\lambda_1(t) = k_1 N_1^2 i_1(t) \tag{7-39}$$

The voltage produced across this coil is given by Eq. 7-21:

$$v_1(t) = k_1 N_1^2 \frac{di_1(t)}{dt} \tag{7-40}$$

The constants $k_1 N_1^2$ are called the self-inductance, or simply the inductance and are referred to by the symbol L_1, that is,

$$v_1(t) = L_1 di_1(t)/dt \tag{7-41}$$

Now suppose that the second coil was close enough so that the flux linkages produced by i_1 intercepted it. A voltage would be generated in the second coil that would be proportional to several things. First of all, it would be proportional to the current in the first coil, $i_1(t)$. Then it would be proportional to the turns in the first coil, N_1. It also would be proportional to the turns in the second coil, N_2. And finally it would be proportional to the coupling between the two coils and a number of secondary effects, k_m. We can write

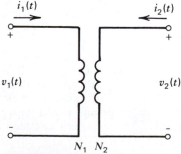

FIGURE 7-23
Mutually coupled coils.

$$v_2(t) = N_1 N_2 k_m \frac{di_1(t)}{dt} \tag{7-42}$$

We call the constant $N_1 N_2 k_m$ the mutual inductance and give it the symbol M. The units of M, just like L_1, are expressed in henrys. We can rewrite Eq. 7-42 as

$$v_2(t) = M \, di_1(t)/dt \tag{7-43}$$

The same reasoning could be followed if the current were flowing in coil two. There would be a self-induced voltage and one voltage induced in coil one via mutual inductance. That is, we would obtain two terms similar to Eqs. 7-41 and 7-43,

$$v_2(t) = k_2 N_2^2 \frac{di_2(t)}{dt} \tag{7-44}$$

which we would write as

$$v_2(t) = L_2 \, di_2(t)/dt \tag{7-45}$$

and

$$v_1(t) = N_1 N_2 k_m \frac{di_2(t)}{dt} \tag{7-46}$$

which we would write as

$$v_1(t) = M \, di_2(t)/dt \tag{7-47}$$

Now suppose *both* currents were flowing as in Figure 7-24a. We could combine the results of Eqs. 7-41, 7-43, 7-45, and 7-47 using superposition to obtain the terminal voltages for each coil.

$$v_1(t) = L_1 \frac{di_1(t)}{dt} + M \frac{di_2(t)}{dt}$$

and $\qquad\qquad\qquad\qquad\qquad\qquad\qquad\qquad$ (7-48)[4]

$$v_2(t) = M \frac{di_1(t)}{dt} + L_2 \frac{di_2(t)}{dt}$$

Up to now we have assumed that the two coils were wound with **positive** polarity. This may not be the case. The way a coil is wound—clockwise or counterclockwise—can make a difference to the polarity of the mutual voltages generated. In Figure 7-24a it was assumed that both coils were

---

[4] We have assumed that the mutual inductance M is the same regardless of which current is inducing the voltage. In some rare cases this will not be true and two different M's will have to be identified. Under these conditions Eq. 7-48 will have to be written as

$$v_1(t) = L_1 \frac{di_1(t)}{dt} + M_{12} \frac{di_2(t)}{dt}$$

and

$$v_2(t) = M_{21} \frac{di_1(t)}{dt} + L_2 \frac{di_2(t)}{dt}$$

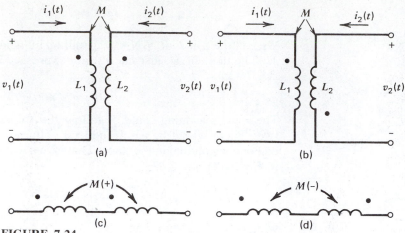

FIGURE 7-24
Coupled coils showing the dot convention.

wound in the same direction. To show that this was true we added two dots to the top of each coil. The result then is Eq. 7-48. Suppose, however, that one coil was wound differently than the other. This would result in a **negative** polarity. We would indicate this winding by the dot convention shown in Figure 7-24b. What the dots refer to is the positive polarity for the **mutually** induced voltage not the self-induced voltage. When the windings are as depicted in Figure 7-24b, Eq. 7-48 must be written as

$$v_1(t) = L_1 \frac{di_1(t)}{dt} - M \frac{di_2(t)}{dt}$$

and
(7-49)

$$v_2(t) = -M \frac{di_1(t)}{dt} + L_2 \frac{di_2(t)}{dt}$$

Let us try to realize why this is so. Recall that whenever a current flows in a conductor, a magnetic field is generated. This field is oriented in a particular direction in accordance with what is often referred to as the **right-hand rule.** Hence if we looked at a coil head on and the windings were clockwise, the north pole of the resultant field would point away from us. If the windings were counterclockwise, the north pole would point toward us. It seems reasonable to assume that the voltages generated by flux linkages created by currents running in opposite directions would have opposite polarities. The dots are a simple way to keep track of these polarities. If both the flux linkage generating current and the mutually induced current enter dotted terminal, the mutual inductance term is " + ." If one of the currents exits a dotted terminal while the other enters one, the mutual inductance term is " − ." If both currents exit dotted terminals, the mutual inductance term is once again " + ." Figures 7-24c and 7-24d are another way to look at additive and subtractive mutual inductance.

Example 7-12

For the coupled circuit of Figure 7-25, find $v_2(t)$.

Since the output is open-circuited, $i_2(t) = 0$. Therefore, substituting for $v_1(t)$, Eq. 7-48 is

$$10 \cos 100t = 0.01 \frac{di_1(t)}{dt} + 0$$

$$v_2(t) = 0.002 \frac{di_1(t)}{dt} + 0$$

Solving for $i_1(t)$ by direct integration yields

$$i_1(t) = (10)(100) \int_0^t \cos 100t \, dt$$

$$i_1(t) = \frac{1000}{100} \sin 100t \bigg|_0^t = 10 \sin 100t$$

so that $v_2(t) = 0.002 \, d(10 \sin 100t)/dt$

$$v_2(t) = 2 \cos 100t \text{ V}$$

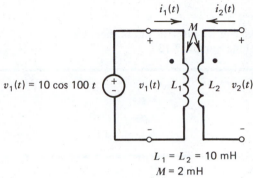

$$L_1 = L_2 = 10 \text{ mH}$$
$$M = 2 \text{ mH}$$

FIGURE 7-25
Coupled coil circuit for Example 7-12.

Example 7-13

Consider the circuit of Figure 7-26. Suppose $v_1(t) = \cos t$ and find $i_2(t)$.

Since the output is short-circuited, $v_2(t) = 0$. Under this constraint and noting the negative polarity of the windings, we can write Eq. 7-49 as

$$\cos t = 2 \frac{di_1(t)}{dt} - 1 \frac{di_2(t)}{dt}$$

$$0 = -1 \frac{di_1(t)}{dt} + 2 \frac{di_2(t)}{dt}$$

This yields

$$\frac{di_1(t)}{dt} = 2 \frac{di_2(t)}{dt}$$

Substituting we find

$$\cos t = 4 \frac{di_2(t)}{dt} - 1 \frac{di_2(t)}{dt} = 3 \frac{di_2(t)}{dt}$$

Integrating both sides we find

$$i_2(t) = \frac{1}{3} \int_0^t \cos t \, dt$$

$$i_2(t) = \frac{1}{3} \sin t \text{ A}$$

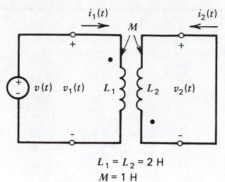

$$L_1 = L_2 = 2 \text{ H}$$
$$M = 1 \text{ H}$$

FIGURE 7-26
Coupled coil circuit with output short-circuited for Example 7-13.

Example 7-14

For the circuit of Figure 7-27, find $v_1(t)$ and $v_2(t)$.

The currents $i_1(t)$ and $i_2(t)$ are fixed because of the current sources driving the system. Therefore Eq. 7-48 yields

$$v_1(t) = 0.2 \frac{d(5 \cos 10t)}{dt} + 1.0 \frac{d(-2 \sin 5t)}{dt}$$

$$v_2(t) = 1.0 \frac{d(5 \cos 10t)}{dt} + 0.5 \frac{d(-2 \sin 5t)}{dt}$$

which immediately results in

$$v_1(t) = -10 \sin 10t - 10 \cos 5t$$
$$v_2(t) = -50 \sin 10t - 5 \cos 5t$$

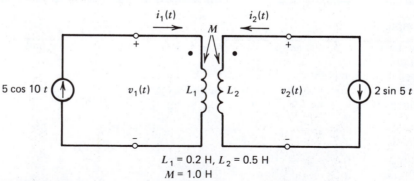

$$L_1 = 0.2 \text{ H}, L_2 = 0.5 \text{ H}$$
$$M = 1.0 \text{ H}$$

FIGURE 7-27
Doubly driven coupled coil circuit for Example 7-14.

Example 7-15

Three coils are mutually coupled as shown in Figure 7-28. Write a set of equation describing $v_1(t)$, $v_2(t)$, and $v_3(t)$ in terms of the currents and the inductances.

Following the example used to develop Eq. 7-48 and 7-49 we can write the results as

$$v_1(t) = L_1 \frac{di_1(t)}{dt} + M_1 \frac{di_2(t)}{dt} - M_3 \frac{di_3(t)}{dt}$$

$$v_2(t) = M_1 \frac{di_1(t)}{dt} + L_2 \frac{di_2(t)}{dt} - M_2 \frac{di_3(t)}{dt}$$

$$v_3(t) = -M_3 \frac{di_1(t)}{dt} - M_2 \frac{di_2(t)}{dt} + L_3 \frac{di_3(t)}{dt}$$

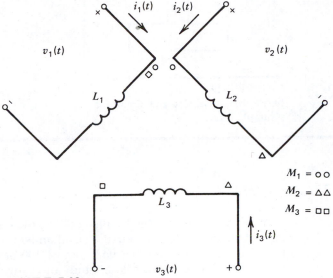

$$M_1 = \circ \circ$$
$$M_2 = \triangle \triangle$$
$$M_3 = \square \square$$

FIGURE 7-28
Three mutually coupled coils for Example 7-15.

Example 7-16

Find an expression for $v_x(t)$ in the circuit of Figure 7-29.

The desired $v_x(t)$ is equal by Kirchoff's voltage law to $v_1(t) - v_2(t)$. Also $i_1(t) = -i_2(t)$ by Kirchhoff's current law. Hence using Eq. 7-49 because the dots are opposing, we can write

$$v_1(t) = L_1 \frac{di_1(t)}{dt} - M \frac{d(-i_1(t))}{dt}$$

$$v_2(t) = M \frac{di_1(t)}{dt} - L_2 \frac{d(-i_1(t))}{dt}$$

Therefore,

$$v_x(t) = v_1(t) - v_2(t)$$

$$v_x(t) = L_1 \frac{di_1(t)}{dt} - M \frac{di_1(t)}{dt} + M \frac{di_1(t)}{dt} + L_2 \frac{di_1(t)}{dt}$$

$$v_x(t) = (L_1 + L_2) \frac{di_1(t)}{dt}$$

The mutual inductances have canceled each other.

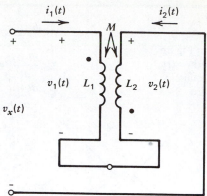

FIGURE 7-29
Series-driven coupled coil circuit for
Example 7-16.

7-7   THE IDEAL TRANSFORMER

A most important use for coupled coils is the **transformer.** Transformers are physically designed so that a small amount of current can generate a lot of flux linkages. Furthermore almost 100 percent of this flux linkage couples both coils. In our brief study of the transformer we discuss the ideal model. The transformer is a very efficient device. Over 99 percent of the power input to the transformer comes out! Transformers are ubiquitous; they find applications in all kinds of power and instrumentation systems. For example, in the home transformers are likely to be found in the doorbell circuit, providing reduced voltages from the powerlines to the home junction box, providing low voltage for thermostats, in various power supplies for television sets, stereos, and radios, in high-intensity lamps, and in the chargers for calculators, to mention just some of the more common applications. Their ubiquitousness derives from their ability to simplify interfacing circuits. Figure 7-30 shows a model of the ideal transformer and sketches of some real devices.

In our development of the ideal transformer we make two idealizations. The first assumes that all the flux linkages produced by the first coil link the second coil as well, that is, $k_1 = k_2 = k_m \simeq 1$. This really is not a far-fetched idealization since in practical iron-core-type transformers $k_1 \simeq k_2 \simeq k_m \simeq 0.99$. This idealization allows us to rewrite Eq. 7-48 as

$$v_1(t) = N_1^2 \frac{di_1(t)}{dt} + N_1 N_2 \frac{di_2(t)}{dt}$$

$$v_2(t) = N_1 N_2 \frac{di_1(t)}{dt} + N_2^2 \frac{di_2(t)}{dt}$$

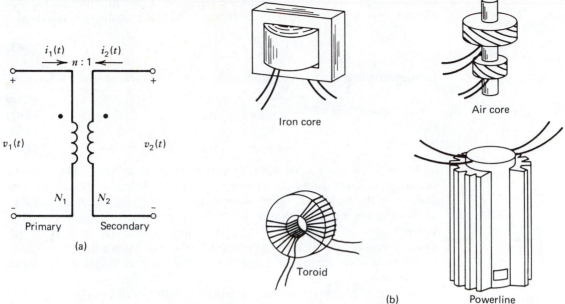

FIGURE 7-30
Transformers. (a) Model. (b) Real transformers.

Factoring N_1 from the first equation and N_2 from the second gives

$$v_1(t) = N_1[N_1 \, di_1(t)/dt + N_2 \, di_2(t)/dt]$$
$$v_2(t) = N_2[N_1 \, di_1(t)/dt + N_2 \, di_2(t)/dt]$$

and therefore the ratio of $v_1(t)$ to $v_2(t)$ yields

$$\frac{v_1(t)}{v_2(t)} = \frac{N_1}{N_2} \equiv n \tag{7-50}$$

where n is called the **turns ratio.**

Thus our first idealization leads to the statement that the ratio of the voltages across the coils is directly proportional to the turns ratio.

The second idealization we must make is that negligible power is lost in the transformer. That essentially means that the instantaneous power input to the transformer equals the instantaneous power output or

$$p_1(t) = p_2(t) \tag{7-51}$$

which we can write as

$$v_1(t) \, i_1(t) = -v_2(t) \, i_2(t) \tag{7-52}$$

The minus sign simply takes into account our referenced direction for the current $i_2(t)$. The transformer *supplies* power at its output; hence the negative sign on the right side of the equation. If we manipulate Eqs. 7-52 and 7-50, we can generate the remaining ideal transformer relation

$$\frac{v_1(t)}{v_2(t)} = -\frac{i_2(t)}{i_1(t)} = n \tag{7-53}$$

In summary we can list the properties of the ideal transformer a follows:

$$n = N_1/N_2 \tag{7-50}$$
$$p_1(t) = p_2(t) \tag{7-51}$$
$$v_1(t) = nv_2(t) \tag{7-54}$$
$$i_1(t) = -i_2(t)/n \tag{7-55}$$

Some vocabulary used with transformers is as follows: The input is generally referred to as the **primary,** while the output is called the **secondary.** If the number of turns of the secondary is larger than the number of turns in the primary, that is, $N_2 > N_1$, then $n < 1$. This means that a voltage $v_2(t)$ will be greater than $v_1(t)$. Thus the voltage is stepped up. Such a transformer is often referred to as a **step up** transformer. If the opposite is true, then $n > 1$ and the transformer is called a **step down** transformer.

A word of caution is in order in working with circuits containing transformers. Our two idealizations mean that *no* current is needed to produce flux linkages and that all the inductances, both self- and mutual, are infinite. Do not try to write coupled-coil equations such as Eqs. 7-48 and 7-49 if the transformer is to be treated as ideal (Eqs. 7-50, 7-51, 7-54, and 7-55).

The real advantage of transformers is in their ability to ease the interfacing of circuits. Consider the circuit of Figure 7-31.

The transformer will alter the way in which the load R_L appears at the input. What we wish to find then is the Thévenin circuit presented at the input. To do this we note that

$$v_2(t) = i_L(t)R_L$$

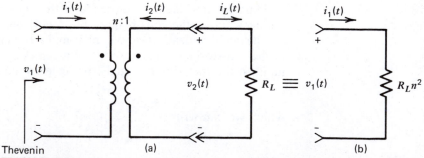

Thevenin
FIGURE 7-31
Transformer used as an interfacing tool. (*a*) Transformer circuit. (*b*) Equivalent.

but

$$i_2(t) = -i_L(t)$$

so that

$$v_2(t) = -i_2(t)R_L$$

Now using the ideal transformer equation, we let

$$v_2(t) = v_1(t)/n$$

and

$$i_2(t) = -i_1(t)n$$

Substituting we find that

$$\frac{v_1(t)}{n} = -(-i_1(t)n)R_L$$

or

$$\frac{v_1(t)}{i_1(t)} = n^2R_L \qquad (7\text{-}56)$$

Hence the load R_L will appear to be n^2 times what it actually is. For example, if R_L is 1 kΩ and $n = 10$, R_L will seem to be 100 kΩ at the input. Transformers are used frequently for matching a source to a load for maximum power transfer.

A few examples will demonstrate some of the transformer's applications.

Example 7-17

Determine the turns ratio of a step-down transformer to drop ordinary house voltage ($v_1(t) = 156 \cos 377t$) to 24 V rms ($v_2(t) = 33.9 \cos 377t$).

The input and output voltages are related by Eq. 7-54. Therefore we will need a step down transformer with the following turns ratio:

$$n = \frac{v_1(t)}{v_2(t)} = \frac{156}{33.9} = 4.6$$

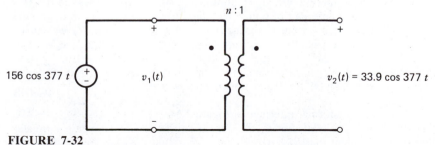

FIGURE 7-32
House voltage step-down transformer for Example 7-17.

Example 7-18

The output stage of a stereo amplifier has a "resistance" of 600 Ω. The speaker (the load) has a "resistance" of 8 Ω. Match the two for maximum power transfer.

We recall from Chapter 3 that for maximum power transfer R_L must equal R_S. We could use a resistive interface circuit, but while maximum power would be transferred from the source, much of what is transferred would be lost in the resistive network. If we use a transformer, essentially all the transferred power will be delivered to R_L, provided we use a transformer with the right turns ratio.

What we wish to do is make our load appear to the source as 600 Ω. We saw from Eq. 7-56 that a load will appear to be n^2 times larger than it actually is at the primary. Therefore we must select a transformer with a turns ratio equal to the following:

$$600 = n^2 8$$

or

$$n = \sqrt{\frac{600}{8}} = 8.66$$

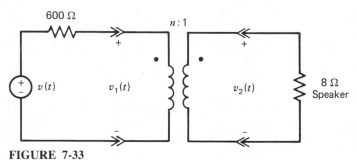

FIGURE 7-33
"Stereo amplifier" output transformer circuit for Example 7-18.

Example 7-19

Design an appropriate interface and select R_L so that $v_L(t) = 15 \cos \omega t$ in the circuit of Figure 7-34a.

A source transformation (Figure 7-34b) shows that a resistive circuit cannot work, since more voltage is needed than the source can provide. We could use an OP AMP, but then we must provide power to it. A better solution, especially if size and weight are not problem, is to use a transformer. (Actually at times the size and weight of a transformer are comparable to those of an OP AMP and associated power supply.)

Consider the transformer interface of Figure 7-34c. We can simplify our choice of R_L if we first find the Thévenin equivalent that R_L "sees."

To find the Thévenin circuit at the R_L–transformer interface we remove R_L and calculate the open-circuit

voltage $v_{OC}(t)$. If we remove R_L, we force $i_2(t)$ to be zero due to the open circuit. If $i_2(t)$ is zero, so must be $i_1(t)$ (Eq. 7-55). If $i_1(t)$ is zero, there is no voltage drop across the 100-Ω resistor. Hence $v_1(t)$ must equal the source voltage $v_s(t)$. $v_{OC}(t)$ then is simply $v_s(t)/n$.

To find R_T we must find $i_{SC}(t)$. If we short-circuit the output, we force $v_2(t)$ to be zero. If $v_2(t)$ is zero, so is $v_1(t)$ (Eq. 7-54). Therefore $i_1(t) = nv_s(t)/100$, so that R_L can be found as

$$R_T = \frac{v_{OC}(t)}{i_{SC}(t)} = \frac{v_s(t)/n}{nv_s(t)/100} = \frac{100}{n^2}$$

Our Thévenin equivalent circuit is as shown in Figure 7-34d. The resulting problem is typical of many engineering problems. There is more than one solution. A good engineer now researches what transformers are available—their size, cost, shape—and the range of R_L's that can be chosen to select which transformer and load are best for the application.

One thing is clear. The engineer must choose a transformer with $n < 1$, otherwise the desired voltage still will not be available. Suppose a transformer is found that meets all the constraints with a turns ratio of 0.1. A proper R_L to achieve the desired $v_L(t)$ could then be found by using a voltage divider (Figure 7-

34e). That is,

$$15 \cos \omega t = \frac{R_L}{R_L + 10{,}000} 100 \cos \omega t$$

so that

$$R_L = 1765 \ \Omega$$

Of course, many other solutions are possible.

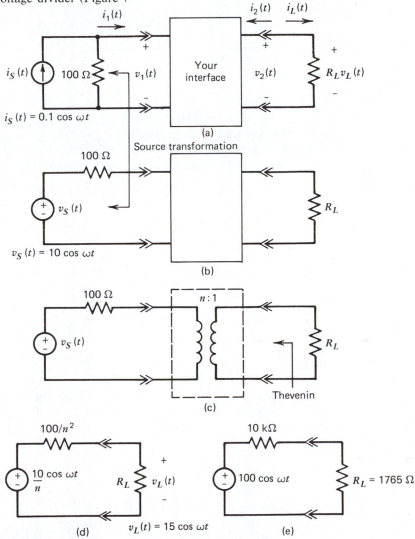

FIGURE 7-34
Circuit and solution technique for Example 7-19. (*a*) Circuit. (*b*) Source-transformed input. (*c*) Transformer interface. (*d*) Thévenized input. (*e*) One solution.

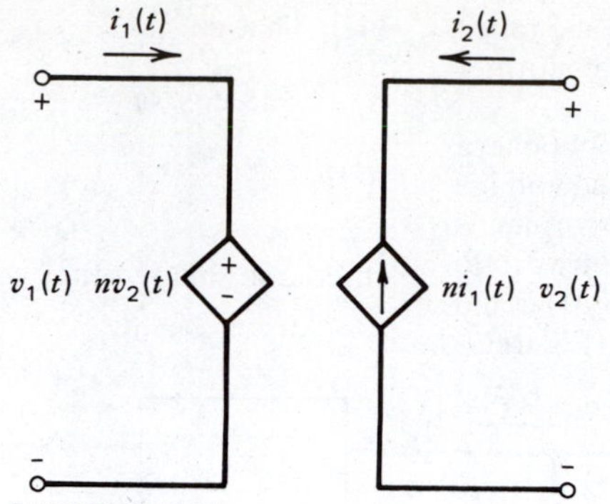

FIGURE 7-35
Model of an ideal transformer.

As with most electric devices we can develop a dependent source model for the ideal transformer that may help us visualize how it operates. A suitable model is shown in Figure 7-35. It is left up to the reader to verify that it indeed represents the ideal transformer we have been studying.

SUMMARY

- The behavior of two-terminal devices can be modeled using the element i-v relationships given in Table 7-3.

Element	Units	Circuit Symbol	i-v Relationship
Resistance (R)	Ohms, Ω		$v = R_i$ $i = Gv$
Capacitance (C)	Farads, F		$v = \dfrac{1}{C} \displaystyle\int_0^t i\,dt + v(0)$ $i = C\dfrac{dv}{dt}$
Inductance (L)	Henrys, H		$v = L\dfrac{di}{dt}$ $i = \dfrac{1}{L}\displaystyle\int_0^t v\,dt + i(0)$

TABLE 7-3

- Capacitance is the proportionality constant relating charge and voltage ($q = Cv_C$). It is a measure of the ability of a device to store energy in an electric field ($w_C = \frac{1}{2} Cv_C^2$).

- Inductance is the proportionality constant relating flux linkages and current ($\lambda = Li_L$). It is a measure of the ability of a device to store energy in a magnetic field ($w_L = \frac{1}{2} Li_L^2$).

- The energy stored in a memory element must be a continuous function of time. As a result the voltage across a capacitance and the current through an inductance cannot change instantaneously.

- Table 7-4 summarizes the rules for combining two-terminal elements connected in series or parallel to obtain an equivalent element.

Element	Series	Parallel
Resistance	$R_{EQ} = R_1 + R_2 + \cdots + R_N$	$\dfrac{1}{R_{EQ}} = \dfrac{1}{R_1} + \dfrac{1}{R_2} + \cdots + \dfrac{1}{R_N}$
Capacitance	$\dfrac{1}{C_{EQ}} = \dfrac{1}{C_1} + \dfrac{1}{C_2} + \cdots + \dfrac{1}{C_N}$	$C_{EQ} = C_1 + C_2 + \cdots + C_N$
Inductance	$L_{EQ} = L_1 + L_2 + \cdots + L_N$	$\dfrac{1}{L_{EQ}} = \dfrac{1}{L_1} + \dfrac{1}{L_2} + \cdots + \dfrac{1}{L_N}$

TABLE 7-4

- Two coupled coils can be described in terms of i-v relationships involving both self- and mutual inductance.

$$v_1 = L_1 \frac{di_1}{dt} \pm M \frac{di_2}{dt}$$

$$v_2 = \pm M \frac{di_1}{dt} + L_2 \frac{di_2}{dt}$$

The appropriate sign for the mutual inductance term is determined by the dot convention shown in Figure 7-36.

- The ideal transformer is a model of two coupled coils that assumes that all the flux linkages produced by each coil couple to the other coil and that the coils are lossless. The *i-v* relationships of an ideal transformer are

$$v_1 = nv_2$$
$$i_1 = -i_2/n$$

where $n = N_1/N_2$ is called the turns ratio.

- The transformer is often used to interface source–load circuits. The resistance seen looking in at the input of an ideal transformer is

$$R_{IN} = n^2 R_L$$

where R_L is the resistance connected across the output terminals.

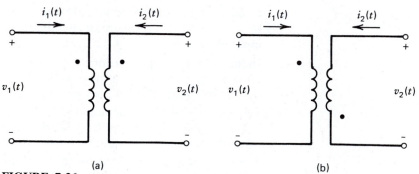

(a)   (b)

FIGURE 7-36
Dot convention for coupled coils. (*a*) Positive sign. (*b*) Negative sign.

EN ROUTE OBJECTIVES
AND RELATED EXERCISES

7-1 MEMORY ELEMENTS (SECS. 7-1 to 7-3, 7-5)

Given the waveform of the current through (voltage across) a capacitor or inductor:
(a) Determine the voltage across (current through) the element.
(b) Determine the energy and power delivered to the element.

Exercises

7-1-1 A 10-μF capacitor is excited at different times by the voltage given below. For each voltage what is the resulting current?
(a) $v_C(t) = t$ V
(b) $v_C(t) = 5 \sin 10t$ V
(c) $v_C(t) = 2 e^{-2t}$ V
(d) $v_C(t) = 10 e^{-t} \cos 100t$ V
(e) $v_C(t) = 5 r(t) - 5 r(t - 2)$ V
(f) $v_C(t) = 5 \cos (100t + 45°)$ V

7-1-2 A 2-pF capacitor has an initial voltage of 2 V across it. At $t = 0$ the current $i_C(t) = 10 e^{-t}$ A is suddenly applied. Draw a sketch of the voltage across the capacitor. Repeat assuming the current is $i_C(t) = 0.5 \cos 10^5 t$ A. Repeat assuming the current is $i_C(t) = 100 \, u(t)$ A.

7-1-3 A 2-mH inductor has, at different times, the currents given below flowing through it. For each current what is the resulting voltage?
(a) $i_L(t) = 2 r(t)$ A
(b) $i_L(t) = 0.5 \cos 10t$ A
(c) $i_L(t) = 0.001 e^{-0.5t}$ A
(d) $i_L(t) = 10 e^{-0.5t} \cos 100t$ mA
(e) $i_L(t) = 500 \cos (100t - 45°) \mu$A
(f) $i_L(t) = 30$ pA

7-1-4 For the circuit of Figure E7-1-4 find the currents $i_1(t)$ and $i_2(t)$.

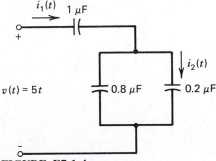

FIGURE E7-1-4
Capacitor circuit for Exercise 7-1-4.

7-1-5   For the circuit of Figure E7-1-4 taken as a whole, find an expression for the input power (is it being absorbed or produced by the circuit?) and the input energy.

7-1-6   For the text example problem 7-4, prove that the maximum power absorbed by the capacitor is given by $I_o^2 T_C / 4C$ and occurs at $t = T_C \ln (2)$ s.

7-1-7   In instrumentation it often is necessary to convert an analog signal to a digital signal. A part of the A/D conversion circuitry consists of the sample-hold circuit. A simplified sample-hold circuit is shown in Figure E7-1-7. Suppose the switch closes every second and remains closed for 1 μs (sufficient time for the capacitor to become fully charged). Draw a sketch of the output for the input given. What purpose does the OP AMP serve?

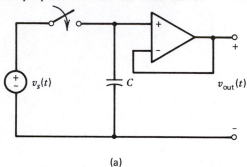

(a)

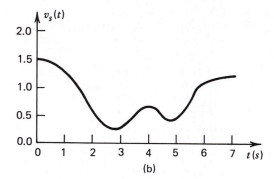

(b)

FIGURE E7-1-7
Sample-hold circuit. (*a*) Circuit. (*b*) Input voltage.

7-1-8 Simplify the circuits of Figure E7-1-8.

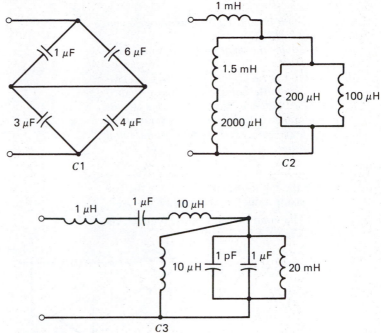

FIGURE E7-1-8
Equivalent inductance and capactiance problems.

7-1-9 Show than an inductor is a linear element.

7-1-10 An unmarked box is known to contain one memory element. A signal $v_{in} = 10 \cos 10t$ is input into the box. A current with a peak amplitude of 100 μA is measured exiting the box. What is in the box and what is its value?

7-1-11 A 10-mH inductor has a current equal to $10[1 - e^{-5t}]$ A flowing through it. Draw a sketch of the voltage across the inductor, the power absorbed or delivered, and the energy stored.

7-1-12 You need a 2.35-mH inductor. Referring to Appendix C for standard values, what would you use if cost was no object? Suppose cost was important, and the cost of ± 5 percent inductors was two times that of ± 10 percent inductors and five times that of ± 20 percent inductors. Would you change your order? Explain.

7-1-13 A 12-V automobile battery is connected to a 10-μF capacitor. How much power will be delivered after 10 minutes? How much energy will be stored?

7-2 FUNCTIONAL OP AMP MEMORY CIRCUITS (SEC. 7-4)

Given a circuit composed of a memory element and an OP AMP:

(a) *Determine the input-output relation for the circuit.*

(b) *Using the functional relation of integration and differentiation and those studied earlier (adders, subtractors and scalar multipliers), devise an analog solution for a first- or second-order differential equation.*

Exercises

7-2-1 Derive the input-output relation for the bottom resistor-inductor circuit of Figure 7-17.

7-2-2 Derive the input-output relation for each circuit of Figure E7-2-2.

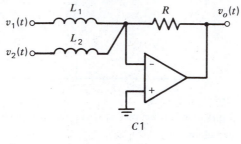

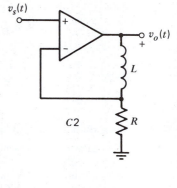

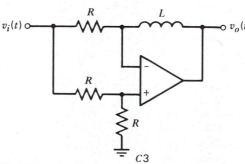

FIGURE E7-2-2
OP AMP—memory element circuits.

7-2-3 Using the functional building blocks of Figure 7-17 and those in Figure 4-27, devise solutions to the following differential equations.

(a) $\quad x = \dfrac{d^2y}{dx^2} + 5\dfrac{dy}{dx} - 6y$

(b) $\quad 0 = \dfrac{d^2y}{dx^2} + 25y$

(c) $e^{-2t} = \dfrac{dv}{dt} - 5v$

(d) $5 \sin 10t = \dfrac{d^2i}{dt^2} + 6\dfrac{di}{dt} + 5i$

7-2-4 Write a differential equation for each of the circuits of Figure E7-2-4.

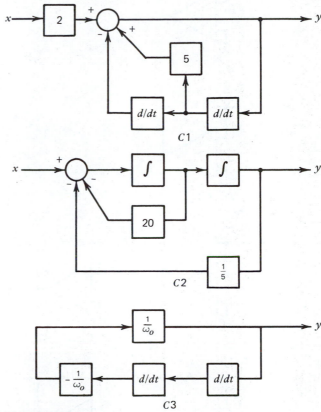

FIGURE E7-2-4
Block diagrams for Exercise 7-2-4.

7-2-5 Show that the following have the dimension of time:
(a) L/R
(b) RC

7-2-6 The signals shown in Figure E7-2-6 are input to each of the circuits shown. Sketch the various outputs.

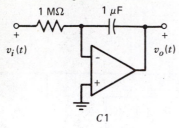

C1

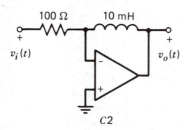

C2

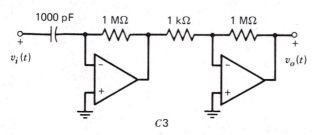

C3

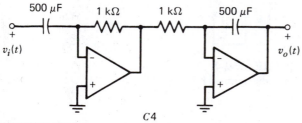

C4

FIGURE E7-2-6
More OP AMP memory element circuits.

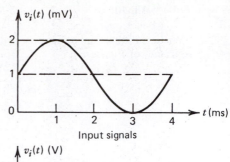

Input signals

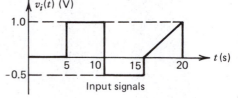

Input signals

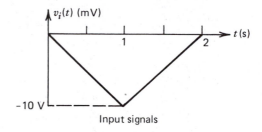

Input signals

7-3 MUTUAL INDUCTANCE (SEC. 7-6)

Given the current through or voltage across both branches of two coupled coils, determine the unspecified currents or voltages.

Exercises

7-3-1 For the coupled circuit of Figure E7-3-1, find $v_2(t)$ and $i_2(t)$.

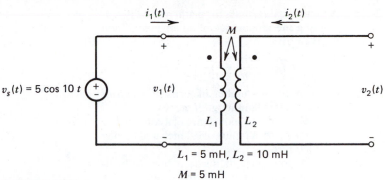

FIGURE E7-3-1
Coupled coil circuit for Exercise 7-3-1.

7-3-2 Repeat Exercise 7-3-1 as assuming a 100-Ω load is connected across the output.

7-3-3 For the coupled coil circuit of Figure E7-3-3, find $v_1(t)$, $i_1(t)$, $v_2(t)$, and $i_2(t)$.

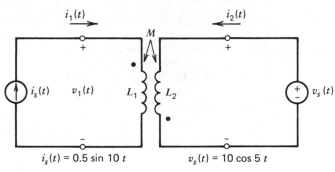

FIGURE E7-3-3
Coupled coil circuit for Exercise 7-3-3.

7-3-4 The three coils of Figure E7-3-4 are mutually coupled. Find an expression for $i_1(t)$, $v_1(t)$, $v_2(t)$, and $v_3(t)$. $L_1 = L_2 = 3$ mH; $L_3 = 5$ mH; $M_{\bullet\bullet} = 0.5$ mH; $M_{\blacktriangle\blacktriangle} = 0.2$ mH; $M_{\blacksquare\blacksquare} = 1$ mH.

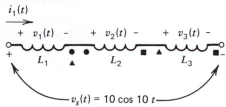

FIGURE E7-3-4
Three mutually couple coil circuit for Exercise 7-3-4.

7-3-5 An empirical formula for mathematically calculating the inductance of an air-wound coil is given by

$$L = \frac{N^2 r \times 10^{-6}}{9r + 101}$$

where N is the number of turns of a coil of radius r inches and length l inches.

(a) Compute the inductance of a coil 2 inches long with a radius of 0.2 inch and having 50 turns.

(b) A coil must fit inside a box with dimensions of ½ inch by ½ inch by 1.5 inches. Design a coil of 10 mH, 100 μH, and 1 H.

(c) What is the maximum thickness the wire can be for the 10-mH design of Exercise 7-3-5b, if the coils cannot overlap, but must lie side by side?

7-3-6 Find $v_x(t)$ in terms of $i_1(t)$ for the circuit of Figure E7-3-6.

7-3-7 The coil of Figure E7-3-1 is excited by $v_S(t) = 10\,e^{-10t}\,u(t)$ rather than $v_S(t) = 5\cos 10t$. Find $v_2(t)$. Repeat if $v_S(t) = [1 - e^{-5t}]u(t)$.

7-3-8 Find the equivalent inductance for each of the circuits of Figure E7-3-8.

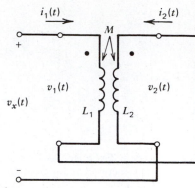

FIGURE E7-3-6
Series-driven coupled coil circuit for Exercise 7-3-6—compare with Example 7-16.

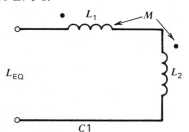
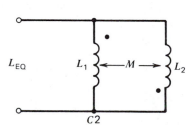

FIGURE E7-3-8
Equivalent inductance exercise.

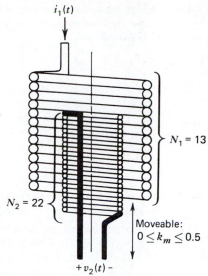

FIGURE E7-3-9
Coupled coils for Exercise 7-3-9.

7-3-9 A coupled pair of coils are designed so that they can slip one inside the other, causing k_m to vary (Figure E7-3-9). Suppose k_m can be changed from $\simeq 0$ to 0.5. Find the range of $v_2(t)$ if $i_1(t) = 10 \cos 10t$.

7-4 IDEAL TRANSFORMERS (SEC. 7-7)

Given a circuit containing an ideal transformer, determine the output current, voltage, and power, or use the ideal transformer to match the source to the load to meet prescribed constraints.

Exercises

7-4-1 Select the turns ratio of a step-up transformer to increase a 12-V signal ($v_1(t) = 12 \cos 377t$) to ordinary house voltage ($v_2(t) \doteq 156 \cos 377t$).

7-4-2 Two amplifier stages are to be coupled using a transformer. The output resistance of the first stage is 300 Ω while the input resistance to the second is 2200 Ω. Determine the turns ratio of an appropriate transformer so that maximum power is transferred.

7-4-3 A typical filament transformer is shown in Figure E7-4-3. Find $V_{2\text{RMS}}$ and $V_{3\text{RMS}}$ if $v_1(t) = 156 \cos 377t$.

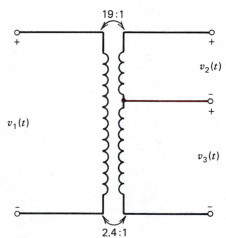

FIGURE E7-4-3
A filament transformer.

7-4-4   Design an interface using a transformer to deliver maximum power to the 1 kΩ resistor in Figure E7-4-4.

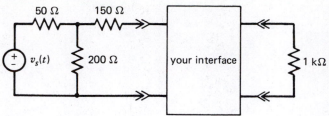

FIGURE E7-4-4
An interface problem.

7-4-5   Show that the dependent source model of an ideal transformer in Figure 7-35 is valid.

7-4-6   Two transformers are connected in cascade as shown in Figure E7-4-6. The first is ideal, the second is not. Find $v_{\text{out}}(t)$.

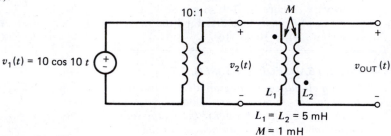

FIGURE E7-4-6
Cascade transformer pair.

7-4-7   A **push-pull** amplifier can be modeled as shown in Figure E7-4-7.

(a)   If $v_{in}(t) \pm 0.1 \cos 2\pi t$, sketch $v_{out}(t)$. The transformers are ideal and $\pm V_{CC} = \pm 15$ V.

(b)   Repeat if $\pm V_{CC} = +15$ V and 0 V, respectively.

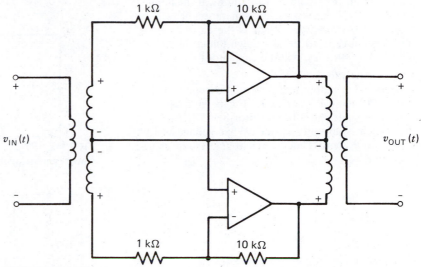

FIGURE E7-4-7
A push-pull amplifier problem.

PROBLEMS

P7-1 (Analysis)
A parallel-plate capacitor is constructed by separating two 6 cm by 6 cm square aluminum plates. The separation between the plates is adjustable and is filled with air ($\epsilon = 8.86 \times 10^{-12}$ F/m). If the applied voltage is 50 V, what must the separation be to store 5×10^{-7} C of charge? What is the capacitance of this capacitor? If air breaks down under an electric field of 3×10^6 V/m, will the capacitor break down?

P7-2 (Analysis)
Two ideal capacitors and a switch are connected as shown in Figure P7-2. The 0.5-F capacitor has 10 V across it, while the 3-F capacitor has 5V. At $t = 0$ the switch is abruptly closed. What voltage will ultimately exist across each capacitor? (Assume charge is conserved.)

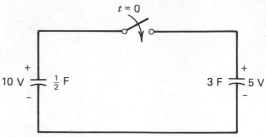

FIGURE P7-2
Conservation problem.

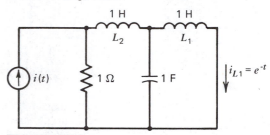

FIGURE P7-3
Circuit for Problem 7-3.

P7-3 (Analysis)
Consider the circuit of Figure P7-3, and answer the following questions.

What must $i(t)$ be to produce $i_{L1} = e^{-t}$? How much energy is stored in the capacitor at $t = 1$ s? How much is stored in the inductor L_1 at $t = 1$ s?

P7-4 (Analysis)
A certain inductor has an inductance that is time dependent, $L(t)$. Suppose $L(t) = e^{-t}$ and $i(t) = e^{-2t}$, what is the voltage $v_L(t)$ across the inductor?

P7-5 (Evaluation)
An OP AMP integrator has been designed for use in a particular application. The form of the input signal is $v_1(t) = 10 \cos \omega t$ where ω can vary from 100 rad/s to 100 krad/s. Part of your analysis involves testing the integrator using an oscilloscope whose most sensitive scale is 0.005 V/cm. Can you effectively test the integrator circuit over the entire frequency range?

P7-6 (Analysis)
Show that the circuit of Figure P7-6 is equivalent to a nonideal transformer. This representation of a transformer is called a π **equivalent.**

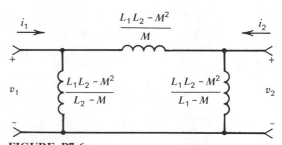

FIGURE P7-6
π-**equivalent model of nonideal transformer.**

Chapter 8
Memory Circuits

The mere formulation of a problem is far more often essential than its solution, which may be merely a matter of mathematical or experimental skill. To raise new questions, new possibilities, to regard old problems from a new angle requires creative imagination and marks real advances in science.

Albert Einstein

The formulation of the differential equations for simple *RC*, *RL*, and *RLC* circuits yields considerable insight into their behavior and use. The ability of these circuits to store energy gives rise to exponential and oscillatory responses that find many applications from timing circuits to filters. In this chapter we treat memory circuits to establish a connection between their mathematical representation, a differential equation, and the physical behavior of the circuit. We begin our study with the relatively simple single-memory circuit, including its response to both step functions and sinusoids. The more complicated double-memory circuit introduces the need for complex numbers to represent the range of possible responses. The step response of memory circuits is so important that it is worth undertaking using the classical techniques introduced in this chapter. Once the behavior of these circuits is understood using classical methods, our understanding can be broadened through use of the Laplace transformation studied in subsequent chapters.

8-1   SINGLE-MEMORY CIRCUITS

The equilibrium in any circuit is always the result of constraints of two types: (1) connection constraints (KVL and KCL) and (2) device constraints (element equations). The major difference with memory circuits is that the equilibrium is dynamic and the formulation process leads to differential equations rather than the algebraic equations that characterize memoryless circuits. Nonetheless the formulation methods studied in the context of memoryless circuits in the first five chapters are quite general, and can easily be extended to the study of circuits with memory. In this chapter we examine the methods of formulating the differential equations describing memory circuits, and some of the classical techniques used to obtain solutions.

The simplest memory circuits are those containing a single capacitor or a single inductor as illustrated in Figure 8-1. The resistors and sources can always be replaced by a Thévenin or Norton equivalent, which leads to the simple series and parallel circuits shown. To formulate the differential equation of the resistor-capacitor (RC) circuit, Figure 8-1a, we note that the Thévenin equivalent source circuit is governed by the constraint

$$v + R_T i = v_T \qquad \text{(8-1)}$$

while the capacitor is characterized by the $i\text{-}v$ constraint

$$i = C \frac{dv}{dt} \qquad \text{(8-2)}$$

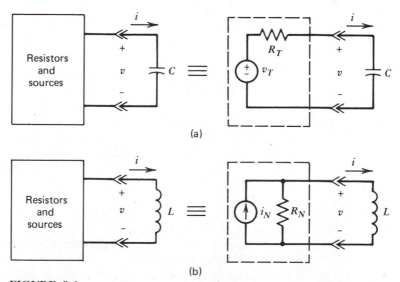

(a)

(b)

FIGURE 8-1
General single-memory circuits. (*a*) Resistor-capacitor circuits. (*b*) Resistor-inductor circuits.

Substituting the element constraint into the source constraint yields

$$R_T C \frac{dv}{dt} + v = v_T \qquad (8\text{-}3)$$

Mathematically this result is called a *first-order, linear differential equation*. The dependent variable (v) is the voltage across the capacitor. In Chapter 7 we showed that the energy stored on a capacitor is proportional to the square of the voltage across it. Thus the differential equation involves the energy, memory, or state variable associated with the capacitor.

With reference to the resistor-inductor (*RL*) circuit in Figure 8-1*b*, we observe that the Norton equivalent source is governed by the constraint

$$i + G_N v = i_N \qquad (8\text{-}4)$$

and the inductor by the constraint

$$v = L \frac{di}{dt} \qquad (8\text{-}5)$$

Combining these constraints yields

$$G_N L \frac{di}{dt} + i = i_N \qquad (8\text{-}6)$$

which is likewise a *linear, first-order differential equation* in the energy, memory, or state variable associated with the inductor.

Moreover we observe that Eqs. 8-3 and 8-6 have the same form. In fact, if we make the following replacements,

$$R_T \leftrightarrow G_N \qquad C \leftrightarrow L \qquad v \leftrightarrow i \qquad v_T \leftrightarrow i_N$$

we can change one equation into the other. This is yet another example of duality. What it means in the present context is that we do not need to study both equations in detail. To have studied one is to have studied the other. Therefore, in what follows, we concentrate on the *RC* circuit.

In dealing with *RC* circuits we will find that the *response* (solution of the differential equation) depends on three kinds of things:

1. The form of the input signal $v_T(t)$.
2. The values of R_T and C.
3. The value of $v(t)$ at the time we initiate our solution ($t = 0$) (i.e., the initial condition).

This last dependence is something new. The response of memoryless circuits depends only on the input signals and the circuit parameters. But here we have those dependencies and a third one, which is related to what has happened in the past. In sum, the circuit has memory. Put differently, the circuit can have a nonzero response even if there is no input signal.

To highlight this difference we first find what is called the **zero-input** response of the circuit. Setting $v_T = 0$ in Eq 8-3 produces

$$R_TC\frac{dv}{dt} + v = 0 \tag{8-7}$$

Mathematically this is called a **homogeneous equation** (the right side is zero). In any case, the equation requires that a linear combination of the voltage and its derivative be zero. This observation suggests that we try a solution in the form of an exponential,

$$v(t) = Ke^{st} \tag{8-8}$$

where k and s are constants to be determined. Substituting this trial solution into Eq. 8-7 yields

$$R_TCKse^{st} + Ke^{st} = 0$$

or

$$Ke^{st}(R_TCs + 1) = 0$$

The condition $K = 0$ satisfies this requirement, but this is a trivial solution. The function e^{st} cannot be zero for all t; hence the general way to meet the requirement is to have

$$R_TCs + 1 = 0 \tag{8-9}$$

The result is called the circuit's **characteristic equation** and has the root $s = -1/R_TC$. Thus the response is of the form

$$v(t) = Ke^{-t/R_TC}$$

The remaining constant K can be evaluated by knowing the value of $v(t)$ at one particular time. In circuit analysis this time is usually taken to be $t = 0$, and the value of $v(0)$ is denoted as V_o. Thus

$$v(0) = Ke^{-0} = V_o$$

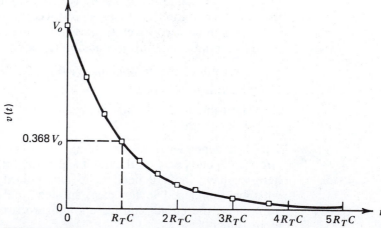

FIGURE 8-2
Zero-input response of the general *RC* circuit.

Therefore $K = V_o$ and we obtain the zero-input response as

$$v(t) = V_o \, e^{-t/R_T C} \qquad \textbf{(8-10)}$$

The zero-input response of the RC circuit is the familiar exponential waveform studied in Chapter 6 and shown in Figure 8-2. The time constant of the waveform is $T_C = R_T C$ and the initial amplitude is V_o. From our previous study we know that the waveform will decay to about 37 percent of its initial amplitude in one time constant, and to essentially zero in about five time constants. In sum, the zero-input response of the capacitor circuit is determined by two quantities: (1) the circuit time constant and (2) the initial value of the voltage across the capacitor.

Example 8-1

The switch in Figure 8-3 is closed at $t = 0$ connecting a capacitor with an initial voltage of 30 V to the resistance shown. Find the circuit response [$v_C(t)$, $i(t)$, $i_1(t)$, and $i_2(t)$].

To do this we must first determine the time constant. The equivalent resistance seen by the capacitor is

$$R_T = 10 + \frac{20 \times 20}{20 + 20} = 10 + 10 = 20 \text{ k}\Omega$$

Hence the circuit time constant is

$$T_C = R_T C = 20 \times 10^3 \times 5 \times 10^{-7}$$
$$= 10^{-2} = 10 \text{ ms}$$

Since the initial capacitor voltage is $V_o = 30$ V, we have

$$v_C(t) = 30e^{-100t} \qquad t > 0$$

This gives us the capacitor voltage from which we can determine any other variable in the circuit. For example, the current i is

$$i(t) = C\frac{dv_C}{dt} = 5 \times 10^{-7} \, (-3000e^{-100t})$$
$$= -1.5 \times 10^{-3} \, e^{-100t}$$

Then by current division we have the other currents as

$$i_1(t) = \frac{20}{40} i(t) = -0.75 \times 10^{-3} e^{-100t}$$
$$i_2(t) = \frac{20}{40} i(t) = -0.75 \times 10^{-3} e^{-100t}$$

Notice the analysis pattern. We first determine the circuit's state variable, that is, the voltage across the capacitor. Armed with that response we can then find every other voltage or current in the circuit by applying the memory element's i-v relation and simple analysis techniques.

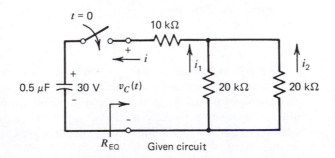

Given circuit

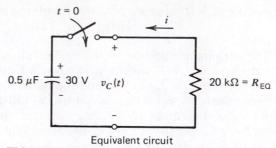

FIGURE 8-3
Circuit and its Thévenin equivalent for Example 8-1.

Although the RL circuit is the dual of the circuit just considered it is instructive to outline briefly the derivation of its zero-input response. With $i_N(t) = 0$ (zero input), Eq. 8-6 reduces to

$$G_N L \frac{di}{dt} + i = 0 \tag{8-11}$$

Again a trial solution in the form of an exponential

$$i(t) = K e^{st}$$

is called for. Substituting this trial response into Eq. 8-11 yields the circuit characteristic equation as

$$G_N L s + 1 = 0 \tag{8-12}$$

with a root $s = -1/G_N L$. Denoting the initial value of the inductor current as

$$i(0) = I_o = K e^{-0}$$

we obtain the zero-input response as

$$i(t) = I_o \, e^{-t/G_N L} \tag{8-13}$$

The RL circuit zero-state response is, as expected, the familiar exponential waveform with a time constant of $T_C = G_N L = L/R_T$, and an initial amplitude of I_o. In sum, the duality is complete. To find the zero-state response of an RL circuit we need two quantities: (1) the circuit time constant and (2) the initial value of the state variable (inductor current in this case).

Example 8-2

Figure 8-4 shows an RL circuit with an initial inductor current I_o. Find the zero-state response.

To find the circuit time constant we observe that the inductors are connected in series and can be replaced by a single equivalent inductor

$$L_{EQ} = L_1 + L_2$$

Likewise, the resistors are connected in parallel, and

hence

$$G_N = G_1 + G_2$$

The circuit time constant is then

$$T_C = G_N L_{EQ} = (G_1 + G_2)(L_1 + L_2)$$

and the zero-state response is

$$i(t) = I_o\, e^{-t/T_C}$$

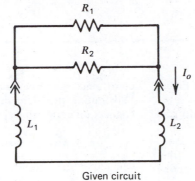

Given circuit

Equivalent circuit

FIGURE 8-4
Circuit and its Thévenin equivalent for Example 8-2.

The previous example illustrates an important point. The circuit studied is a single-memory circuit since the two inductors can be replaced by a single equivalent inductor. In general then, a single-memory circuit is one that contains the equivalent of a single inductor or a single capacitor. The general form of the differential equation describing a single-memory circuit can be written as

$$T_C \frac{dx}{dt} + x = f(t) \tag{8-14}$$

where $x(t)$ is a current or a voltage, T_C is the circuit time constant, and $f(t)$ is the equivalent source term.

In some cases it may not be convenient to formulate this differential equation in terms of the state variable (capacitor voltage or inductor current). That is, it may be inconvenient to determine the Thévenin or Norton equivalent seen by the energy storage element. In such cases we can always appeal to the basic concepts of device and connection constraints to derive the differential equation in terms of any signal variable that is convenient. The next example illustrates this principle using an OP AMP circuit.

Example 8-3

Find the zero-input response of the circuit of Figure 8-5.

The OP AMP circuit contains a single capacitor and it is somewhat difficult to determine the Thévenin equivalent seen by this element. From our previous experience with memoryless circuits we know that only one node voltage equation is needed to describe the circuit. We begin by writing a KCL equation at the inverting input node,

$$i_1 + i_2 + i_C = i_N$$

and the element equations as

$$i_1 = G_1 (v_S - v_N)$$
$$i_2 = G_2 (v_o - v_N)$$
$$i_C = C \frac{d(v_o - v_N)}{dt}$$
$$i_N = 0$$
$$v_N = v_P = 0$$

The last two element equations represent the OP AMP

constraints with the noninverting input grounded. Substituting the element constraints into the KCL connection constraint yields

$$G_1 v_S + G_2 v_o + C \frac{dv_o}{dt} = 0$$

which can be written as

$$R_2 C \frac{dv_o}{dt} + v_o = - \frac{R_2}{R_1} v_S$$

This is in the form given in Eq. 8-14. By inspection the circuit time constant is $T_C = R_2 C$, and the zero-input response is

$$v_o(t) = V_o \, e^{-t/R_2 C}$$

where V_o is the initial value of the output voltage $v_o(t)$. In this case the dependent variable is the OP AMP output, which is usually the signal quantity of interest rather than the capacitor voltage.

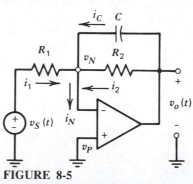

FIGURE 8-5
OP AMP circuit for Example 8-3.

8-2   SINGLE-MEMORY STEP RESPONSE

The concept of step response is one of the fundamental notions in circuit and system theory. By step response we mean the response of a circuit or system to a step-function input. The step function is one of the two premier test inputs used to study the dynamics of linear systems. The other standard input is the sinusoid, which we study later in this chapter, and more extensively in later chapters.

In developing the step response of a single-memory circuit we first treat the general RC circuit in detail and then summarize the corresponding results for its dual, the RL circuit. If the input to the general RC circuit in Figure 8-1 is a step function $v_T(t) = A\, u(t)$, then the circuit differential equation (Eq. 8-3) becomes

$$R_TC\,\frac{dv}{dt} + v = A\, u(t) \tag{8-15}$$

For $t > 0$ the step function $u(t) = 1$, and we can write the differential equation as

$$R_TC\,\frac{dv}{dt} + v = A \qquad t > 0 \tag{8-16}$$

There are a number of ways to solve this equation, including separation of variables and integrating factors. However, because the circuit is linear we chose an approach that divides the solution into two parts as

$$v(t) = v_N(t) + v_F(t) \tag{8-17}$$

In this expression $v_N(t)$ is called the **natural response** and is the general solution of the homogeneous equation (input set to zero). The component $v_F(t)$ is called the **forced response** and is a particular solution of Eq. 8-16.

Dealing first with the natural response, we are required to find the general solution of Eq. 8-16 with the input set to zero. That is,

$$R_TC\,\frac{dv_N}{dt} + v_N = 0$$

But this is the same equation used to obtain the zero-input response (Eq. 8-7), so we know that the natural response takes the form

$$v_N(t) = Ke^{-t/R_TC} \tag{8-18}$$

This is a general solution of the homogeneous equation since it contains an arbitrary constant K. But we cannot evaluate K at this point from the initial conditions as we did for the zero-input response. The initial condition applies to the *total* response (natural plus forced) and we still have to determine the forced response.

Turning now to the forced response, we seek a particular solution of the equation

$$R_TC\,\frac{dv_F}{dt} + v_F = A \qquad t > 0 \tag{8-19}$$

It seems clear by inspection that a particular solution is $v_F = A$, since

$$\frac{dv_F}{dt} = \frac{dA}{dt} = 0$$

so that

$$R_T C(0) + A = A$$

and Eq. 8-19 reduces to the identity $A = A$.

Combining the forced and natural responses as prescribed in Eq. 8-17, we obtain

$$v(t) = v_N(t) + v_F(t)$$
$$= K e^{-t/R_T C} + A$$

This result is a general solution for the step response since it satisfies Eq. 8-16 and contains an arbitrary constant K. This constant now can be evaluated if we make use of the initial condition as

$$v(0) = V_o = Ke^{-0} + A = K + A$$

which means that $K = (V_o - A)$. If we substitute this into our total solution, the step response can be expressed as

$$v(t) = (V_o - A) e^{-t/R_T C} + A \qquad \text{(8-20)}$$

A general plot of this response waveform is shown in Figure 8-6.

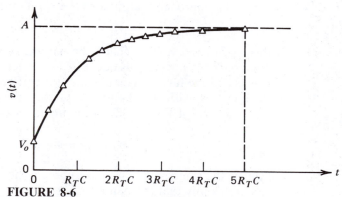

FIGURE 8-6
Step response of the general *RC* circuit.

It is interesting to contrast the difference between the zero-input response and the step response. If left to its own devices (zero input), the capacitor voltage simply exponentially decays to zero with a time constant determined by the circuit. When the circuit is driven by a step function, the response departs from the initial condition and asymptotically approaches a constant value determined by the amplitude of the step input. That is, if we evaluate the response $v(t)$ at $t = 0$, the time we start our event, and $t = \infty$, a time long after our circuit has quieted down, we obtain

$$v(0) = (V_o - A)e^{-0} + A = V_o - A + A = V_o$$

and

$$v(\infty) = (V_o - A)e^{-\infty} + A = (V_o - A) \times 0 + A = A$$

The path between these two values is an exponential rise waveform whose time constant is none other than the circuit's time constant. From our study of exponential signals in Chapter 6, we know that the step response will essentially reach (within less than 1 percent) its final value in a time period of five time constants ($5\ T_C$). In summary then, the step response of an RC circuit depends on three things:

1. The circuit time constant.
2. The initial ($t = 0$) capacitor voltage.
3. The amplitude of the step function applied at $t = 0$.

Example 8-4

Find the step response of the circuit in Figure 8-7.
   The circuit is a *single*-memory RC circuit since the two capacitors can be replaced by an equivalent capacitor

$$C_{EQ} = \frac{C_1 C_2}{C_1 + C_2} = \frac{(0.1)(0.5)}{0.6} = 0.0833 \ \mu\text{F}$$

Since the two capacitors are connected in series, the initial condition on this equivalent capacitor is by KVL

$$V_o = V_{o1} + V_{o2} = 15 \text{ V}$$

The Thévenin equivalent seen by the equivalent capacitor can be found by first determining the open-circuit voltage using voltage division:

$$v_{OC} = v_T = \frac{R_2 A u(t)}{R_1 + R_2} = \frac{10}{40} 100u(t) = 25u(t)$$

The Thévenin resistance looking back into the source

circuit with the voltage driver replaced by a short circuit consists of R_1 and R_2 in parallel. Hence

$$R_T = \frac{R_1 R_2}{R_1 + R_2} = \frac{(30)(10)}{40} = 7.5 \text{ k}\Omega$$

The circuit time constant is then found as

$$T_C = R_T C_{EQ} = 7.5 \times 10^3 \times 8.33 \times 10^{-8}$$
$$= \frac{1}{1600} = 0.625 \text{ ms}$$

The step response of the voltage $v(t)$ is of the form

$$v(t) = (V_o - A) e^{-t/T_C} + A$$
$$= (15 - 25) e^{-1600t} + 25$$
$$= 25 - 10 e^{-1600t} - \qquad , t > 0$$

The initial ($t = 0$) value of the waveform is $25 - 10 = 15$ V as required. The final value is 25 V, which is obtained after about $5\ T_C = 3.125$ ms.

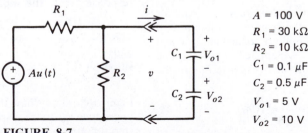

FIGURE 8-7
RC **Circuit demonstrating step response.**

$A = 100$ V
$R_1 = 30$ kΩ
$R_2 = 10$ kΩ
$C_1 = 0.1 \ \mu$F
$C_2 = 0.5 \ \mu$F
$V_{o1} = 5$ V
$V_{o2} = 10$ V

A special case of the step response occurs when the initial condition is zero. In this event the step response in Eq. 8-20 reduces to

$$v(t) = A(1 - e^{-t/R_TC}) \qquad t > 0 \tag{8-21}$$

We observe that the response is proportional to the amplitude of the input step function, which reminds us that the circuit is linear and has the proportionality property discussed in Chapter 3. However, if the initial condition is not zero, then Eq. 8-20 (the general case) points out that the response is not simply proportional to the input step amplitude. This does not mean that the circuit with a nonzero initial condition is nonlinear. Rather it simply means that in the general case the response is due to two causes: (1) the input step function applied at $t = 0$ and (2) the initial condition reflecting what has happened prior to $t = 0$. Once again we see that the circuit has memory: Its response for $t > 0$ depends *both* on what happened at $t = 0$ and on what happened prior to $t = 0$!

Among the consequences of memory is that the response of a circuit due to two or more inputs applied at $t = 0$ cannot be determined by superposition unless the initial conditions are zero. Again this does not mean that a circuit with nonzero initial conditions is nonlinear. It means that the circuit has memory and its response is influenced by all inputs, including those that occurred prior to $t = 0$, that lead to the initial conditions. However, superposition can be used when the initial conditions are zero, as illustrated in the next example.

Example 8-5

Find the response of the circuit of Figure 8-8a.

The circuit shown is a simple RC network with zero initial capacitor voltage. The applied input is a rectangular pulse of amplitude A and duration T.

$$v_{IN}(t) = Au(t) - Au(t - T)$$

Since the initial condition is zero, we will account for all inputs if we think of the response as due to two inputs: (1) a step of amplitude A applied at $t = 0$ and (2) a step of amplitude $-A$ applied at $t = T$. The differential equation giving the response due to the first input is

$$RC \frac{dv_1}{dt} + v_1 = Au(t)$$

whose solution for the zero initial condition is

$$v_1(t) = A(1 - e^{-t/RC})u(t)$$

The differential equation for the second pulse input is

$$RC \frac{dv_2}{dt} + v_2 = - Au(t - T)$$

This has the same form except that the amplitude of the step input is negative and it occurs at $t = T$. The response due to this input has the form

$$v_2(t) = - A(1 - e^{-(t - T)/RC})u(t - T)$$

This response goes negative and is delayed by T seconds. The complete response is the sum (superposition) of the two responses.

$$v(t) = v_1(t) + v_2(t)$$

Figure 8-8c shows how these two responses combine to produce the pulse response of the circuit. The first step function initiates a transient that begins at zero and eventually would reach an amplitude of $+A$. However, at $t = T$, an equal but opposite transient occurs that eventually cancels the first transient after about $T + 5_{RC}$ seconds.

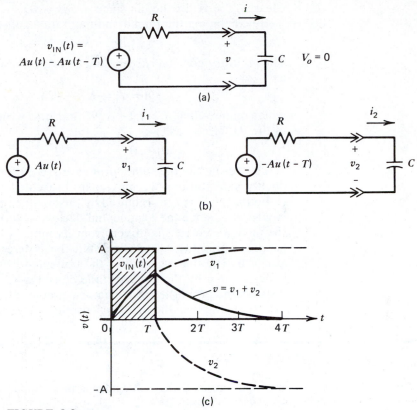

FIGURE 8-8
Circuits and pulse response for Example 8-5. (*a*) Given circuits. (*b*) Superposition applied. (*c*) Pulse response.

The step response of the general *RL* circuit follows the same pattern discussed since it is the dual of the *RC* circuit. If the Norton equivalent input to the *RL* circuit in Figure 8-1 is $Au(t)$, a step of current applied at $t = 0$, then the differential equation for the circuit (see Eq. 8-6) becomes

$$G_N L \frac{di}{dt} + i = A \qquad t > 0 \qquad \text{(8-22)}$$

The solution of this equation can be partitioned into a natural and a forced component. The natural response is the solution of the homogeneous equation (right side of Eq. 8-22 set to zero), and takes the same form as the zero input response.

$$i_N(t) = Ke^{-t/G_N L}$$

In this result K is an arbitrary constant that can be evaluated from the initial condition once the forced response is found. The forced response is a particular solution of the equation

$$G_N L \frac{di_F}{dt} + i_F = A$$

Clearly $i_F = A$ fits the bill since $dA/dt = 0$.

By combining the forced and natural responses we obtain

$$i(t) = Ke^{-t/G_N L} + A$$

The arbitrary constant K can now be evaluated from the initial condition as

$$i(0) = I_o = Ke^{-0} + A = K + A$$

which requires that $K = I_o - A$. We can now write down the step response of the RL circuit as

$$i(t) = (I_o - A)\, e^{-t/G_N L} + A \qquad t > 0 \tag{8-23}$$

This result has the same form as Eq. 8-20 for the *RC* circuit with the duality substitutions we previously presented. The initial value of the response is $i(0) = I_o$ as required by the boundry condition. The final value is $v(\infty) = A$ since the exponential decays to zero for large time. In sum, the inductor current is driven from its initial value determined by what happened prior to $t = 0$ to a final state that is determined by what happened at $t = 0$. The time needed for the transition from one state to the other is determined by the circuit's time constant $(G_N L)$. Thus the *RL* circuit displays the properties of a memory circuit, where the memory, energy, or state variable is the inductor current.

Example 8-6

The circuit in Figure 8-9a is a single-memory *RL* network driven by a step-function current input. Determine the step response.

To solve this problem we must first find the Norton equivalent of the source-resistor circuit. The short-circuit current at the interface is, by current division,

$$i_{SC}(t) = \frac{R_1}{R_1 + R_2} A\, u(t)$$

Looking back into the source-resistor circuit with the current source off (replaced by an open circuit), we see a resistance that is simply R_1 and R_2 in series. Hence

$$R_N = R_1 + R_2$$

The Norton equivalent just derived is then shown in Figure 8-9b. The time constant of the circuit is

$$T_C = G_N L = L/(R_1 + R_2)$$

The differential equation describing the step response is

$$\frac{L}{R_1 + R_2} \frac{di}{dt} + i = \frac{AR_1}{R_1 + R_2} \qquad t > 0$$

The forced and natural responses are then

$$i_F = AR_1/(R_1 + R_2)$$
$$i_N = Ke^{-(R_1 + R_2)t/L}$$

Combining these responses yields

$$i(t) = Ke^{-(R_1 + R_2)t/L} + \frac{AR_1}{(R_1 + R_2)}$$

The constant K can be evaluated from the initial condition as

$$i(0) = I_o = K + \frac{R_1 A}{(R_1 + R_2)}$$

which implies that

$$K = I_o - \frac{AR_1}{(R_1 + R_2)}$$

Finally we can write down the step response in all its glory as

$$i(t) = \left[I_o - \frac{AR_1}{R_1 + R_2} \right] e^{-(R_1 + R_2)t/L}$$
$$+ \frac{AR_1}{R_1 + R_2} \qquad t > 0$$

The general form of this response waveform is shown in Figure 8-9b. Notice that the final value of the inductor current is determined by the amplitude of the Norton equivalent step function and not the amplitude of the current step function in the original circuit shown in Figure 8-9a.

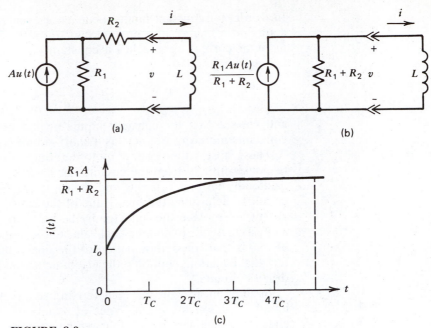

FIGURE 8-9
Circuits and pulse response for Example 8-6. (*a*) Given circuit. (*b*) Norton equivalent. (*c*) Step response.

8-3 INITIAL AND FINAL CONDITIONS

If we review the step responses of the last section carefully, we find that the capacitor voltage of an *RC* circuit can be written as

$$v_C(t) = (IC - FC)\, e^{-t/T_C} + FC \tag{8-24}$$

Similarly, the inductor current in an *RL* circuit has a step response, which is

$$i_L(t) = (IC - FC)\, e^{-t/T_C} + FC \tag{8-25}$$

In both of these expressions *IC* stands for the initial condition ($t = 0$) and *FC* for the final condition ($t = \infty$). Thus to determine the step response of any single-memory circuit, we need to find three quantities: *IC*, *FC*, and T_C. We know how to determine the time constant of a circuit. Clearly it would be useful if we had a simple way to determine *IC* and *FC*.

The key to accomplishing this goal is to observe that as time approaches infinity both responses approach a constant value (*FC*). Thus for large time $v_C = FC$ and hence

$$i_C = C\frac{dv_C}{dt} = 0$$

In words, for constant inputs a capacitor eventually (after five time constants or so) acts as an open circuit ($i_C = 0$). By similar reasoning the inductor current approaches a constant value (*FC*) for large time, and hence

$$v_L = L\frac{di_L}{dt} = 0$$

and we see that for constant inputs an inductor eventually (five time constants more or less) acts as a short circuit ($v_L = 0$).

These observations allow us to determine the final value of the step response directly from the circuit. For an *RC* circuit we replace the capacitor by an open circuit, and then analyze the resulting memoryless circuit to determine the final value of the capacitor voltage. For the *RL* circuit we replace the inductor by a short circuit and use memoryless analysis methods to find the final value of the inductor current. The beauty of this is that the determination of the final value does not involve differential equations, but only the algebraic methods used to analyze memoryless circuits.

In addition, it turns out that we can use this method to determine the initial condition in certain types of situations. One common case is a circuit containing a switch that remains in one state, say closed, for a period of time that is long compared with the circuit time constant. If the circuit input is constant, then the capacitor voltage (or inductor current) will approach a constant value, which is the final condition for the circuit with the switch closed. If the switch is now opened at a time that we choose to call $t = 0$, then a transient will ensue in which the capacitor voltage (or inductor current) is driven to a new and excitingly different final condition. But note, the final condition that existed for $t < 0$ is the initial condition at $t = 0$ when the transient begins. The reason for this is the continuity conditions discussed in Chapter 7.

$$v_C(0^+) = v_C(0^-) \quad \text{and} \quad i_L(0^+) = i_L(0^-)$$

Here $x(0^-)$ implies the value of the variable x, a whisker before the epoch-making event occurs. $x(0^+)$, on the other hand, is the value of the variable an instant after the event occurred. The event is said to occur in zero time (instantaneously).

The switching action thus cannot cause an instantaneous change in the value of the *capacitor voltage* or the *inductor current* at $t = 0$. What this means in the present context is that

$$FC(0^+) = IC(0^-)$$

In other words, the switch is an epoch-making-event maker. The final conditions of the memory variables from the previous epoch are their initial conditions for the subsequent epoch.

The usual way to state such analysis problems is to say that a switch has been (say) closed for a "long time." In this context a "long time" means something more than five time constants. Since electric circuits rarely have time constants that exceed a few hundred milliseconds, we do not need great patience to wait the requisite "long time" before opening the switch. The next four examples illustrate the use of these concepts in single-memory circuits with constant inputs and a switch that is activated at $t = 0$.

Example 8-7

The circuit in Figure 8-10a has a constant input and a switch. The switch has been closed long enough for the circuit to reach equilibrium (a "long time") and is then opened at $t = 0$. Determine the capacitor voltage $v(t)$ for $t > 0$.

To accomplish this we need three things: IC, FC, and T_C. The initial condition can be found from the initial value circuit in Figure 8-10b. In this circuit the capacitor has been replaced by an open circuit since the circuit has reached equilibrium with the switch closed. The initial condition for our problem is then found by voltage division:

$$v(0^-) = IC = \frac{R_2 V_o}{R_1 + R_2}$$

The final value is found from the circuit in Figure 8-10c. Since this circuit is undriven (the switch is open), the final value is clearly zero. In addition the final condition circuit gives us the appropriate time constant. In looking back at the interface we see a Thévenin resistance of R_2 since R_1 is connected to an open switch. Thus the time constant for $t > 0$ is R_2C. We can now write the capacitor voltage for $t > 0$ as

$$v(t) = (IC - FC)\, e^{-t/T_C} + FC$$
$$= \frac{R_2 V_o}{R_1 + R_2}\, e^{-t/R_2C} \qquad t > 0$$

This looks remarkably like a zero-input response, which indeed it is for $t > 0$. But now we see how to establish the initial condition for such a problem by simply opening a switch that has been closed for a "long time." To complete the picture we can determine the capacitor current by appealing to its element constraint as

$$i_C(t) = C\,\frac{dv}{dt} = -\frac{V_o}{R_1 + R_2}\, e^{-t/R_2C} \qquad t > 0$$

This is the capacitor current for $t > 0$. For $t < 0$ the initial condition circuit points out that $i_C(0^-) = 0$ since the capacitor is an open circuit. Both of these responses are plotted in Figure 8-11. Observe that the capacitor voltage is continuous at $t = 0$ as re-

quired by the continuity condition, but that the capacitor current has a jump at $t = 0$. This jump does not violate the continuity condition since the capacitor current is not the energy, memory, or state variable. In other words, nonstate variables can have discontinuities at $t = 0$. For this reason it is usually desirable first to solve for the state variable since it is constrained by continuity, and then to solve for other variables in the circuit as required using element or connection constraints.

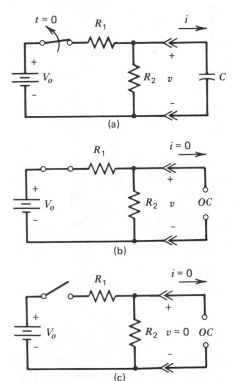

FIGURE 8-10
Circuits for demonstrating initial and final conditions of Example 8-7. (a) Given circuit. (b) Initial value circuit. (c) Final value circuit.

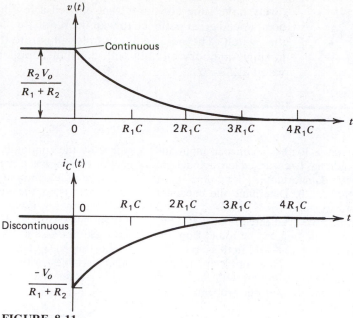

FIGURE 8-11
Step-response waveforms for Example 8-7.

Example 8-8

The circuit in Figure 8-12a contains a switch that has been open for a "long time" and then is closed at $t = 0$. Determine the inductor current, the state variable, for $t > 0$.

We must first find the initial condition from the circuit in Figure 8-12b. By simple series equivalence

$$i(0^-) = i(0^+) = IC = \frac{V_o}{R_1 + R_2}$$

The final condition and the time constant are determined from the circuit in Figure 8-12c. With the switch

closed and the inductor replaced by a short circuit, it is easy to see that

$$i(\infty) = FC = V_o/R_1 \qquad \text{and} \qquad G_N = 1/R_1$$

since R_2 is short-circuited by the switch. Consequently we now write the step response of the inductor current as

$$i(t) = (FC - IC)\, e^{-t/T_C} + FC$$
$$= \left[\frac{V_o}{R_1 + R_2} - \frac{V_o}{R_1}\right] e^{-R_1 t/L} + \frac{V_o}{R_1} \qquad t > 0$$

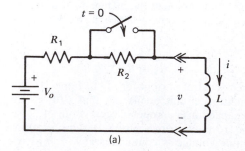

(a)

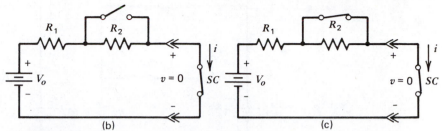

(b)

(c)

FIGURE 8-12
Circuits for demonstrating initial and final conditions of Example 8-8. (*a*) Given circuits. (*b*) Initial value circuit. (*c*) Final value circuit.

Example 8-9

The circuit in Figure 8-13*a* is slightly different in that we are asked to find the output voltage $v_o(t)$, which is not the voltage across the capacitor (that is, it is not the state or memory variable). But we have previously noted that once we find the state variable (capacitor voltage in this case), we can then determine any other signal. So our first task, regardless of what is desired, is to determine the capacitor voltage. From the initial value circuit (Figure 13*b*) and voltage division,

$$v(0^-) = v(0^+) = IC = \frac{R_1 V_o}{R_1 + R_2}$$

From the final value circuit (Figure 13*c*) we see that

$$v(\infty) = FC = V_o$$

since no current flows. Likewise we see that the time constant is R_2C for $t > 0$. Hence the capacitor voltage is

$$v(t) = (IC - FC) e^{-t/T_C} + FC$$

$$v(t) = \left[\frac{R_1 V_o}{R_1 + R_2} - V_o\right] e^{-t/R_2 C} + V_o$$

$$= V_o - \frac{R_2 V_o}{R_1 + R_2} e^{-t/R_2 C} \qquad t > 0$$

Now we can determine the voltage $v_o(t)$ by writing a KVL equation around the perimeter of the original circuit:

$$-V_o + v(t) + v_o(t) = 0$$

or

$$v_o(t) = V_o - v(t)$$

$$= \frac{R_2 V_o}{R_1 + R_2} e^{-t/R_2 C} \qquad t > 0$$

This looks like a zero-input response even though there is a nonzero input for $t > 0$. The reason is that the response $v_o(t)$ is not the capacitor voltage but the voltage across the resistor R_2.

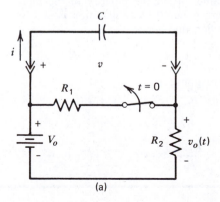

(a)

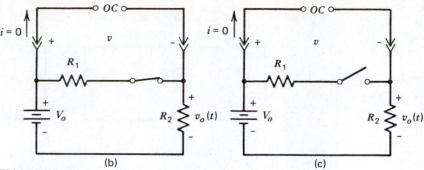

FIGURE 8-13
Circuits for demonstrating initial and final conditions of Example 8-9. (*a*) Given circuit. (*b*) Initial value circuit. (*c*) Final value circuit.

Example 8-10

Our final switching example is the OP AMP circuit in Figure 8-14. The switch has been in position *A* for "quite awhile." At $t = 0$ it is moved to position *B*. Determine the output voltage $v_o(t)$.

In Example 8-3 we found the differential equation of this circuit as

$$R_2 C \frac{dv_o}{dt} + v_o = - \frac{R_2}{R_1} V_S$$

Therefore we know the circuit time constant is $R_2 C$. The initial value circuit is a simple inverting amplifier, and using our knowledge from Chapter 4 we write

$$v_o(0^-) = v_o(0^+) = IC = - \frac{R_2}{R_1} V_S$$

In the final value circuit the polarity of the input source is reversed but the configuration is still an inverting amplifier, and hence

$$v_o(\infty) = FC = + \frac{R_2}{R_1} V_S = - IC$$

Thus the step response is

$$v_o(t) = (IC - FC) e^{-t/T_C} + FC$$

$$= - 2 \frac{R_2}{R_1} V_S e^{-t/R_2 C} + \frac{R_2}{R_1} V_S$$

$$= \frac{R_2 V_S}{R_1} [1 - 2 e^{-t/R_2 C}] \quad t > 0$$

Since v_o is not a state variable we might wonder if there could be a discontinuity at $t = 0$. But a KVL equation around the perimeter of the OP AMP in the circuit shows that

$$- v_o + v_C + v_N = 0$$

But $v_N = 0$ since the OP AMP input constraint requires $v_N = v_P$ and $v_P = 0$ because the noninverting input is grounded. Thus from the KVL equation $v_o = v_C$. In words, the output voltage equals the capacitor voltage, which is the state variable and there cannot be a jump at $t = 0$.

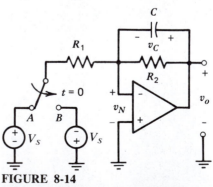

FIGURE 8-14
OP AMP circuit for Example 8-10.

8-4 SINGLE-MEMORY SINUSOIDAL RESPONSE

The response of linear circuits to sinusoidal inputs is one of the central themes of electrical engineering. In this introduction to the concept we treat the general RC circuit by means of its differential equation. In later chapters we see that sinusoidal response can found by means of the Laplace transformation, and ultimately from the circuit itself using the concept of a phasor. But for now, to attain an initial physical insight we concentrate on the use of the classical method of finding solutions to differential equations.

If the input to the general RC circuit in Figure 8-1 is a sinusoid, then the circuit differential equation (Eq. 8-3 again) can be written as

$$R_T C \frac{dv}{dt} + v = u(t) A \cos \omega t \qquad (8\text{-}26)$$

Now the input here (right side of Eq. 8-26) is *not* the eternal sine wave studied in Chapter 6, but a sinusoid that is initiated at $t = 0$ through some action such as the closing of a switch. We seek a solution function $v(t)$ that satisfies Eq. 8-26 for $t > 0$, and that meets a prescribed initial condition $v(0) = V_o$.

As with the step response, we find the solution in two parts: natural response and forced response. The natural response is of the form

$$v_N(t) = K e^{-t/R_T C} \qquad (8\text{-}27)$$

This follows from the fact that the natural response is a general solution of the homogeneous equation (input set to zero), and is therefore independent of the nature of the input. It represents what the circuit naturally tends to do with zero input, and thus is called the *natural response*.

The forced response is a particular solution of Eq. 8-26. Since the input is a sinusoid, we are naturally led to try a solution in the form of a sinusoid.

$$v_F = a \cos \omega t + b \sin \omega t \qquad (8\text{-}28)$$

In this expression the Fourier coefficients a and b are unknown. The technique we are using is known mathematically as the **method of undetermined coefficients,** in this case, undetermined *Fourier* coefficients. To show that Eq. 8-28 is indeed a solution, we substitute into Eq. 8-26:

$$R_T C \frac{dv_F}{dt} + v_F = A \cos \omega t \qquad t > 0$$

or

$$R_T C(-\omega a \sin \omega t + \omega b \cos \omega t) + a \cos \omega t + b \sin \omega t = A \cos \omega t$$

Now gather all sine and cosine terms on one side of the equation:

$$(a + R_T C \omega b - A)\cos \omega t + (b - R_T C \omega a)\sin \omega t = 0$$

This equation can be valid for all $t > 0$ only if the coefficients of the cosine and sine terms are identically zero. From this observation we obtain two linear equations in a and b:

$$a + R_T C \omega b = A$$
$$-R_T C \omega a + b = 0$$

which yield solutions

$$a = \frac{A}{1 + (\omega R_T C)^2} \qquad b = \frac{\omega R_T C A}{1 + (\omega R_T C)^2}$$

The undetermined Fourier coefficients are now determined. That is, a and b have been expressed in terms of known circuit parameters ($R_T C$) and known input signal parameters (ω and A).

We now combine the forced and natural responses as

$$v(t) = K e^{-t/R_T C} + \frac{A}{1 + (\omega R_T C)^2} (\cos \omega t + \omega R_T C \sin \omega t) \qquad \textbf{(8-29)}$$

The constant K can now be evaluated from the prescribed initial condition

$$v(0) = V_o = K + \frac{A}{1 + (\omega R_T C)^2}$$

which yields

$$K = V_o - \frac{A}{1 + (\omega R_T C)^2}$$

We can now write down the function $v(t)$ that satisfies the differential equation and the given initial condition.

$$v(t) = \underbrace{\left[V_o - \frac{A}{1 + (\omega R_T C)^2} \right] e^{-t/R_T C}}_{\text{Natural response}} +$$

$$\underbrace{\frac{A}{1 + (\omega R_T C)^2} [\cos \omega t + \omega R_T C \sin \omega t]}_{\text{Forced response}} \qquad \textbf{(8-30)}$$

This result may seem somewhat less formidable if we convert the forced response to an amplitude and phase-angle format.

$$v(t) = \underbrace{\left[V_o - \frac{A}{1 + (\omega R_T C)^2} \right] e^{-t/R_T C}}_{\text{Natural response}} + \underbrace{\frac{A}{\sqrt{1 + (\omega R_T C)^2}} \cos (\omega t - \theta)}_{\text{Forced response}} \qquad \textbf{(8-31)}$$

where

$$\theta = \tan^{-1} (\omega R_T C).$$

This result is worthy of several comments. The natural response is a decaying exponential whose time constant is none other than the circuit time constant. This component of the response essentially vanishes after about five time constants. Thereafter there remains the forced response, which is a sinusoid with the same frequency (ω) as the input, but with a different amplitude and phase angle. The amplitude of the forced response is proportional to the amplitude of the input (A), reminding us that the circuit is linear and has the proportionality property. However, the amplitude and phase angle of the forced response are not linear functions of the input frequency. In other words, the circuit will respond to different input frequencies in quite different ways. This frequency-selective characteristic of linear circuits gives rise to a signal-processing method called **filtering.** We do not discuss this matter further here, but simply note that an entire chapter (Chapter 13) is devoted to filter characteristics of linear circuits.

In the terminology of electrical engineering, the forced response is usually called the **sinusoidal steady-state response.** The choice of the words "steady state" is perhaps unfortunate since it seems to imply a constant, "steady" value, whereas the forced response is actually a sustained oscillation. Nonetheless, this terminology so permeates the literature and folklore of electrical engineering that we are compelled to use it. Hereafter we use the terms *sinusoidal steady-state response* and *forced response due to a sinusoidal input* interchangeably.

Example 8-11

The switch in Figure 8-15a is closed at $t = 0$ connecting a source

$$v_S(t) = 20 \sin 1000t$$

to the circuit shown. Find the resulting response $v(t)$, the voltage across the capacitor.

The Thévenin voltage seen by the capacitor is, by voltage division,

$$v_T = \frac{4}{4 + 4} v_S = 10 \sin 1000t$$

The Thévenin resistance (switch closed and source off) is simply two 4-kΩ resistors in parallel, or 2 kΩ. Hence the circuit time constant is

$$R_T C = 2 \times 10^3 \times 1 \times 10^{-6} = 2 \times 10^{-3} = 1/500$$

Therefore the differential equation of the circuit is

$$2 \times 10^{-3} \frac{dv}{dt} + v = 10 \sin 1000t \qquad t > 0$$

Note that the right side of this equation is the Thévenin equivalent input and not the original source input $v_S(t)$. The natural response is of the form

$$v_N(t) = Ke^{-500t}$$

and the forced response with undetermined Fourier coefficients is

$$v_F(t) = a \cos 1000t + b \sin 1000t$$

Substituting the forced response into the differential equation produces

$$2 \times 10^{-3}(-1000a \sin 1000t + 1000b \cos 1000t)$$
$$+ a \cos 1000t + b \sin 1000t = 10 \sin 1000t$$

Collecting all sine and cosine terms on one side of this equation yields

$$(2b + a) \cos 1000t$$
$$+ (-2a + b - 10) \sin 1000t = 0$$

which by the method of undetermined coefficients requires that

$$a + 2b = 0 \qquad and \qquad -2a + b = 10$$

whose solutions are $a = -4$ and $b = 2$. Now combining the forced and natural responses,

$$v(t) = Ke^{-500t} - 4 \cos 1000t + 2 \sin 1000t$$

The constant K can be determined from the initial condition

$$v(0) = V_o = K - 4$$

The initial condition has not been prescribed, but can be determined by simple physical arguments. If the switch was open for a "long time" prior to $t = 0$, then any zero-input response would have decayed to zero. Hence it is reasonable to assume that $V_o = 0$, and therefore $K = 4$. Incidentally, a "long time" in this case is $5T_C = 10^{-2} = 10$ ms, a mere blink of the eye in the great sweep of history. We can now write the complete solution as

$$v(t) = 4e^{-500t} - 4 \cos 1000t + 2 \sin 1000t$$

in Fourier coefficient form, or

$$v(t) = 4e^{-500t} + 4.47 \cos (1000t - 153°)$$

in magnitude and phase-angle form. Plots of this response and its components are shown in Figure 8-16. Notice that the natural response decays to zero and that the total response merges into the forced response—a sustained oscillation. Thus after five time constants or so the circuit settles down to the sinusoidal steady state.

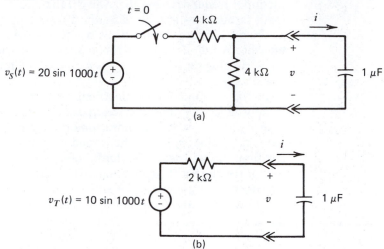

(a)

(b)

FIGURE 8-15
Circuit and equivalent driven by sudden application of a sinusoidal signal (Example 8-11). (a) Given circuit. (b) Equivalent circuit.

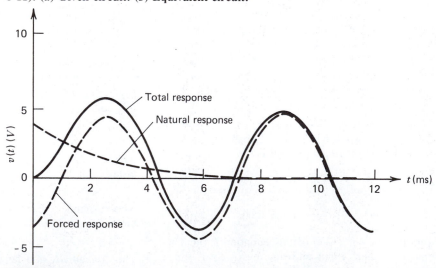

FIGURE 8-16
Response for Example 8-11.

Example 8-12

The circuit in Figure 8-17 is operating in the sinusoidal steady state. Find the output voltage $v_o(t)$.

Our approach is first to find the state variable (inductor current) and then to use it to determine the output. The differential equation of the circuit in terms of the inductor current is

$$GL \frac{di}{dt} + i = A \cos \omega t$$

The circuit is operating in the steady state and so we only need to determine the forced component,

$$i_F(t) = a \cos \omega t + b \sin \omega t$$

Substituting into the differential equation yields

$$GL(-\omega a \sin \omega t + \omega b \cos \omega t) + a \cos \omega t$$
$$+ b \sin \omega t = A \cos \omega t$$

Collecting terms produces

$$(a + GL\omega b - A)\cos \omega t$$
$$+ (-GL\omega a + b)\sin \omega t = 0$$

For this equation to be identically zero for all t requires that

$$a + \omega GLb = A \quad \text{and} \quad -GL\omega a + b = 0$$

whose solutions are

$$a = \frac{A}{1 + (\omega GL)^2} \quad \text{and} \quad b = \frac{\omega GLA}{1 + (\omega GL)^2}$$

Hence the forced response of the inductor current is

$$i_F(t) = \frac{A}{1 + (\omega GL)^2} (\cos \omega t + \omega GL \sin \omega t)$$

We can now determine the output voltage since it is the voltage across the inductor. Hence

$$v_o(t) = L \frac{di}{dt}$$
$$= L \frac{A}{1 + (\omega GL)^2} \frac{d}{dt} [\cos \omega t + \omega GL \sin \omega t]$$
$$= \frac{A}{1 + (\omega GL)^2} [-\omega L \sin \omega t + G(\omega L)^2 \cos \omega t]$$

Thus the output voltage is a sinusoid with the same frequency as the input signal, but with a different amplitude and phase angle. In fact, in the sinusoidal steady state every voltage and current in the circuit is sinusoidal *with the same frequency*.

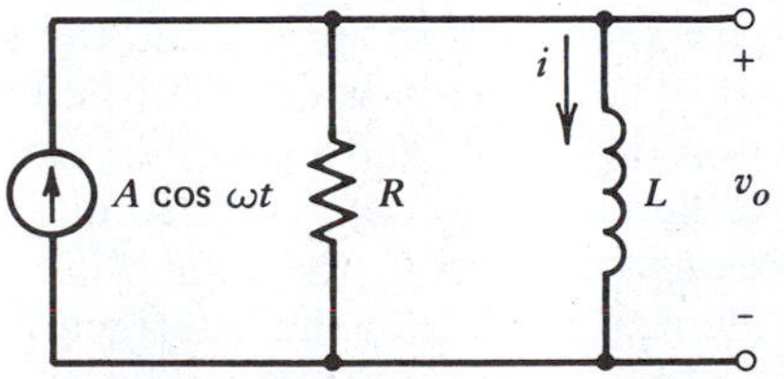

FIGURE 8-17
RL circuit driven by sinusoidal signal (Example 8-12).

8-5 DOUBLE-MEMORY CIRCUITS

By double memory we mean a circuit that contains at least two memory elements that cannot be combined to produce a single equivalent memory element. Although the number of such circuits is endless, we concentrate our attention on two classical, canonical forms, the series *RLC* circuit and the parallel *RLC* circuit. These two circuits illustrate almost all of the basic concepts of double-memory circuits and serve as vehicles for studying the solutions of second-order differential equations. In subsequent chapters we use the Laplace transformation to develop techniques for handling any type of double-memory circuit.

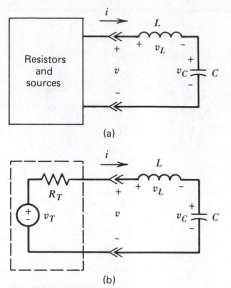

FIGURE 8-18
The series *RLC* circuit. (*a*) Given circuit.
(*b*) Equivalent circuit.

The general series *RLC* circuit is shown in Figure 8-18. The inductor and capacitor are connected in series, and the source-resistor circuit can be reduced to its Thévenin equivalent (R_T and v_T). The result is a circuit in which resistor, inductor, and capacitor are connected in series, and hence we have the name "series *RLC* circuit." Our first task is to write down the equations that describe the equilibrium in this circuit. The source-resistor equivalent to the left of the interface introduces the constraint

$$v + iR_T = v_T \qquad\qquad \textbf{(8-32)}$$

Writing a KVL equation around the loop on the load side of the interface yields

$$v = v_L + v_C \qquad\qquad \textbf{(8-33)}$$

Finally, we can write down the *i-v* characteristics of the two memory elements:

$$v_L = L \frac{di}{dt} \qquad\qquad \textbf{(8-34)}$$

$$i = C \frac{dv_C}{dt} \qquad\qquad \textbf{(8-35)}$$

We have written four equations in four unknowns (i, v, v_L, v_C). This indicates that we should be able to eliminate all but one variable, and our problem is to select an appropriate solution variable. For reasons that will become apparent, the most convenient solution variable is the capacitor voltage. To isolate v_C we first substitute Eqs. 8-33 and 8-35 into Eq. 8-32.

$$v_L + v_C + R_T C \frac{dv_C}{dt} = v_T \qquad (8\text{-}36)$$

This substitution eliminated everything except v_L. To eliminate the inductor voltage we substitute Eq. 8-35 into Eq. 8-34 to obtain

$$v_L = LC \frac{d^2 v_C}{dt^2}$$

When this result is substituted into Eq. 8-36, there results

$$LC \frac{d^2 v_C}{dt^2} + R_T C \frac{dv_C}{dt} + v_C = v_T \qquad (8\text{-}37)$$

This is a *second-order, linear differential equation* in which the dependent variable is the capacitor voltage. Once we have solved this equation for v_C, we can work backward through our analysis to determine every other voltage or current in the circuit if need be. In fact, we could have derived a differential equation in which the dependent variable is any one of the other signal variables. If we had done so, the left side of the equation would have the same form as Eq. 8-37, but the right side would involve derivatives of the input v_T. Since this could be a bit awkward at times, we choose the capacitor voltage as our solution variable.

In dealing with Eq. (8-37) we will find that the solution depends on three kinds of things:

1. The nature of the input signal (v_T).

2. The values of the circuit parameters (R_T, L, C).

3. The values of the capacitor voltage and the inductor current at $t = 0$, which will be denoted as V_o and I_o respectively.

The first two categories are not surprising since these things also influence the response of memoryless circuits. The last category points out that the circuit has memory. That is, its response is influenced by what has happened in the past, as represented by the energy stored in the capacitor $(\frac{1}{2})CV_o^2$ and the inductor $(\frac{1}{2})LI_o^2$ at $t = 0$, the time at which we initiated our solution.

Among the consequences of memory is that the circuit can have a response even though the input signal is identically zero for $t > 0$. To highlight this characteristic we first deal with the zero-input response. That is, we make $v_T = 0$. Equation 8-37 then becomes

$$LC \frac{d^2 v_C}{dt^2} + R_T C \frac{dv_C}{dt} + v_C = 0 \qquad (8\text{-}38)$$

This is, of course, a homogeneous differential equation. In words, it requires that a linear combination of a dependent variable and its first and second derivative somehow sum to zero for all $t > 0$. This suggests, as it did in the single-memory case, that we try a solution in the form

$$v_C(t) = Ke^{st}$$

where s can be real, imaginary, or complex. When this trial solution is inserted in Eq. 8-38, there results

$$Ke^{st}(LCs^2 + R_TCs + 1) = 0$$

The possibility that $K = 0$ is called the trivial solution. The quantity e^{st} cannot be zero for all $t > 0$. Hence the general condition is that

$$LCs^2 + R_TCs + 1 = 0 \tag{8-39}$$

This is the circuit's characteristic equation. It is a quadratic because the circuit contains two noncombinable memory elements. If we were to set $L = 0$, the equation would reduce to the characteristic equation of the RC circuit. In fact, if we replace the inductor in Figure 8-18 by a short circuit, we see that the circuit reduces to a single-memory RC circuit.

Thus the presence of two memory elements leads to a quadratic characteristic equation that necessarily has two roots:

$$s_1, s_2 = \frac{-R_TC \pm \sqrt{(R_TC)^2 - 4LC}}{2LC} \tag{8-40}$$

Apparently then there are two possible solutions: $K_1 e^{s_1t}$ and $K_2 e^{s_2t}$. But we seek the general solution and we also know that we must satisfy two initial conditions (V_o and I_o). Consequently the general solution of the homogeneous equation must be

$$v_C(t) = K_1 e^{s_1t} + K_2 e^{s_2t} \tag{8-41}$$

We can evaluate the constants K_1 and K_2 from the initial conditions. If we let $t = 0$, we can find the first equation:

$$v_C(0) = V_o = K_1(1) + K_2(1) \tag{8-42}$$

To obtain a second equation we must use the initial value of the inductor current. Looking back at Eq. 8-35 we see that

$$C\frac{dv_C(t)}{dt}\bigg|_{t=0} = Cv'_C(0) = i(0)$$

But $i(t)$ is the current through the inductor, and hence

$$v'_C(0) = I_o/C$$

Thus the initial inductor current specifies the value of the first derivative of the capacitor voltage at $t = 0$. To use this result we differentiate Eq. 8-41 as

$$\frac{dv_C(t)}{dt} = K_1s_1 e^{s_1t} + K_2s_2 e^{s_2t}$$

and hence

$$v'_C(0) = I_o/C = K_1s_1 + K_2s_2 \tag{8-43}$$

Equations 8-42 and 8-43 provide two equations in two unknowns:

$$K_1 + K_2 = V_o$$
$$s_1K_1 + s_2K_2 = I_o/C$$

whose solutions are

$$K_1 = \frac{s_2V_o - I_o/C}{s_2 - s_1} \quad \text{and} \quad K_2 = \frac{-s_1V_o + I_o/C}{s_2 - s_1}$$

Putting these results back into Eq. 8-41 yields the zero-input response as

$$v_C(t) = \frac{s_2V_o - I_o/C}{s_2 - s_1} e^{s_1t} + \frac{-s_1V_o + I_o/C}{s_2 - s_1} e^{s_2t} \qquad \textbf{(8-44)}$$

This result is indeed the general form of the zero-input response of the series *RLC* circuit. The response depends on two initial conditions (V_o and I_o), and on the circuit parameters (R_T, L, and C) since the roots s_1 and s_2 derive from the characteristic equation. However, there is more here than meets the eye. The nature of the roots can dramatically change the basic form of the response.

Looking back at Eq. 8-40 defining the roots

$$s_1, s_2 = \frac{-R_TC \pm \sqrt{(R_TC)^2 - 4LC}}{4LC}$$

we find that there are three distinct possibilities:

CASE A If $(R_TC)^2 - 4LC > 0$, the radicand is positive and there will be two real, unequal roots ($s_1 \neq s_2$).

CASE B If $(R_TC)^2 - 4LC = 0$, the radicand vanishes and there will be two real, equal roots ($s_1 = s_2$).

CASE C If $(R_TC)^2 - 4LC < 0$, the radicand will be negative, which leads to two complex conjugate roots ($s_1 = s_2^*$).

In what follows we deal with all three of these cases. To set the stage for our discussion, let us consider a concrete example.

Example 8-13

The circuit in Figure 8-19 is a series *RLC* circuit in which the switch is closed at $t = 0$. The initial capacitor voltage is 15 V and the inductor current is initially zero. The values of the capacitance and inductance are fixed. Find the roots of the characteristic equation for three different values of resistance: 8.5 kΩ, 4 kΩ, and 1 kΩ.

From Eq. 8-38 we know that the differential equation for the circuit is

$$LC \frac{d^2v_C}{dt^2} + RC \frac{dv_C}{dt} + v_C = 0 \qquad t > 0$$

The trial solution $v_C = Ke^{st}$ then yields the characteristic equation

$$LCs^2 + RCs + 1 = 0$$

What we need to do is to find the roots of this equation for the three values of resistance given in the foregoing. For the first value (8.5 kΩ) the characteristic equation is

$$0.25 \times 10^{-6}s^2 + 2.125 \times 10^{-3}s + 1 = 0$$

or

$$s^2 + 8500s + 4 \times 10^6 = (s + 500)(s + 8000) = 0$$

The solution contains *two real, unequal* roots ($s_1 = -500$ and $s_2 = -8000$), which is an example of Case A. For the second value of R, 4 kΩ, the equation becomes

$$0.25 \times 10^{-6}s^2 + 10^{-3}s + 1 = 0$$

or

$$s^2 + 4000s + 4 \times 10^6 = (s + 2000)^2 = 0$$

For this value of resistance there are *two real, equal* roots ($s_1 = s_2 = -2000$), which illustrates Case B. For the final value of 1 kΩ the characteristic equation is

$$0.25 \times 10^{-6}s^2 + 0.25 \times 10^{-3}s + 1 = 0$$

whose roots are

$$s_1, s_2 = -500 \pm 500 \sqrt{-15}$$

The quantity under the radical is negative, illustrating the Case C situation. Defining the symbol $j = \sqrt{-1}$, the roots can be written as

$$s_1, s_2 = -500 \pm j500 \sqrt{15}$$

This is a *pair of complex, conjugate* roots. We will return to this example later and calculate the responses corresponding to each of these cases.

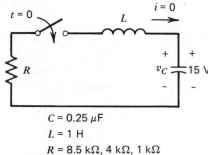

$C = 0.25\ \mu F$

$L = 1\ H$

$R = 8.5\ k\Omega,\ 4\ k\Omega,\ 1\ k\Omega$

FIGURE 8-19
RLC circuit for Example 8-13.

In the foregoing example, and in what follows, we use the symbol j to stand for $\sqrt{-1}$ rather than the symbol i used by mathematicians to avoid confusion with our symbol for current in electrical engineering. In any case, we have not introduced complex numbers simply to make things——well, complex. The need to deal with complex numbers arises quite naturally in physical situations, from something as simple as factoring a quadratic equation. The ability to deal with complex numbers is essential to our study. For those who need a review of such matters there is a brief discussion in Appendix E.

We now return to our development and consider each of these cases in detail. For Case A the roots are *real* and *unequal*. Using the notation

$$s_1 = \alpha_1 \quad \text{and} \quad s_2 = \alpha_2$$

we can write the general form of the response from Eq. 8-44 as

$$v_C(t) = \frac{(\alpha_2 V_o - I_o/C)\, e^{-\alpha_1 t} + (-\alpha_1 V_o + I_o/C)\, e^{-\alpha_2 t}}{\alpha_2 - \alpha_1} \tag{8-45}$$

There is nothing particularly surprising here. The response is called the **overdamped** case and simply consists of two exponential waveforms. So to speak, the response has two time constants. The two time constants can be greatly different, or nearly equal, but they cannot be exactly equal because we would then have Case B.

For Case B the roots are *real* and *equal*. Using the notation

$$s_1 = s_2 = \alpha$$

the general form in Eq. 8-44 becomes

$$v_C(t) = \frac{(\alpha V_o - I_o/C)\, e^{-\alpha t} + (-\alpha V_o + I_o/C)\, e^{-\alpha t}}{\alpha - \alpha}$$

We immediately see a problem with this result since the denominator vanishes. However, a careful examination of the numerator shows that it vanishes as well. Our solution reduces to the indeterminant form 0/0! To investigate this further we let

$$s_1 = \alpha \qquad \text{and} \qquad s_2 = \alpha + x$$

and explore the situation as x approaches zero. Using this notation in Eq. 8-44 produces

$$v_C(t) = V_o e^{\alpha t} + \frac{\alpha V_o - I_o/C}{x} e^{\alpha t} + \frac{-\alpha V_o + I_o/C}{x} e^{\alpha t} e^{xt} .$$

which can be arranged as

$$v_C(t) = e^{\alpha t}\left[V_o + (\alpha V_o - I_o/C)\frac{1 - e^{xt}}{x}\right] \qquad \textbf{(8-46)}$$

In this form we see that our problem comes from the term $(1 - e^{xt})/x$, which is indeed indeterminant as x approaches zero. But the application of l'Hospital's rule reveals

$$\underset{x \to 0}{\text{Limit}}\ \frac{1 - e^{xt}}{x} = \underset{x \to 0}{\text{Limit}}\ \frac{-te^{xt}}{1} = -t$$

Thus the indetermancy is resolved, and for x approaching zero ($s_1 = s_2 = \alpha$) Eq. 8-46 becomes

$$v_C(t) = e^{\alpha t}\,[V_o - (\alpha V_o - I_o/C)t] \qquad \textbf{(8-47)}$$

This special form of the response is called the **critically damped** case. A special form is required because the two roots are identical.

In Case C we have *complex conjugate* roots, which we denote as

$$s_1 = \alpha - j\beta \qquad \text{and} \qquad s_2 = \alpha + j\beta$$

Inserting these roots directly into Eq. 8-44 yields

$$v_C(t) = \frac{(\alpha + j\beta)V_o - I_o/C}{j2\beta} e^{\alpha t} e^{-j\beta t} + \frac{(-\alpha + j\beta)V_o + I_o/C}{j2\beta} e^{\alpha t} e^{j\beta t}$$

which can be arranged as

$$v_C(t) = e^{\alpha t}\left[V_o\frac{e^{j\beta t} + e^{-j\beta t}}{2} - (\alpha V_o - I_o/C)\frac{e^{j\beta t} - e^{-j\beta t}}{2j}\right] \qquad \textbf{(8-48)}$$

The complex parts of this result have been arranged in a special way. Using Euler's relationships for the complex exponential,

$$e^{jx} = \cos x + j \sin x$$

and

$$e^{-jx} = \cos x - j \sin x$$

We add and subtract these relationships to show that

$$\cos x = \frac{e^{jx} + e^{-jx}}{2} \quad \text{and} \quad \sin x = \frac{e^{jx} - e^{-jx}}{2j}$$

Comparing these expressions for $\sin x$ and $\cos x$ with the complex terms in Eq. 8-48 reveals that we can write it as

$$v_C(t) = e^{\alpha t} [V_o \cos \beta t - (\alpha V_o - I_o/C) \sin \beta t] \tag{8-49}$$

Thus, even though the roots are complex, we have ended up with a real function of time. In fact, it is the damped sinusoidal waveform studied in Chapter 6. The real part of the roots (α) ends up as the damping term in the exponential, while the imaginary part (β) ends up as the frequency of oscillation of the resulting damped sinusoidal waveform. This case is referred to as the **underdamped** response.

We now return to Example 8-13 to apply these results to the circuit introduced there.

Example 8-14

In Example 8-13 we found the roots of the characteristic equation of the series RLC circuit shown in Figure 8-19 for the three values of resistance shown in the figure. The initial conditions were given as $V_o = 15$ V and $I_o = 0$. We are now in a position to apply the results derived for each of the types of responses.

Case A: For $R = 8.5$ kΩ, we found $s_1 = -500$ and $s_2 = -8000$. Applying Eq. 8-45 yields

$$v_C(t) = 16e^{-500t} - e^{-8000t}$$

This is the overdamped response and consists of two exponential waveforms.

Case B: For $R = 4$ kΩ, we found $s_1 = s_2 = -2000$. Applying Eq. 8-47 yields

$$v_C(t) = 15(1 + 2000t)e^{-2000t}$$

This special form of the response occurs because the roots are equal and is called the critically damped case.

Case C: For $R = 1$ kΩ, we found $s_1, s_2 = -500 \pm j500 \sqrt{15}$. Applying Eq. 8-49 yields

$$v_C(t) = e^{-500t} (15 \cos 500 \sqrt{15}\, t + 7500 \sin 500 \sqrt{15}\, t)$$

This is a damped sinusoidal response characteristic of the underdamped case. All three of these responses are shown in Figure 8-20. All start out at 15 V (the initial condition) and eventually decay to zero. However, the nature of the response waveforms is dramatically different. The overdamped response (Case A) is relatively sluggish and unexciting. The critically damped response (Case B) decays rapidly to zero but does not undershoot its final value. The underdamped response passes through zero rapidly but undershoots the mark, and eventually decays to zero in a sequence of damped oscillations.

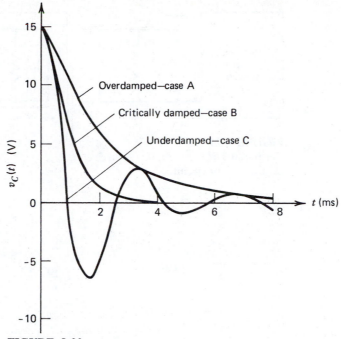

FIGURE 8-20
Possible responses for the series *RLC* circuit (Example 8-13).

So far we have seen that the zero-input response of a double-memory circuit is somewhat more complicated than the single-memory case. The response can take quite different forms, depending on the nature of the roots of the characteristic equation. Our vehicle for demonstrating this has been the series *RLC* circuit. We now turn to the parallel *RLC* circuit to see whether these results have general applicability.

The general case is shown in Figure 8-21. The inductor and capacitor are connected in parallel in the given circuit and the source-resistor circuit can be replaced by its Norton equivalent. The result is a circuit in which the resistance, inductance, and capacitance are connected in parallel, and hence the name "parallel *RLC* circuit." Our first task is to develop the differential equation of this circuit, which we expect will be second order because of two uncombinable memory elements.

The source-resistor side of the equivalent circuit introduces the constraint

$$i + G_N v = i_N \tag{8-50}$$

The application of KCL to the load side of the interface produces

$$i_L + i_C = i \tag{8-51}$$

and the *i-v* characteristics of the two memory elements are

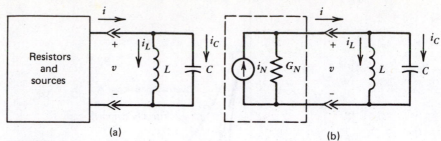

FIGURE 8-21
The parallel *RLC* circuit. (*a*) Given circuit.
(*b*) Equivalent circuit.

$$i_C = C \frac{dv}{dt} \qquad (8\text{-}52)$$

$$v = L \frac{di_L}{dt} \qquad (8\text{-}53)$$

We have now written four equations in four unknowns (i, v, i_C, i_L), so it should be possible to eliminate all but one to obtain the circuit differential equation. In this case the most convenient solution variable is the inductor current. Hence we begin by substituting Eqs. 8-51 and 8-53 into Eq. 8-50:

$$i_L + i_C + G_N L \frac{di_L}{dt} = i_N \qquad (8\text{-}54)$$

The capacitor current can be eliminated from this result by substituting Eq. 8-53 into Eq. 8-52 to yield

$$i_C = LC \frac{d^2 i_L}{dt^2}$$

When this is inserted in Eq. 8-54 we obtain

$$LC \frac{d^2 i_L}{dt^2} + G_N L \frac{di_L}{dt} + i_L = i_N \qquad (8\text{-}55)$$

This is a second-order, linear differential equation of the same form we obtained for the series *RLC* circuit (Eq. 8-37). In fact, if we make the replacement

$$L \leftrightarrow C \qquad R_T \leftrightarrow G_N \qquad v_T \leftrightarrow i_N \qquad v_C \leftrightarrow i_L$$

we can change one into the other. Thus duality triumphs again, which means that we need not study the parallel *RLC* circuit in detail since all of the results obtained for the series *RLC* case apply here.

However, we do need to summarize the major features of the parallel *RLC* circuit. The initial conditions in the circuit are, of course, the initial inductor current (I_o) and capacitor voltage (V_o). The first of these works directly to provide one boundry condition on the differential equation:

$$i_L(0) = I_o$$

The capacitor voltage specifies the initial rate of change of the inductor current since, from Eq. 8-53,

$$L\left.\frac{di_L(t)}{dt}\right|_{t=0} = Li'_L(0) = v(0)$$

But from the circuit in Figure 8-21 we see that $v(t)$ is the voltage across the capacitor. Hence

$$i'_L(0) = V_o/L$$

These results are the dual of those obtained for the series *RLC* circuit.

Turning now to the zero-input response, we set $i_N = 0$ in Eq. 8-55 to obtain

$$LC\frac{d^2i_L}{dt^2} + G_N L\frac{di_L}{dt} + i_L = 0$$

The solution of this homogeneous equation yields the zero-input response. The equation requires that a linear combination of a signal and its first two derivatives must sum to zero, suggesting a trial solution of the form $i_L = Ke^{st}$. When this trial is inserted into the homogeneous equation, we are led to the characteristic equation:

$$LCs^2 + G_N Ls + 1 = 0 \qquad\qquad \textbf{(8-56)}$$

This is a quadratic equation because of double memory. The characteristic equation has two roots:

$$s_1, s_2 = \frac{-G_N L \pm \sqrt{(G_N L)^2 - 4LC}}{2LC}$$

and, as before, we have three distinct types of roots.

CASE A If $(G_N L)^2 > 4LC$, the radicand is positive and there will be two, unequal, real roots (α_1 and α_2). The zero-input response will be of the form

$$i_L(t) = K_1 e^{\alpha_1 t} + K_2 e^{\alpha_2 t}$$

which is the overdamped case.

CASE B If $(G_N L)^2 = 4LC$, the radicand vanishes and there will be two real, equal roots $s_1 = s_2 = \alpha$. The zero-input response will be of the form

$$i_L(t) = (K_1 + K_2 t)e^{\alpha t}$$

which is the critically damped case.

CASE C If $(G_N L)^2 < 4LC$, the radicand is negative and there will be a pair of complex, conjugate roots $s_1, s_2 = \alpha \pm j\beta$. The zero-input response will then take the form

$$i_L(t) = e^{\alpha t}(K_1 \cos \beta t + K_2 \sin \beta t)$$

which is the underdamped case.

In sum, all of the results derived for the series *RLC* circuit apply to the parallel *RLC* case with the appropriate duality replacements. In particular, the concept of overdamped, critically damped, and underdamped response applies to both cases. Our next two examples show that this concept applies to other types of double-memory circuits as well.

Example 8-15

The OP AMP circuit in Figure 8-22 contains three equal resistors and two unequal capacitors. It is a double-memory circuit since the two capacitors are neither in series nor in parallel, and therefore cannot be replaced by a single equivalent capacitor. To formulate the differential equation for this circuit, we use the basic concepts of device and connection equations. From our experience in Chapter 4 we know that we need only two node equations to describe the circuit. We formulate these by writing KCL connection constraints at nodes Ⓐ and Ⓑ.

Node Ⓐ $i_1 + i_2 + i_3 - i_4 = 0$
Node Ⓑ $-i_2 + i_5 - i_N = 0$

and the device constraints as

$$i_1 = G(v_S - v_A)$$
$$i_2 = G(v_B - v_A)$$
$$i_3 = G(v_o - v_A)$$
$$i_4 = C_1 \frac{dv_A}{dt}$$
$$i_5 = C_2 d\frac{(v_o - v_B)}{dt}$$
$$i_N = 0$$
$$v_B = 0$$

The last two device equations are the OP AMP constraints with the noninverting input connected to ground.

Substituting the element equations into the connection (KCL) equations yields

$$-3Gv_A - C_1\frac{dv_A}{dt} + Gv_o + Gv_S = 0$$
$$Gv_A + C_2\frac{dv_o}{dt} = 0$$

Solving the second of these equations for v_A,

$$v_A = -RC_2\frac{dv_o}{dt}$$

and inserting this result into the first equation yields

$$R^2C_1C_2\frac{d^2v_o}{dt^2} + 3RC_2\frac{dv_o}{dt} + v_o = -v_S$$

This is a second-order, linear differential equation of the same form obtained for the *RLC* circuits, although it is not a dual of those equations. A trial solution of the form $v_o = Ke^{st}$ leads to the characteristic equation

$$R^2C_1C_2s^2 + 3RC_2s + 1 = 0$$

whose roots are

$$s_1, s_2 = \frac{-3RC_2 \pm \sqrt{R^2C_2(9C_2 - 4C_1)}}{2R^2C_1C_2}$$

and we see that there are three distinct types of roots:

CASE A If $9C_2 > 4C_1$ we have the overdamped case.

CASE B If $9C_2 = 4C_1$ we have the critically damped case.

CASE C If $9C_2 < 4C_1$ we have the underdamped case.

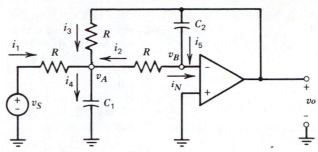

FIGURE 8-22
OP AMP Double-memory circuit (Example 8-15).

Example 8-16

The non-ideal transformer circuit in Figure 8-23 is a double-memory circuit. Show that it cannot exhibit an underdamped response.

To show this we first need to derive a differential equation describing the circuit. We begin by writing the i-v characteristics of the two coupled coils using the results studied in Chapter 7. For the indicated polarities these are

$$v_1(t) = L_1 \frac{di_1}{dt} + M \frac{di_2}{dt}$$

$$v_2(t) = M \frac{di_1}{dt} + L_2 \frac{di_2}{dt}$$

These are two equations in four unknowns. The resistor (R_1)-source connection at the input and the load (R_2) connection at the output provide two additional equations:

$$v_1(t) = v_S - R_1 i_1 \quad \text{and} \quad v_2(t) = -R_2 i_2$$

The minus sign in the last equation results from the reference marks assigned to v_2 and i_2 in the coupled-coil equations. Substituting the source and load constraints into the coupled-coil equations leads to

$$L_1 \frac{di_1}{dt} + R_1 i_1 + M \frac{di_2}{dt} = v_S$$

$$M \frac{di_1}{dt} + L_2 \frac{di_2}{dt} + R_2 i_2 = 0$$

We now have two equations in two unknowns (i_1 and i_2). Solving the second of these equations for the derivative of i_1 yields

$$\frac{di_1}{dt} = -\frac{L_2}{M} \frac{di_2}{dt} - \frac{R_2}{M} i_2 \qquad \textbf{(8-57)}$$

Inserting this result into the first of the two foregoing equations yields

$$-\frac{(L_1 L_2 - M^2)}{M} \frac{di_2}{dt} - \frac{R_2 L_1}{M} i_2 + R_2 i_1 = v_S$$

We have not yet eliminated i_1 and the only available equation involves the derivative of i_1. Hence we differentiate this equation to obtain

$$-\frac{(L_1 L_2 - M^2)}{M} \frac{d^2 i_2}{dt^2} - \frac{R_2 L_1}{M} \frac{di_2}{dt} + R_2 \frac{di_1}{dt} = \frac{dv_S}{dt}$$

We can now eliminate di_1/dt using Eq. 8-57 to obtain

$$(L_1 L_2 - M^2) \frac{d^2 i_2}{dt^2} + (R_1 L_2 + R_2 L_1) \frac{di_2}{dt} + R_1 R_2 i_2$$
$$= -M \frac{dv_S}{dt}$$

This is a second-order, linear differential equation in the current i_2. A trial solution of the homogeneous equation in the form $i_2 = K e^{st}$ leads to the characteristic equation

$$(L_1 L_2 - M^2)s^2 + (R_1 L_2 + R_2 L_1)s + R_1 R_2 = 0$$

whose roots are

$$s_1, s_2 = \frac{-(R_1 L_2 + R_2 L_1)}{2(L_1 L_2 - M^2)}$$
$$\pm \frac{\sqrt{(R_1 L_2 + R_2 L_1)^2 - 4R_1 R_2 (L_1 L_2 - M^2)}}{2(L_1 L_2 - M^2)}$$

The key question in this rather formidable result is the sign of the radicand:

$$\text{Radicand} = (R_1L_2)^2 + 2R_1R_2L_1L_2 + (R_2L_1)^2$$
$$- 4R_1R_2L_1L_2 + 4R_1R_2M^2$$
$$= (R_1L_2)^2 - 2R_1R_2L_1L_2 + (R_2L_1)^2$$
$$+ 4R_1R_2M^2$$
$$= (R_1L_2 - R_2L_1)^2 + 4R_1R_2M^2$$

In the final form we see that the radicand can never be negative. Generally it will be positive, indicating the overdamped response. It can be zero only in the special case $R_1 = R_2 = 0$, which yields the critically damped case. (But this can never happen in reality since there will always be some resistance associated with our circuit.) However, for no combination of element values can the radicand be negative, which means that the circuit cannot exhibit the underdamped response.

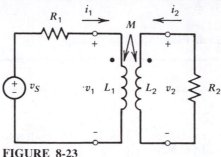

FIGURE 8-23
Transformer circuit for Example 8-16.

8-6 DOUBLE-MEMORY STEP RESPONSE

The step response is an important characterization of linear circuits, and so we are naturally led to investigate this matter for double-memory circuits. In a later chapter we develop general techniques for determining the step response of any type of memory circuit. However, in this introductory development we use the series RLC circuit to demonstrate the results employing classical differential equation methods. As we now know, similar results would be obtained for the parallel RLC circuit and, in fact, for any double-memory circuit.

If the input to a series RLC circuit is a step function $v_T(t) = A\ u(t)$, then the differential equation (see Eq. 8-37) is

$$LC\frac{d^2v_C}{dt^2} + R_TC\frac{dv_C}{dt} + v_C = A\ u(t) \tag{8-58}$$

We seek a solution of this equation for $t > 0$, subject to the initial conditions V_o and I_o, which are the initial capacitor voltage and inductor current respectively. As we found with single-memory circuits, this can be done by dividing the solution into forced and natural components.

$$v_C(t) = v_{CN}(t) + v_{CF}(t) \tag{8-59}$$

The natural response is the general solution of the homogeneous equation (input set to zero), while the forced response is a particular solution of

$$LC\frac{d^2v_{CF}}{dt^2} + R_TC\frac{dv_{CF}}{dt} + v_{CF} = A \qquad t > 0$$

Again it is not hard to see that $v_{CF} = A$ is a particular solution since dA/dt and d^2A/dt^2 are both zero.

Turning now to the natural response, we seek a general solution of the homogeneous equation. But we know that the natural response will take one of the three possible forms of the zero input response: overdamped, critically damped, or underdamped. To highlight these possibilities we introduce two new parameters, ω_0 (omega zero) and ζ (zeta), and define them as

$$\frac{1}{\omega_0^2} \equiv LC \quad \text{and} \quad \frac{2\zeta}{\omega_0} \equiv R_T C \tag{8-60}$$

The parameter ω_0 is called the **undamped natural frequency,** or simply the **natural frequency,** and ζ is called the **damping ratio.** In terms of these two parameters the homogeneous equation is

$$\frac{1}{\omega_0^2} \frac{d^2v_C}{dt^2} + \frac{2\zeta}{\omega_0} \frac{dv_C}{dt} + v_C = 0$$

The trial solution of the form $v_C = Ke^{st}$ then leads to the characteristic equation

$$\frac{1}{\omega_0^2} s^2 + \frac{2\zeta}{\omega_0} s + 1 = 0 \quad \text{or} \quad s^2 + 2\zeta\omega_0 s + \omega_0^2 = 0$$

whose roots are

$$s_1, s_2 = -\zeta\omega_0 \pm \omega_0 \sqrt{\zeta^2 - 1}$$

We can now begin to see the virtue of these two parameters. The radicand defining the roots depends only on ζ, the damping ratio. We can write down the three possible types of roots in terms of that parameter.

CASE A   For $\zeta > 1$, the radicand is positive and there are two unequal, real roots

$$\alpha_1 = -\zeta\omega_0 + \omega_0 \sqrt{\zeta^2 - 1}$$

and

$$\alpha_2 = -\zeta\omega_0 - \omega_0 \sqrt{\zeta^2 - 1}$$

and the natural response will be of the form

$$v_{CN}(t) = K_1 e^{\alpha_1 t} + K_2 e^{\alpha_2 t}$$

CASE B   For $\zeta = 1$, the radicand vanishes and there are two real, equal roots

$$s_1 = s_2 = -\zeta\omega_0$$

and the natural response has the form

$$v_{CN}(t) = (K_1 + K_2 t) e^{-\zeta\omega_0 t}$$

CASE C   For $\zeta < 1$, the radicand is negative leading to two complex, conjugate roots $s_1, s_2 = \alpha \pm j\beta$, where

$$\alpha = -\zeta\omega_0 \quad \text{and} \quad \beta = \omega_0 \sqrt{1 - \zeta^2}$$

and the natural response has the form

$$v_{CN}(t) = e^{\alpha t}(K_1 \cos \beta t + K_2 \sin \beta t)$$

In other words, for $\zeta > 1$ we have the overdamped response, for $\zeta = 1$ we have the critically damped response, and for $\zeta < 1$ we have the underdamped response.

We can now at least write down the general form of the step response of the series RLC circuit:

$$v_C(t) = A + v_{CN}(t) \qquad \text{(8-61)}$$

where $v_{CN}(t)$ takes one of the three forms listed. The appropriate form is determined by the value of the damping ratio ζ. The arbitrary constants in each form can be evaluated from the initial conditions applied to Eq. 8-61. In other words, it takes three kinds of parameters to determine the step response:

1. The amplitude of the input step function (A).
2. The values of the initial conditions (V_o and I_o)
3. The natural frequency and damping ratio (ζ and ω_0).

In this regard the natural frequency and damping ratio play the same role for double-memory circuits that the time constant plays for single-memory circuits. That is, they are parameters that completely describe the form of the natural response, just as the circuit time constant completely defines the form of the natural response for a single-memory circuit. That it should take two parameters to do this should not be surprising since the circuit has two memory elements.

Example 8-17

The circuit in Figure 8-24 is a series RLC circuit with zero initial conditions and a step-function input. The differential equation for the circuit is

$$LC \frac{d^2 v_C}{dt^2} + RC \frac{dv_C}{dt} + v_C = A \qquad t > 0$$

Upon inserting the numerical values given in the figure,

$$10^{-6} \frac{d^2 v_C}{dt^2} + \frac{10^{-3}}{2} \frac{dv_C}{dt} + v_C = 10$$

from which it is apparent that $v_{CF} = 10$ V. Comparing this equation to the standard form yields

$$\frac{1}{\omega_0^2} = 10^{-6}$$

And

$$\frac{2\zeta}{\omega_0} = \frac{10^{-3}}{2}$$

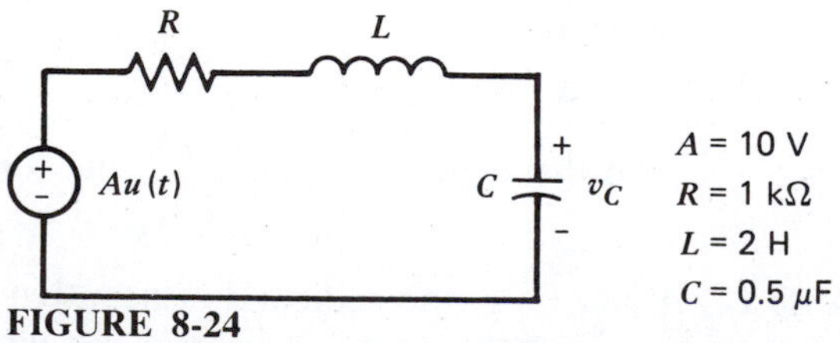

FIGURE 8-24
Double-memory circuit for Example 8-17.

In the figure:
R   L
Au(t)
$+$
C v_C
$-$
A = 10 V
R = 1 kΩ
L = 2 H
C = 0.5 μF

Hence

$$\omega_0 = 10^3 \text{ and } \zeta = 0.25$$

Since the damping ratio is less than unity we now know that the response is underdamped. Thus

$$\alpha = -\zeta\omega_0 = -250$$

and

$$\beta = \omega_0 \sqrt{1 - \zeta^2} = 500 \sqrt{3}$$

and the step response has the form

$$v_C(t) = 10 + e^{-250t}(K_1 \cos 500 \sqrt{3} \, t$$
$$+ K_2 \sin 500 \sqrt{3} \, t)$$

which is the underdamped form. The constants K_1 and K_2 can be determined now from the initial conditions (both zero):

$$v_C(0) = 10 + K_1 = 0$$
$$v_C'(0) = -250 K_1 + 500 \sqrt{3} K_2 = 0$$

which yields $K_1 = -10$ and $K_2 = -5/\sqrt{3}$. Thus the step response is

$$v_C(t) = 10 - e^{-250t} (10 \cos 500 \sqrt{3} \, t$$
$$+ \frac{5}{\sqrt{3}} \sin 500 \sqrt{3} \, t)$$

This response is plotted in Figure 8-25. The waveform is zero at $t = 0$ as required by the initial condition, and eventually settles down on the final value of 10 V. The final value is the amplitude of the forced response since the natural response decays to zero. However, between zero and the steady-state value the response initially overshoots the mark, then undershoots, and gradually decays to the steady response. This damped sinusoidal behavior is, of course, the result of the underdamped natural response.

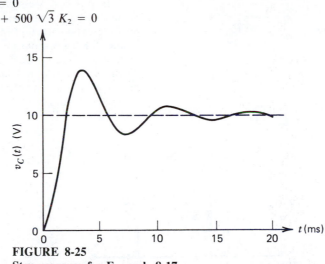

FIGURE 8-25
Step response for Example 8-17.

Example 8-18

The circuit in Figure 8-26 contains two equal resistors (R), two equal capacitors (C), and a voltage-controlled voltage source (μv_B). We develop the differential equation for this circuit by writing node equations at nodes Ⓐ and Ⓑ. The KCL equations are

Node Ⓐ $i_1 - i_2 + i_3 = 0$

Node Ⓑ $i_2 - i_4 = 0$

The element constraints are

$$i_1 = G(v_S - v_A) \qquad i_2 = G(v_A - v_B)$$
$$i_3 = \frac{Cd(v_o - v_A)}{dt} \qquad i_4 = Cdv_B/dt \qquad v_o = \mu v_B$$

The last of these equations comes from the controlled source. Substituting the element equations into the connection equations yields the node equations as

$$Gv_S - 2Gv_A + Gv_B + \mu C \frac{dv_B}{dt} - C \frac{dv_A}{dt} = 0$$

$$Gv_A - Gv_B - C\frac{dv_B}{dt} = 0$$

Solving the second node equation for v_A,

$$v_A = v_B + RC\frac{dv_B}{dt}$$

and substituting this result into the first equation yields the differential equation of the circuit as

$$(RC)^2\frac{d^2v_B}{dt^2} + RC(3 - \mu)\frac{dv_B}{dt} + v_B = v_S$$

This is a second-order differential equation of the same form we have studied. Comparison with the standard form yields

$$\frac{1}{\omega_0^2} = (RC)^2$$

Hence

$$\omega_0 = \frac{1}{RC} \quad \text{and} \quad \frac{2\zeta}{\omega_0} = RC(3 - \mu)$$

Hence

$$\zeta = \frac{3 - \mu}{2}$$

Thus the natural frequency is determined by the product RC and the damping ratio by the gain of the controlled source. Specifically:

CASE A   If $\mu < 1$, then $\zeta > 1$, and we have the overdamped response.

CASE B   If $\mu = 1$, then $\zeta = 1$, and we have the critically damped response.

CASE C   If $3 > \mu > 1$, then $0 < \zeta < 1$, and we have the underdamped response.

There are two other interesting cases for this circuit that are not available in the passive RLC circuits. If $\mu = 3$, then $\zeta = 0$, and we have zero damping. This corresponds to a sustained or eternal oscillation. The energy required to sustain the oscillation comes from the controlled source, which is a model of an active device that must have a power supply to operate. Even more dramatic is the case $\mu > 3$, and hence $\zeta < 0$. In this case we have negative damping, which means that the exponentials in the response have positive exponents and the response grows without bound. Such a response is called **unstable** and is generally considered undesirable. We treat the question of circuit stability in a later chapter. The essential message of this example is the great value of ζ and ω_0 as descriptors of the response of double-memory circuits.

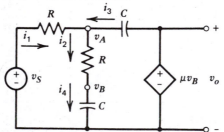

FIGURE 8-26
Dependent-source double-memory circuit
(Example 8-18).

SUMMARY

- Single-memory circuits consist of linear memoryless elements and the equivalent of one capacitor or one inductor. The response of a single-memory circuit can be determined by solving a first-order differential equation of the form

$$T_C \frac{dx}{dt} + x = f(t)$$

where $x(t)$ is a current or voltage, $f(t)$ is the equivalent input source, and T_C is the circuit time constant.

- The zero-input response of a single-memory circuit is of the form

$$x(t) = X_o \, e^{-t/T_C}$$

where X_o is the value of $x(t)$ at $t = 0$.

- The response for nonzero inputs can be found by:
  (1) Finding the natural response that is the general solution of the differential equation with zero input.
  (2) Finding the forced response that is a particular solution of the differential equation for the specified input.
  (3) Adding the natural and forced responses and determining the unknown constants in the natural response from the initial conditions.

- The step response of a single-memory circuit has the form

$$x(t) = (IC - FC)e^{-t/T_C} + FC \qquad t > 0$$

where IC is the initial condition of $x(t)$ and FC is the final condition of $x(t)$.

- For constant inputs the initial and final values can be determined by replacing capacitors by open circuits and inductors by short circuits, and analyzing the resulting memoryless circuit.

- The sinusoidal response of a single-memory circuit has the form

$$x(t) = K \, e^{-t/T_C} + A \cos(\omega t) + B \sin(\omega t)$$

The constants A and B can be determined by the method of undetermined coefficients and the constant K from the initial conditions.

- Double-memory circuits consist of linear memoryless elements and the equivalent of two memory elements (capacitors and/or inductors). The response of a double-memory circuit can be determined by solving a second-order differential equation of the form

$$\frac{1}{\omega_0^2} \frac{d^2x}{dt^2} + \frac{2\zeta}{\omega_0} \frac{dx}{dt} + x = f(t)$$

where $x(t)$ is a current or a voltage, $f(t)$ is the equivalent input, ω_0 is the circuit undamped natural frequency, and ζ is the circuit damping ratio.

- The zero-input response of a double-memory circuit takes one of three possible forms, depending on the nature of the roots of the characteristic equation

$$\frac{1}{\omega_0^2} s^2 + \frac{2\zeta}{\omega_0} s + 1 = 0$$

Case A (overdamped): $\zeta > 1$, two real, unequal roots α_1 and α_2:

$$x_N(t) = K_1 e^{-\alpha_1 t} + K_2 e^{-\alpha_2 t}$$

Case B (critically damped): $\zeta = 1$, two real, equal roots α:

$$x_N(t) = (K_1 + K_2 t)e^{-\alpha t}$$

Case C (underdamped): $\zeta < 1$, two complex, conjugate roots $\alpha \pm j\beta$:

$$x_N(t) = e^{-\alpha t}(K_1 \cos \beta t + K_2 \sin \beta t)$$

- The step response of a double-memory circuit is of the form

$$x(t) = x_N(t) + FC$$

where $x_N(t)$ is the natural response and FC is the final condition of the circuit.

- Memory circuits containing active devices can produce zero or even negative damping. The former corresponds to a sustained oscillation while the latter case causes the output to grow without bound and is called unstable.

EN ROUTE OBJECTIVES
AND RELATED EXERCISES

8-1 SINGLE-MEMORY CIRCUITS (SECS. 8-1 to 8-4)

Given a circuit consisting of linear memoryless elements and the equivalent of one capacitor or one inductor:

(a) *Determine the circuit differential equation and the initial conditions (if not given).*

(b) *Solve for the natural and forced components of selected signal variables.*

Exercises

8-1-1 Solve the following differential equations.

(a) $\dfrac{dx}{dt} + 10x = 0$, $x(0) = 5$

(b) $10^{-4}\dfrac{dx}{dt} + x = 10\,u(t)$, $x(0) = -10$

(c) $\dfrac{dx}{dt} - 10x = -10\,u(t)$, $x(0) = 0$

(d) $10^{-3}\dfrac{dx}{dt} + 10\,x = -10\,u(t)$,

$x(0) = -2$

(e) $\dfrac{dx}{dt} + 5x = 10\cos 5t$, $x(0) = 0$

(f) $10^{-4}\dfrac{dx}{dt} + x = 10\sin 100t$, $x(0) = 3$

(g) $10^{-2}\dfrac{dx}{dt} + x = 5\cos(100t - 60°)$,

$x(0) = 0$

8-1-2   For each of the circuits in Figure E8-1-2 write
an expression for the circuit time constant or
determine the numerical value of the time
constant.

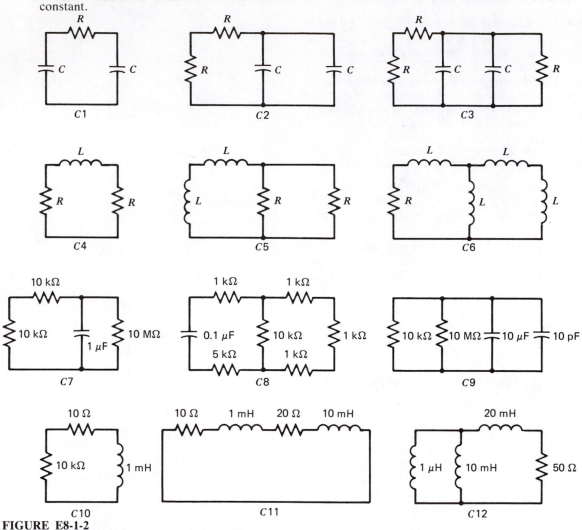

FIGURE E8-1-2
Circuits for determining time constants.

8-1-3 In each of the circuits in Figure E8-1-3 the switch has been open for a "long time." At $t = 0$ the switch is closed. Determine $v_C(t)$ or $i_L(t)$, sketch its waveform, and identify the forced and natural components.

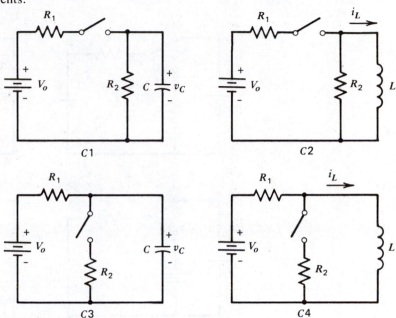

FIGURE E8-1-3
RL and RC circuits for determining step responses.

8-1-4 In each of the circuits in Figure E8-1-3 the switch has been closed for a "long time." At $t = 0$ the switch is opened. Determine $v_C(t)$ or $i_L(t)$, sketch its waveform, and identify the forced and natural components.

8-1-5 In each of the circuits in Figure E8-1-5 the switch has been in position A for a "long time." At $t = 0$ the switch is moved to position B. Determine $v_C(t)$ or $i_L(t)$, sketch its waveform, and identify the forced and natural components.

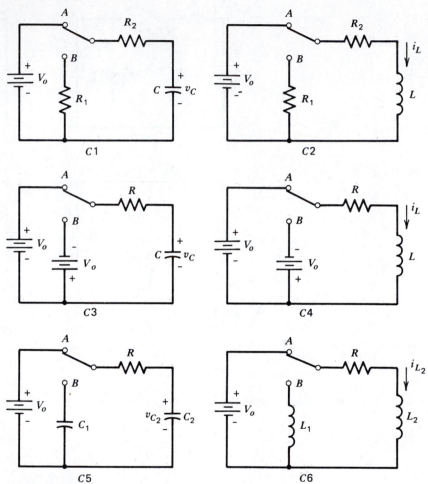

FIGURE E8-1-5
RL and *RC* circuits for determining forced and natural responses.

8-1-6 In each of the circuits in Figure E8-1-5 the switch has been in position B for a "long time." At $t = 0$ the switch is moved to position A. Determine $v_C(t)$ or $i_L(t)$, sketch its waveform, and identify the forced and natural components.

8-1-7 In each of the circuits in Figure E8-1-7 the
initial conditions are zero and a step function
$A\,u(t)$ is applied where A is either V_o or I_o.
Determine the output voltage $v_o(t)$, sketch its
waveform, and identify the forced and natural
components.

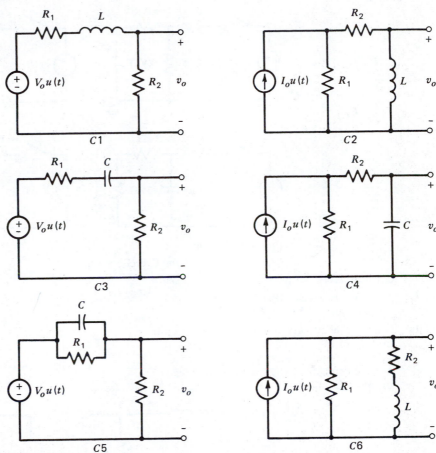

FIGURE E8-1-7
RL and *RC* circuits for finding step responses.

8-1-8   In each of the circuits in Figure E8-1-8 the switch has been open for a "long time" and is closed at $t = 0$. For $v_S(t) = 10\ u(t)$ solve for the indicated signal variable, sketch its waveform, and identify the forced and natural components.

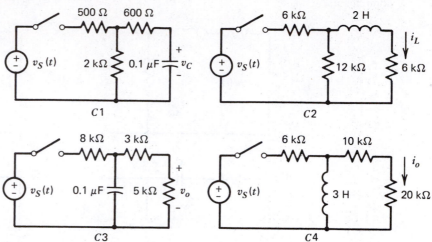

FIGURE E8-1-8
RL and *RC* circuits for Exercise 8-1-8.

8-1-9   Repeat Exercise 8-1-8 for $v_S(t) = 10 \cos 5000t$.

8-1-10   Repeat Exercise 8-1-8 for $v_S(t) = 10 \cos 5000t + 5 \sin 5000t$.

8-1-11   Each of the circuits in Figure E8-1-11 is operating in the sinusoidal steady state with $v_S(t) = A \cos \omega t$. Determine the output voltage $v_o(t)$ and state whether the output leads or lags the input.

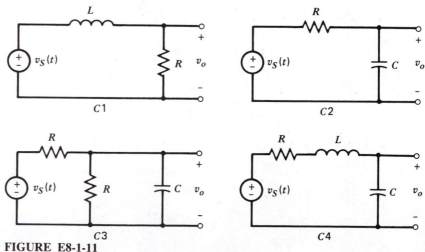

FIGURE E8-1-11
RL and *RC* circuits driven by sinusoidal signals.

8-1-12 Show that each of the circuits in Figure 8-1-12 has a time constant $T_C = RC$. Then solve for the output voltage $v_o(t)$ if $v_S(t) = A\,u(t)$ and the initial conditions are zero.

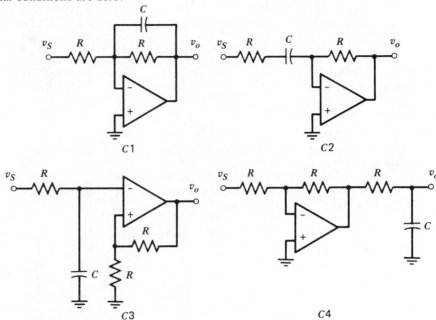

FIGURE E8-1-12
OP AMP circuits for Exercise 8-1-12.

8-2 DOUBLE-MEMORY CIRCUITS (SECS. 8-5 and 8-6)

Given a circuit consisting of linear memoryless elements and the equivalent of two memory elements (capacitors and/or inductors):
(a) Determine the circuit differential equation and the initial conditions (if not given).
(b) Solve for the forced and natural components of selected signal variables for step function inputs.

Exercises

8-2-1 Solve the following differential equations.

(a) $\dfrac{d^2x}{dt^2} + 7\dfrac{dx}{dt} + 10x = 0$

$x(0) = 0,\; x'(0) = 15$

(b) $\dfrac{d^2x}{dx^2} + 4\dfrac{dx}{dt} + 4x = 0$

$x(0) = 0,\; x'(0) = 2$

(c) $\dfrac{d^2x}{dx^2} + 2\dfrac{dx}{dt} + 5x = 0$

$x(0) = 0,\ x'(0) = 2$

(d) $\dfrac{d^2x}{dt^2} + 11\dfrac{dx}{dt} + 10x = 100u(t)$

$x(0) = 5,\ x'(0) = 5$

(e) $\dfrac{d^2x}{dt^2} + 10\dfrac{dx}{dt} + 25x = 100u(t)$

$x(0) = 5,\ x'(0) = 25$

(f) $\dfrac{d^2x}{dt^2} + 2\dfrac{dx}{dt} + 10x = 100u(t)$

$x(0) = 5,\ x'(0) = 5$

8-2-2   For each of the circuits in Figure E8-2-2 determine the numerical value of the circuit damping ratio (ζ) and the undamped natural frequency (ω_0). Indicate whether the circuit is overdamped, underdamped, or critically damped.

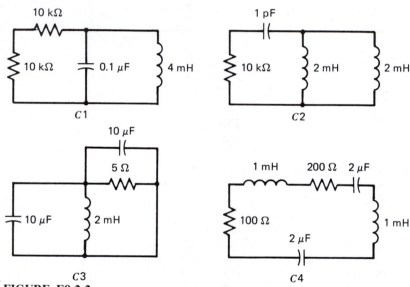

FIGURE E8-2-2
Double-memory (second-order) circuits for Exercise 8-2-2.

8-2-3   For each of the circuits in Figure E8-2-3 write an expression for the circuit damping ratio (ζ) and the undamped natural frequency (ω_0) in terms of the circuit parameters. Then derive an expression for the value of R that will produce critical damping.

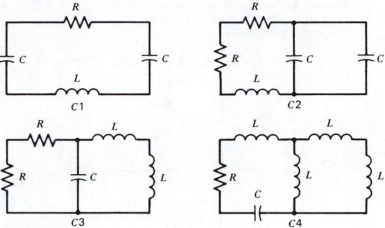

FIGURE E8-2-3
Circuits for determining ζ, ω, and critical resistance.

8-2-4   In each of the circuits in Figure E8-2-4 the switch has been open for a "long time." At $t = 0$ the switch is closed. For $L = 0.4$ H, $C = 0.1$ μF, $R = 3$ kΩ, and $V_0 = 10$ V, determine $v_C(t)$ or $i_L(t)$ as indicated, identify the forced and natural components of the response, and sketch the response waveform. Indicated whether the response is over-damped or underdamped.

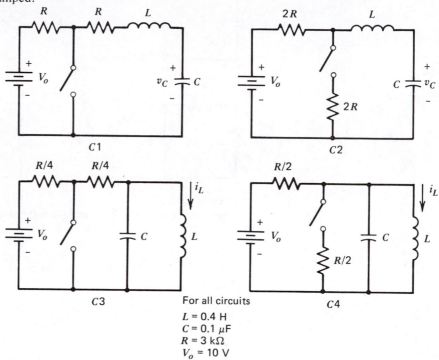

FIGURE E8-2-4
Step-response double-memory circuits.

For all circuits
$L = 0.4$ H
$C = 0.1$ μF
$R = 3$ kΩ
$V_o = 10$ V

8-2-5   Repeat Exercise 8-2-4 assuming the switch is first closed for a "long time" and then is opened at $t = 0$.

8-2-6 In each of the circuits in Figure E8-2-6 the switch has been in position *A* for a "long time." At $t = 0$ it is moved to position *B*. For $R_1 = 40$ kΩ, $R_2 = 2.5$ kΩ, $L = 2$ H, $C = 0.02$ μF, and $V_o = 15$ V, determine $v_o(t)$ or $i_L(t)$ as indicated, identify the forced and natural components of the response, and sketch the response waveform. Indicate whether the response is overdamped or underdamped.

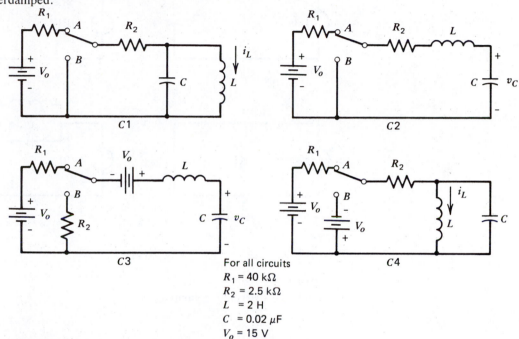

For all circuits
$R_1 = 40$ kΩ
$R_2 = 2.5$ kΩ
$L = 2$ H
$C = 0.02$ μF
$V_o = 15$ V

FIGURE E8-2-6
Double-memory *RLC* circuits for Exercise 8-2-6.

8-2-7 Repeat Exercise 8-2-6 assuming the switch first is in position *B* for a "long time" and then is moved to position *A*.

8-2-8   Each of the circuits in Figure E8-2-8 has a
step-function input and the initial conditions
are zero. Determine the output voltage $v_o(t)$
for $L = 50$ mH, $C = 0.05$ μF, $R = 4$ kΩ,
and $A = 100$ V. Indicate whether the response is overdamped, underdamped, or critically damped.

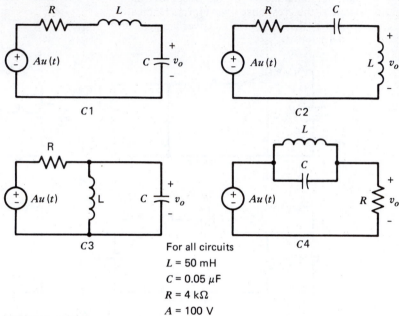

For all circuits
$L = 50$ mH
$C = 0.05$ μF
$R = 4$ kΩ
$A = 100$ V

FIGURE E8-2-8
Circuits for determining damping.

8-2-9 Show that each of the circuits in Figure E8-2-9 is critically damped regardless of the values of R and C.

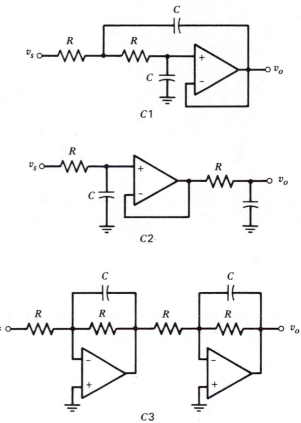

FIGURE E8-2-9
Critically damped OP AMP circuits.

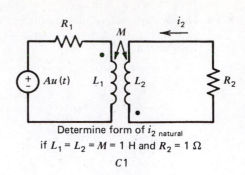

Determine form of $i_{2 \text{ natural}}$
if $L_1 = L_2 = M = 1$ H and $R_2 = 1$ Ω

C1

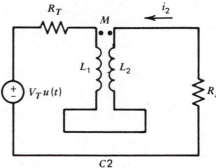

C2

FIGURE E8-2-10
Transformer circuits for Exercise 8-2-10.

8-2-10   For each transformer circuit of Figure E8-2-10, solve for $i_2(t)$.

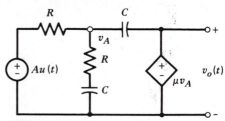

FIGURE E8-2-11
Dependent-source stability problem.

8-2-11   What would you do to the circuit of Figure E8-2-11 so that it remains stable?

PROBLEMS

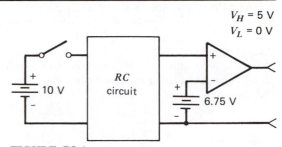

$V_H = 5$ V
$V_L = 0$ V

FIGURE P8-1
RC **timing circuit design.**

P8-1 (Design)

The timing circuit in Figure P8-1 is to be designed so that the output of the comparator (see Sec. 4-8) goes to the high level exactly 1.25 s after the switch is closed. Design an *RC* circuit to accomplish this objective using any of the standard resistors and capacitors described in Appendix C.

P8-2 (Design)

Figure P8-2 shows a sample-hold circuit often used in analog-to-digital signal conversion. In the sample mode, the converter commands the analog switch to close (ON) and the capacitor charges to the current value of the analog input signal. In the hold mode, the converter commands the analog switch to open (OFF) and the capacitor stores the sampled value which feeds to the converter through the OP AMP voltage follower. When the conversion process is completed, the switch is again commanded to close and the sample-hold cycle is repeated. The entire process is subject to two limitations:

(1) The number of sample-hold cycles per second must be at least twice the highest frequency contained in the input signal.

(2) The hold mode must be about ten times as long as the sample mode.

The design problem is to select a suitable value of capacitance. The value must be small enough so that the capacitor voltage can change rapidly during the sample mode and large enough so that the capacitor voltage does not change appreciably during the hold

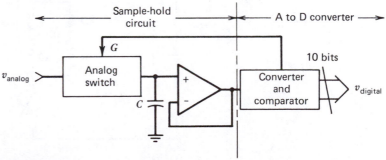

FIGURE P8-2
Analog-to-digital converter sample-hold problem.

mode. For an input signal $v(s) = 5 + 5 \cos 2\pi 1000t$ and an analog switch with $R_{ON} = 200\Omega$ and $R_{OFF} = 500\,k\Omega$; determine the range of acceptable capacitance and select a suitable capacitor from Appendix C.

P8-3 (Analysis)
For the circuit shown in Figure P8-3, find a differential equation relating the desired output $v_o(t)$ and the input $i_s(t)$.

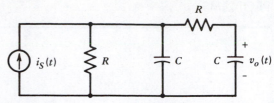

FIGURE P8-3
Differential equation problem.

P8-4 (Analysis)
The circuit of Figure P8-4 is excited by $i_s(t) = 3.3e^{-400t}A$. Find $i_L(t)$ if $i_L(0) = 20$ A.

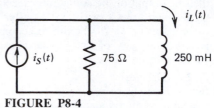

FIGURE P8-4
RL circuit driven by exponential source.

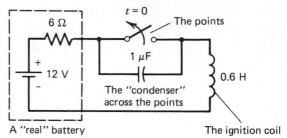

P8-5 (Analysis)
The circuit in Figure P8-5 represents the ignition system of an automobile with a standard ignition (with points). If the switch (the points) opens suddenly, how would you sketch the voltage across the inductor v_L? Assume the system has reached equilibrium with the points closed.

FIGURE P8-5
Automobile ignition system problem.

P8-6 (Analysis)
A circuit often used to avoid arching of the contacts of a relay is shown in Figure P8-6a. Qualitatively explain how the diode achieves this small feat.

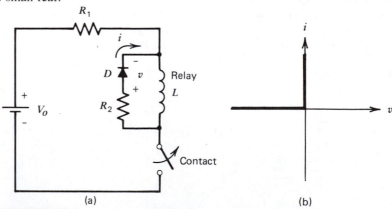

FIGURE P8-6
Diode bypass RL circuit. (a) Circuit. (b) Diode i-v characteristics.

Chapter 9
Laplace Transforms

Shall I refuse my dinner because I do not fully understand the process of digestion.

Oliver Heaviside

The Laplace transformation is used to alter the form of signals and systems to simplify the study of signal processing. The method provides a viewpoint and a terminology that pervade the study of linear circuits. The central concept of the chapter is the idea of a transformation as a mathematical process that alters the form of data, but not the data themselves. The basic pattern, properties, and pairs of the transformation are presented in the first two sections. The third section introduces a new way of looking at signals via poles and zeros. The fourth and fifth sections treat the inverse transformation. Transform methods are applied to single-memory circuit responses in the final two sections.

9-1 SIGNAL WAVEFORMS AND TRANSFORMS

A transformation is the process of changing the form of data in accordance with a specified rule. The conversion of decimal numbers into binary numbers at the input to a computer is an example of a transformation. In the present context, transformations are used to obtain alternative representations of a signal or a system. They are used because they often simplify certain signal-processing operations, or because they provide a different perspective that can be quite useful or even essential. Modern signal processing uses a host of transformation methods, including Fourier transforms, fast Fourier transforms (FFT), discrete Fourier transforms (DFT), Z-transforms, and Laplace transforms. In every case, these methods involve a specific transformation rule, make certain analysis techniques more manageable, and provide a useful viewpoint for system design.

In this book, we concern ourselves only with the Laplace transformation. Our initial exposure to Laplace ($\mathscr{L}$) transforms will follow the pattern shown in Figure 9-1. The process begins with a linear circuit and the differential equation that characterize its behavior. This equation is transformed into the complex frequency domain, where it is changed into an algebraic equation. Simple algebraic techniques are then used to obtain the solution in the frequency domain. The inverse Laplace ($\mathscr{L}^{-1}$) transform changes the s-domain solution back into the time domain, yielding the solution of the original differential equation. The figure also points out that there is another route to the solution using the classical techniques studied in Chapter 8. For simple circuits, this route may be easier and more direct. But for more complex circuits, the advantages of the Laplace transformation can be quite significant.

An advantage of the transform approach is that solving a differential equation becomes a purely algebraic process. However, Laplace transforms are more than just another way to solve differential equations. They offer a new and different perspective toward circuit behavior. As we shall see in subsequent chapters, we will not transform the circuit differential equation, but rather the circuit itself. Put differently, eventually the reader must learn to think about signals and systems in both the familiar time domain and the arena of Laplace transforms called the **complex frequency domain.**

Symbolically we represent the Laplace transformation operation as

$$\mathscr{L}\,[v(t)] \,=\, V(s) \tag{9-1}$$

In words, this expression says that $V(s)$ is the Laplace transform of the waveform $v(t)$. The transformation operation involves two domains: (1) the time domain in which the signal is characterized by its waveform $v(t)$ and (2) the complex frequency domain in which the signal is represented by its transform $V(s)$. The variable s is called the complex frequency variable and, being complex, is equal to $\sigma + j\omega$, where σ is the attenuation

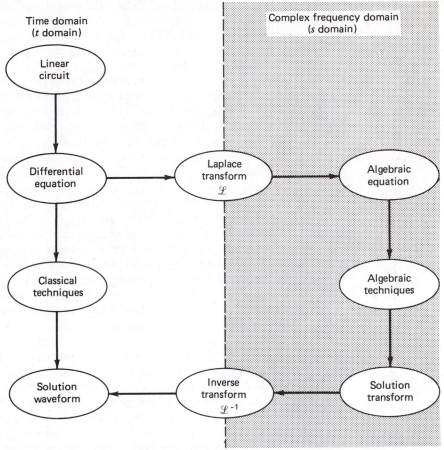

FIGURE 9-1
Laplace transformation pattern.

constant and ω is the radian frequency, variable in the time domain.[1] In sum, a signal can uniquely be expressed as a waveform or a transform. Collectively $v(t)$ and $V(s)$ are called a **transform pair,** meaning alternative representations of the same thing, the signal. Note that a lowercase letter is used to represent the signal waveform and an uppercase letter for its transform.

The Laplace transform of a signal waveform $v(t)$ is defined as

$$V(s) = \int_{0-}^{\infty} v(t)\, e^{-st}\, dt \qquad\qquad \textbf{(9-2)}$$

This definition involves an improper integral (the upper limit is infinite), and so we must discuss the conditions under which the integral converges.

---

[1] j is used in electrical engineering to represent the square root of negative one ($\sqrt{-1}$); i is not used, to avoid confusion with the symbol for electric current.

Briefly, the integral will exist if the waveform $v(t)$ is piecewise continuous and of exponential order. Piecewise continuous means that in any finite interval, $v(t)$ has a finite number of steplike discontinuities. Exponential order means that there exists constants K and b such that $|v(t)| < Ke^{bt}$ for all t greater than some value T. Fortunately essentially all signals encountered in real systems meet these two conditions.

Equation 9-2 used a lower limit denoted $t = 0-$ to indicate a time just a whisker before $t = 0$. This is done because in dealing with system response we divide all of history into two eras as a result of an epoch-making event that occurs at $t = 0$, the movement of a switch, for example. Occasionally the epoch-making event leads to a discontinuity in $v(t)$ at $t = 0$, that is, $v(0-) \neq v(0)$. To capture this discontinuity in the integration process, we set the lower limit at $t = 0-$, just prior to the event. Fortunately, in most situations, there is no discontinuity, so we will not distinguish between $t = 0-$ and $t = 0$, unless it is crucial.

Finally it is important to note at the outset that the Laplace transformation that we will deal with in this text inherently involves waveforms that are *identically zero for all negative t*. This is not obvious from the definition of the direct transformation in Eq. 9-2 since the integral does not involve the values of $v(t)$ for negative t. However, the inverse transformation, which we study in a later section, always produces a waveform that is zero for $t < 0$. Hence to preserve *uniqueness* of the transformation we must assume that all waveforms are identically zero for $t < 0$.[2]

Equally fortunate is the fact that the number of different signal waveforms encountered in linear systems is relatively small. The list includes the basic waveforms studied in Chapter 6 (step, exponential, sinusoid), as well as a number of composite waveforms such as the exponential rise, damped ramp, impulse, ramp and damped sinusoid. Since the total number is relatively small, we do not need to appeal to the definition repeatedly in using transform methods. Once the transform of a waveform has been found, it can be cataloged in a Table of Transform pairs for future reference and use. A brief table of important pairs is provided in this chapter (see Table 9-2). More extensive tables are available in other references such as in *Standard Mathematical Tables*, published yearly by the CRC Press.

In the remainder of this section, we will derive the transform of some important waveforms using the definition in Eq. 9-2.

Example 9-1

For the step function $v(t) = u(t)$, Eq. 9-2 becomes

$$v(s) = \int_0^\infty u(t)e^{-st}\, dt$$

Since $u(t) = 1$ throughout the range of integration,

$$V(s) = \int_0^\infty e^{-st}dt = \left. -\frac{1}{s}e^{-st}\right|_0^\infty$$

$$= \frac{1}{s} \text{ provided } \sigma = > 0$$

[2] The particular Laplace transform we use is called the *one-sided* transform since it deals with time ≥ 0. In advanced systems texts other Laplace transforms occur such as the *double-sided* transform covering all time from $-\infty$ to $+\infty$.

Whenever we find the transform of a waveform, we tacitly assume a $u(t)$, even if the waveform appears to extend from $-\infty$. This occurs because our definition of the Laplace transform is single sided from 0 to $+\infty$. Hence, the transform of $Ku(t)$ and K, a constant, for example, appear to be the same, K/s. This occurs since we have tacitly assumed the constant K to start at $t = 0$. If we wanted to consider some function occurring before $t = 0$, we cannot use our one-sided Laplace transform.

Example 9-2

For the exponential waveform $v(t) = e^{-at}$, we have

$$V(s) = \int_0^\infty e^{-at} e^{-st} dt = \int_0^\infty e^{-(s + a)t} dt$$

$$V(s) = \frac{-e^{-(s + a)t}}{s + a} \bigg|_0^\infty = \frac{1}{s + a}$$

provided $\sigma + a > 0$.

Example 9-3

For the impulse function $v(t) = \delta(t)$,

$$V(s) = \int_0^\infty \delta(t) e^{-st} dt$$

In this case the difference between $t = 0-$ and $t = 0$ is important since the impulse exists only at $t = 0$. To capture the impulse in the integration, we must take the lower limit at $t = 0-$. Since the impulse is zero, except at $t = 0$, the upper limit can be taken as $t = 0+$.

$$V(s) = \int_{0-}^{0+} \delta(t) e^{st} dt = \int_{0-}^{0+} \delta(t) dt = 1$$

The impulse is the waveform whose transform is 1.

9-2 LAPLACE TRANSFORM PROPERTIES AND PAIRS

The previous section gave the definition of the Laplace transformation and showed that it could be used directly to obtain the transform of some simple waveforms. In this section, we develop some of the general properties of the Laplace transformation and show that these can be used to obtain additional transform pairs.

The foremost feature of the Laplace transformation is that it is a linear operation. If

$$\mathcal{L}\{v_1(t)\} = V_1(s) \quad \text{and} \quad \mathcal{L}\{v_2(t)\} = V_2(s) \tag{9-3}$$

and if A and B are constants, then

$$\mathcal{L}\{Av_1(t) + Bv_2(t)\} = \int_0^\infty [Av_1(t) + Bv_2(t)] e^{-st} dt$$

$$= A \int_0^\infty v_1(t) e^{-st} dt + B \int_0^\infty v_2(t) e^{-st} dt$$

Hence

$$\mathcal{L}\{Av_1(t) + Bv_2(t)\} = AV_1(s) + BV_2(s) \tag{9-4}$$

The linearity property is a very important feature. Among its many implications is that Kirchhoff's laws are valid in both the time domain and the complex frequency domain, but that is getting ahead of our story. The next two examples show how this property can be used to obtain the transform of the exponential rise and the sinusoid.

Example 9-4

The exponential rise waveform $v(t) = u(t)(1 - e^{-at})$ is the difference between a step function and an exponential. Therefore, we can use the linearity property of Laplace transforms to write

$$\mathcal{L}\{v(t)\} = \mathcal{L}\{u(t)(1 - e^{-at})\}$$
$$= \mathcal{L}\{u(t)\} - \mathcal{L}\{u(t)e^{-at}\}$$

Using the transform pairs for the step and exponential functions found in Examples 9-1 and 9-2,

$$V(s) = \frac{1}{s} - \frac{1}{s+a} = \frac{a}{s(s+a)}$$

Example 9-5

If $v(t) = \sin(\beta t)$, then using Euler's relationship

$$v(t) = \frac{e^{j\beta t}}{2j} - \frac{e^{-j\beta t}}{2j}$$

The transform pair $\mathcal{L}\{e^{-at}\} = 1/(s+a)$ obtained in Example 9-2 applies even if a is complex. Hence,

using linearity, we have

$$\mathcal{L}\{\sin(\beta t)\} = \frac{1}{2j}\mathcal{L}[e^{j\beta t}] - \frac{1}{2j}\mathcal{L}\{e^{-j\beta t}\}$$
$$= \frac{1}{2j}\left[\frac{1}{s-j\beta} - \frac{1}{s+j\beta}\right] = \frac{\beta}{s^2+\beta^2}$$

Since memory element i-v relationships involve the time-domain operations of integration and differentiation, their frequency-domain equivalents are important. Beginning with integration, we write as follows:

$$\text{If } \mathcal{L}\{v(t)\} = V(s) = \int_0^\infty v(t)e^{-st}dt, \tag{9-5}$$

$$\text{then } \mathcal{L}\left\{\int_0^t v(t)dt\right\} = \int_0^\infty \left\{\int_0^t v(t)dt\right\}e^{-st}dt \tag{9-6}$$

This expression can be integrated by parts using

$$y = \int_0^t v(t)dt, \; dx = e^{-st}dt$$

Hence

$$dy = v(t) \qquad x = \frac{-e^{-st}}{s}$$

Using these factors reduces Eq. 9-6 to

$$\mathcal{L}\left\{\int_0^t v(t)dt\right\} = \left[\frac{-e^{-st}}{s}\int_0^t v(t)dt\right]_0^\infty + \frac{1}{s}\int_0^\infty v(t)e^{-st}dt \tag{9-7}$$

The first term on the right in Eq. 9-7 is zero. It vanishes at the lower limit because $\int_0^0 f(t)dt = 0$. It vanishes at the upper limit if $\sigma = Re\{s\}$ is sufficiently large so that $e^{-st} \to 0$ at $t \to \infty$. Using Eq. 9-5 in the second term yields

$$\mathcal{L}\left\{ \int_0^t v(t)dt \right\} = \frac{V(s)}{s} \tag{9-8}$$

Integration of a waveform $v(t)$ in the time domain can be accomplished by the algebraic process of dividing its transform $V(s)$ by s in the frequency domain. The next example applies this property to obtain the transform of the ramp function.

Example 9-6

From Example 9-1, we have $\mathcal{L}\{u(t)\} = 1/s$. From our study of signals, we know that the ramp waveform can be obtained from $u(t)$ as

$$r(t) = \int_0^t u(t)dt$$

Hence

$$\mathcal{L}\{r(t)\} = \mathcal{L}\left\{ \int_0^t u(t)dt \right\}$$

$$= \frac{1}{s}\mathcal{L}\{u(t)\} = \frac{1}{s^2}$$

The differentiation operation is transformed into the frequency domain as follows:

$$\text{If } \mathcal{L}\{v(t)\} = V(s) = \int_0^\infty v(t)e^{-st}dt \tag{9-9}$$

then

$$\mathcal{L}\left\{ \frac{dv}{dt} \right\} = \int_0^\infty \frac{dv}{dt} e^{-st}dt \tag{9-10}$$

Again, integrating by parts using

$$y = e^{-st} \qquad dx = \frac{dv}{dt} dt$$

Hence

$$dy = -se^{-st} \qquad x = v$$

Using these factors reduces Eq. 9-10 to

$$\mathcal{L}\left\{ \frac{dv}{dt} \right\} = v(t)e^{-st} \Big|_0^\infty + (s) \int_0^\infty v(t)e^{-st}dt \tag{9-11}$$

For $\sigma = Re\{s\}$ sufficiently large, the first term on the right in this equation vanishes at the upper limit, leaving

$$\mathcal{L}\left\{ \frac{dv}{dt} \right\} = sV(s) - v(0-) \tag{9-12}$$

Time differentiation of the waveform $v(t)$ transforms into multiplication of $V(s)$ by s to within an additive constant $v(0-)$. Note that in this case we must use $t = 0-$ to account for any discontinuities in $v(t)$ that occur at $t = 0$. If none exist, the difference between 0 and $0-$ can be ignored. By repeated application of the differentiation rule in Eq. 9-12, we obtain

$$\frac{d^2v}{dt^2} = s^2V(s) - sv(0-) - v'(0-) \tag{9-13}$$

$$\frac{d^3v}{dt^3} = s^3V(s) - s^2v(0-) - sv'(0-) - v''(0-) \tag{9-14}$$

where $v' = dv/dt$ and $v'' = d^2v/dt^2$.

The fact that time integration and differentiation change into algebraic processes in the s-domain is a hallmark feature of the Laplace transformation. The next example shows how the differentiation rule can be used to obtain the transform of $v(t) = \cos \beta t$.

Example 9-7

From Example 9-5, we have $\mathscr{L}\{\sin \beta t\} = \beta/(s^2 + \beta^2)$. By differentiation, we can write

$$\frac{1}{\beta}\frac{d}{dt}\sin \beta t = \cos \beta t$$

$$\mathscr{L}\{\cos \beta t\} = \frac{1}{\beta}s\,\frac{\beta}{s^2 + \beta^2} - \sin 0$$

$$= \frac{s}{s^2 + \beta^2}$$

Hence

$$\mathscr{L}\{\cos \beta t\} = \frac{1}{\beta}\mathscr{L}\left\{\frac{d}{dt}\sin \beta t\right\}$$

Property	Time Domain	Frequency Domain
Independent variable	Time t	Complex frequency $s = \sigma + j\omega$
Signal representation	Waveform $v(t)$	Transform $V(s)$
Linearity property	$Av_1(t) + Bv_2(t)$	$AV_1(s) + BV_2(s)$ A and B are constants
Integration property	$\int_0^t v(t)dt$	$\dfrac{V(s)}{s}$
Differentiation property	$\dfrac{dv(t)}{dt}$	$sV(s) - v(0-)$
	$\dfrac{d^2v}{dt^2}$	$s^2V(s) - sv(0-) - v'(0-)$
	$\dfrac{d^3v}{dt^3}$	$s^3V(s) - s^2v(0-) - sv'(0-) - v''(0-)$

TABLE 9-1
Basic Laplace Transformation Properties

Signal	Waveform	Transform
Impulse	$\delta(t)$	1
Step or constant	$u(t)$	$\dfrac{1}{s}$
Ramp	$r(t) = tu(t)$	$\dfrac{1}{s^2}$
Exponential	e^{-at}	$\dfrac{1}{s+a}$
Damped ramp	te^{-at}	$\dfrac{1}{(s+a)^2}$
Sinusoid	$\cos \beta t$	$\dfrac{s}{s^2+\beta^2}$
Sinusoid	$\sin \beta t$	$\dfrac{\beta}{s^2+\beta^2}$
Damped sinusoid	$e^{-\alpha t}\cos \beta t$	$\dfrac{(s+\alpha)}{(s+\alpha)^2+\beta^2}$
Damped sinusoid	$e^{-\alpha t}\sin \beta t$	$\dfrac{\beta}{(s+\alpha)^2+\beta^2}$

TABLE 9-2
Basic Laplace Transform Pairs

The basic properties of the Laplace transformation are summarized in Table 9-1. Table 9-2 lists a basic set of transform pairs, including all of those derived in these two sections. The final example in this section shows how these properties and pairs can be used to obtain a transform when $v(t)$ is a linear combination of basic waveforms.

Example 9-8

Given the waveform $v(t) = e^{-2t} + \cos 2t - \sin 2t$, then by the linearity property,

$\mathcal{L}\{v(t)\} = V(s)$
$= \mathcal{L}\{e^{-2t}\} + \mathcal{L}\{\cos 2t\} - \mathcal{L}\{\sin 2t\}$

Each of these individual transforms is available in Table 9-2.

$$V(s) = \frac{1}{s+2} + \frac{s}{s^2+4} - \frac{2}{s^2+4}$$

Normally transforms are not left as linear combinations but are written as quotients of polynomials, that is, as rational functions. Rationalizing the above sum yields

$$V(s) = \frac{2s^2}{(s+2)(s^2+4)}$$

9-3 POLE-ZERO DIAGRAMS

If we look at the transforms of various signals in Table 9-2, we see that they are represented by rational functions in s. We can write each signal as a polynomial fraction in the following form.

$$V(s) = \frac{b_m s^m + b_{m-1} s^{m-1} + \cdots + b_1 s + b_0}{a_n s^n + a_{n-1} s^{n-1} + \cdots + a_1 s + a_0} \qquad (9\text{-}15)$$

The two polynomials can now be factored in terms of their roots:

$$V(s) = K \frac{(s - z_1)(s - z_2) \cdots (s - z_n)}{(s - p_1)(s - p_2) \cdots (s - p_n)} \qquad (9\text{-}16)$$

We define a **pole** as a value of s that makes the denominator zero, and therefore the function $V(s)$ undefined. Hence $p_1, p_2, \ldots p_n$ are all poles. Further, we define a **zero** as a value of s that makes the numerator zero, and therefore $V(s)$ zero. We find that $z_1, z_2, \ldots z_m$ are all zeros.

In general, poles and zeros are complex numbers, that is, $p = \alpha + j\beta$. It turns out to be beneficial to plot these values on a complex s-plane. In plotting our values of poles and zeros on the s-plane, we indicate a pole by a small "$\times$" and a zero by a small "$\circ$" at the proper location on the complex plane. For example, consider a simple exponential signal $v(t) = e^{-t}$. The transform of this signal is readily found from Table 9-2 to be $V(s) = 1/(s + 1)$. This signal has a single pole at $s = -1$ and no observable zeros. We can plot a representation of this signal on a pole-zero diagram as shown in Figure 9-2a.

The damped sinusoid, $e^{-\alpha t} \cos \beta t$, has a pole at $s = -\alpha - j\beta$, another at $s = -\alpha + j\beta$, and a zero at $s = -\alpha$. Its pole-zero plot is shown in Figure 9-2b. A unit step function, $u(t)$ has a transform equal to $1/s$, which means that it has no zeros but one pole at the origin. This is shown plotted in the s-plane in Figure 9-2c.

It should be apparent that, if one is given a pole-zero diagram, the waveform of the signal represented by the pole-zero diagram can be readily identified, even without transforming the signal back into the time domain. The utility of a pole-zero diagram will become clearer in subsequent chapters.

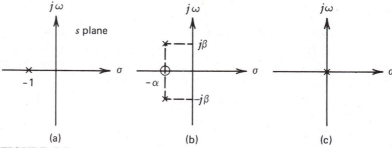

FIGURE 9-2
Pole-zero diagrams. (*a*) Decaying exponential. (*b*) Damped sinusoid. (*c*) Step function.

Example 9-9

In Example 9-8 we found the Laplace transform of the waveform

$$v(t) = e^{-2t} + \cos 2t - \sin 2t$$

to be

$$V(s) = \frac{2s^2}{(s + 2)(s^2 + 4)}$$

This transform has two zeros at $s = 0$, a pole at $s = -2$, and poles at the roots of $s^2 + 4 = 0$, which are $s = j2$ and $s = -j2$. The resulting pole-zero diagram is shown in Figure 9-3. We can attribute the poles of $V(s)$ to specific components of $v(t)$. The pole at $s = -2$ came from the component e^{-2t}, while the two conjugate poles on the imaginary axis came from the sinusoid $\cos 2t - \sin 2t$.

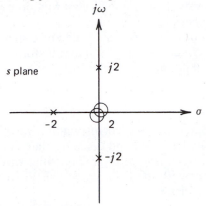

FIGURE 9-3
Pole- zero diagram for Example 9-9.

9-4   INVERSE LAPLACE TRANSFORMS

So far, we have been concerned with the process of obtaining the transform of a signal from its waveform. But, as the pattern in Figure 9-1 shows, we eventually will be confronted with the need to perform the inverse process. The process of recovering the waveform from a transform is called the *inverse Laplace transformation*. The important feature of this process is that the Laplace transformation is unique. Symbolically, we can state the uniqueness property as follows:

$$\text{If } \mathscr{L}\{v(t)\} = V(s), \text{ then } \mathscr{L}^{-1}\{V(s)\} (=) u(t)v(t)$$

The symbol $\mathscr{L}^{-1}$ stands for the inverse Laplace transform. The notation $(=)$ means equal almost everywhere. The only points at which equality may not hold are at the discontinuities of $v(t)$.

Finally, note that the waveform recovered by the inverse transformation is zero for $t < 0$. The mathematical justification for this is beyond the scope of our treatment. But to preserve uniqueness, that is, that one and only one signal waveform for each signal transform exists and vice-versa, we must assume that all waveforms involved are zero for negative time. This is a very important point. For example, when we write

$$V(s) = \mathscr{L}\{\sin \beta t\} = \frac{\beta}{s^2 + \beta^2}$$

and

$$v(t) = \mathcal{L}^{-1}\left\{\frac{\beta}{s^2 + \beta^2}\right\} = \sin \beta t$$

we do *not* mean an eternal sine wave, but a waveform $v(t) = u(t) \sin \beta t$ that starts at $t = 0$. The step function required is not always shown explicitly. However, an implied step function must be understood, otherwise the process is not unique.

The upshot of all this is that a table of Laplace transform pairs can be used in either direction: waveform to transform or transform to waveform. For example, if we are given the transform

$$V(s) = \frac{10}{s + 2}$$

and need to determine the corresponding waveform, we scan the transform column in Table 9-2 and discover that $V(s)$ is of the form of $1/(s + a)$. This is the transform of the exponential waveform. Using this observation, together with the linearity property, we conclude that the corresponding waveform is

$$v(t) = 10u(t)e^{-2t}$$

This is an exponential signal with an amplitude of 10 and a time constant of 0.5. Note that these data are contained in the transform. In other words, the transform and the waveform are equivalent representations of the exponential signal. Stated differently, the data needed to describe the signal are available in either the transform or the waveform. The Laplace transformation alters the form of the data, but not the data themselves.

Example 9-10

To determine the waveform for $V(s) = 15/(s^2 + 9)$, we scan the transform column in Table 9-2 and observe that $V(s)$ is nearly of the form of $\beta/(s^2 + \beta^2)$, which is the transform of $\sin \beta t$. Rearranging $V(s)$ as

$$V(s) = 5\frac{3}{s^2 + 3^2}$$

and using the linearity property again, we have

$$v(t) = 5\ u(t) \sin 3t$$

Again, all of the data needed to describe the signal are contained in either the transform or the waveform.

It is a simple matter to go from a transform to a waveform if we can find $V(s)$ in a table of pairs. Unfortunately it does not take a very complicated signal (or system) before we exceed the capability of Table 9-2, or even the more extensive tables available in the literature. However, all is not lost, because the general method of performing the inverse transformation is based on decomposing $V(s)$ into a linear combination of terms, each of which is available in Table 9-2.

Transforms of the basic signal waveforms are all of the form of a ratio of polynomials in the complex frequency variable s. A quick scan of the transform column in Table 9-2 will convince the reader that this is true, so far. A function that can be written as a quotient of polynomials is called a **rational function,** and all transforms of interest to us are rational functions. Consequently we write $V(s)$ as

$$V(s) = \frac{r(s)}{q(s)} = \frac{b_m s^m + b_{m-1} s^{m-1} + \cdots + b_1 s + b_0}{a_n s^n + a_{n-1} s^{n-1} + \cdots + a_1 s + a_0} \tag{9-17}$$

where $r(s)$ and $q(s)$ are polynomials in the complex frequency variable s, and a_k and b_k are real numbers. The transforms of interest to us are always quotients of polynomials with real coefficients. The two polynomials can be expressed in factored form in terms of their roots. Hence an alternative representation of $V(s)$ is

$$V(s) = K \frac{(s - z_1)(s - z_2) \cdots (s - z_m)}{(s - p_1)(s - p_2) \cdots (s - p_n)} \tag{9-18}$$

where z_k, $k = 1, 2, \ldots, m$ are the roots of the polynomial $r(s)$, and p_k, $k = 1, 2, \ldots n$ are the roots of the polynomial $q(s)$. The z_k's are called the zeros of the rational function $V(s)$, because when $s = z_k$, the function $V(s)$ vanishes. The p_k's are called poles, because when $s = p_k$, then $V(s)$ blows up. Collectively the poles and zeros are called critical frequencies because they represent values of the complex frequency variable s at which $V(s)$ does interesting things, such as vanish, or blow up.

If there are more poles than zeros ($n > m$), and if none of the poles are repeated roots of $q(s)$, then $V(s)$ can be decomposed by partial fractions as

$$V(s) = \frac{k_1}{s - p_1} + \frac{k_2}{s - p_2} + \cdots + \frac{k_n}{s - p_n} \tag{9-19}$$

Thus $V(s)$ can be expressed as a linear combination of terms, one term for each of its n distinct poles. The k's associated with each term are called residues. They are simply the amplitude or weight of each term in the decomposition. Each of the terms in the partial fraction decomposition of $V(s)$ is of the form of the transform of an exponential signal. Since we recognize each term, we can easily write down the corresponding waveform as

$$v(t) = u(t) \{k_1 e^{p_1 t} + k_2 e^{p_2 t} + \cdots + k_n e^{p_n t}\} \tag{9-20}$$

The poles end up as the exponents in the exponential waveform and the residues become the amplitudes.

Assuming that the poles of $V(s)$ are known, the algorithm for finding $v(t)$ reduces to finding the residues. Let us consider the case where $V(s)$ has three simple (unequal) poles and one zero.

$$V(s) = \frac{K(s - z_1)}{(s - p_1)(s - p_2)(s - p_3)} \tag{9-21}$$

$$= \frac{k_1}{s - p_1} + \frac{k_2}{s - p_2} + \frac{k_3}{s - p_3}$$

The residue k_1 can be found by first multiplying this equation through by the factor $(s - p_1)$:

$$(s - p_1)V(s) = \frac{K(s - z_1)}{(s - p_2)(s - p_3)} = k_1 + \frac{k_2(s - p_1)}{(s - p_2)} + \frac{k_3(s - p_1)}{(s - p_3)}$$

If we now set $s = p_1$, the last two terms on the right are annihilated, leaving

$$k_1 = (s - p_1)V(s) \Big|_{s = p_1} = \frac{K(s - z_1)}{(s - p_2)(s - p_3)} \Big|_{s = p_1} \tag{9-22}$$

The same approach applied to determining the residue k_2 yields

$$k_2 = (s - p_2)V(s) \Big|_{s = p_2} \tag{9-23}$$

The technique clearly generalizes to the case where k_i is the residue at any simple pole p_i, as

$$k_i = (s - p_i)V(s) \Big|_{s = p_i} \tag{9-24}$$

The process of determining the residue at any simple pole is sometimes called the **cover-up algorithm.** In essence, what we do is temporarily remove (cover up) the factor $(s - p_i)$ in $V(s)$, and then evaluate the remainder at $s = p_i$.

Example 9-11

The transform

$$V(s) = \frac{2(s + 3)}{s(s + 1)(s + 2)}$$

has simple poles at $s = 0$, $s = -1$, $s = -2$. Its partial fraction decomposition reads

$$V(s) = \frac{k_1}{s} + \frac{k_2}{s + 1} + \frac{k_3}{s + 2}$$

Using the cover-up algorithm to determine the residues,

$$k_1 = sV(s) \Big|_{s = 0} = \frac{2(s + 3)}{(s + 1)(s + 2)} \Big|_{s = 0} = 3$$

$$k_2 = (s + 1)V(s) \Big|_{s = -1} = \frac{2(s + 3)}{s(s + 2)} \Big|_{s = -1} = -4$$

$$k_3 = (s + 2)V(s) \Big|_{s = -2} = \frac{2(s + 3)}{s(s + 1)} \Big|_{s = -2} = 1$$

Hence the waveform $v(t)$ is

$$v(t) = u(t)[3e^{0t} - 4e^{-t} + e^{-2t}]$$

or

$$v(t) = 3u(t) - 4u(t)e^{-t} + u(t)e^{-2t}$$

An important consideration occurs when $V(s)$ has a simple, complex pole. As a result of the fact that $V(s)$ is a quotient of polynomials with real coefficients, it follows that a complex pole $p = \alpha + j\beta$ must be accompanied by a pole $p^* = \alpha - j\beta$. In other words, complex poles of $V(s)$ must occur in conjugate pairs. As a consequence, the partial fraction decomposition of $V(s)$ will contain the terms

$$V(s) = \cdots + \frac{k}{s - \alpha - j\beta} + \frac{k^*}{s - \alpha + j\beta} + \cdots \qquad \text{(9-25)}$$

The residues at the conjugate poles are, in turn, conjugates because $V(s)$ is a rational function with real coefficients. The residues at complex poles can be calculated using the cover-up algorithm. In general, they turn out to be complex numbers. Writing the residues in polar form

$$k = |k|\, e^{j\theta}$$

and hence

$$k^* = |k|\, e^{-j\theta}$$

produces the corresponding waveform as

$$v(t) = u(t)\, [\ldots|k|\, e^{j\theta}\, e^{\alpha t} e^{j\beta t} + |k|\, e^{-j\theta}\, e^{\alpha t} e^{-j\beta t} \ldots]$$

which can be rearranged as

$$v(t) = u(t)\, [\ldots 2|k|e^{\alpha t}\left[\frac{e^{j(\beta t + \theta)} + e^{-j(\beta t + \theta)}}{2}\right] \ldots]$$

The expression inside the bracket is of the form

$$\cos x = \frac{(e^{jx} + e^{-jx})}{2}$$

Consequently we can combine these terms as the damped sinusoid

$$v(t) = u(t)\, [\ldots + 2\,|k|e^{\alpha t} \cos\,(\beta t + \theta) + \ldots] \qquad \text{(9-26)}$$

What all of this means is that when $V(s)$ has a complex pole, there must be an accompanying conjugate pole. These two poles then combine to produce a damped sinusoidal waveform. The residues at these poles must be the conjugates of each other; therefore we only need to compute one of them. The amplitude of the resulting damped sinusoid is equal to twice the magnitude of the residue at the pole with a positive imaginary part. The phase angle is equal to the angle of this residue. The exponent of the exponential factor equals the real part of the pole, while the frequency of the sinusoid equals the imaginary part of the pole. Again we see that all of the data needed to recover the waveform are available in the transform. Transforms and waveforms are simply two sides of the same coin—the signal.

Example 9-12

The transform

$$V(s) = \frac{20(s + 3)}{(s + 1)(s^2 + 2s + 5)}$$

has a simple pole at $s = -1$ and a pair of complex poles located at the roots of the quadratic factor in the denominator:

$(s^2 + 2s + 5) = (s + 1 - j2)(s + 1 + j2)$

Consequently the partial fraction decomposing of $V(s)$ is

$$V(s) = \frac{k_1}{s + 1} + \frac{k_2}{s + 1 - j2} + \frac{k_2^*}{s + 1 + j2}$$

The residues at each of these poles can be obtained from the cover-up algorithm as

$$k_1 = \left.\frac{20(s + 3)}{s^2 + 2s + 5}\right|_{s = -1} = 10$$

$$k_2 = \left.\frac{20(s + 3)}{(s + 1)(s + 1 + j2)}\right|_{s = -1 + j2}$$
$$= 5\sqrt{2}\, e^{+j5\pi/4}$$

Hence

$$k_2^* = 5\sqrt{2}\, e^{-j5\pi/4}$$

We now have all of the data needed to construct the waveform as

$$v(t) = u(t)\,[10e^{-t} + 10\sqrt{2}\, e^{-t} \cos(2t + 5\pi/4)]$$

9-5 SPECIAL CASES

The previous section treated the basic process of performing the inverse Laplace using partial fraction expansions. That process assumed that the transform had more poles than zeros, that is, $n > m$ in Eq. 9-17. Such a function is called a proper rational function, and can always be expanded by partial fractions. However, it is quite possible to have transforms that are not proper rational functions.

For example, the transform

$$v(s) = \frac{s^2 + 4s + 6}{s^2 + 3s + 2}$$

has exactly as many finite poles as zeros ($n = m = 2$), and hence is not proper. It can be reduced to a proper function by the simple process of long division:

$$
\begin{array}{r}
1 \\
s^2 + 3s + 2 \overline{)\, s^2 + 4s + 6} \\
\underline{s^2 + 3s + 2} \\
s + 4
\end{array}
$$

Hence

$$v(s) = 1 + \frac{s + 4}{s^2 + 3s + 2} \qquad (9\text{-}27)$$

Impulse   Proper rational function

The first term in Eq. 9-27 is an impulse and the remainder term is a proper rational function that can be expanded by partial fractions to obtain its inverse.

If the transform has more zeros than poles ($m > n$), long division will yield additive terms such as $s, s^2, \ldots s^{m-n}$ before a proper remainder function is obtained. These powers of s correspond to higher order singularity functions called doublets, triplets, and so on. While it is theoretically possible to obtain these rather pathological waveforms, they do not actually occur in real circuits. Therefore we will not treat cases in which $m > n$.

Our second special case is an alternative way of handling complex poles that avoids the manipulation of complex numbers, but at the cost of additional algebra. In the previous section we showed that if $V(s)$ has complex poles at $s = -\alpha + j\beta$ and $s = -\alpha - j\beta$, then its partial fraction will contain two terms of the form

$$\frac{k}{s + \alpha - j\beta} + \frac{k^*}{s + \alpha + j\beta}$$

An alternative way to write these two terms is

$$\frac{K_1(s + \alpha)}{(s + \alpha)^2 + \beta^2} + \frac{K_2\beta}{(s + \alpha)^2 + \beta^2} \tag{9-28}$$

In this form the complex conjugate poles have been combined to produce two quadratic terms with real constants K_1 and K_2. To obtain the inverse transform of Eq. 9-28 we observe that the first term is a damped cosine and the second a damped sine. That is,

$$\mathscr{L}^{-1}\left\{\frac{K_1(s + \alpha)}{(s + \alpha)^2 + \beta^2} + \frac{K_2\beta}{(s + \alpha)^2 + \beta^2}\right\} \tag{9-29}$$
$$= K_1 e^{-\alpha t}\cos \beta t + K_2 e^{-\alpha t}\sin \beta t$$

This yields the damped sinusoid in Fourier coefficient form, which can be converted to amplitude-phase form by the methods discussed in Chapter 6.

$$K_1 e^{-\alpha t}\cos \beta t + K_2 e^{-\alpha t}\sin \beta t = A e^{-\alpha t}\cos (\beta t - \theta) \tag{9-30}$$

where

$$A = \sqrt{K_1^2 + K_2^3}$$
$$\theta = \tan^{-1} K_2/K_1$$

Thus the problem of finding the damped sinusoid corresponding to a pair of complex poles reduces to finding the two real constants K_1 and K_2. There are several ways to do this, but the general method is to expand the given transform with the unknown constants left in symbolic form. Then rationalize the expansion (find a common denominator) and equate like powers of s in the numerator. This yields a set of linear equations whose solution gives us the unknown constants. This method is completely general and works for real poles, complex poles, and even repeated poles. However it involves solving n linear equations, where n is the total number of poles, and usually requires more algebraic manipulation than the cover-up algorithm. (See Example 9-13 on page 410.)

The final special case we must treat occurs when $V(s)$ has equal or multiple poles. For example,

$$V(s) = \frac{K(s - z_1)}{(s - p_1)(s - p_2)^2} \tag{9-31}$$

Example 9-13

In Example 9-12 we found the inverse transform of

$$V(s) = \frac{20(s + 3)}{(s + 1)(s^2 + 2s + 5)}$$

using partial fractions and the cover-up algorithm. To apply the alternate method discussed, we first complete the square in the quadratic factor

$$s^2 + 2s + 5 = (s + 1)^2 + 4^2$$

and expand $V(s)$ as

$$V(s) = \frac{K_1}{s + 1} + \frac{K_2(s + 1)}{(s + 1)^2 + 2^2} + \frac{K_3 2}{(s + 1)^2 + 2^2}$$

We now find the common denominator of the expansion

$$V(s) = \frac{(K_1 + K_2)s^2 + (2K_1 + 2K_2 + 2K_3)s + 5K_1 + K_2 + 2K_3}{(s + 1)(s^2 + 2s + 5)}$$

Comparing like powers of s in the numerator with those in the given $V(s)$ yields the equations

s^2 terms	$K_1 + K_2$	$= 0$
s^1 terms	$2K_1 + 2K_2 + 2K_3$	$= 20$
s^0 terms	$5K_1 + K_2 + 2K_3$	$= 60$

Using Cramer's rule,

$$K_1 = \frac{D_1}{D} = \frac{\begin{vmatrix} 0 & 1 & 0 \\ 20 & 2 & 2 \\ 60 & 1 & 2 \end{vmatrix}}{\begin{vmatrix} 1 & 1 & 0 \\ 2 & 2 & 2 \\ 5 & 1 & 2 \end{vmatrix}} = \frac{80}{8} = 10$$

$$K_2 = \frac{D_2}{D} = \frac{\begin{vmatrix} 1 & 0 & 0 \\ 2 & 20 & 2 \\ 5 & 60 & 2 \end{vmatrix}}{8} = \frac{-80}{8} = -10$$

$$K_3 = \frac{D_3}{D} = \frac{\begin{vmatrix} 1 & 1 & 0 \\ 2 & 2 & 20 \\ 5 & 1 & 60 \end{vmatrix}}{8} = \frac{80}{8} = 10$$

which yields the inverse transform as

$$v(t) = 10e^{-t} - 10e^{-t} \cos 2t + 10e^{-t} \sin 2t$$

The reader can now show that the two damped sinusoids can be combined to produce the same result obtained in Example 9-12.

---

has a simple pole at $s = p_1$ and a pole of order 2 at $s = p_2$. The partial fraction expansion of this transform is

$$V(s) = \frac{k_1}{s - p_1} + \frac{k_{21}}{s - p_2} + \frac{k_{22}}{(s - p_2)^2} \tag{9-32}$$

Note that the decomposition contains two terms for the pole at $s = p_2$ because it is of order 2. The corresponding waveform is

$$v(t) = u(t)\,(k_1 e^{p_1 t} + k_{21} e^{p_2 t} + k_{22} t\, e^{p_2 t}) \tag{9-33}$$

As before, the residues turn out to be the amplitudes, but the double pole leads to a damped ramp waveform.

The residues k_1 and k_{22} can be obtained using the cover-up algorithm. The constant k_{21} can be found by first multiplying the expansion through by $(s - p_2)^2$:

$$(s - p_2)^2 V(s) = k_1 \frac{(s - p_2)^2}{s - p_1} + k_{21}(s - p_2) + k_{22}$$

If we differentiate this result with respect to s, and then evaluate the result at $s = p_2$, the only term on the right side that remains is k_{21}, so that

$$k_{21} = \frac{d}{ds}(s - p_2)^2 V(s) \Big|_{s = p_2} \tag{9-34}$$

In sum, the multiple poles of $V(s)$ lead to multiple exponentials. In general, if $V(s)$ has a pole of order r, $(s - p_i)^r$, then $v(t)$ will contain all of the multiples up to $t^{r-1}e^{p_it}$. The residues for each of these terms can be obtained from the general formula

$$k_{ij} = \frac{1}{(r - j)!} \frac{d^{r-j}}{ds^{r-j}} \left[(s - p_i)^r V(s) \right] \bigg|_{s = pi} \qquad (9\text{-}35)$$

Example 9-14

The transform

$$V(s) = \frac{4(s + 3)}{s(s + 2)^2}$$

has a simple pole at $s = 0$ and a double pole at $s = -2$. Its partial fraction decomposition is

$$V(s) = \frac{k_1}{s} + \frac{k_{21}}{s + 2} + \frac{k_{22}}{(s + 2)^2}$$

The residues k_1 and k_{22} are obtained from the cover-up algorithm.

$$k_1 = sV(s) \bigg|_{s = 0} = 3$$

$$k_{22} = (s + 2)^2 V(s) \bigg|_{s = -2} = -2$$

k_{21} is found using Eq. 9-32:

$$k_{21} = \frac{d}{ds} (s + 2)^2 V(s) \bigg|_{s = -2}$$

$$= \frac{-12}{s^2} \bigg|_{s = -2} = -3$$

Hence $v(t)$ is

$$v(t) = u(t) \{3 - 3e^{-2t} - 2te^{-2t}\}$$

9-6 STEP RESPONSE OF SINGLE-MEMORY CIRCUITS

In using the Laplace transformation to study single-memory circuits we can either (1) transform the circuit differential equation or (2) transform the circuit itself. In this section, we use the differential equation approach. The latter viewpoint is developed in the next chapter.

The single-memory RC circuit in Figure 9-4 is driven by a constant source and a switch that closes at $t = 0$. For the purpose of analysis, we can replace the switch and constant source by a voltage source that delivers a signal $V_iu(t)$, a step function. In sum, the circuit input signal is a step function, and so the resulting circuit response is called the **step response.**

From our previous study of single-memory circuits, we can predict that the step response will depend on three things:

1. The amplitude of the input step function (V_i)

2. The circuit time constant (RC)

3. The voltage across the capacitor at $t = 0$,

We also know that the response is governed by the circuit differential equation

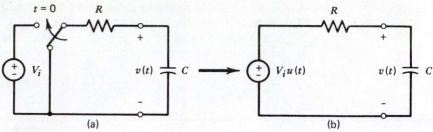

FIGURE 9-4
Transformation of a single-memory *RC* circuit. (*a*) Original circuit. (*b*) Equivalent step function.

$$RC \frac{dv}{dt} + v = V_i u(t) \qquad (9\text{-}36)$$

with the initial condition $v(0) = v_C(0)$.

Our objective is to determine the waveform $v(t)$ that satisfies this differential equation and the initial condition. Simply put, we must determine the step response. To achieve this objective, we follow the pattern shown in Figure 9-1 by transforming the differential equation into the s-domain, solving for the transform of the response, and then performing the inverse transformation to obtain the response waveform $v(t)$.

To begin this process, we transform Eq. 9-36 as

$$\mathscr{L} \left\{ RC \frac{dv}{dt} + v \right\} = \mathscr{L} \left\{ V_i u(t) \right\}$$

$$RC \, \mathscr{L} \left\{ \frac{dv}{dt} \right\} + \mathscr{L} \left\{ v(t) \right\} = V_i \mathscr{L} \left\{ u(t) \right\}$$

$$RC[sV(s) - v(0)] + V(s) = \frac{V_i}{s} \qquad (9\text{-}37)$$

To achieve this transformation, we have used the linearity property, the differentiation property, and the transform pair for a unit step function. The beauty of this result is that it is now an algebraic equation in $V(s)$, the transform of the response we seek.

We now algebraically separate $V(s)$ in Eq. 9-37.

$$(RCs + 1) \, V(s) = \frac{V_i}{s} + RCv(0)$$

Solving for $V(s)$ yields

$$V(s) = \frac{V_i/RC}{s(s + 1/RC)} + \frac{v(0)}{(s + 1/RC)} \qquad (9\text{-}38)$$

The function $V(s)$ in this equation is the transform of the waveform $v(t)$ that satisfies the differential equation and meets the initial conditions. Note that the initial condition first appeared explicitly as a result of applying the differentiation rule to obtain Eq. 9-37.

To recover the waveform $v(t)$, we must apply the inverse transformation to Eq. 9-38. The partial fraction expansion of the first term on the right is

$$\frac{V_i/RC}{s(s + 1/RC)} = \frac{k_1}{s} + \frac{k_2}{s + 1/RC}$$

The cover-up algorithm yields the residues as

$$k_1 = \frac{V_i/RC}{s + 1/RC}\bigg|_{s\,=\,0} = V_i \qquad k_2 = \frac{V_i/RC}{s}\bigg|_{s\,=\,-1/RC} = -V_i$$

Hence Eq. 9-38 can be written as

$$V(s) = \frac{V_i}{s} - \frac{V_i}{s + 1/RC} + \frac{v(0)}{s + 1/RC} \qquad \text{(9-39)}$$

Each term in this expansion is now recognizable. The first term on the right is a step function, and each of the next two terms is an exponential. Thus the waveform $v(t)$ corresponding to $V(s)$ is

$$v(t) = u(t)\,[V_i - V_i e^{-t/RC} + v(0)e^{-t/RC}] \qquad \text{(9-40)}$$

or

$$v(t) = V_i + [v(0) - V_i]\,e^{-t/RC} \qquad t \geq 0 \qquad \text{(9-41)}$$

This is a familiar result from our previous study of single-memory RC circuits. The first term on the right is the forced response due to the input source, while the second is the natural response due to the characteristic of the circuit itself. The complete response, as predicted, depends on three parameters: the input amplitude V_i, the time constant RC, and the initial condition $v(0)$. Note, the Laplace transformation method yields the complete response (forced plus natural) in one process. The pattern is the one outlined in flow graph form in Figure 9-1. The reader is advised to begin with Eq. 9-36, and to relate each step leading to Eq. 9-41 to this flow graph.

Example 9-15

The step response of the single-memory RL circuit in Figure 9-5 is governed by the differential equation

$$L\frac{di}{dt} + Ri = V_i u(t)$$

with the initial condition $i(0) = i_L(0)$. Transforming this equation into the s-domain yields

$$L[sI(s) - i(0)] + RI(s) = \frac{V_i}{s}$$

The function $I(s)$ is the transform of the waveform $i(t)$, which satisfies the differential equation and the initial condition. Solving for $I(s)$ yields

$$I(s) = \frac{V_i/L}{s(s + R/L)} + \frac{i(0)}{s + R/L}$$

The partial fraction expansion of this equation is

$$I(s) = \frac{V_i/R}{s} - \frac{V_i/R}{s + R/L} + \frac{i(0)}{s + R/L}$$

The first term in this expansion is a step function and the other terms are exponentials. Hence the waveform $i(t)$ is

$$i(t) = \frac{V_i}{R} + [i(0) - \frac{V_i}{R}]e^{-Rt/L} \qquad t \geq 0$$

The first term on the right is the forced response due to the step input. The second term is the familiar natural response.

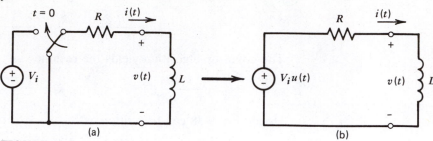

(a)   (b)

FIGURE 9-5
Transformation of a single-memory *RL* circuit. (*a*) Constant source and switch. (*b*) Step function.

Example 9-16

The Laplace transformation method can be used to determine the step response when the governing circuit equation is not the familiar first-order differential equation. For example, the single-memory *RC* circuit in Figure 9-6 can be analyzed by first writing a KVL equation around the one mesh in the circuit.

$$- V_i u(t) + v_R(t) + v_C(t) = 0$$

The element voltages can then be expressed in terms of the mesh circuit $i(t)$ as

$$v_R = Ri, \qquad v_C = \frac{1}{C} \int_0^t i \, dt + v_C(0)$$

Substituting these relations into the KVL equation yields

$$Ri + \frac{1}{C} \int_0^t i \, dt + v_C(0) = V_i u(t)$$

This equation describes the step response of the circuit in terms of the current $i(t)$. It is not a differential equation, but rather what is called an integrodifferential equation since it contains integral and derivative terms. The equation can be transformed directly into the *s*-domain using the integration property of the Laplace transformation rather than first having to differentiate out the integrals as would be necessary in a classical solution approach:

$$RI(s) + \frac{I(s)}{Cs} + \frac{v_C(0)}{s} = \frac{V_i}{s}$$

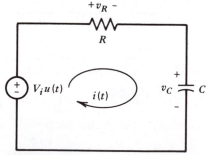

FIGURE 9-6
Single-memory *RC* circuit analyzed using a mesh current.

The integrodifferential equation has been transformed into an algebraic equation in the current transform $I(s)$. Solving for $I(s)$ yields

$$\frac{RCs + 1}{Cs} I(s) = \frac{V_i - v_C(0)}{s}$$

or

$$I(s) = \frac{[V_i - v_C(0)]}{R(s + 1/RC)}$$

This result is simply the transform of an exponential, and hence

$$i(t) = \frac{[V_i - v_C(0)]}{R} e^{-t/RC} \qquad t \geq 0$$

In the case of the current $i(t)$, the step response does not contain a forced component, only a natural component.

9-7 RESPONSE TO GENERAL INPUTS

It is comforting to discover that the Laplace transformation yields results that agree with those obtained by classical methods. However, the Laplace transformation is more than just another way to find the step response of a circuit. It can also be used to find the response attributable to a wide variety of input signals.

For example, the single-memory RC circuit in Figure 9-7 is driven by an input signal denoted $v_A(t)$, where the input signal is not necessarily a step function, but any time-varying waveform. The differential equation describing the voltage across the capacitor is

$$RC \frac{dv}{dt} + v = v_A(t) \tag{9-42}$$

The only difference here is that the right side of the differential equation, the driving force, can be any time-varying waveform. The classical methods of dealing with this situation involve such concepts as variation of parameters, integrating factors, and undetermined coefficients.

However, with the Laplace transform method we can proceed without specifying the exact nature of the input signal. We first transform Eq.

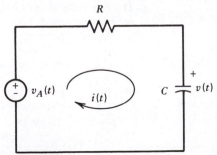

FIGURE 9-7
Single-memory RC circuit with a general input signal.

9-42 into the s-domain as

$$RC[sV(s) - v(0)] + V(s) = V_A(s) \qquad \textbf{(9-43)}$$

The only assumption necessary is that the input $v_A(t)$ be Laplace transformable, a condition met by all signals of interest to us. We now solve for the response $V(s)$ as

$$(RCs + 1)V(s) = V_A(s) + RCv(0)$$

or

$$V(s) = \frac{V_A(s)/RC}{s + 1/RC} + \frac{v(0)}{s + 1/RC} \qquad \textbf{(9-44)}$$

The function $V(s)$ is the transform of the response of the circuit due to any input signal $v_A(t)$. Notice that we can proceed this far in the solution process without knowing the exact form of the input signal. In a sense we have found the general solution—in the s-domain, of course—of the differential equation (Eq. 9-42) for any input signal. All of the necessary ingredients are present in Eq. 9-44: the input signal $V_A(s)$, the circuit time constant RC, and the initial condition $v(0)$. However, to recover the waveform $v(t)$, we must have a particular input in mind. The rest of this section is devoted to several examples that illustrate the remainder of the solution process.

Example 9-17

If the input to the circuit in Figure 9-7 is $v_A(t) = Ve^{-At}$, find the output, $v(t)$.

Using the transform of an exponential from Table 9-2, we obtain

$$V_A(s) = \frac{V}{s + A}$$

and the response in Eq. 9-42 becomes

$$V(s) = \frac{V/RC}{(s + A)(s + 1/RC)} + \frac{v(0)}{s + 1/RC}$$

To obtain the waveform $v(t)$ we expand the first term on the right by partial fractions:

$$\frac{V/RC}{(s + A)(s + 1/RC)} = \frac{k_1}{s + A} + \frac{k_2}{s + 1/RC}$$

The cover-up algorithm gives the residues as

$$k_1 = \left.\frac{V/RC}{s + 1/RC}\right|_{s=-A} = \frac{V}{1 - ARC}$$

$$k_2 = \left.\frac{V/RC}{s + A}\right|_{s=-1/RC} = \frac{V}{ARC - 1}$$

Hence the waveform $v(t)$ is

$$v(t) = \underbrace{\frac{V}{1 - ARC}\, e^{-At}}_{\text{Forced}} +$$

$$\underbrace{\frac{V}{ARC - 1}\, e^{-t/RC} + v(0)\, e^{-t/RC}}_{\text{Natural}} \qquad t \geq 0$$

The first term is the forced response and the rest the natural response. The forced component is exponential because the input is an exponential. The natural response is also exponential, but one whose time constant (RC) depends on the circuit rather than the input. We have tacitly assumed that $A \neq 1/RC$ in finding this solution. If $A = 1/RC$, then the forced and natural response have the same time constant and $V(s)$ can be written as

$$V(s) = \frac{V/RC}{(s + 1/RC)^2} + \frac{v(0)}{s + 1/RC}$$

Using Table 9-2, we recognize the first term on the right as the transform of a damped ramp, and hence

$$v(t) = \frac{V}{RC}\, te^{-t/RC} + v(0)\, e^{-t/RC}$$

In this case we cannot separate the response into forced and natural components since the two time constants are equal and therefore combine to produce a single waveform, the damped ramp.

Example 9-18

If the input signal to the circuit of Figure 9-7 is $v_A(t) = V \cos \beta t$, find $v(t)$.

From Table 9-2 we find that

$$V_A(s) = \frac{sV}{s^2 + \beta^2}$$

and the response in Eq. 9-44 becomes

$$V(s) = \frac{sV/RC}{(s^2 + \beta^2)(s + 1/RC)} + \frac{v(0)}{s + 1/RC}$$

The input signal in this case introduces a pair of imaginary poles located at $s = j\beta$ and $s = -j\beta$. Expanding the first term on the right by partial fractions yields

$$\frac{sV/RC}{(s - j\beta)(s + j\beta)(s + 1/RC)}$$
$$= \frac{k_1}{s - j\beta} + \frac{k_1{}^*}{s + j\beta} + \frac{k_2}{s + 1/RC}$$

The cover-up algorithm yields the residues as

$$k_1 = \frac{sV/RC}{(s + j\beta)(s + 1/RC)} \bigg|_{s=j\beta}$$
$$= \frac{V/2}{1 + j\beta RC} = |k_1|e^{+j\theta}$$

where

$$|k_1| = \frac{V/2}{\sqrt{1 + (\beta RC)^2}}$$

$$\theta = \tan^{-1}(-\beta RC)$$

and

$$k_1{}^* = |k_1|\, e^{-j\theta}$$
$$k_2 = \frac{sV/RC}{(s^2 + \beta^2)} \bigg|_{s=-1/RC} = \frac{-V}{1 + (\beta RC)^2}$$

Using the results in Eq. 9-26 for complex poles, we obtain

$$v(t) = 2|k_1| \cos(\beta t + \theta) + k_2 e^{-t/RC} + v(0)e^{-t/RC}$$

or

$$v(t) = \underbrace{\frac{V}{\sqrt{1 + (\beta RC)^2}} \cos(\beta t + \theta)}_{\text{Forced}}$$
$$\underbrace{- \frac{Ve^{-t/RC}}{1 + (\beta RC)^2} + v(0)e^{-t/RC}}_{\text{Natural}}$$

The forced response is a sinusoidal waveform because the input is sinusoidal. The natural response is the familiar exponential with a time constant of RC.

Example 9-19

The concept of impulse response is often used in the study of circuits. The term *impulse response* means that the input is a unit impulse and implies that all initial conditions are zero. Find the impulse response of the RC circuit in Figure 9-7.

We are given that

$$v_A(t) = \delta(t)$$

Hence

$$V_A(s) = 1$$

and we assume $v(0) = 0$.

Under these conditions Eq. 9-44 reduces to

$$V(s) = \frac{1/RC}{s + 1/RC}$$

and hence

$$v(t) = \frac{1}{RC} e^{-t/RC}$$

The impulse response contains only a natural component and *no* forced term. This is the reason that impulse response is often used in the study of circuits and systems.

SUMMARY

- A transformation is the operation of changing the form of data using a mathematical rule. The Laplace transformation changes signals and systems from the time domain to the complex frequency domain. The signal representation in the time domain is called a waveform. In the frequency domain it is called a transform.

- The Laplace transformation (waveform to transform) is defined as

$$\mathcal{L}\{v(t)\} = V(s) = \int_0^\infty v(t)e^{-st}dt$$

  The basic properties of this transformation are listed in Table 9-1.

- Waveform–transform pairs are unique and can be derived using the definition of the Laplace transformation. Pairs are also available in catalog form in Table 9-2. Waveforms and transforms contain the same data but in different form.

- The inverse Laplace transformation (transform to waveform) can be obtained from the partial fraction expansion of the transform. If $V(s)$ is a proper rational function with simple poles $p_1, p_2, \ldots, p_n$, then

$$V(s) = \frac{k_1}{s - p_1} + \frac{k_2}{s - p_2} + \cdots + \frac{k_n}{s - p_n}$$

  where

$$k_1 = (s - p_i)V(s)\bigg|_{s=p_i}$$

  is called the residue at p_i. Multiple poles of $V(s)$ require special treatment.

- The waveform $v(t)$ recovered from the partial fraction expansion of $V(s)$ consists of a linear combination of exponential terms. The form of each term is determined by the nature of the poles. The amplitude depends on the residues at the poles. See Table 9-3.

Pole Factor	$V(s)$ Term	$v(t)$ Term
Real, simple $(s + a)$	$\dfrac{k}{s + a}$	ke^{-at}
Real, double $(s + a)^2$	$\dfrac{k_1}{s + a} + \dfrac{k_2}{(s + a)^2}$	$k_1e^{-at} + k_2te^{-at}$
Complex, conjugate $(s + \alpha)^2 + \beta^2$	$\dfrac{k}{s + \alpha - j\beta} + \dfrac{k^*}{s + \alpha + j\beta}$	$2|k|e^{-\alpha t}\cos(\beta t + \underline{/k})$

TABLE 9-3

- The Laplace transformation can be used to study circuit response by:
    - **(1)** Find the circuit differential equation and initial conditions.
    - **(2)** Transform the differential equation into the s-domain.
    - **(3)** Solve for the response transform.
    - **(4)** Perform the inverse transformation to find the response waveform.
- The Laplace transformation is a convenient method for finding the response of a circuit for any input signal. The method yields the forced and natural response in one process.
- Pole-zero diagrams are a way of expressing a signal's characteristic in the s-domain.

EN ROUTE OBJECTIVES
AND RELATED EXERCISES

9-1 *LAPLACE TRANSFORM (SECS. 9-1 to 9-3)*

Given a signal waveform, determine its Laplace transform and construct its pole-zero diagram.

Exercises

9-1-1 For each of the signal waveforms in Exercise 6-1-1, determine $V(s)$ and construct its pole-zero diagram.

9-1-2 Repeat Exercise 9-1-1 for the following waveforms:
(a) $v(t) = 1 - e^{-at}$
(b) $v(t) = (1 + at)e^{-at}$
(c) $v(t) = e^{-at} - e^{-bt}$
(d) $v(t) = \sin(bt - \phi)$
(e) $v(t) = e^{-at}\cos(bt + \phi)$
(f) $v(t) = e^{-at}(1 - \cos bt)$

9-1-3 Repeat Exercise 9-1-1 for each of the following waveforms.
(a) $v(t) = Ae^{-\zeta\omega_0 t}\cos(\omega_0\sqrt{1-\zeta^2}\,t)$
(b) $v(t) = A[1 - e^{-\zeta\omega_0 t}\cos(\omega_0\sqrt{1-\zeta^2}t)]$

9-1-4 Repeat Exercise 9-1-1 for each of the following waveforms.
(a) $v(t) = 10$
(b) $v(t) = 10e^{-t}$
(c) $v(t) = 10e^{-10t}$
(d) $v(t) = 10 - 10e^{-10t}$
(e) $v(t) = 10 - 10e^{-10t}$
(f) $v(t) = 10e^{-t} - 10e^{-10t}$

9-1-5 Repeat Exercise 9-1-1 for each of the following waveforms.
(a) $v(t) = 10\cos t$
(b) $v(t) = 10\cos 10t$
(c) $v(t) = 10\sin 10t$
(d) $v(t) = 10e^{-t}\cos t$
(e) $v(t) = 10e^{-t}\cos 10t$
(f) $v(t) = 10e^{-10t}\cos t$

9-1-6 For each of the following waveforms, construct a sketch in the s-plane of the location of the poles (poles only) of $V(s)$.
(a) $v(t) = 1 - e^{-t} + e^{-3t} - e^{-6t}$
(b) $v(t) = 10 + 3e^{-t} + 4e^{-3t} + 8e^{-6t}$
(c) $v(t) = 3\cos 6t - 10\sin 10t + 7\cos 10t$
(d) $v(t) = 3\cos 6t - 10\sin 10t + 7e^{-t}\cos 10t$

(e) $v(t) = 3 \cos 6t - 10e^{-t} \sin 10t + 6e^{-t} \cos 10t$

(f) $v(t) = 3e^{-t} \cos 6t - 10e^{-2t} \sin 10t + 6e^{-t} \cos 10t$

9-1-7 The initial value theorem is a relationship between a waveform $v(t)$ at $t = 0^+$ and the corresponding transform at $s \to \infty$.

$$\underset{t \to 0+}{\text{Limit}}\ v(t) = \underset{s \to \infty}{\text{Limit}}\ s\ V(s)$$

Verify the initial value theorem for each of the signals in Table 9-2.

9-2 INVERSE TRANSFORMS (SECS. 9-4 and 9-5)

Given a signal transform, plot its pole-zero map, determine the coresponding waveform, and sketch the waveform.

Exercises

9-2-1 For each of the following transforms $V(s)$, sketch a pole-zero map, determine $v(t)$, and sketch the waveform.

(a) $V(s) = \dfrac{1}{(s + a)(s + b)}$

(b) $V(s) = \dfrac{1}{s(s + a)(s + b)}$

(c) $V(s) = \dfrac{s}{(s + a)(s + b)}$

(d) $V(s) = \dfrac{s}{s^2 - a^2}$

(e) $V(s) = \dfrac{s}{s^2 + a^2}$

(f) $V(s) = \dfrac{s}{(s + a)^2 + b^2}$

9-2-2 Repeat Exercise 9-2-1 for the following transforms for three conditions: (1) $\zeta > 1$, (2) $\zeta = 1$, and (3) $\zeta < 1$.

(a) $V(s) = \dfrac{\omega_0^2}{s^2 + 2\zeta\omega_0 s + \omega_0^2}$

(b) $V(s) = \dfrac{s + \zeta\omega_0}{s^2 + 2\zeta\omega_0 s + \omega_0^2}$

9-2-3 For each of the following transforms $V(s)$, plot the pole-zero map, determine the waveform $v(t)$, and plot the waveform.

(a) $V(s) = \dfrac{4}{s^2 + 5s + 4}$

(b) $V(s) = \dfrac{4}{s^2 + 4s + 4}$

(c) $V(s) = \dfrac{4}{s^2 + 2s + 4}$

(d) $V(s) = \dfrac{4(s + 1)}{s^2 + 5s + 4}$

(e) $V(s) = \dfrac{4(s + 1)}{s^2 + 4s + 4}$

(f) $V(s) = \dfrac{4(s + 1)}{s^2 + 2s + 4}$

(g) $V(s) = \dfrac{4}{s^2 + 4}$

(h) $V(s) = \dfrac{s^2}{s^2 + 4}$

(i) $V(s) = \dfrac{4(s + 1)}{s(s^2 + 4)}$

9-2-4 Verify that the following partial fraction expansions are correct, determine the unknown residue(s), and then determine $v(t)$.

(a) $V(s) = \dfrac{6s^2 + 24s + 18}{(s + 2)(s + 4)(s + 5)} =$

$\dfrac{-1}{s + 2} + \dfrac{K}{s + 4} + \dfrac{16}{s + 5}$

(b) $V(s) = \dfrac{s^2 + 5s + 6}{(s + 2)(s + 4)(s + 5)} =$

$\dfrac{K}{s + 2} + \dfrac{-1}{s + 4} + \dfrac{2}{s + 5}$

(c) $V(s) = \dfrac{8s + 16}{s(s^2 + 4s + 8)} = \dfrac{2}{s} +$

$\dfrac{K}{s + 2 - j2} + \dfrac{-1 + j}{s + 2 + j2}$

(d) $V(s) = \dfrac{4}{(s + 1)^2(s^2 + 1)} = \dfrac{A}{s + 1} +$

$\dfrac{2}{(s + 1)^2} + \dfrac{-1 - j}{s - j} + \dfrac{B}{s + j}$

(e) $V(s) = \dfrac{8(s^2 + 1)}{s(s^2 + 4)} = \dfrac{2}{s} + \dfrac{A}{s - j2} +$

$\dfrac{B}{s + j2}$

9-2-5 The following statements describe the location of the poles and zeros of $V(s)$. Construct a pole-zero map of $V(s)$ and qualitatively describe the types of waveforms that will appear in $v(t)$.

(a) Poles: $p_1 = -2$, $p_2 = -10$, $p_3 = -1 + j10$, $p_4 = -1 - j10$
Zeros: $z_1 = 0$, $z_2 = -j10$, $z_3 = +j10$

(b) Poles: $p_1 = -100$, $p_2 = -10^4$, $p_3 = -200$

Zeros: $z_1 = -50$, $z_2 = -150$, $z_3 = -1000$

(c) Poles: $p_1 = -5$, $p_2 = -j8$, $p_3 = +j8$, $p_4 = -60$

Zeros: $z_1 = 0$, $z_2 = -60$

(d) Poles: $p_1 = -10$, $p_2 = -10^4$, $p_3 = -10$

Zeros: None

(e) Poles: $p_1 = -j4$, $p_2 = -j9$, $p_3 = +j4$, $p_4 = +j9$

Zeros: $z_1 = 0$, $z_2 = -j5$, $z_3 = +j5$

9-3 DIFFERENTIAL EQUATIONS (SECS. 9-6 and 9-7)

Given a circuit consisting of linear memoryless elements and not more than two memory elements:

(a) *Determine the circuit differential equation and the initial conditions (if not given).*

(b) *Transform the differential equation into the s-domain and solve for the response transform.*

(c) *Determine the response waveform from its transform.*

(d) *Identify the forced and natural components of the response in both its waveform and its transform.*

Exercises

9-3-1 Repeat Exercise 8-1-1 using the Laplace transformation method outlined in Objective 9-3.

9-3-2 Repeat Exercise 8-1-3 using the Laplace transformation method outlined in Objective 9-3.

9-3-3 Repeat Exercise 8-1-5 using the Laplace transformation method outlined in Objective 9-3.

9-3-4 Repeat Exercise 8-1-7 using the Laplace transformation method outlined in Objective 9-3.

9-3-5 Repeat Exercise 8-2-1 using the Laplace transformation method outlined in Objective 9-3.

9-3-6 Repeat Exercise 8-2-4 using the Laplace transformation method outlined in Objective 9-3.

9-3-7 Solve the following differential equations using the Laplace transformation method outlined in Objective 9-3.

(a) $\dfrac{dx}{dt} + 10x = 5e^{-5t}$ $x(0) = 1$

(b) $\dfrac{dx}{dt} + 10x = 10e^{-10t}$ $x(0) = 0$

(c) $\dfrac{dx}{dt} + 10x = 5e^{-5t}$ $x(0) = 0$

(d) $\dfrac{dx}{dt} + 10x = 10\,\delta(t)$ $x(0) = 0$

(e) $\dfrac{d^2x}{dt^2} + 10\dfrac{dx}{dt} + 20x = 100\,\delta(t)$

$x(0) = 0$ and $\dfrac{dx}{dt}(0) = 0$

9-3-8 For each of the input signals given, determine $v(t)$ and $i(t)$ in the circuit of Figure E9-3-8 by transforming the circuit differential equation. Identify the forced and natural components of $v(t)$ and $i(t)$.

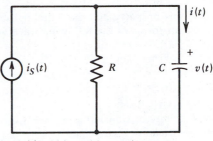

$i_S(t) = \delta(t),\ Au(t),\ Ae^{-at},\ A\cos\beta t$
$v(0) = 0$ for all inputs.

FIGURE E9-3-8
RC circuit for Exercise 9-3-8.

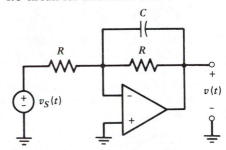

$v_S(t) = \delta(t),\ Au(t),\ Ae^{-at},\ A\cos\beta t$
$v(0) = 0$

FIGURE E9-3-9
OP AMP circuit for Exercise 9-3-9.

9-3-9 For each of the input signals given, determine $v(t)$ in the circuit of Figure E9-3-9 by transforming the circuit differential equation. Identify the forced and natural components of $v(t)$.

PROBLEMS

P9-1 (Analysis)
The complex differentiation property of the Laplace transformation states that:

$$\text{If } \mathcal{L}\{f(t)\} = F(s), \text{ Then } \mathcal{L}\{tf(t)\} = -\frac{dF(s)}{ds}$$

Use this property and the pairs in Table 9-2 to find the transforms of the following waveforms.
(a) $v(t) = t\, e^{at}$
(b) $v(t) = t \sin bt$
(c) $v(t) = t^2$
(d) $v(t) = t^2 e^{-at}$

P9-2 (Analysis)
The complex translation property of the Laplace transformation states that:

$$\text{If } \mathcal{L}\{f(t)\} = F(s), \text{ Then } \mathcal{L}\{e^{-at}f(t)\} = F(s + a)$$

Use this property and the pairs in Table 9-2 to find the transforms of the following waveforms.
(a) $v(t) = t\, e^{-at}$
(b) $v(t) = e^{-at} \sin bt$
(c) $v(t) = [(b - a)\, t + 1]\, e^{-at}$
(d) $v(t) = t\, e^{-at} \cos bt$

P9-3 (Analysis)
For each of the following transforms select the parameter K so that the corresponding waveform has the form $v(t) = e^{-t}$.

(a) $V(s) = \dfrac{s + 1}{s^2 + (K + 1)\, s + K}$

(b) $V(s) = \dfrac{s + K}{s^2 + 3s + K}$

(c) $V(s) = \dfrac{s^2 + K}{s^3 + s^2 + 9s + K}$

P9-4 (Analysis)
If $V(s)$ is a rational function with only simple poles and

$$\underset{s \to \infty}{\text{Limit}} \; sV(s) = 0$$

show that the sum of the residues at the poles is zero. Construct the partial fraction expansion of

$$V(s) = \frac{4}{(s + 2)(s^2 + 2s + 2)}$$

and use this property to check your result.

P9-5 (Analysis)
The poles of a transform can be widely separated so that poles nearer the origin dominate the waveform

and the remote poles can be neglected. For example, the transform

$$V_1(s) = \frac{100}{(s + 1)(s + 100)}$$

can be approximated by neglecting the remote pole (set $s = 0$ in the term $s + 100$) to obtain the approximation

$$V_2(s) = \frac{1}{s + 1}$$

Show that these two transforms have approximately the same waveform for $t > 0.05$ s. Then explain why the dominant pole approximation cannot be applied to the transform

$$V_3(s) = \frac{100(s + 2)}{(s + 1)(s + 100)}$$

P9-6   (Analysis)
The final value theorem of Laplace transforms states that:

If $\mathcal{L}\{v(t)\} = V(s)$,

$$\text{Then } FV = \underset{t\to\infty}{\text{Limit}}\ v(t) = \underset{s\to 0}{\text{Limit}}\ sV(s)$$

Use this property to find the final value of the waveforms corresponding to the following transforms. Check your results by finding $v(t)$ and determining its final value.

(a) $V(s) = \dfrac{1}{s(s + a)} \qquad a > 0$

(b) $V(s) = \dfrac{1}{(s + a)(s + b)} \qquad a > 0,\ b > 0$

(c) $(V(s) = \dfrac{s + a}{(s + a)^2 + b^2} \qquad a > 0$

(d) $V(s) = \dfrac{a^2}{s(s + a)^2} \qquad a > 0$

P9-7   (Analysis)
Find the waveform corresponding to the transform

$$V(s) = \frac{1}{(s + a)(s + b)}$$

Then show that

$$\underset{b\to a}{\text{Limit}}\ v(t) = t\, e^{-at}$$

Does the transform of $t\, e^{-at}$ correspond to $V(s)$ with $a = b$? Discuss what happens to the waveform as the poles of $V(s)$ approach each other.

P9-8 (Analysis)

Use Laplace transforms to solve for $v_1(t)$ and $v_2(t)$, which satisfy the differential equations

$$\frac{dv_1}{dt} + 2\,v_1 - v_2 = u(t)$$

$$\frac{dv_2}{dt} + v_2 - v_1 = 0$$

with initial conditions $v_1(0) = 1$ and $v_2(0) = 0$.

Chapter 10
S-Domain Circuit Analysis

Transformed Circuits
Circuit Analysis Techniques
Node-Voltage Equations
Mesh-Current Equations
Solution of Linear Equations
Transfer Functions
Poles and Zeros Revisited

Be not conformed to this world but be ye transformed.

Romans 12:24A
King James Version

The Laplace transformation is a rule for carrying signals or circuits from the time domain to the complex frequency domain. This chapter begins the study of the process of transforming circuits into the s domain. We find that Kirchhoff's laws are unchanged by the transformation. Circuit equilibrium equations in the s domain can be written using either node voltages or mesh currents as variables. The resulting linear equations can be solved for the response transform using algebraic methods. Circuit analysis in the s domain is closely related to the analysis of memoryless circuits in the t domain. This leads to the concept of viewing the response in transform form. The last section of the chapter introduces response poles as the means of describing the response of a circuit to signal inputs.

429

10-1 TRANSFORMED CIRCUITS

So far we have used the Laplace transformation to change waveforms into transforms, and to change circuit differential equations into algebraic equations. These operations are useful and provide insight into the nature of the s domain. However, the utility of the Laplace transformation is not just solving differential equations, but providing an alternative representation of signals and systems. The real advantage of the Laplace transformation shows up when we transform the circuit itself and analyze its behavior in the s domain.

The pattern of s-domain circuit analysis is outlined in Figure 10-1. We begin with a circuit described in the t domain in the usual way. We transform the circuit into the s domain, write the circuit equilibrium equations directly in that domain as algebraic equations, and then solve for current or voltage transforms using algebraic techniques. If necessary, we can obtain current or voltage waveforms by performing the inverse Laplace transformation. The illustration also points out that there is an-

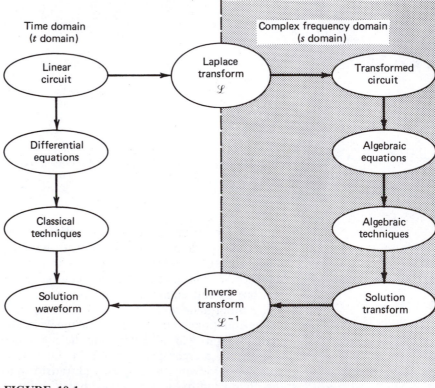

FIGURE 10-1
Pattern of s-domain circuit analysis.

other route to the solution waveform, using the circuit differential equation and classical time-domain techniques.

From the flow diagram in Figure 10-1 it may appear that s-domain circuit analysis is just a way to circumvent the classical differential equation technique. But it is really more than this. It provides a new viewpoint, a different way of thinking about signals and circuits. The solution transform is actually just another representation of a signal. The solution transform contains the same data as the solution waveform, otherwise we could not recover the waveform from the transform by using the inverse transformation. We can now begin to think of the solution transform as *the* solution; to think of circuit behavior in terms of signal transforms rather than signal waveforms. In sum, we can begin to think about circuits in the s domain.

But how are we to transform the circuit itself? As we have seen many times, the behavior of a circuit is rooted in an equilibrium established by constraints of two types: (1) connection constraints and (2) device constraints. The connection constraints are represented mathematically by equations obtained using KCL and KVL. The device constraints are mathematically expressed by the i-v relationships of the elements used to model the devices in the circuit. In other words, connection equations and element equations are the foundation of circuit analysis. How are we to transform circuits? We must see how connection and element equations are altered by the Laplace transformation.

The connection equations are based on Kirchhoff's laws. A typical current-law equation in the time domain would be

$$i_1(t) + i_2(t) + i_3(t) + i_4(t) = 0 \tag{10-1}$$

In words, this equation says that the sum of current waveforms at a node is zero for all values of t. If we take the Laplace transform of this equation, then because of the linearity property of the transformation we obtain

$$I_1(s) + I_2(s) - I_3(s) + I_4(s) = 0 \tag{10-2}$$

In words, this equation says that the sum of current transforms is zero for all values of s. Clearly this idea generalizes to any number of currents at any node. Just as clearly, it applies to Kirchhoff's voltage law as well. In sum, KCL and KVL are not altered by the Laplace transformation. They apply in either the t domain to waveforms or the s domain to transforms. Hence the connection constraints in a circuit are the same in the s domain.

Turning now to the device constraints, we deal first with the signal voltage source in Figure 10-2. The i-v relationships for this element are

$$v(t) = v_S(t) \quad \text{and} \quad i(t) = \text{Depends} \tag{10-3}$$

In words, this says that a voltage source produces a prescribed waveform at its terminals and can deliver any current waveform that may be de-

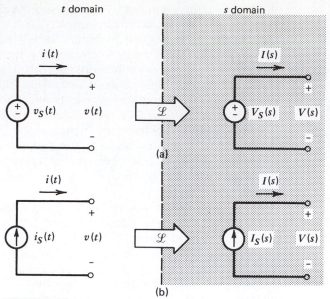

FIGURE 10-2
S-domain models of signal sources. (*a*) Voltage source. (*b*) Current source.

manded by the circuit to which it is connected. Taking the Laplace transform of this relationship produces

$$V(s) = V_S(s) \qquad \text{and} \qquad I(s) = \text{Depends} \qquad \textbf{(10-4)}$$

This says that the voltage source produces a prescribed transform and can deliver whatever current transform may be required. Clearly the same idea applies to the current source in Figure 10-2. In sum, signal sources behave in exactly the same way in the *s* domain, except that we think of them as producing a transform rather than a waveform.

Next we consider the three linear passive circuit elements shown in Figure 10-3. In the time domain the *i-v* relationships are

$$\text{Resistor} \quad v_R(t) = Ri_R(t)$$
$$\text{Inductor} \quad v_L(t) = L\frac{di_L(t)}{dt}$$
$$\text{Capacitor} \quad v_C(t) = \frac{1}{C}\int_0^t i_C(t)dt + v_C(0)$$

$$\textbf{(10-5)}$$

These relationships can be transformed into the *s* domain using the linearity, differentiation, and integration properties of the Laplace transformation:

$$\text{Resistor} \quad V_R(s) = RI_R(s)$$
$$\text{Inductor} \quad V_L(s) = (Ls)I_L(s) - Li_L(0)$$
$$\text{Capacitor} \quad V_C(s) = (1/Cs)I_C(s) + (1/s)v_C(0)$$

$$\textbf{(10-6)}$$

As might be expected, the *i-v* relationships in the *s* domain for all three elements turn out to be linear algebraic equations. In particular we see that for the resistor. Ohm's law is the same in the *s* domain.

These three-element equations lead to the *s*-domain circuit models in Figure 10-3. the initial conditions associated with the two memory elements are modeled as voltage sources connected in series with the elements. The element symbols in the *s* domain stand for what is called the element impedance. The concept of impedance is based on writing Eqs. 10-6 in the form

$$\begin{aligned} \text{Resistor} \quad & V_R = Z_R I_R \\ \text{Inductor} \quad & V_L = Z_L I_L - L i_L(0) \\ \text{Capacitor} \quad & V_C = Z_C I_C + v_C(0)/s \end{aligned} \qquad \textbf{(10-7)}$$

The symbol Z stands for the element **impedance,** which can be defined as the proportionality factor in the *s*-domain relationship between the current

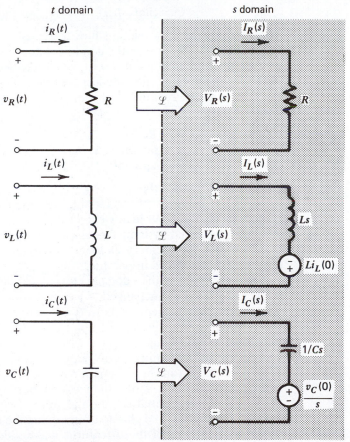

FIGURE 10-3
***S*-domain models for passive elements used in mesh-current analysis.**

transform and the voltage transform. In comparing Eqs. 10-6 and 10-7 we identify the three-element impedances as

$$\begin{aligned} \text{Resistor} \quad & Z_R = R \\ \text{Inductor} \quad & Z_L = Ls \\ \text{Capacitor} \quad & Z_C = 1/Cs \end{aligned} \tag{10-8}$$

Impedance is inherently an s-domain concept since it is based on a proportionality between a current transform and a voltage transform. It is a generalization of the concept of resistance, and hence we have the name "impedance." The inpedance of a resistor is a constant, its resistance. The impedances of the two memory elements are not constants, but depend on the complex frequency variable s.

Alternatively the i-v relationships in Eq. 10-6 can be solved for the current transform.

$$\begin{aligned} \text{Resistor} \quad & I_R(s) = GV_R(s) \\ \text{Inductor} \quad & I_L(s) = (1/Ls)V_L(s) + (1/s)i_L(0) \\ \text{Capacitor} \quad & I_C(s) = (Cs)V_C(s) - Cv_C(0) \end{aligned} \tag{10-9}$$

This form of the $i-v$ relationships leads to the s-domain circuit models in Figure 10-4. Here the initial conditions for the memory elements appear as current sources in parallel with what is called the element **admittance**, Y. The concept of admittance is a generalization of conductance and can be defined as the reciprocal of impedance ($Y = 1/Z$). The three-element admittances are

$$\begin{aligned} \text{Resistor} \quad & Y_R = 1/Z_R = 1/R = G \\ \text{Inductor} \quad & Y_L = 1/Z_L = 1/Ls \\ \text{Capacitor} \quad & Y_C = 1/Z_C = Cs \end{aligned} \tag{10-10}$$

Alternatively we see that the element admittances are the proportionality factor in the linear i-v characteristics relating transform voltage to transform current, Eq. 10-9. In any event, in the s domain the passive elements can be represented by impedances, with initial-condition voltage sources in series with the memory elements, or by admittances with initial-condition current sources in parallel.

Circuit analysis in the s domain closely parallels t-domain analysis since KCL and KVL are unchanged. To transform a circuit into the s domain, we replace each element by its s-domain model. For sources and resistors this is really no change at all. For capacitors and inductors we replace the element by its impedance in series with an initial-condition voltage source, or by its admittance in parallel with an initial-condition current source. Note that in the s domain the circuit initial conditions appear as sources in the circuit, rather than as boundary conditions on the solution of the circuit differential equation. The i-v relationships in the s domain have an Ohm's law-like character involving the impedance or admittance of the element. All of these features make s-domain circuit analysis an algebraic process similar to the analysis of memoryless resistance circuits in the t domain.

t domain   *s* domain

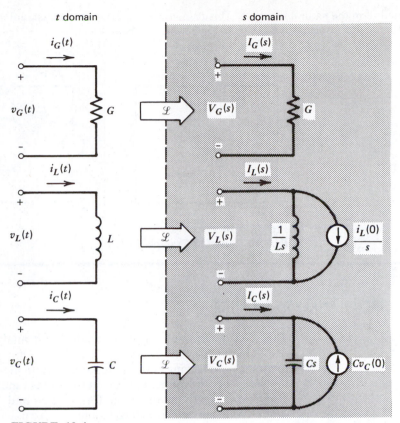

FIGURE 10-4
S-domain model for passive elements as used in node-voltage analysis.

Example 10-1

Figure 10-5a shows the series *RL* circuit studied in Example 9-15. Figure 10-5b shows the circuit transformed into the *s* domain. The sum of voltage transforms around the one loop in this circuit is

$$-\frac{V}{s} + V_R(s) + V_L(s) = 0$$

The voltages $V_R(s)$ and $V_L(s)$ can be written in terms of the mesh current $I(s)$ using the *s*-domain element equations

$$V_R(s) = RI(s) \quad \text{and} \quad V_L(s) = LsI(s) - Li_L(0)$$

Substituting these relationships into the foregoing KVL equation and collecting terms yields

$$(Ls + R)I(s) = \frac{V}{s} + Li_L(0)$$

Solving for $I(s)$,

$$I(s) = \frac{V/L}{s(s + R/L)} + \frac{i_L(0)}{s + R/L}$$

This is the transform of the mesh current for a step-function input. If necessary, we can determine the corresponding waveform. First we expand the transform by partial fractions:

$$I(s) = \frac{V/R}{s} - \frac{V/R}{s + R/L} + \frac{i_L(0)}{s + R/L}$$

Performing the inverse transformation yields

$$i(t) = \frac{V}{R} + \left[i_L(0) - \frac{V}{R}\right]e^{-Rt/L} \qquad t \geq 0$$

The first term on the right is the forced response and

the second term the natural response. This response is the same as that obtained in Example 9-15 by trans-forming the circuit differential equation into the s domain.

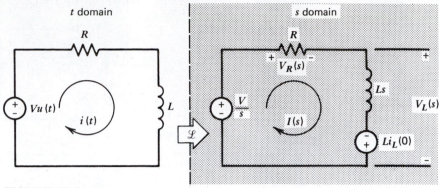

FIGURE 10-5
RL circuit used in Example 10-1.

10-2   CIRCUIT ANALYSIS TECHNIQUES

This section focuses on the special circuit analysis techniques that are useful in dealing with fairly simple circuits. Of particular interest here are techniques involving series and parallel equivalence, voltage and current division, the Thévenin and Norton theorems, and superposition. The general, formal methods involving formulation and solution of node-voltage or mesh-current equilibrium equations are discussed in subsequent sections.

The analysis methods we have lumped together as circuit-reduction techniques can be extended to the s domain. For example, the two elements Z_1 and Z_2 in Figure 10-6 are connected in series, and hence the same current $I(s)$ flows through both. Using the element equations in the s domain, we write

$$V_1(s) = Z_1 I(s) \quad \text{and} \quad V_2(s) = Z_2 I(s) \quad \text{(10-11)}$$

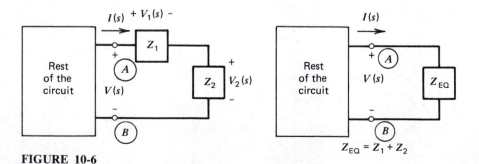

FIGURE 10-6
Series equivalence and voltage division.

But according to KVL, $V(s) = V_1(s) + V_2(s)$, and hence

$$V(s) = Z_1 I(s) + Z_2 I(s) \qquad (10\text{-}12)$$
$$= (Z_1 + Z_2) I(s)$$

This result indicates that $V(s)$ and $I(s)$ are unchanged if Z_1 and Z_2 are replaced by an equivalent impedance Z_{EQ} defined as

$$Z_{EQ} = Z_1 + Z_2 \qquad (10\text{-}13)$$

In other words, Z_{EQ} is the equivalent impedance of the series connection insofar as behavior at terminals Ⓐ and Ⓑ is concerned. It is easy to see that this result generalizes as follows:

The equivalent impedance of two or more elements connected in series equals the sum of the individual element impedances.

By combining Eqs. 10-12 and 10-13, we can express the current $I(s)$ as

$$I(s) = \frac{V(s)}{Z_{EQ}} \qquad (10\text{-}14)$$

By now combining Eqs. 10-11 and 10-14, we can express each of the element voltages as

$$V_1 = \frac{Z_1}{Z_{EQ}} V(s) \quad \text{and} \quad V_2 = \frac{Z_2}{Z_{EQ}} V(s) \qquad (10\text{-}15)$$

These equations are the *s*-domain version of the voltage-division relationships. The following statement summarizes that concept:

In a series connection the voltage across any one element is proportional to the ratio of its impedance to the equivalent impedance of the connection.

The dual situation occurs in the parallel circuit in Figure 10-7. The two elements Y_1 and Y_2 are connected in parallel and thus the same voltage

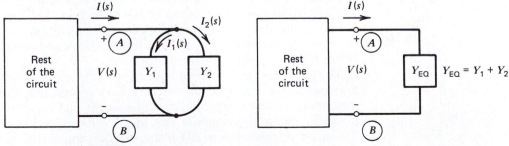

FIGURE 10-7
Parallel equivalence and current division.

$V(s)$ appears across both. The individual element currents can be written as

$$I_1 = Y_1 V(s) \qquad \text{and} \qquad I_2 = Y_2 V(s) \tag{10-16}$$

But KCL at node Ⓐ demands that $I(s) = I_1(s) + I_2(s)$, and hence

$$I(s) = Y_1 V(s) + Y_2 V(s) \tag{10-17}$$
$$= (Y_1 + Y_2)V(s)$$

This equation indicates that the pair (V,I) would be unchanged if Y_1 and Y_2 were replaced by an equivalent admittance Y_{EQ} defined as

$$Y_{EQ} = Y_1 + Y_2 \tag{10-18}$$

This result clearly generalizes as follows:

The equivalent admittance of two or more elements connected in parallel equals the sum of the individual element admittances.

Continuing with the duality, we combine Eq. 10-17 and 10-18 to write $V(s)$ as

$$V(s) = \frac{I(s)}{Y_{EQ}} \tag{10-19}$$

and using Eq. 10-16 we obtain the individual element currents as

$$I_1 = \frac{Y_1}{Y_{EQ}} I(s) \qquad \text{and} \qquad I_2 = \frac{Y_2}{Y_{EQ}} I(s) \tag{10-20}$$

These equations are the s-domain version of the current division principle, which can be summarized as follows:

In a parallel connection the current through any element is proportional to the ratio of its admittance to the equivalent admittance of the connection.

From all of this we see that s-domain circuit analysis involves concepts that closely parallel the analysis of resistance circuits in the t domain. Repeated or successive application of equivalence and division lead to analysis by what we have called circuit reduction. The major difference is that in the s domain we use impedance and admittances rather than resistance and conductance, and determine voltage and current transforms rather than waveforms.

The superposition principle is often a useful circuit analysis tool. As previously used, the principle indicates that any response of a linear circuit containing several voltage or current sources can be obtained by algebraically adding the response due to all voltages acting together, then finding the response due to all current sources, and obtaining the complete response by algebraically adding (superimposing) these two responses.

When circuits are transformed into the s domain, the diagram fairly bristles with current and voltage sources. However, all of the sources can be grouped as either input signal sources or initial conditions sources. The superposition principle then tells us that any response can be found as

$$
\text{Response} = \underbrace{\begin{bmatrix} \text{Response due to} \\ \text{input signal} \\ \text{sources} \end{bmatrix}}_{\text{Zero-state response}} + \underbrace{\begin{bmatrix} \text{Response due} \\ \text{to initial} \\ \text{condition sources} \end{bmatrix}}_{\text{Zero-input response}} \qquad \textbf{(10-21)}
$$

Thus any response can be expressed as the sum of two components. The first component is called the **zero-state** response. It is caused by the input signal sources and is found by setting all initial conditions to zero (zero state). The second component is called the **zero-input** response. It is caused by the initial condition sources and is found by turning off all input signal sources (zero input).

In sum, any voltage or current response transform can be written as

$$
V = V_{zs} + V_{zi} \qquad \text{or} \qquad I = I_{zs} + I_{zi} \qquad \textbf{(10-22)}
$$

where the subscript zs stands for zero state and zi for zero input. The next three examples illustrate the use of circuit reduction and superposition in s-domain circuit analysis.

Example 10-2

In this example we use the circuit from Example 10-1. The transformed circuit is shown in Figure 10-8. The circuit contains two sources: one due to the input signal (a step function) and one due to the initial inductor current. The resistor and inductor are connected in series, hence

$$
Z_{EQ} = Ls + R
$$

By turning the initial condition source off (voltage source replaced by a short circuit), we obtain the zero-state response as

$$
I_{zs} = \frac{V/s}{Z_{EQ}} = \frac{V}{s(Ls + R)} = \frac{V/L}{s(s + R/L)}
$$

Next, by turning the input source off, we obtain the zero-input response as

$$
I_{zi} = \frac{Li_L(0)}{Z_{EQ}} = \frac{i_L(0)}{s + R/L}
$$

Hence, by superposition,

$$
I(s) = I_{zs} + I_{zi}
$$

$$
I(s) = \frac{V/L}{s(s + R/L)} + \frac{i_L(0)}{s + R/L}
$$

In expanding the first term on the right by partial fraction,

$$
I(s) = \frac{V/R}{s} - \frac{V/R}{s + R/L} + \frac{i_L(0)}{s + R/L}
$$

we obtain the waveform $i(t)$ as

$$
i(t) = \overbrace{\underbrace{\frac{V}{R}}_{\text{Zero state}}}^{\text{Forced}} - \overbrace{\frac{V}{R} e^{-Rt/L} + \underbrace{i_L(0) e^{-Rt/L}}_{\text{Zero input}}}^{\text{Natural}} \qquad t \geq 0
$$

This is the same result as obtained in Example 10-1. Superposition partitions the response in a slightly different way. The forced response is contained in the zero-state component; the natural response is contained in both the zero-state and zero-input components.

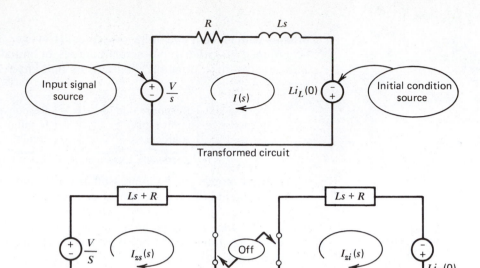

Transformed circuit

Zero state

Zero input

FIGURE 10-8
Applying superposition to find zero-state and zero-input responses of circuit in Example 10-2.

Example 10-3

The switch in the circuit shown in Figure 10-9a has been closed for a "long time" and is opened at $t = 0$. Determine the voltage $v(t)$.

To determine the initial capacitor voltage, we use the circuit in Figure 10-9b. With the capacitor replaced by an open circuit, voltage division yields

$$v_C(0) = \frac{R_2}{R_1 + R_2} V$$

Now, in transforming the circuit ($t > 0$) into the s domain, we obtain Figure 10-9c, which has two voltage sources. Turning the initial condition source off and using s-domain voltage division on the circuit in Figure 10-9(d), we write

$$V_{zs} = \frac{1/Cs}{Z_{EQ}} \times \frac{V}{s} = \left(\frac{1/Cs}{R_1 + 1/Cs}\right)\frac{V}{s}$$

$$= \frac{V/R_1C}{s(s + 1/R_1C)}$$

Turning the input source off and using s-domain voltage division in Figure 10-9e, we write

$$V_{zi} = \frac{R_1}{Z_{EQ}}\frac{V_C(0)}{s} = \frac{R_1}{R_1 + 1/Cs}\frac{v_C(0)}{s} = \frac{v_C(0)}{s + 1/R_1C}$$

Hence by superposition,

$$V(s) = V_{zs} + V_{zi}$$

$$= \frac{V/R_1C}{s(s + 1/R_1C)} + \frac{v_C(0)}{s + 1/R_1C}$$

Expanding the zero-state component by partial fractions:

$$V(s) = \frac{V}{s} - \frac{V}{s + 1/R_1C} + \frac{v_C(0)}{s + 1/R_1C}$$

And we recover the waveform $v(t)$ as

$$
\overbrace{\phantom{\text{Forced}}}^{\text{Forced}} \quad \overbrace{\phantom{\text{Natural..........}}}^{\text{Natural}}
$$

$$v(t) = \underbrace{V - V\,e^{-t/R_1C}}_{\text{Zero state}} + \underbrace{v_C(0)\,e^{-t/R_1C}}_{\text{Zero input}} \qquad t \geq 0$$

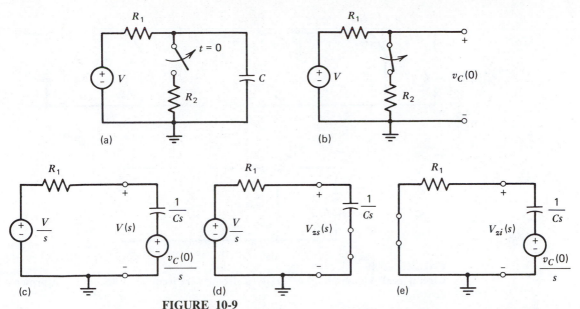

FIGURE 10-9
Finding zero-state and zero-input responses using superposition of circuit in Example 10-3. (*a*) Switch opens at $t = 0$. (*b*) Circuit at $t=0^-$. Used to find $v_C(0)$. (*c*) Transformed circuit. (*d*) Zero-state circuit. (*e*) Zero-input circuit.

Example 10-4

The switch in Figure 10-10*a* has been in position *A* for a "long time" and is moved to position *B* at $t = 0$. Determine the output voltage $v(t)$.

To solve this problem, we first determine the initial condition using Figure 10-10*b*. With the switch in position *B* and the capacitor replaced by an open circuit, we concluded that $v_C(0) = 0$. The transformed circuit is shown in Figure 10-10*c*. There is no initial condition source associated with the capacitor since the initial condition is zero. The circuit is in the zero state at $t = 0$, and therefore the zero-input response is zero. Using *s*-domain voltage division:

$$V_{zs}(s) = \frac{RV/s}{Z_{EQ}} = \frac{RV/s}{R + 1/Cs}$$

$$= \frac{V}{s + 1/RC}$$

Hence

$$v(t) = \underbrace{Ve^{-t/RC}}_{\substack{\text{Zero} \\ \text{state}}} + \underbrace{0}_{\substack{\text{Zero} \\ \text{input}}} \qquad t \geq 0$$

Note that the zero-state response contains no forced component.

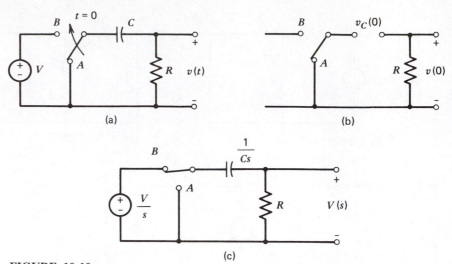

FIGURE 10-10
Circuit used in Example 10-4. (*a*) Given circuit. (*b*) Circuit at $t = 0^-$. (*c*) Transformed circuit.

The use of Thévenin and Norton equivalent circuits at source–load interfaces also applies in the *s* domain as illustrated in Figure 10-11. The general proposition is that the signals at the interface will be unchanged if the source circuit is replaced by its Thévenin or Norton equivalent. The important difference here is that the interface signals are represented as transforms rather than waveforms. Likewise the Thévenin source $V_T(s)$ and Norton source $I_N(s)$ are also transforms. Finally the "elements" Z_T and Z_N are not just resistance but the *s*-domain generalization impedance.

With these important differences, the interface analysis proceeds in much the same way as presented in Chapter 3. The Thévenin and Norton circuits are both equivalent to the given source circuit, and thus are equivalent to each other. We establish relationships among the circuits by first finding the open-circuit voltage as

$$V_{OC}(s) = V_T(s) = I_N(s)Z_N(s) \tag{10-23}$$

When a short circuit is connected at the interface we obtain

$$I_{SC}(s) = V_T(s)/Z_T = I_N(s) \tag{10-24}$$

Together Eqs. 10-23 and 10-24 tell us that

$$\begin{aligned} V_T &= V_{OC} \\ I_N &= I_{SC} \\ Z_N &= Z_T = V_{OC}/I_{SC} \end{aligned} \tag{10-25}$$

There results point out that to find the Thévenin or Norton equivalent at an interface, we need only find the open-circuit voltage and the short-circuit current. Algebraically Eqs. (10-25) are identical to the results ob-

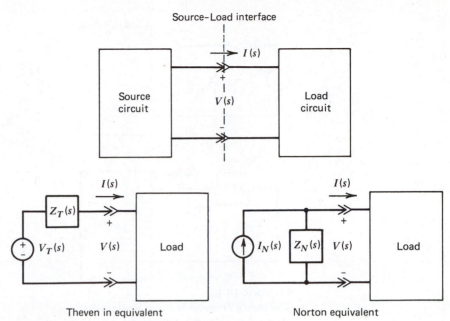

FIGURE 10-11
S-domain Thévenin and Norton equivalent circuits for source–load interface.

tained in Chapter 3. The important difference is that we are here dealing with transforms and impedances rather than waveforms and resistances.

Example 10-5

Find the Thévenin equivalent of the circuit given in Figure 10-12.

The circuit is given in the s domain with zero as the capacitor initial condition. The open-circuit voltage at the interface can be found by voltage division.

$$V_{OC} = \frac{1/Cs}{R + 1/Cs} \frac{V_O}{s}$$

$$= \frac{V_O}{s(RCs + 1)}$$

With a short circuit connected at the interface, the capacitor is effectively removed from the circuit. Hence

$$I_{SC} = \frac{V_O}{sR}$$

and hence

$$Z_T = \frac{V_{OC}}{I_{SC}} = \frac{R}{RCs + 1}$$

Notice that the poles of V_{OC} come from both the input source (V_O/S) and the circuit ($RCs + 1$). The Thévenin impedance could also be found by turning off the input source and finding the equivalent impedance at the interface "looking back" into the source circuit.

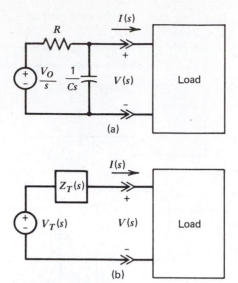

FIGURE 10-12
Thévenin's Theorem in s-domain. (a) Given circuit. (b) Thévenin equivalent.

10-3 NODE-VOLTAGE EQUATIONS

The methods of linear resistance circuit analysis previously studied in the t domain can be applied in the s domain with only minor changes. The formal analysis process can be divided into two major phases: (1) formulation of equilibrium equations and (2) solution of linear algebraic equations. This section deals with the formulation of equilibrium equations using node-voltage variables. The following section treats in a parallel fashion the classical mesh-current method. And in Sec. 10-5 we look at the solution process.

For most circuits, node voltages are a convenient set of formulation and solution variables. These variables are defined by selecting a reference (or ground) node, and then using the voltages at all other nodes as the unknowns. We previously developed a systematic algorithm for formulating the node-voltage equations for resistance circuits. The same steps can be used here after the circuit has been transformed into the s domain:

STEP 1   Transform the circuit into the s domain using the current source representation of the initial condition sources.

STEP 2   Select a reference node and identify a node voltage at every other node and an element current for every nonsource branch.

STEP 3   Write KCL at each nonreference node using the source currents and the element currents defined in step 2.

STEP 4   Write the element equations relating each element current to the node voltages using the element admittances.

STEP 5   Substitute the element equations from step 4 into the KCL connection equations from step 3 and collect terms.

We represent the initial conditions as current sources in step 1 since current sources are more convenient in node analysis. The next example illustrates the node-voltage equation formulation process for a transformed circuit.

Example 10-6

Figure 10-13 shows a transformed circuit (step 1) with node voltages and element currents identified (step 2). Note that the initial condition sources are represented as current sources. The KCL equations (step 3) at each of the two nonreference nodes are

Node Ⓐ $I_S - \dfrac{1}{s} i_L(0) - I_1(s) - I_2(s) = 0$

Node Ⓑ $Cv_C(0) + \dfrac{1}{s} i_L(0) + I_1(s) - I_3(s) = 0$

The element equations can be written in terms of the node voltage (step 4) as

$$I_1 = \frac{1}{Ls}(V_A - V_B)$$
$$I_2 = GV_A$$
$$I_3 = CsV_B$$

Substituting these element constraints into the KCL connection constraints (step 5) yields

$$I_S - \frac{1}{s} i_L(0) - \frac{1}{Ls}(V_A - V_B) - GV_A = 0$$

$$Cv_C(0) + \frac{1}{s} i_L(0) + \frac{1}{Ls}(V_A - V_B) - CsV_B = 0$$

Collecting terms, we obtain the two node-voltage equations that describe the circuit:

$$\left(G + \frac{1}{Ls}\right) V_A - \frac{1}{Ls} V_B = I_S - \frac{1}{s} i_L(0)$$

$$-\frac{1}{Ls} V_A + \left(\frac{1}{LS} + Cs\right) V_B = Cv_C(0) + \frac{1}{s} i_L(0)$$

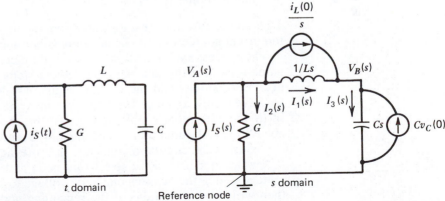

FIGURE 10-13
Circuit for Example 10-6.

The final form of the node equations in Example 10-6 have a very familiar ring to them. First of all, they are two linear algebraic equations in the two unknown node-voltage transforms, $V_A(s)$ and $V_B(s)$. As advertised, equilibrium equations in the s domain are linear algebraic equations, even for memory circuits. The terms on the right side of the equations are all due to current sources—both the input signal source and the initial condition sources. The factor multiplying V_A in the first equation (node Ⓐ) is the sum of the admittances connected at node Ⓐ. The factor multiplying V_B in the same equation is the admittance connected between node Ⓐ and node Ⓑ. This is a familiar pattern from our previous study of node equations. Is this pattern always repeated in s-domain circuit analysis?

If an admittance Y has one terminal connected at a node, say node Ⓐ, then there are but two possibilities. Either the other terminal of admittance is connected to the reference node, in which case the current leaving node Ⓐ through this admittance is

$$I = Y(V_A - 0) = YV_A \tag{10-26}$$

or else the other terminal of the admittance is connected to a nonreference node, say node Ⓑ, in which case the current leaving the node through the admittance is

$$I = Y(V_A - V_B) = YV_A - YV_B \tag{10-27}$$

The pattern of the final result is quite easy to see. The sum of the currents leaving node Ⓐ through admittances is

1. V_A times the sum of admittances directly connected to node Ⓐ
2. minus V_B times the sum of admittances connected between node Ⓐ and node Ⓑ
3. minus similar terms for all other nodes.

This sum of currents must equal the sum of currents entering node Ⓐ due to current sources—both signal sources and initial condition sources.

The process outlined allows us to write node equations by inspection without passing through the intermediate steps involving the KCL connection equations and the element equations. In describing the method we have tacitly assumed that there are no voltage sources in the circuit. In the case of initial condition sources, the choice is ours since we can always select the admittance-current-source representation of the memory elements. If the circuit contains one or more signal voltage sources with a common node, then we select that node as the reference node. Each voltage source then causes one node voltage to be a known quantity. We then write node equations at the remaining nodes in the usual way and transfer the known node voltages to the right side of the equations.

The next example illustrates the formulation of node equations by inspection, including accounting for a signal voltage source.

Example 10-7

For the circuit of Figure 10-14(a), formulate a set of node equations.

The circuit in Figure 10-14 contains a signal voltage source. For the selected reference node this source causes the voltage at node Ⓐ to be known. Hence we need only write equations at nodes Ⓑ and Ⓒ. The sum of the admittances connected at node Ⓑ is $G_1 + G_2 + C_1s$. The admittance connected between nodes Ⓑ and Ⓐ is G_1. The admittance connected between nodes Ⓑ and Ⓒ is G_2. The current source sum at node Ⓑ is $C_1v_{C1}(0)$. Consequently the node equation at node Ⓑ is

Node Ⓑ
$$-G_1V_A + (G_1 + G_2 + C_1s)V_B - G_2V_C = C_1v_{C1}(0)$$

Similarly the equation at node Ⓒ is

Node Ⓒ
$$-G_3V_A - G_2V_B + (G_2 + G_3 + C_2s)V_C = C_2v_{C2}(0)$$

But V_A is known since $V_A = V_S$, the signal source voltage, and all of the terms involving V_A can be transferred to the right side of the node-voltage equation.

$$(G_1 + G_2 + C_1s)V_B - G_2V_C = G_1V_S + C_1v_{C1}(0)$$
$$-G_2V_B + (G_2 + G_3 + C_2s)V_C = G_3V_S + C_2v_{C2}(0)$$

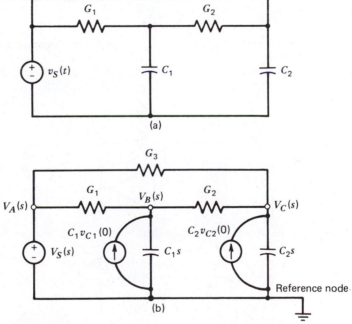

(a)

(b)

FIGURE 10-14
Circuit for Example 10-7. (*a*) *t* domain. (*b*) *s* domain.

The process used in Example 10-7 will not work if the circuit contains two or more voltage sources that do not have a common node. In that event we must use the modified node analysis algorithm discussed in

Chapter 5. The extension of this algorithm to s-domain circuit analysis is left as an exercise for the reader.

An important application of node equations occurs in OP AMP circuits. Happily the OP AMP *i-v* constraints are algebraically unchanged in the *s* domain.

$$V_P(s) = V_N(s)$$
$$I_N(s) = 0 \qquad\qquad (10\text{-}28)$$
$$I_P(s) = 0$$

Of course the constraints now apply to transforms rather than wave-forms. To formulate node equations for OP AMP circuits we proceed in the manner described in Chapter 5. Basically we write node equations at all nodes except the OP AMP output node. We then invoke the OP AMP constraints to reduce the number of unknowns. Our final node-voltage example illustrates the process.

Example 10-8

Formulate a set of node-voltage equations for the OP AMP circuit of Figure 10-15(a).

Figure 10-15(b) shows the transformed OP AMP circuit. To formulate node equations we need only consider nodes Ⓑ and Ⓒ since the input source forces $V_A = V_S$ and V_D is the OP AMP output.

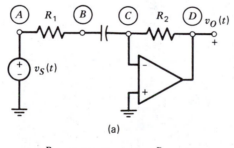

(a)

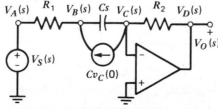

(b)

FIGURE 10-15
OP AMP circuit for Example 10-8. (*a*) Given time-domain circuit. (*b*) Transformed circuit.

Node Ⓑ
$$-G_1V_A + (G_1 + Cs)V_B - CsV_C = Cv_C(0)$$
Node Ⓒ
$$-CsV_B + (G_2 + Cs)V_C - G_2V_D = -Cv_C(0)$$

But $V_A = V_S$ and $V_D = V_o$ as noted above. The OP AMP constraint $V_P = V_N$ then dictates that $V_C = 0$

since the noninverting input is grounded. Inserting these observations yields

$$(G_1 + Cs)\,V_B = Cv_C(0) + G_1V_S$$
$$-Cs\,V_B - G_2V_0 = Cv_C(0)$$

We now have two equations in two unknowns (V_B and V_O). The equations are not symmetrical since the OP AMP is a unilateral device.

10-4 MESH-CURRENT EQUATIONS

We have seen that in the s domain the node-voltage method applies with only minor changes due to initial condition sources and the use of element admittances. We naturally would expect to be able to apply the mesh-current method as well. To review, the mesh-current method is not completely general. It works only if the circuit is planar, which basically means that the circuit can be drawn on a flat surface without having any branch crossings. A mesh is a special type of loop, one that contains no branches within it. The mesh-current variables are the loop currents assigned to each of the special loops (meshes) in a planar circuit. The process of obtaining mesh-current equations in the s domain can be outlined in five steps:

STEP 1   Transform the circuit into the s domain using the voltage source representation of the initial condition sources.

STEP 2   Identify a mesh current with each mesh and an element voltage with each nonsource branch.

STEP 3   Write KVL around each mesh using the source voltages and the element voltages defined in step 2.

STEP 4   Write the element equations relating the element voltages to the mesh currents using the element impedances.

STEP 5   Substitute the element equations from step 4 into the KVL connection equation from step 3 and collect terms.

We use the voltage source representation of the initial conditions since voltage sources are more convenient in mesh analysis. The following example illustrates mesh-current-equation formulation.

Example 10-9

Figure 10-16 shows a transformed circuit (step 1) with the mesh currents and element voltages identified (step 2) and the initial conditions represented as voltage sources. Step 3 calls for a KVL connection equation around each mesh.

Mesh A
$$-V_S(s) + V_1(s) - L_1i_{L_1}(0) + V_2(s) = 0$$
Mesh B
$$-V_2(s) + V_3(s) - L_2i_{L1}(0) + V_4(s) = 0$$

Step 4 requires each element current to be written in terms of mesh currents and element impedances.

$$V_1 = L_1 s I_A$$
$$V_2 = R_1(I_A - I_B)$$
$$V_3 = L_2 s I_B$$
$$V_4 = R_2 I_B$$

Substituting these element equations into the foregoing KVL connection equations and collecting terms (step 5) yields

$$(L_1 s + R_1)I_A - R_1 I_B = V_S + L_1 i_{L1}(0)$$
$$-R_1 I_A + (L_2 s + R_1 + R_2)I_B = L_2 i_{L2}(0)$$

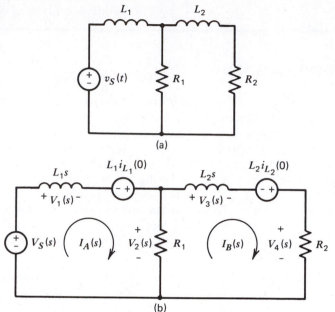

(a)

(b)

FIGURE 10-16
Circuit for Example 10-9. (a) t domain. (b) s domain.

The mesh-current equations in Example 10-9 have a familiar pattern. The terms on the right side of the equations are due to the signal and initial-condition voltage sources in each mesh. The factor multiplying I_A in the mesh A equation is the sum of impedances in that mesh. Likewise the factor multiplying I_B in the same equation is the impedance common to meshes A and B. This familiar result is a consequence of the manner in which mesh currents are defined.

If an impedance Z is contained within a mesh, say mesh A, then either it is contained in that mesh only, in which case the voltage across it is

$$V = Z(I_A - 0) = Z I_A \qquad \textbf{(10-29)}$$

or it is contained in another mesh as well, say mesh B, in which case its voltage is

$$V = Z(I_A - I_B) = Z I_A - Z I_B \qquad \textbf{(10-30)}$$

The general pattern follows from these results. The sum of voltages across impedances in mesh A is

1. I_A times the sum of impedances in mesh A
2. minus I_B times the sum of impedances common to meshes A and B
3. minus similar terms for all other meshes.

This sum of voltages must equal the algebraic sum of source voltages around the mesh.

The algorithm outlined allows us to write mesh-current equations by inspection without the need first to write connection and element equations. We have assumed that no current sources are present in the circuit. Initial condition sources are always selected as voltage sources if mesh-current equations are to be used. If a signal current source is contained in only one mesh, then that mesh current is known. We can then write mesh-current equations around the remaining meshes and transfer the known terms to the right side of the equations in the final step. If a current source is contained in more than one mesh, we must use the modified mesh-current method.

In Example 10-7 we wrote the node equations for the circuit in Figure 10-14. In Example 10-10 we obtain the mesh-current equations for the

---

Example 10-10

For the circuit of Figure 10-17 formulate a set of mesh equations.

The circuit in Figure 10-17 contains only voltage sources. The total impedance in mesh A is $R_1 + 1/C_1s$; the impedance common to meshes A and B is $1/C_1s$; the impedance common to meshes A and C is R_1; and the sum of voltage sources in mesh A is $V_S - (1/s(vc_1(0)))$. Therefore the mesh A equation is

$$\left(R_1 + \frac{1}{C_1s} \right) I_A - \frac{1}{C_1s} I_B - R_1 I_C = V_S - \frac{1}{s} v_{C1}(0)$$

Similarly the equations for meshes B and C are

$$- \frac{1}{C_1s} I_A + \left(R_2 + \frac{1}{C_1s} + \frac{1}{C_2s} \right) I_B - R_2 I_C$$
$$= \frac{1}{s} v_{C1}(0) - \frac{1}{s} v_{C2}(0)$$

$$-R_1 I_A - R_2 I_B + (R_1 + R_2 + R_3) I_C = 0$$

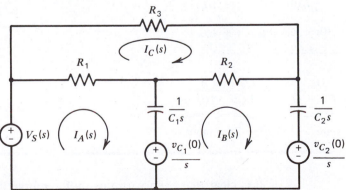

FIGURE 10-17
Applying mesh-analysis techniques to three-mesh circuit (Example 10-10).

same circuit. The appropriate transformed circuit is shown in Figure 10-17 with impedances and voltage sources to represent the two capacitors.

10-5 SOLUTION OF LINEAR EQUATIONS

The process of formulating node-voltage or mesh-current equations in the s domain leads to systems of linear equations. We now turn to the second phase of s-domain circuit analysis—solving these equations. In principle, this is a straightforward mathematical process. Among the standard techniques at our disposal are Cramer's rule and systematic elimination methods such as Gauss' reduction as discussed in Chapter 5. However, the algebra involved can become somewhat tedious because we must manipulate polynominals in the complex frequency variable. The next example illustrates the solution process using Cramer's rule for solving problems in the s domain.

Example 10-11

The node-voltage equations from Example 10-6 are

$$(G + 1/Ls)V_A - (1/Ls)v_B = I_S - (1/s)i_L(0)$$
$$- (1/Ls)V_A + (Cs + 1/Ls)V_B = (1/s)i_L(0) + Cv_C(0)$$

To use Cramer's rule we first form the determinant

$$D = \begin{vmatrix} G + 1/Ls & - 1/Ls \\ - 1/Ls & Cs + 1/Ls \end{vmatrix}$$

$$= (G + 1/Ls)(Cs + 1/Ls) - (1/Ls)^2$$

$$= \frac{GLCs^2 + Cs + G}{Ls}$$

Solving for V_A:

$$V_A = \frac{D_A}{D} = \frac{\begin{vmatrix} I_S - \dfrac{1}{s}i_L(0) & - 1/Ls \\ \dfrac{1}{s}i_L(0) + Cv_C(0) & Cs + 1/Ls \end{vmatrix}}{D}$$

$$= \frac{(LCs^2 + 1)I_S}{GLCs^2 + Cs + G} + \frac{- LCsi_L(0) + Cv_C(0)}{GLCs^2 + Cs + G}$$

$$\underbrace{\phantom{\frac{(LCs^2 + 1)I_S}{GLCs^2 + Cs + G}}}_{\text{Zero state}} \quad \underbrace{\phantom{\frac{- LCsi_L(0) + Cv_C(0)}{GLCs^2 + Cs + G}}}_{\text{Zero input}}$$

Solving for V_B:

$$V_B = \frac{D_B}{D} = \frac{\begin{vmatrix} G + 1/Ls & I_B - \dfrac{1}{s}i_L(0) \\ - 1/Ls & \dfrac{1}{s}i_L(0) + Cv_C(0) \end{vmatrix}}{D}$$

$$= \underbrace{\frac{I_S}{GLCs^2 + Cs + G}}_{\text{Zero state}}$$

$$+ \underbrace{\frac{(GLs + 1)Cv_C(0) + GLi_L(0)}{GLCs^2 + Cs + G}}_{\text{Zero input}}$$

It is clear that even with this rather simple two-node circuit Cramer's rule leads to a fair amount of algebraic drudgery. Nonetheless, the example does suggest that, at least in principle, we can formulate and solve the s-domain equilibrium equations for linear circuits. As compensation for our effort we obtain both the zero-state and zero-input responses in the process.

The example also points out that response transforms are quotients of polynominals in the complex frequency variable, that is, rational functions. Moreover the poles of the response transforms are zeros of the circuit determinant, that is, roots of the equation $D = 0$. These observations are revealed by our use of Cramer's rule, which involves forming quotients of determinants such as D_A/D.

Still the thought of tackling a circuit with many more than two or three nodes (or meshes) is not an idea of irresistible appeal. As a practical matter, however, there are several considerations that frequently reduce the amount of labor involved. First, we are usually not interested in finding every node-voltage (or mesh-current) response, but only the one identified as the circuit output. Second, we often do not need both the zero-state and zero-input components of the response. Most commonly we need only the zero-state term. The next example illustrates the simplification resulting from these considerations.

Example 10-12

The mesh-current equilibrium equations from Example 10-9 are

$$(L_1s + R_1)I_A - R_1I_B = V_S + L_1i_{L1}(0)$$
$$-R_1I_A + (L_2s + R_1 + R_2)I_B = L_2i_{L2}(0)$$

To solve for the zero-state component of I_B we set the initial conditions to zero,

$$i_{L1}(0) = i_{L2}(0) = 0$$

and apply Cramer's rule:

$$(I_B)_{zs} = \frac{D_B}{D} = \frac{\begin{vmatrix} L_1s + R_1 & V_S \\ -R_1 & 0 \end{vmatrix}}{\begin{vmatrix} L_1s + R_1 & -R_1 \\ -R_1 & L_2s + R_1 + R_2 \end{vmatrix}}$$

$$= \frac{R_1V_S}{L_1L_2s^2 + (L_1R_1 + L_2R_1 + L_1R_2)s + R_1R_2}$$

Note that the response transform is a rational function whose poles are roots of the equation $D = 0$.

The algebraic process of determining just the zero-state component of one response variable is somewhat less trying. Our next subject, *transfer functions*, capitalizes on this and focuses on the zero-state response of a circuit with a single input signal.

10-6 TRANSFER FUNCTIONS

An important application of circuit analysis is the processing of a signal in its passage from input to output. In the s domain the signal processing involved is described by a rational function of the complex frequency variable called a **transfer function**. A transfer function is defined as

$$\text{Transfer function} = \frac{(\text{Zero-state response transform})}{(\text{Input signal transform})} = T(s) \qquad \textbf{(10-31)}$$

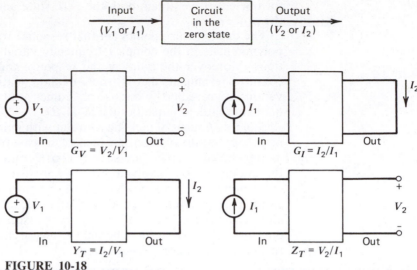

FIGURE 10-18
Signal transfer through a circuit.

Note that the formal definition applies only to the zero-state response and implies that the circuit has only one input. Both of these conditions simplify the process of finding and using transfer functions.

The pattern of signal transfer from input to output is illustrated in Figure 10-18. There is a single input signal, either a current I_1 or voltage V_1, and a single output, either a voltage V_2 or a current I_2. Since the input and output signals can take one of two possible forms, it is possible to define four different kinds of transfer functions (V_2/V_1, I_2/I_1, V_2/I_1, I_2/V_1). We will concentrate our attention on the two dimensionless functions.

$$\text{Voltage transfer function} = T(s) = \frac{V_2(s)}{V_1(s)}$$

$$\text{Current transfer function} = T(s) = \frac{I_2(s)}{I_1(s)}$$

(10-32)

The distinguishing feature of a transfer function is that it involves an input at one place in a network and a response that occurs somewhere else. In this regard the transfer function $T(s)$ is the s-domain generalization of the input-output relationship $y = Kx$, which we studied for memoryless circuits. As such we see that it is the result of the circuit being linear. Strictly speaking only linear circuits have transfer functions, although it is fairly common for practicing engineers to apply the term loosely (and usually incorrectly) to nonlinear circuits.

Certain simple circuits occur so frequently that it is worth our time to work out their transfer functions for the general case. First, for the voltage divider in Figure 10-19 we use s-domain voltage division to write

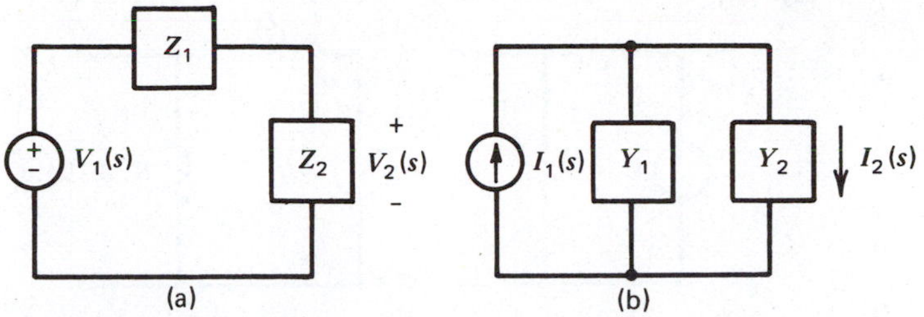

FIGURE 10-19
Transfer functions using voltage and current division. (*a*) Voltage divider. (*b*) Current divider.

$$V_2 = \frac{Z_2}{Z_1 + Z_2} V_1$$

Hence

$$T(s) = \frac{V_2}{V_1} = \frac{Z_2}{Z_1 + Z_2} \tag{10-33}$$

Using *s*-domain current division in the second circuit we have

$$I_2 = \frac{Y_2}{Y_1 + Y_2} I_1 \tag{10-34}$$

Hence

$$T(s) = \frac{I_2}{I_1} = \frac{Y_2}{Y_1 + Y_2}$$

The next two examples apply voltage and current division to obtain the transfer functions of some simple but useful circuits.

Example 10-13

Figure 10-20 shows several simple voltage and current dividers.

For *C*1

$$Z_1 = R, \ Z_2 = 1/Cs$$

$$T(s) = \frac{1/Cs}{R + 1/Cs} = \frac{1}{RCs + 1}$$

For *C*2

$$Z_1 = Ls, \ Z_2 = R$$

$$T(s) = \frac{R}{Ls + R} = \frac{1}{GLs + 1}$$

For *C*3

$$Y_1 = Cs, \ Y_2 = G$$

$$T(s) = \frac{G}{Cs + G} = \frac{1}{Cs + G} = \frac{1}{RCs + 1}$$

Note that all three transfer functions have the form $1/(Ms + 1)$. Thus two or more circuits can have the same transfer function. Put another way, a desired transfer function often can be obtained in several ways.

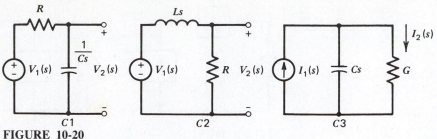

FIGURE 10-20
Circuits for finding transfer functions $T(s)$ **(Example 10-13).**

Example 10-14

A somewhat more complicated voltage divider is shown in Figure 10-21. Find the circuit transfer function.

The two elements in parallel make up the impedance Z_1, hence

$$Z_1 = \frac{1}{Y_C + Y_R} = \frac{1}{C_1 s + 1/R_1} = \frac{R_1}{R_1 C_1 s + 1}$$

The two elements in series combine to produce Z_2, and hence

$$Z_2 = Z_R + Z_C = R_2 + 1/C_2 s = \frac{R_2 C_2 s + 1}{C_2 s}$$

and

$$Z_1 + Z_2$$
$$= \frac{R_1 R_2 C_1 C_2 s^2 + (R_1 C_1 + R_2 C_2 + R_1 C_2)s + 1}{C_2 s(R_1 C_1 s + 1)}$$

So finally,

$$T(s) = \frac{Z_2}{Z_1 + Z_2}$$

$$= \frac{(R_1 C_1 s + 1)(R_2 C_2 s + 1)}{R_1 R_2 C_1 C_2 s^2 + (R_1 C_1 + R_2 C_2 + R_1 C_2)s + 1}$$

This rather formidable appearing result is actually the quotient of two quadratic polynomials in the complex frequency variable of the form

$$T(s) = \frac{(M_1 s + 1)(M_2 s + 1)}{M_1 M_2 s^2 + (M_1 + M_2 + M_3)s + 1}$$

where

$$M_1 = R_1 C_1$$
$$M_2 = R_2 C_2$$
$$M_3 = R_1 C_2$$

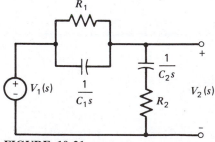

FIGURE 10-21
Circuit for finding transfer function via
voltage division (Example 10-14).

Two other very common simple circuits are the inverting and noninverting OP AMP configurations shown in Figure 10-22. To determine the

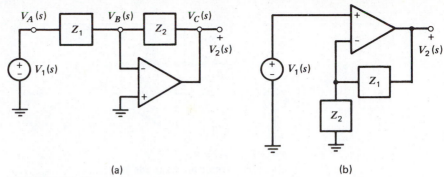

(a)   (b)

FIGURE 10-22
Generalized transfer function of two Simple OP AMP circuits. (*a*) Inverting circuit.
(*b*) Noninverting circuit.

transfer function of the inverting configuration we need only write a single
node equation at node $\circledB$.

$$-Y_1 V_A + (Y_1 + Y_2)V_B - Y_2 V_C = 0 \qquad (10\text{-}35)$$

But the OP AMP constraint requires that $V_B = 0$ since the noninverting
input is grounded. Since $V_A = V_1$ and $V_C = V_2$ we obtain $T(s)$ as

$$T(s) = \frac{V_2}{V_1} = -\frac{Z_2}{Z_1} \qquad (10\text{-}36)$$

The reader should readily recognize that this is simply an *s*-domain generalization of the result obtained in Chapter 4. It should not be much of
a leap of faith to note that the transfer function of the noninverting circuit
is

$$T(s) = \frac{V_2}{V_1} = \frac{Z_1 + Z_2}{Z_2} \qquad (10\text{-}37)$$

which is the voltage-divider relationship upside down.
    The results greatly simplify the analysis and design of OP AMP circuits.
The next example shows an analysis application.

Example 10-15

Find the transfer function of the OP AMP circuit
shown in Figure 10-23.
    The circuit in Figure 10-23 is an inverting configu-
ration with

$$Z_1 = R_1 + \frac{1}{C_1 s} = \frac{R_1 C_1 s + 1}{C_1 s}$$

and

$$Z_2 = \frac{1}{C_2 s + 1/R_2} = \frac{R_2}{R_2 C_2 s + 1}$$

Hence

$$T(s) = \frac{-Z_2}{Z_1}$$

$$= -\frac{R_2 C_1 s}{(R_1 C_1 s + 1)(R_2 C_2 s + 1)}$$

Notice that the circuit is easy to analyze compared
with the voltage divider in Example 10-14 and that
the denominator of $T(s)$ is in factored form.

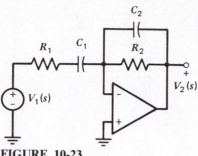

FIGURE 10-23
OP AMP circuit for Example 10-15.

The methods of voltage current division and the simple OP AMP relationships are extremely useful in finding the transfer functions in many practical situations. However, as the circuit complexity increases, these basic concepts become cumbersome or impossible to apply. A more general approach involves formulating either node-voltage or mesh-current equilibrium equations for the circuit. We then solve these equations, by Cramer's rule if necessary, to determine the output response. The ratio of this response to the input is then the transfer function. The algebra can be somewhat tedious at times, but the process is smoothed since we are concerned only with the zero-state response to a single input signal. The next three examples illustrate this general method, including dealing with an active circuit containing a dependent source.

Example 10-16

Find the transfer function of the circuit in Figure 10-24. To determine the voltage transfer function of the circuit in Figure 10-24, we first write mesh-current equations.

$$(Ls + R)I_A - RI_B = V_S$$
$$-RI_A + (2R + 1/Cs)I_B = 0$$

To determine the desired transfer function we must find the voltage V_2. Since $V_2 = RI_B$ it is clear that we must first solve for I_B. Using Cramer's rule:

$$I_B = \frac{D_B}{D} = \frac{\begin{vmatrix} Ls + R & V_S \\ -R & 0 \end{vmatrix}}{\begin{vmatrix} Ls + R & -R \\ -R & 2R + 1/Cs \end{vmatrix}}$$

$$= \frac{CsV_1}{2LCs^2 + (RC + GL)s + 1}$$

Hence

$$T(s) = \frac{V_2}{V_1} = \frac{RI_B}{V_1} = \frac{RCs}{2LCs_2 + (RC + GL)s + 1}$$

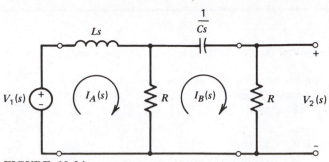

FIGURE 10-24
Circuit finding its transfer function via mesh analysis.

Example 10-17

Find the transfer function of the dependent source circuit of Figure 10-25.

To find the voltage transfer function of the circuit in Figure 10-25, first note that the input signal source and the dependent voltage source share a common node. If this node is selected as the reference node, then two of the four remaining node voltages are known.

$$V_A = V_1 \quad \text{and} \quad V_D = KV_C$$

Hence we need write a node-voltage equation only at the remaining two unknown nodes.

Node Ⓑ

$$-GV_A + (2G + Cs)V_B - GV_C - CsV_D = 0$$

Node Ⓒ

$$-GV_B + (G + Cs)V_C = 0$$

Substituting the node Ⓒ equation, together with the two known node voltages, into the node Ⓑ equation yields

$$-GV_1 + [(2G + Cs)(G + Cs)/G$$
$$- (G + KCs)]V_C = 0$$

or

$$[(RCs)^2 + (3 - K)RCs + 1]V_C = V_1$$

But $V_2 = V_D = KV_C$, and thus $T(s)$ is

$$T(s) = \frac{K}{(RCs)^2 + (3 - K)RCs + 1}$$

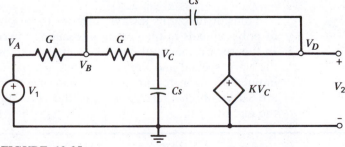

FIGURE 10-25
Dependent source circuit for Example 10-17.

Example 10-18

Find the transfer function of the OP AMP circuit in Figure 10-26.

To determine the transfer function, we need only write node equations at nodes Ⓑ and Ⓒ since $V_A = V_1$ and V_D is the OP AMP output.

Node Ⓑ

$$(G_1 + G_2 + C_1s + C_2s)V_B - G_1V_A \\ - C_1sV_C - C_2sV_D = 0$$

Node Ⓒ

$$(G_3 + C_1s)V_C - C_1sV_B - G_3V_D = 0$$

But $V_A = V_1$ as noted above and the OP AMP constraint $V_P = V_N$ dictate that $V_C = 0$ since the non-inverting input is grounded. Hence the equations can be written as

$$(G_1 + G_2 + C_1s + C_2s)V_B - C_2sV_D = G_1V_1 \\ - C_1sV_B - G_3V_D = 0$$

Using Cramer's rule we write the output $V_2 = V_D$ as

$$V_D = \frac{D_D}{D} = \frac{\begin{vmatrix} G_1 + G_2 + C_1s + C_2s & G_1V_1 \\ - C_1s & 0 \end{vmatrix}}{\begin{vmatrix} G_1 + G_2 + C_1s + C_2s & - C_2s \\ - C_1s & - G_3 \end{vmatrix}}$$

$$= \frac{-G_1C_1sV_1}{C_1C_2s^2 + G_3(C_1 + C_2)s + G_3(G_1 + G_2)}$$

Hence the transfer function is

$$T(s) = \frac{V_2}{V_1}$$

$$= \frac{-G_1C_1s}{C_1C_2s^2 + G_3(C_1 + C_2)s + G_3(G_1 + G_2)}$$

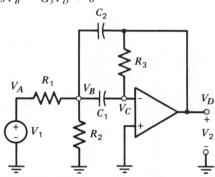

FIGURE 10-26
OP AMP circuit for Example 10-18.

Transfer functions always turn out to be rational functions: quotients of polynominals in the complex frequency variable of the form $r(s)/q(s)$. In this regard they look just like signal transforms. However, except for the very special and very important special case when the input is an impulse $\delta(t)$, transfer functions are not, in general, signal waveforms. They are s-domain characterizations of the zero-state signal transfer through the circuit.

Example 10-19

Determine the transfer function $T(s) = V_2(s)/V_S(s)$ of the transformer circuit in Figure 10-27.

To solve this problem we must first deal with the s-domain representation of two coupled inductors. In the t domain the i-v characteristics of coupled inductors are

$$L_1 \frac{di_1}{dt} + M\frac{di_2}{dt} = v_1(t)$$

and

$$M \frac{di_1}{dt} + L_2 \frac{di_2}{dt} = v_2(t)$$

For zero initial conditions (circuit in the zero state) these transform into the s domain as

$$L_1 s I_1(s) + M s I_2(s) = V_1(s)$$

and

$$M I_1(s) + L_2 s I_2(s) = V_2(s)$$

The source resistor (R_1) at the input and the load resistor (R_2) introduce the s-domain constraints

$$V_1(s) = V_S(s) - R_1 I_1(s)$$

and

$$V_2(s) = - R_2 I_2(s)$$

When these constraints are substituted into the transformer i-v equations we have

$$(L_1 s + R_1) I_1(s) + M s I_2(s) = V_S(s)$$
$$M s I_1(s) + (L_2 s + R_2) I_2(s) = 0$$

Solving for $I_2(s)$:

$$I_2(s) = \frac{D_2}{D} = \frac{\begin{vmatrix} L_1 s + R_1 & V_S \\ M s & 0 \end{vmatrix}}{\begin{vmatrix} L_1 s + R_1 & M s \\ M s & L_2 s + R_2 \end{vmatrix}}$$

$$= \frac{- M s V_S}{(L_1 L_2 - M^2)s^2 + (R_1 L_2 + R_2 L_1)s + R_1 R_2}$$

But $V_2(s) = - R_2 I_2(s)$, and thus

$$T(s) = \frac{V_2(s)}{V_S(s)}$$

$$= \frac{R_2 M s)}{(L_1 L_2 - M^2)s^2 + (R_1 L_2 + R_2 L_1)s + R_1 R_2}$$

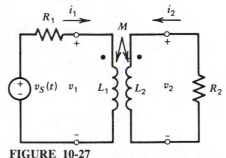

FIGURE 10-27
Transformer circuit for Example 10-19.

10-7 POLES AND ZEROS REVISITED

In Figure 10-1 we outlined the pattern of s-domain circuit analysis. The reader should refer to this flow diagram as we review our progress. We have shown that a circuit specified in the t domain can be transformed into the s domain (Sec. 10-1). Simple transformed circuits can then be analyzed using circuit reduction. Thévenin-Norton theorems, and super-position (Sec. 10-2). Alternatively the equilibrium equations can be formulated directly in the s domain using either node voltages (Sec. 10-3) or mesh currents (Sec. 10-4). The resulting linear algebraic equations can be solved using Cramer's rule (Sec. 10-5). Either the reduction approach or the more formal node or mesh methods reveal the existence of zero-state and zero-input components in a response transform. Using the zero-state component, we then defined a transfer function as an input-output relationship in the s domain (Sec. 10-6). In terms of the flow diagram in Figure 10-1, we have progressed to the block labeled *response transform*. To obtain the response waveform we now must perform the inverse transformation.

A response transform must be a rational function—fundamentally because the connection equations in the s domain are unchanged, and the element i-v characteristics are all linear algebraic equations. Algebraically we know that response transforms are rational because they can be obtained from Cramer's rule as quotients of determinants that are themselves rational functions. Thus if we have solved for a node-voltage transform, say at node ④, then it can always be written in the form

$$V_A(s) = \frac{r(s)}{q(s)} = \frac{b_m s^m + b_{m-1}s^{m-1} + \cdots + b_1 s + b_0}{a_n s^n + a_{n-1}s^{n-1} + \cdots + a_1 s + a_0} \tag{10-38}$$

where $r(s)$ and $q(s)$ are polynomials in the complex frequency variable in which a_k and b_k are the coefficients. The two polynomials can be factored in terms of their roots as

$$V_A(s) = K\frac{(s - z_1)(s - z_2)\ldots(s - z_m)}{(s - p_1)(s - p_2)\ldots(s - p_n)} \tag{10-39}$$

where z_k, with $k = 1, 2, \ldots m$, are called the zeros of $V_A(s)$ and p_j, with $j = 1, 2, \ldots n$, are called poles. If all of the poles are simple (unequal), and if $n > m$, as is usually the case, then V_A can be decomposed by partial fractions as

$$V_A(s) = \frac{k_1}{s - p_1} + \frac{k_2}{s - p_2} + \cdots + \frac{k_n}{s - p_n} \tag{10-40}$$

where k_j, $j = 1, 2, \ldots n$, are the residues at the poles. The response waveform is then available as

$$V_A(t) = u(t)(k_1 e^{p_1 t} + k_2 e^{p_2 t} + \cdots + k_n e^{p_n t}) \tag{10-41}$$

In sum, we can recover the response waveform from its transform if need be. The mechanics of this inversion process were discussed in Chapter 9 (Sec. 9-4) and will not be repeated here.

If we absolutely must have the response waveform, then there is no option but to perform the formal inversion step. But as we have said several times, because it is worth repeating, waveforms and transforms are simply two different ways of representing a signal. All of the data needed to construct the waveform are available in the transform, so why not examine the transform to see how it carries these data?

The prime source of data in a transform is the location of its poles. For example, if a transform has a simple pole at $p = -2$, then we know that the corresponding waveform must contain a term $u(t)(ke^{-2t})$. Note that we can draw this inference without formally constructing the partial fraction expansion.

This observation clearly generalizes. If a response transform has a simple pole as $p = a$, where a is real, then the response waveform must contain a component $u(t)$ (ke^{at}), where k is the residue at the pole. If a < 0, then the component is a decaying exponential. If $a = 0$, then we have a step function $ku(t)$. Finally if $a > 0$, we will have a growing exponential. These observations are summarized in Figure 10-28. Thus the knowledge of the location of a real pole is equivalent to knowing the form of the corresponding component of the waveform. The location of the pole tells us the time constant of the exponential waveform, but not its amplitude. The amplitude is determined by the residue. The size of a residue is affected by all of the poles of the transform, and the zeros as well.

Not only is a response transform rational, but it is a rational function with real coefficients. This means that all of the b_k's and a_k's in Eq. 10-38 are real numbers. Basically this is so because these coefficients are determined by the circuit parameters (R's, L's, and C's) or by the parameters of the input signals or the initial conditions. All of these quantities

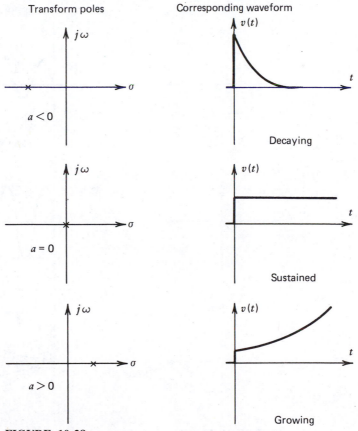

FIGURE 10-28
Real poles and exponential waveforms.

are real numbers. Put differently, the only way that complex numbers creep into the response transform is through the complex frequency variable.

The result is that if a response transform has a simple pole at $p = \alpha + j\beta$, then it is accompanied by its conjugate $p^* = \alpha - j\beta$ because complex roots of polynomials with real coefficients must occur in conjugate pairs. In treating this case in Chapter 9 we found that a pair of conjugate poles led to a waveform component $2|k|e^{\alpha t} \cos(\beta t + \underline{/k})$, where k is the residue at the pole $p = \alpha + j\beta$. Thus complex poles always lead to a composite exponential-sinusoidal waveform whose frequency of oscillation equals the imaginary part of the pole. If $\alpha < 0$, the waveform is a damped or decaying sinusoid. If $\alpha = 0$, we obtain an undamped or sustained sinusoid. Finally if $\alpha > 0$, the waveform is a growing oscillation. These observations are summarized in Figure 10-29. Again the pole locations give us the nature of the waveform. The imaginary parts determine the frequency and the real parts the nature of the damping. In other words, the real part of the pole determines the time constant of the exponential

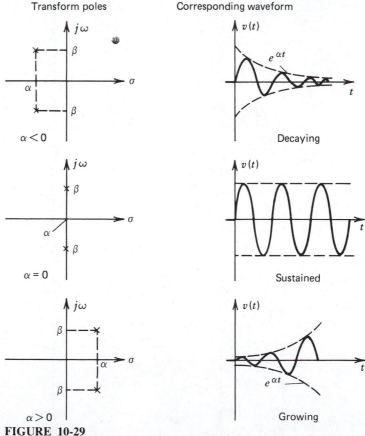

FIGURE 10-29
Complex poles and sinusoidal waveforms.

part of the waveform. But note that we cannot predict the amplitude and phase angle of the waveform from the pole locations alone. These parameters depend on the residue at the pole. To summarize:

The poles of a response transform are either real or complex conjugates. Real poles lead to exponential waveforms. Complex conjugate poles combine to produce damped sinusoidal waveforms.

The locations of the poles of a response transform tell us a great deal about the nature of the corresponding response waveform. If we plot these poles in the s plane, as in Figures 10-28 and 10-29, we can predict the corresponding waveform components. Thus we might just as well think of the response in terms of pole locations instead of step, exponential, and sinusoidal waveforms. We might just as well learn to think about circuit responses in the s domain.

Example 10-20

Classify the poles of the transfer function of Example 10-14.

The voltage transfer function from Example 10-14 is

$$T(s) = \frac{V_2}{V_1}$$

$$= \frac{(R_1C_1s + 1)(R_2C_2s + 1)}{R_1R_2C_1C_2s^2 + (R_1C_1 + R_2C_2 + R_1C_2)s + 1}$$

For $R_1C_1 = R_2C_2 = R_1C_2 = 1$ and $v_1(t) = 10\cos 2t$, and hence $V_1(s) = 10s/(s^2 + 4)$, the zero-state component of the response transform $V_2(s)$ is

$$V_2 = T(s)V_1 = \frac{(s + 1)^2\, 10s}{(s^2 + 3s + 1)(s^2 + 4)}$$

The poles of $V_2(s)$ are roots of the polynominals

$$(s^2 + 3s + 1) = (s + 0.382)(s + 2.62) = 0$$
$$(s^2 + 4) = (s - j2)(s + j2) = 0$$

The form of these poles can be classified as

Real	Complex
$s = -0.382$	$s = j2$
$s = -2.62$	$s = -j2$

Therefore we know that the response waveform will contain two exponential waveforms and a sinusoidal component.

Poles in a response transform are caused either by the circuit itself or by the nature of the driving force. For example, the two real poles in Example 10-20 were caused by the circuit while the complex poles came from the input signal. Poles introduced by the input signal are called **forced poles** since they give rise to the forced response. Similarly we call poles contributed by the circuit **natural poles** since they lead to the natural response. In the s domain we think of the input signals and the circuit as the source of poles.

Poles of a response transform are introduced either by the circuit (natural poles) or by the input signals (forced poles). Natural poles give rise to the natural response and forced poles to the forced response.

Example 10-21

In Example 10-11 we found the node-voltage response transforms

$$V_A(s) = \underbrace{\frac{(LCs^2 + 1)I_S}{GLCs^2 + Cs + G}}_{\text{Zero state}}$$

$$+ \underbrace{\frac{- LCsi_L(0) + Cv_C(0)}{GLCs^2 + Cs + G}}_{\text{Zero input}}$$

$$V_B(s) = \underbrace{\frac{I_S}{GLCs^2 + Cs + G}}_{\text{Zero state}}$$

$$+ \underbrace{\frac{(GLs + 1)Cv_C(0) + GLi_L(0)}{GLCs^2 + Cs + G}}_{\text{Zero input}}$$

The natural poles in both responses are roots of the polynomial

$$GLCs^2 + Cs + G = 0$$

since the form of these roots depend only on the circuit elements R, L, and C. In addition there could be one or more forced poles in the zero-state components, depending on the nature of the input signal I_S. For example,

$$\text{If } i_S(t) = A \cos \beta t, \text{ then } I_S = \frac{As}{s^2 + \beta^2}$$

and the input signal introduces forced poles at $s = \pm j\beta$. Note that the forced poles appear only in the zero-state component while the natural poles appear in both the zero state and the zero input.

The preceding example illustrates the following general principle.

The zero-input response contains only natural poles. The zero-state response contains both forced and natural poles.

The zero-input component can be calculated by turning off all of the input signal sources. Clearly the inputs cannot affect this component of the response and so do not introduce any forced poles therein. On the other hand, the zero-state component reflects the influence of all input sources and the circuit with no initial conditions. It therefore contains natural poles and all of the forced poles. As a corollary of this principle we note that a transfer function contains only natural poles. This follows since a transfer function is defined as the ratio of the zero-state response to the input signal. The forced poles are canceled in this process, leaving only the natural poles.

Stability is a key concept in circuit and system analysis. In the present context we would say that stability is a property of a circuit that suggests that its response eventually will settle down. This heuristic idea, while appealing, involves a fair amount of complication. The *IEEE Standard Dictionary of Electrical and Electronic Terms* contains two pages of definitions of the term *stability* and related concepts. In a way, there is nothing quite as unstable as the definition of stability!

For our purposes we define stability in terms of the zero-input response of a circuit. Assume that a circuit has some nonzero initial conditions. After the passage of some time, does the zero-input response decay to zero? If it does, we say the circuit is stable. If not, we say it is unstable. Note the binary nature of this definition. We would like to be able to say that a circuit either is stable or it is not.

How can we relate this definition of stability to the poles? Figures 10-28 and 10-29 show the relationship between pole locations and the corresponding waveforms. The key to stability is to be found in the real part of the pole. If the real part is negative, the waveforms decay to zero. If the real part is positive, the waveforms do not decay to zero; in fact they increase without bound. If the real part is zero, then as Figures 10-28 and 10-29 show, the waveforms are sustained and neither decay to zero nor increase without bound.

The result is that poles located in the left half of the s plane have negative real parts and clearly imply stability. Those in the right half of the s plane have positive real parts and are unstable. Single poles that lie on the boundary (the j axis in the s plane) have zero real parts and are said to be **marginally** stable. Since the zero-input response contains only natural poles we conclude that:

A circuit is stable if all of its natural poles are located in the left half of the s *plane.*

Note the requirement that all poles be in the left-half plane. If even one is in the right-half plane, or if they are multiple poles on the boundary, the circuit is unstable.

If the natural pole locations are all known, then we need only examine their real parts to determine circuit stability. If all are negative, the circuit is stable. If the polynomial defining the natural poles is not available in factored form, then the pole locations must be determined by inference. For our purposes we need only consider quadratic polynomials of the form $a_2 s^2 + a_1 s + a_0$. The roots of a quadratic equation have negative real parts if and only if the coefficients a_2, a_1, and a_0 are all positive or all negative. This simple test will suffice to illustrate most of the basic issues of circuit stability.

Example 10-22

The voltage transfer function from Example 10-14

$$T(s) = \frac{(R_1 C_1 s + 1)(R_2 C_2 s + 1)}{R_1 R_2 C_1 C_2 s^2 + (R_1 C_1 + R_2 C_2 + R_1 C_2)s + 1}$$

yields the natural poles of the circuit as roots of the denominator polynomial

$$R_1 R_2 C_1 C_2 s^2 + (R_1 C_1 + R_2 C_2 + R_1 C_2)s + 1 = 0$$

All of the polynomial coefficients depend on resistances and capacitances that are positive. Hence all of the coefficients are positive, all of the roots have negative real parts, and the circuit is stable. However, the voltage transfer function from Example 10-17 is

$$T(s) = \frac{K}{(RCs)^2 + (3 - K)RCs + 1}$$

The natural poles are the roots of the denominator polynomial. The product RC is positive, but the factor $(3 - K)$ is negative if K is greater than 3. Hence the circuit is unstable for $K > 3$. For $K = 3$ the denominator polynominal is $(RCs)^2 + 1 = 0$, which has roots at $s = -j/RC$ and $s = +j/RC$. These poles lie on the j axis in the s plane and are marginally stable.

Since we must determine the location of all of the natural poles of a circuit we are naturally led to inquire as to the total number available. The following principle answers this question.

The number of natural poles does not exceed the number of memory elements in a circuit.

Note carefully the bound placed by this statement. The number of natural poles can be *less,* but *never more,* than the total number of capacitors and inductors in a circuit.

We offer the following plausibility argument to support this principle. In the s domain each element of a circuit is represented by its impedance. The impedance of the memory elements are Ls and $1/Cs$, and so each contributes one s to the algebraic process of solving for the circuit response. The process can be performed using Cramer's rule, and the roots of the denominator polynomial, the source of poles, are zeros of a node or mesh determinant. Since each memory element contributes one s to the circuit determinant, it seems reasonable to suggest that the number of zeros of a circuit determinant does not exceed the number of memory elements in the circuit.

So far we have not discussed the zeros of a response transform because they do not play the same role as poles. The zeros determine the amplitude, but not the basic form of the response. A quick review of the cover-up algorithm in Chapter 9 will convince the reader that if a zero is located close to a pole, then the residue at the pole will be small. In the special case that a zero happens to equal a pole, then the residue will be zero. In effect the zero cancels the pole. In such a case the canceled pole is not **observable** in the response transform. The next example illustrates the cancellation of poles by zeros.

Example 10-23

The circuit in Figure 10-30 has two memory elements. To determine the number of natural poles we calculate the admittance Y_{EQ}.

$$Y_{EQ} = Y_{RL} + Y_{RC} = \frac{1}{R + Ls} + \frac{1}{1/Cs + R}$$

$$= \frac{LCs^2 + 2RCs + 1}{(RCs + 1)(Ls + R)}$$

The circuit has two natural poles located at $s = -1/RC$ and $s = -1/GL$. However, if $RC = GL$, and we make use of the obvious fact that $RG = R/R = 1$, then Y_{EQ} can be written as

$$Y_{EQ} = G\frac{LCs^2 + 2RCs + 1}{(RCs + 1)(GLs + 1)}$$

$$= G\frac{LCs^2 + 2RCs + 1}{LCs^2 + 2RCs + 1} = G$$

For $RC = GL$ the two poles are canceled by two zeros that happen to lie at the same points. Thus under special circumstances a response transform may contain no natural poles even though the circuit has the potential of contributing several poles. The principle is that the number of natural poles cannot be

greater than the number of memory elements. The number can be less, including zero. In this case we say that the two poles of Y_{EQ} are unobservable.

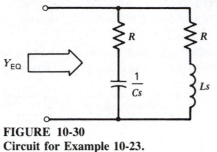

FIGURE 10-30
Circuit for Example 10-23.

SUMMARY

- Kirchhoff's laws are valid in either the t domain or the s domain. They apply to waveforms in the t domain and to transforms in the s domain.

- The element i-v relationships in the s domain are linear algebraic equations based on the concepts of impedance, admittance, and initial condition sources.

Element	Mesh Analysis	Node Analysis
Resistor	$V_R = Z_R I_R$	$I_R = Y_R V_R$
Inductor	$V_L = Z_L I_L - L i_L(0)$	$I_L = Y_L V_L + \dfrac{1}{s} i_L(0)$
Capacitor	$V_C = Z_C I_C + \dfrac{1}{s} v_C(0)$	$I_C = Y_C V_C - C v_C(0)$

- Impedance and admittance are s-domain extensions of the concepts of resistance and conductance. The impedance and admittance of the three passive elements are as follows:

Element	Impedance	Admittance
Resistor	$Z_R = R$	$Y_R = G = 1/R$
Inductor	$Z_L = Ls$	$Y_L = 1/Ls$
Capacitor	$Z_C = 1/Cs$	$Y_C = Cs$

- Node-voltage or mesh-current equilibrium equations can be written by inspection in the s domain. These equations are linear algebraic equations that can be solved for response transforms using Cramer's rule.

- The response of a linear circuit can always be partitioned into zero-state and zero-input components. Most of the important topics in s-domain circuit analysis are based on the zero state response.

- The poles of a response transform determine the form of the corresponding term in the response transform.

  Real pole $p = a$

  $a < 0$ decaying exponential
  $a = 0$ step function
  $a > 0$ growing exponential

  Complex poles $p = \alpha + j\beta$ and $p^* = \alpha - j\beta$

  $\alpha < 0$ damped sinusoid with $\omega = \beta$
  $\alpha = 0$ sustained sinusoid with $\omega = \beta$
  $\alpha > 0$ growing sinusoid with $\omega = \beta$

- Poles of the response transform introduced by signal inputs are called forced poles and give rise to the forced response. Poles introduced by the circuit are called natural poles and lead to the natural response.

- All of the natural poles of a circuit are zeros of a node or mesh determinant. Some of the circuit poles may not be observable in a response due to cancellation by a zero of the response.

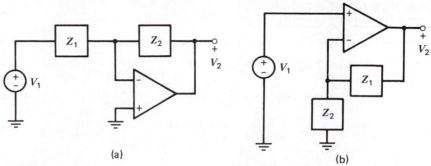

FIGURE 10-31
OP AMP transfer functions in the *s*-domain. (*a*) Inverting. (*b*) Noninverting.

- The transfer function of an inverting OP AMP, Figure 10-31*a*, with impedance elements is given as

$$T(s) = \frac{-Z_2}{Z_1}$$

and of a noninverting OP AMP, Figure 10-31*b*, as

$$T(s) = \frac{Z_2 + Z_1}{Z_2}$$

EN ROUTE OBJECTIVES
AND RELATED EXERCISES

10-1 CIRCUIT ANALYSIS TECHNIQUES (SECS. 10-1 AND 10-2)

Given a circuit consisting of input sources and linear elements:

(a) *Transform the circuit into the s domain.*
(b) *Solve for the equivalent impedance at specified terminals.*
(c) *Solve for the zero-state and/or zero-input responses of selected signal transforms using s-domain circuit reduction, the Thévenin-Norton theorem, and superposition.*

Exercises

10-1-1 Repeat Exercise 8-1-3 using *s*-domain circuit analysis techniques.

10-1-2 Repeat Exercise 8-1-4 using *s*-domain circuit analysis techniques.

10-1-3 Repeat Exercise 8-2-4 using *s*-domain circuit analysis techniques.

10-1-4 Repeat Exercise 8-2-8 using *s*-domain circuit analysis techniques.

10-1-5 Transform each of the circuits in Figure E10-1-5 into the *s* domain and find the equivalent impedance Z_{EQ} at the indicated interface.

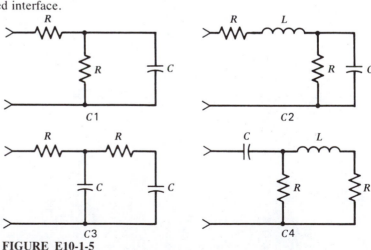

FIGURE E10-1-5
Circuits for finding Z_{EQ}.

10-1-6   Transform each of the circuits in Figure E10-1-6 into the *s* domain and find the Thévenin equivalent at the indicated interface including the effects of the initial conditions shown. Then use the Thévenin equivalent to determine the zero-state and zero-input response transforms of the indicated interface signal variable.

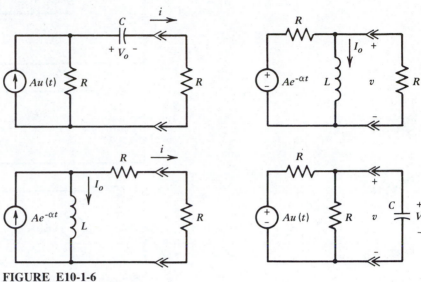

FIGURE E10-1-6
Circuits for finding Thévenin equivalents.

10-1-7 Transform each of the circuits in Figure E10-1-7 into the *s* domain and solve for the zero-state response transform of the indicated signal variables.

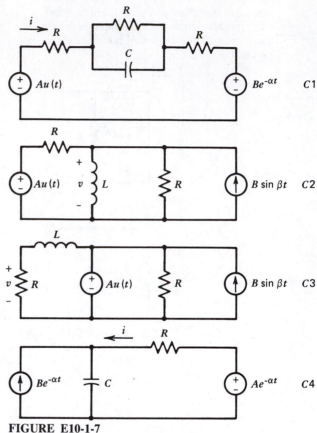

FIGURE E10-1-7
Multiple-source circuits for Exercise 10-1-7.

10-2 NODE AND MESH EQUATIONS IN THE s DOMAIN (SECS. 10-3 AND 10-4)

Given a circuit consisting of linear elements and signal sources:

(a) *Transform the circuit into the s domain.*

(b) *Formulate node-voltage or mesh-current equilibrium equations.*

(c) *Solve for the zero-state and/or zero-input components of selected voltage or current transforms.*

Exercises

10-2-1 For each circuit in Figure E10-2-1 solve for the zero-state and zero-input components of the voltage $V_2(s)$ and the current $I_2(s)$.

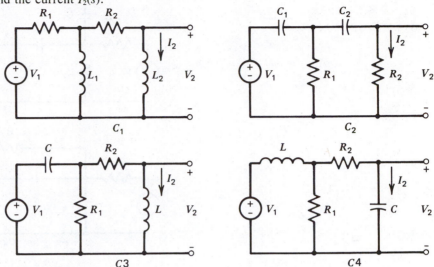

FIGURE E10-2-1
Circuits for finding zero-state and zero-input responses.

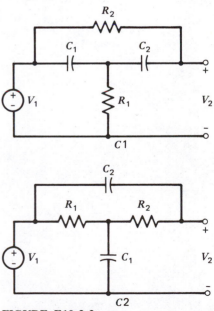

FIGURE E10-2-2
Circuits for finding zero-state and zero-input responses.

10-2-2 For each circuit in Figure E10-2-2 solve for the zero-state and zero-input components of the voltage transform $V_2(s)$.

10-2-3 For each circuit in Figure E10-2-3 solve for the zero-state component of the output voltage transform $V_2(s)$.

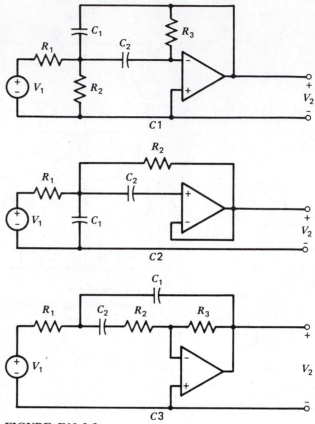

FIGURE E10-2-3
OP AMP circuits for finding zero-state response.

10-2-4 Repeat Exercise 10-2-3 for the circuits in Figure E10-2-4.

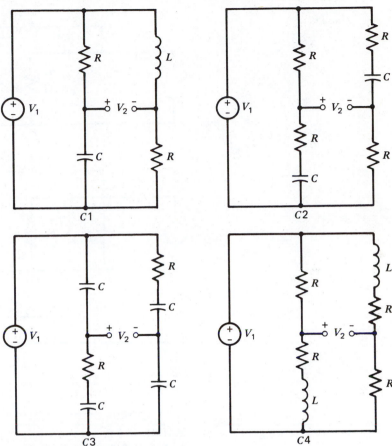

FIGURE E10-2-4
Bridge circuits for Exercise 10-2-4.

10-2-5 Repeat Exercise 10-2-3 for the circuits in Figure E10-2-5.

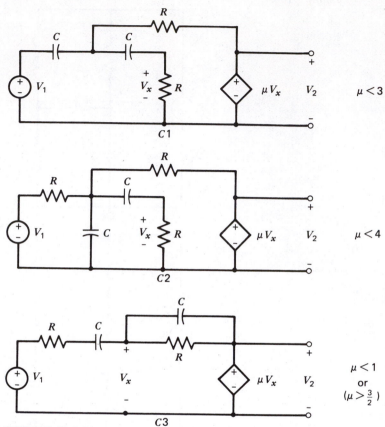

FIGURE 10-2-5
Dependent source circuits for Exercise 10-2-5.

10-2-6   Repeat Exercise 10-2-3 for the circuits in
Figure E10-2-6.

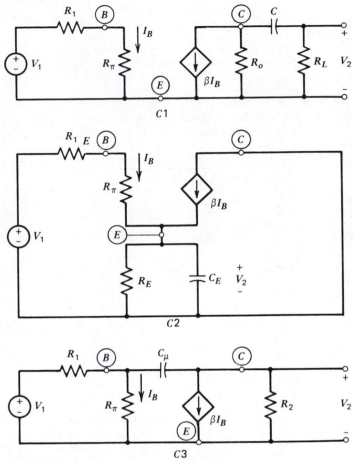

FIGURE E10-2-6
"Transistor model" circuits for Exercise 10-2-6.

10-3 TRANSFER FUNCTIONS (SEC. 10-5 AND 10-6)

Given a linear circuit with a specified output signal and a single input signal source:
(a) Transform the circuit into the s domain.
(b) Solve for the transfer function relating the input and output signals.

Exercises

The following set of exercises uses different combinations of the branches $B0$ through $B6$ and the circuits $C1$ through $C4$ shown in Figure E10-3-1. For each of the exercises connect the identified branches into the circuit indicated and solve for the transfer function

Exercise	Circuit	Branch A	Branch B	T(s)
10-3-1	C1	B0	B2	V_2/V_1
10-3-2	C2	B0	B3	I_2/I_1
10-3-3	C2	B0	B1	I_2/I_1
10-3-4	C3	B0	B4	V_2/V_1
10-3-5	C4	B0	B1	V_2/V_1
10-3-6	C4	B2	B0	V_2/V_1
10-3-7	C2	B2	B3	I_2/I_1
10-3-8	C3	B3	B2	V_2/V_1
10-3-9	C1	B1	B4	V_2/V_1
10-3-10	C2	B4	B1	I_2/I_1
10-3-11	C1	B5	B0	V_2/V_1
10-3-12	C3	B3	B4	V_2/V_1
10-3-13	C1	B0	B5	V_2/V_1
10-3-14	C4	B2	B1	V_2/V_1
10-3-15	C1	B3	B5	V_2/V_1
10-3-16	C2	B2	B6	I_2/I_1

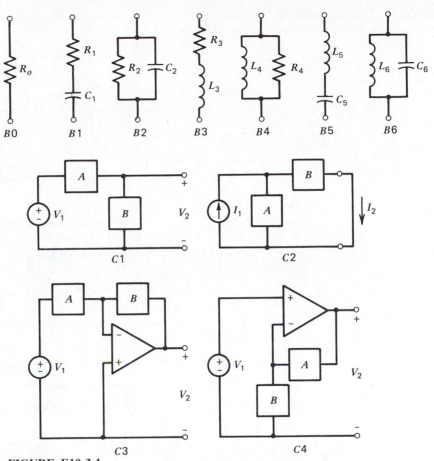

FIGURE E10-3-1
Modular elements and circuits for Exercise 10-3-1.

10-3-17 Find the transfer function $T(s) = V_2/V_1$ for each of the OP AMP circuits in Figure E10-3-17.

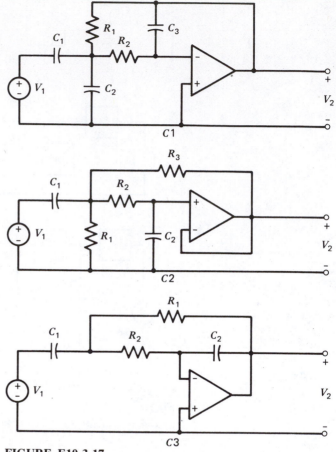

FIGURE E10-3-17
OP AMP circuits for Exercise 10-3-17.

10-3-18 Repeat Exercise 10-3-17 for the circuits in Exercise 10-2-3.

10-3-19 Show that circuits $C1$ and $C2$ in Exercise 10-2-2 have identical transfer functions $T(s) = V_2/V_1$ if $R_1 = R_2 = R$ and $C_1 = C_2 = C$.

10-3-20 Find the transfer function $T(s) = V_2/V_1$ for the circuits in Exercise 10-2-4.

10-3-21 Find the transfer function $T(s) = V_2/V_1$ for the circuits in Exercise 10-2-5.

10-3-22 Find the transfer function $T(s) = V_2/V_1$ for the circuits in Exercise 10-2-6.

10-4 *POLES AND ZEROS (SEC. 10-7)*

Given the zero-state and/or zero-input response transform of a linear circuit, determine the location of the response poles and zeros, and classify the poles as to:

(a) *Form (real or complex).*

(b) *Origin (forced or natural).*

(c) *Stability (stable, unstable, or marginally stable).*

(d) *Number (observable or unobservable).*

Exercises

10-4-1 Using the transfer function found in Exercise 10-3-3, determine the poles and zeros of the zero-state output $I_2(s)$ for a step-function input. Classify the poles as to form, origin, stability, and number.

10-4-2 Using the transfer function found in Exercise 10-3-8, determine the poles and zeros of the zero-state output $V_2(s)$ for a step-function input. Classify the poles as to form, origin, stability, and number.

10-4-3 Using the transfer function found in Exercise 10-3-12, determine the poles and zeros of the zero-state output $V_2(s)$ for a step-function input. Classify the poles as to form, origin, stability, and number.

10-4-4 Using the transfer function found in Exercise 10-3-13, determine the poles and zeros of the zero-state output $V_2(s)$ for an input defined as

$$v_1(t) = A \sin (t/\sqrt{L_5 C_5})$$

Classify the poles as to form, origin, stability, and number.

10-4-5 Using the transfer function found in Exercise 10-3-15, determine the poles and zeros of the zero-state output $V_2(s)$ for a step-function input. Classify the poles as to form, origin, stability, and number.

10-4-6 Using the transfer function found in Exercise 10-3-19, determine the poles and zeros of the response transform $V_2(s)$ for an input defined as

$$v_1(t) = Ae^{-t/RC}$$

Classify the poles as to form, origin, stability, and number.

10-4-7   Using the transfer functions found in Exercise 10-3-20, show that the zero-state response transform $V_2(s)$ in circuit $C1$ has unobservable poles if $RC = GL$.

10-4-8   Using the transfer functions found in Exercise 10-3-21, determine the poles and zeros of the zero-state response transform $V_2(s)$ for a step-function input and with the gain $\mu = 5$. Classify the poles as to form, origin, stability, and number.

PROBLEMS

P10-1   (Analysis)
The unit output method presented in Chapter 3 can
be extended to circuits in the s domain as follows:
(1)   Transform the circuit into the s domain with
      all initial conditions set to zero.
(2)   Assume that the circuit output transform is un-
      ity (i.e., an impulse).
(3)   By successive application of Kirchhoff's laws
      and the element impedance relationships, de-
      termine the input required to produce the unit
      output.
(4)   Then determine the transfer function as

$$T(s) = \frac{\text{Output}}{\text{Input}} = \frac{1}{\text{Input transform}}$$

Use this method to determine the voltage transfer
function and input impedance of the circuit in Figure
P10-1.

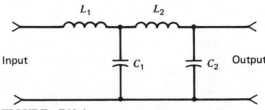

FIGURE   P10-1
Unit output method problem.

P10-2   (Analysis)
The source transformation techniques presented in
Chapter 3 can be extended to circuits in the s domain.
Transform each of the voltage sources in Figure P10-
2 into an equivalent current source and verify that
equivalent circuit has the same i-v relationships. Does
the equivalent current source have the same waveform as the voltage source? If not, explain.

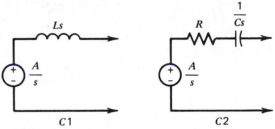

FIGURE P10-2
Source transformation in s-domain problem.

P10-3   (Analysis)
Formulate a set of node-voltage equations for the
circuit in Figure P10-3 and solve for the transfer
function

$$T(s) = \frac{V_2(s)}{V_1(s)}$$

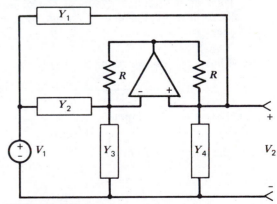

FIGURE P10-3
Unusual OP AMP circuit transfer-function problem.

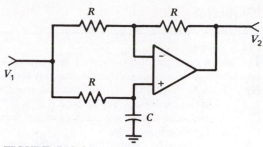

P10-4 (Analysis)
Solve for the transfer function

$$T(s) = V_2/V_1$$

of the circuit in Figure P10-4. Construct a pole-zero
diagram of $T(s)$ and discuss any unique features. Is
the circuit stable?

FIGURE P10-4
OP AMP transfer-function problem.

P10-5 (Analysis, Design)
Circuits with two inputs (V_1 and V_2) and a single
output (V_o) are often characterized in terms of a common-mode transfer function

$$T_{CM} = \frac{V_o}{V_1} \quad \text{with} \quad V_2 = V_1$$

and a difference-mode transfer function

$$T_{DM} = \frac{V_o}{V_1} \quad \text{with} \quad V_2 = -V_1$$

In many applications it is desirable for the circuit to
suppress the common-mode inputs and process only
the difference mode. The ability of the circuit to do
this is expressed as the common-mode rejection ratio
(CMRR).

$$\text{CMRR} = \frac{T_{DM}}{T_{CM}}$$

Find T_{CM}, T_{DM}, and CMRR for the circuit in Figure
P10-5. Then draw circuit diagrams to shown how
physically to measure T_{CM} and T_{DM} using only one
input voltage source.

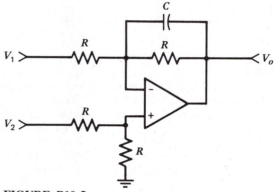

FIGURE P10-5
Common- and difference-mode OP AMP problem.

P10-6 (Analysis)
A circuit has a transfer function

$$T(s) = \frac{s}{(s + 1)(s + 2)}$$

For a certain input the output is

$$y(t) = -e^{-t} + \cos t + \sin t$$

What is the input waveform?

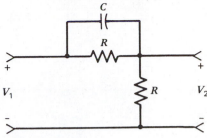

FIGURE P10-7
Circuit for Problem 10-7.

P10-7   (Analysis, Design)
For the circuit in Figure P10-7
(a)   Determine $T(s) = V_2/V_1$.
(b)   Select values of R and C so that an input e^{-100t} produces no forced component in the output.
(c)   Select values of R and C so that an input e^{-10t} cos $200t$ produces no natural component in the output.

P10-8   (Analysis)
Table P10-8 lists transfer functions and input waveforms. For each case answer the following questions without formally determining the output waveform.
(a)   Is the circuit stable?
(b)   Is the forced response observable in the output?
(c)   Does the response settle down to a constant final value? If it does, what is the final value and about how long is the settling time? Is the settling time determined by the circuit or the input?
(d)   Does the response contain a sinusoidal oscillation? If it does, what is the period of the oscillation? Is the period determined by the input or the circuit?

Transfer Functions	Input Waveforms
$\dfrac{s + 2}{s + 1}$	$u(t)$
$\dfrac{s}{s + 1}$	$u(t)$
$\dfrac{s^2 + 1}{(s + 1)(s + 2)}$	$u(t)$
$\dfrac{1}{(s + 1)(s + 2)}$	$\sin 4t$
$\dfrac{s - 1}{s + 1}$	$e^{-t/2}$
$\dfrac{s + 1}{s - 1}$	$e^{-t} \sin 3t$
$\dfrac{1}{s^2 + 1}$	$u(t)$

TABLE P10-8

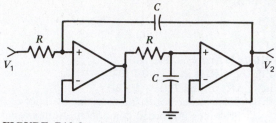

FIGURE P10-9
Cascaded OP AMP circuit with complex poles.

P10-9   (Analysis)
Determine the transfer function $T = V_2/V_1$ of the circuit in Figure P10-9. Then show that the poles of $T(s)$ are always complex regardless of the values of R and C.

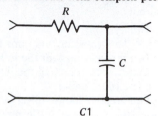

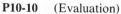

P10-10   (Evaluation)
The response time of circuit $C1$ in Figure P10-10 is considered to be too slow. It is claimed that by adding the ''compensator'' shown in $C2$ the response time at the output of $C2$ will be one-half the response time of $C1$. Prove or disprove this claim.

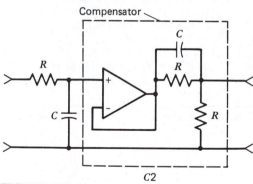

FIGURE P10-10
Response-time ''compensator'' circuit.

BLOCK III
APPLICATIONS

BLOCK OBJECTIVES

ANALYSIS

Given a linear circuit containing one or more memory elements, determine its step or impulse response, including its damping and natural frequency; its sinusoidal response, including the average power delivered at a prescribed interface; and its frequency response, including a sketch of its Bode plot.

DESIGN

Devise an active or passive circuit containing one or more memory elements if necessary or modify an existing circuit to obtain a specified step, sinusoidal, or frequency response.

EVALUATION

Given two or more circuits that perform the same signal-processing function, select the best circuit based on given criteria such as performance, cost, parts count, power dissipation, and simplicity.

Chapter 11
Step Response

Network Theory is essentially the theory of linear systems.

E. A. Guillemin

This chapter begins the study of transfer functions as tools for probing the nature of electric circuits. These functions are an *s*-domain description of the zero-state characteristics of a circuit. They can be used to predict the output of the circuit for almost any type of input signal. A common test input used to study circuit response is the pulse. In the first section we find that the impulse and step functions are reasonable models of pulse signals under certain conditions. We also discover that the impulse and step responses are related by differentiation and integration in the time domain. In the following sections we concentrate on the step response of one-pole and two-pole transfer functions. In both cases we find that the nature of the step response depends on the location of the poles. We also discover that the initial and final values of the step response are related to the value of the transfer function at infinity and zero respectively. In the next section we relate the location of poles to the concepts of circuit damping and natural frequency first introduced in Chapter 8. The final section presents the parameters used to characterize the step response of a circuit.

11-1   IMPULSE AND STEP RESPONSE

In this chapter we begin to use the circuit transfer function to predict the response of circuits. The definition of this function is the ratio of the zero-state response to the input signal transform. As such, the transfer function is an *s*-domain description of the zero-state characteristics of a circuit. However, to use the transfer function to predict responses, we write the defining relationship as

$$\begin{bmatrix} \text{Zero-state} \\ \text{response} \end{bmatrix} = \begin{bmatrix} \text{Transfer} \\ \text{function} \end{bmatrix} \times \begin{bmatrix} \text{Input signal} \\ \text{transform} \end{bmatrix}$$

In symbols this relationship is

$$Y(s) = T(s) \times X(s) \tag{11-1}$$

where $T(s)$ is the circuit transfer function, $X(s)$ is the input signal transform, and $Y(s)$ is the zero-state response or output signal.

The block diagram representation of the input-output relationship is shown in Figure 11-1.

This viewpoint indicates that the transfer function is the rational function, which is multiplied by the input to obtain the output. Since this input-output relationship involves transferring the signal through the circuit, it is reasonable to call the circuit's contribution the transfer function. Likewise it is reasonable to think of the input as the excitation or cause and the output as the response or effect. The input-output or cause–effect relationship is imbedded in the transfer function. A major role of circuit analysis is to deal with the problem of finding the output when given the transfer function and the input signal. Contrast this with the problem of design. In design the excitation or input is known and the output desired for that input is known. The designer's task is to produce a circuit, usually within some equipment, component, and monetary constraints, that has a suitable transfer function.

While the transfer function is relatively easy to come by, it is clear that we cannot find the circuit response until we are given an input signal. Here we encounter a central paradox of circuit analysis. In practice the input signal is a carrier of information and is therefore unpredictable. We could spend a lifetime studying a circuit for various inputs and still not treat all possible signals that might be encountered in practice. What we must do is calculate the responses due to certain simple test signals.

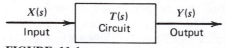

FIGURE 11-1
Block diagram of the input, output, and transfer function.

Although these test signals may never occur as real input signals, their responses tell us enough to begin to understand how the circuit will react to other signals.

The two premier test signals used are the **pulse** and the **sinusoid.** These waveforms have the recommendation that they are relatively easy to describe mathematically and can be easily generated in the laboratory when we need to test our mathematical predictions against the reality of hardware. Moreoever the pulse is in some sense representative of the signals produced by digital transducers such as keyboards and teletypewriters. Likewise the sinusoid is characteristic of the signals produced by the human voice and muscial instruments. Furthermore the study of linear systems is the study of the differential equations that represent those systems. Differential equations have two-part solutions, the transient and the forced or steady-state response. A circuit's transient behavior is alluded to by its pulse response and its forced behavior by its sinusoidal steady-state response. In this chapter we concentrate on pulse response. The study of sinusoidal response is reserved for Chapters 12 and 13.

The study of pulse response of a circuit divides into the two extreme cases shown in Figure 11-2. When the pulse is very short compared with the response time of the circuit, the input can be modeled as an impulse. This all-or-nothing pulse is so brief that the circuit barely has time to begin to respond before the input returns to zero. In essence we get a zero-input response because the sluggish circuit sees essentially no input except for one brief, but spectacular excursion.

At the other extreme is the more sedate ''long'' pulse whose duration exceeds the inherent response time of the circuit. Here the circuit has more than enough time to respond and to reach a steady output before

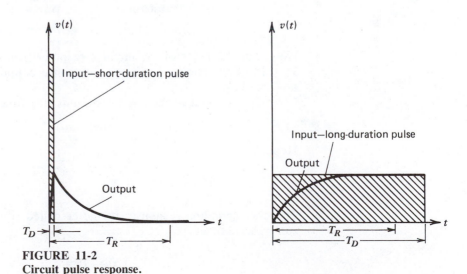

FIGURE 11-2
Circuit pulse response.

the input changes again. The leading edge of this pulse can be treated as a step function as far as the circuit is concerned.

In sum, the two extreme models for a pulse input are the step function $u(t)$ and the impulse $\delta(t)$. We first introduce some notation and then show that there is a simple set of relationships between the two responses.

We first write the transfer function as

$$T(s) = \frac{b_m s^m + b_{m-1} s^{m-1} + \ldots + b_1 s + b_o}{s^n + a_{n-1} s^{n-1} + \ldots + a_1 s + a_o} \tag{11-2}$$

The natural poles, or natural frequencies, are the roots of the denominator polynomial in this expression. Now, if the input is a unit impulse, then

$$x(t) = \delta(t)$$

Hence

$$X(s) = 1$$

The response due to a unit impulse input, hereafter called simply the **impulse response**, is

$$Y(s) = T(s) \times 1 = T(s) \tag{11-3}$$

The remarkable result is that the impulse response transform is none other than the transfer function itself! Only in this very special, very important, case of an impulse input, can we treat a transfer function as a signal transform. To avoid possible confusion between transfer functions $T(s)$ and signal transforms, we denote the impulse response as $H(s)$, the signal transform, and $h(t)$, the corresponding signal waveform. Using this notation, we then write

Impulse response

Transform	Waveform	
$H(s) = T(s)$	$h(t) = \mathscr{L}^{-1}\{H(s)\}$	(11-4)

Note that the poles of the impulse response transform are the poles of $T(s)$. In other words, the impulse response contains only natural poles, no forced poles.

When the input to the circuit is a unit step function, then

$$x(t) = u(t)$$

Hence

$$X(s) = \frac{1}{s}$$

The corresponding response, hereafter called simply the **step response**, can be written as

$$Y(s) = T(s) \times \frac{1}{s} \tag{11-5}$$

The step responses will be denoted as $G(s)$ and $g(t)$, respectively, and hence we write

Step response

Transform	**Waveform**	

$$G(s) = \frac{T(s)}{s} \qquad g(t) = \mathcal{L}^{-1}\{G(s)\} \qquad \textbf{(11-6)}$$

Note that the step response contains natural poles contributed by the circuit through the transfer function $T(s)$, plus a forced pole at $s = 0$ due to the nature of the input signal.

Using this notation, we now show that there are simple relationships between the impulse and step responses. First, combining Eqs. 11-4 and 11-6, we write

$$G(s) = \frac{T(s)}{s} = \frac{H(s)}{s} \qquad \textbf{(11-7)}$$

In words, tbe step-response transform is obtained by dividing the impulse-response transform by s. But division by s in the s domain corresponds to integration in the time domain. Hence

$$g(t) = \int_0^t h(t)dt \qquad \textbf{(11-8)}$$

By the fundamental theorem of calculus then,

$$h(t) \; (=) \; \frac{dg(t)}{dt} \qquad \textbf{(11-9)}$$

where the symbol $(=)$ means equal almost everywhere. The almost-everywhere restriction excludes only those points at which $g(t)$ has a jump or discontinuity. Thus *the step-response waveform is the integral of the impulse-response waveform*, and conversely, *the impulse-response waveform is the derivative of the step-response waveform*. To complete the picture, we note that Eq. 11-7 indicates that

$$H(s) = sG(s) \qquad \textbf{(11-10)}$$

In sum, the five quantities $T(s)$, $H(s)$, $h(t)$, $G(s)$, and $g(t)$ are all related. If we have any one of these quantities, we can obtain all of the others by relatively simple mathematical operations. A summary of the relationships is given in Figure 11-3.

Examples 11-1 and 11-2 (pages 496 and 497) illustrate that the concepts of transfer function, impulse response, and step response are all inter-related. If we have any one of these quantities, we can find the others by the various routes outlined in Figure 11-3. This demonstrates a very important principle of linear systems called **convolution.** Basically convolution says that if we find the response of a circuit to a particular input, we can use the result to calculate the response to almost any other input.

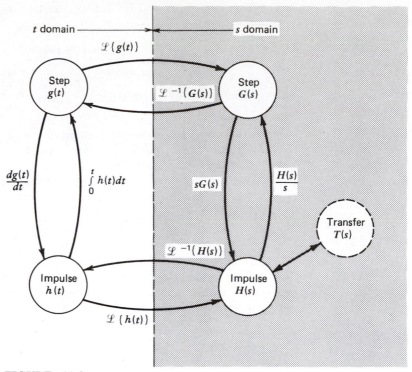

FIGURE 11-3
Relationships between step and impulse responses, and the transfer function in the *t* and *s* domains.

Example 11-1

For the circuit of Figure 11-4, find $T(s)$, $H(s)$, $h(t)$, $g(t)$ and $G(s)$.

The transfer function of the circuit in Figure 11-4 can be obtained by voltage division as

$$T(s) = \frac{V_2(s)}{V_1(s)} = \frac{1/Cs}{R + 1/Cs} = \frac{1/RC}{s + 1/RC}$$

Hence the impulse response waveform is

$$h(t) = \mathscr{L}^{-1}\{H(s)\} = \mathscr{L}^{-1}\{T(s)\} = \frac{1}{RC} e^{-t/RC}$$

The step response can be found by integrating $h(t)$.

$$g(t) = \int_0^t h(t) = \int_0^t \frac{e^{-t/RC}}{RC} dt = -e^{-t/RC} \Big|_0^t$$

$$= (1 - e^{-t/RC})$$

Thus, beginning with $T(s)$, we first obtained $h(t)$ by the inverse transformation and then found $g(t)$ by integrating $h(t)$. Alternatively we could obtain $g(t)$ directly from $T(s)$ as

$$G(s) = \frac{T(s)}{s} = \frac{1/RC}{s(s + 1/RC)} = \frac{1}{s} - \frac{1}{s + 1/RC}$$

Hence

$$g(t) = (1 - e^{-t/RC})$$

Then $h(t)$ is obtained by differentiating $g(t)$.

$$h(t) = \frac{dg(t)}{dt} = \frac{d}{dt}(1 - e^{-t/RC}) = \frac{1}{RC} e^{-t/RC}$$

In sum, we can find $g(t)$ from $h(t)$ or $h(t)$ from $g(t)$.

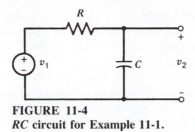

FIGURE 11-4
RC circuit for Example 11-1.

Example 11-2

Suppose the step response of a circuit is

$$g(t) = 5e^{-t^2} \sin 2t$$

Find the impulse response and the circuit's transfer function.

First we obtain $h(t)$ from $g(t)$ as

$$h(t) = \frac{dg(t)}{dt} = \frac{d}{dt}(5e^{-t} \sin 2t)$$
$$= 5e^{-t}(2 \cos 2t - \sin 2t)$$
$$= 10e^{-t} \cos 2t - 5e^{-t} \sin 2t$$

Hence $H(s)$ is

$$H(s) = \mathcal{L}\{h(t)\}$$

$$= \frac{10(s + 1)}{(s + 1)^2 + 4} - \frac{10}{(s + 1)^2 + 4}$$
$$= \frac{10s}{(s + 1)^2 + 4}$$

But since $T(s) = H(s)$ we have the transfer function. The same result can be obtained by transforming $g(t)$.

$$G(s) = \mathcal{L}\{g(t)\} = \frac{10}{(s + 1)^2 + 4}$$

But

$$T(s) = H(s) = sG(s) = \frac{10s}{(s + 1)^2 + 4}$$

and we have found the transfer function by another route.

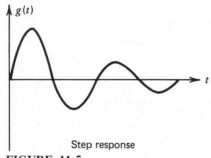

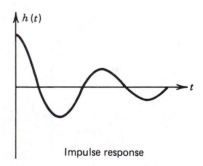

FIGURE 11-5
Step and impulse responses for Example 11-2.

In the present context, this means that we do not need to study both the impulse and step responses of circuits as separate problems. If we have one, we can obtain the other by differentiation or integration. Although the impulse response offers several mathematical advantages and is used extensively in classical solution techniques, we will concentrate on the step response since it is easier to understand.

We have spent considerable effort explaining the difference between the system's transfer function $T(s)$ and the same system's impulse response $H(s)$. To reiterate, the system's transfer function $T(s)$ represents the system's makeup—*it is not a signal*. The system's impulse response $H(s)$, on the other hand, is the output generated by the system when the input is an impulse $\delta(t)$. The impulse response *is a signal*. Notwithstanding this important difference, the forms of $T(s)$ and $H(s)$ are identical. For this reason, and since most advanced texts use $H(s)$ to mean *both* the impulse response of a system *and* its transfer function, $H(s)$ will be used to mean both for the remainder of this text.

11-2   ONE-POLE STEP RESPONSE

Since the transfer function is the mediator between the input and output, it must contain all of the data needed to construct the step response. Studying the step response in terms of transfer functions amounts to discovering how the transfer function carries these data. We begin our study by examining transfer functions with but one pole.

The most general form of one-pole transfer function is

$$T(s) = \frac{b_1 s + b_0}{s + \alpha} = H(s) \tag{11-11}$$

where the one pole is located at $s = -\alpha$ and the zero at $s = -b_0/b_1$. Figure 11-6 shows several circuits that exhibit a voltage ratio transfer function of this form. All of these circuits contain a single memory element, which accounts for the one pole in the transfer function.

The step response of such a circuit is given by

$$G(s) = \frac{H(s)}{s} = \frac{b_1 s + b_0}{s(s + \alpha)} = \frac{k_0}{s} + \frac{k_1}{s + \alpha} \tag{11-12}$$

From the partial fraction expansion of $G(s)$ we obtain $g(t)$ as

$$g(t) = k_0 + k_1 e^{-\alpha t} \tag{11-13}$$

Of particular interest to us is the final value (FV) and the initial value (IV) of this response.

$$
\begin{aligned}
\text{FV} &= \underset{t \to \infty}{\text{Limit}}\; g(t) = g(\infty) = k_0 \\
\text{IV} &= \underset{t \to 0^+}{\text{Limit}}\; g(t) = g(0) = k_0 + k_1
\end{aligned}
\tag{11-14}
$$

The existence of a final value assumes that the system is stable, that is, that the pole is in the left-half plane ($\alpha > 0$). The existence of a nonzero initial value does not contradict the assumption that the circuit is in the zero state since $g(t)$ may jump at $t = 0$. That is, it may be that $g(0+) \neq g(0-)$. In any event it is clear from Eq. 11-14 that

$$k_0 = \text{FV} \qquad \text{and} \qquad k_1 = (\text{IV} - \text{FV})$$

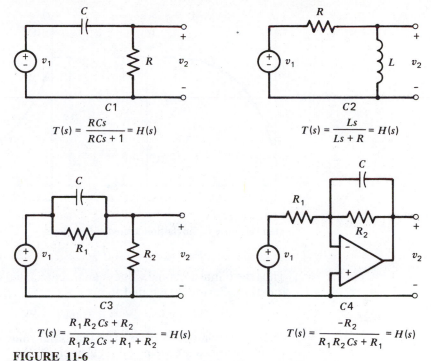

$$T(s) = \frac{RCs}{RCs + 1} = H(s)$$

$$T(s) = \frac{Ls}{Ls + R} = H(s)$$

$$T(s) = \frac{R_1 R_2 Cs + R_2}{R_1 R_2 Cs + R_1 + R_2} = H(s)$$

$$T(s) = \frac{-R_2}{R_1 R_2 Cs + R_1} = H(s)$$

FIGURE 11-6
Some examples of circuits with one-pole transfer functions.

Thus the step response can always be written as Eq. 11-15. This form is the same as we obtained in our classical study of step response in Chapter 8.

$$g(t) = FV + (IV - FV)e^{-\alpha t} \tag{11-15}$$

Figure 11-7 shows two possible forms of the step response. Again the jump at $t = 0$ does not violate the zero-state condition. In fact the initial value or the final value could be zero, or even negative.

There remains the need to determine the initial and final values from the transfer function $H(s)$. From the cover-up algorithm we calculate the two residues as

$$k_0 = b_0/\alpha \tag{11-16}$$
$$k_1 = b_1 - b_0/\alpha$$

If we now calculate the IV and FV of the step response, we find that

$$IV = k_0 + k_1 = b_1 = g(0)$$
$$FV = k_0 = b_0/\alpha = g(\infty)$$

If we cleverly observe that the value of the transfer function $H(s)$ is equal to the IV when $s = \infty$ and the FV when $s = 0$, we find

$$k_0 = FV = sG(s)\big|_{s = 0} = H(s)\big|_{s = 0} = H(0) \tag{11-17}$$
$$k_0 + k_1 = IV = sG(s)\big|_{s = \infty} = H(s)\big|_{s = \infty} = H(\infty)$$

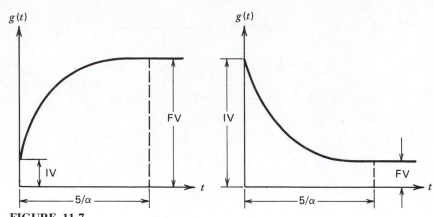

FIGURE 11-7
Step response of single-pole transfer functions.

Thus the initial and final values can be calculated directly from the transfer function. Curiously we find that

$$\text{IV} = g(0) = H(\infty) \qquad \text{and} \qquad \text{FV} = g(\infty) = H(0) \qquad \textbf{(11-18)}$$

These two relations are based on the initial value and final value theorems. Proofs of these theorems are in Appendix D.

Finally note that the time constant of the exponential is determined by the location of the pole.

$$\text{Time constant} = \frac{1}{\alpha} = \frac{1}{|\text{pole}|} \qquad \textbf{(11-19)}$$

Since it takes about five time constants for the exponential to decay to less than 1 percent of its initial value, the circuit response time is $5/\alpha$.

In sum, all of the ingredients necessary to construct the step response are contained in the transfer function itself. The values of $H(s)$ at $s = 0$ and $s = \infty$ give us the final and initial values respectively. The location of the pole gives us the time constant. We can now evaluate the step response directly from the transfer function using Eqs. 11-18 and 11-19.

Example 11-3

Find the step response of circuit C2 in Figure 11-6. Sketch your result.

The transfer function is found by voltage division as

$$H(s) = \frac{Ls}{Ls + R}$$

Hence

$$\text{FV} = H(0) = 0$$

and

$$\text{IV} = H(\infty) = \frac{L}{L} = 1$$

The location of the pole is found from $Ls + R = 0$, and hence $\alpha = R/L$. The step response is then

$$g(t) = 0 + (1 - 0)e^{-\alpha t} = e^{-Rt/L}$$

This response is shown graphically in Figure 11-8.

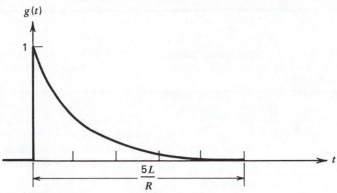

FIGURE 11-8
Step response for Example 11-3.

Example 11-4

Find and sketch the step response of circuit C3 in Figure 11-6.

The transfer function is

$$H(s) = \frac{R_1 R_2 Cs + R_2}{R_1 R_2 Cs + R_1 + R_2}$$

Hence

$$FV = H(0) = \frac{R_2}{R_1 + R_2}$$

and

$$IV = H(\infty) = \frac{R_1 R_2 C}{R_1 R_2 C} = 1$$

The pole is the root of the $R_1 R_2 Cs + R_1 + R_2 = 0$. Hence $\alpha = (R_1 + R_2)/R_1 R_2 C$ and the step response is

$$g(t) = \frac{R_2}{R_1 + R_2} + \frac{R_1}{R_1 + R_2} e^{-\alpha t}$$

This response is shown graphically in Figure 11-9.

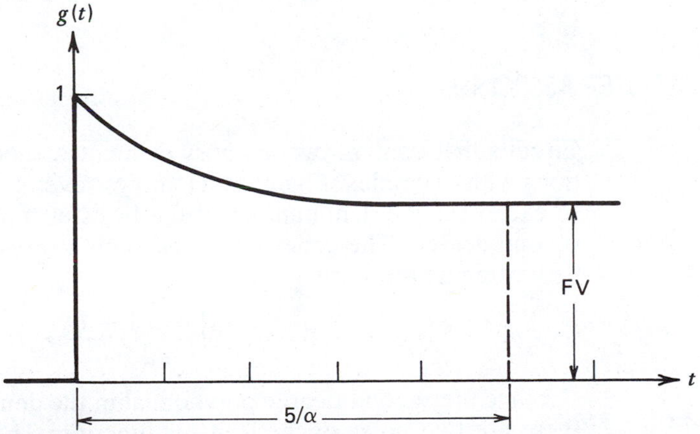

FIGURE 11-9
Step response for Example 11-4.

Example 11-5

Show that the initial and final values do not always turn out to be positive by finding the step response of circuit $C4$ in Figure 11-6.

The transfer function is

$$H(s) = \frac{-R_2}{R_1 R_2 Cs + R_1}$$

Hence

$$H(0) = \text{FV} = \frac{-R_2}{R_1}$$

$$H(\infty) = \text{IV} = 0$$

The pole location is determined from $R_1(R_2 Cs + 1) = 0$, and thus $\alpha = 1/R_2 C$. The step response is

$$g(t) = -\frac{R_2}{R_1} + \frac{R_2}{R_1} e^{-\alpha t}$$

This waveform is shown in Figure 11-10. In this example a positive input produces a negative output. This illustrates the signal inversion provided by the inverting OP AMD circuit.

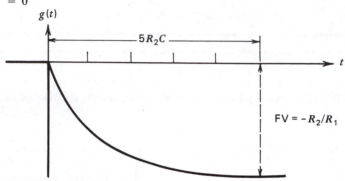

FIGURE 11-10
Step response for Example 11-5.

11-3   TWO-POLE STEP RESPONSE

Circuits that contain two memory elements generally yield transfer functions with two poles. Figure 11-11 shows several double-memory circuits. In each case the denominator of the transfer function is a polynomial of second degree. The general form of such a transfer function can always be written as the form

$$T(s) = \frac{b_2 s^2 + b_1 s + b_0}{s^2 + a_1 s + a_0} = H(s) \qquad \textbf{(11-20)}$$

Since the second-degree polynomial in the denominator has two roots, there are two poles of the transfer function. The coefficients of the polynomial all depend on circuit parameters (i.e., R, L, C), which are real numbers. Consequently there are two important cases to consider.

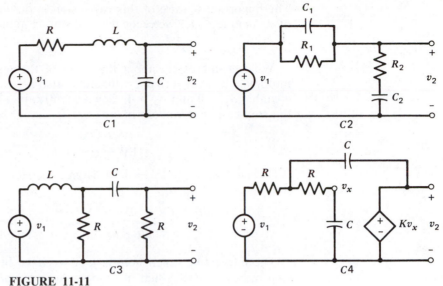

FIGURE 11-11
Some examples of circuits with two-pole transfer functions.

CASE A—Real Roots

$$s^2 + a_1s + a_0 = (s + \alpha_1)(s + \alpha_2) \quad \text{with } |\alpha_2| > |\alpha_1|$$

CASE B—Complex Roots

$$s^2 + a_1s + a_0 = (s + \alpha)^2 + \beta^2 = (s + \alpha - j\beta)(s + \alpha + j\beta)$$

In either case, we assume that the circuit is stable, which means that the poles of $T(s)$ lie in the left-half plane. In symbols, this means that

CASE A

$$\alpha_2 > \alpha_1 > 0$$

CASE B

$$\alpha > 0$$

This in turn means that the step response of the circuit is bounded and approaches some constant final value as $t \to \infty$.

For Case A the step-response transform can be expanded as

$$G(s) = \frac{H(s)}{s} = \frac{b_2s^2 + b_1s + b_0}{s(s + \alpha_1)(s + \alpha_2)} = \frac{k_0}{s} + \frac{k_1'}{s + \alpha_1} + \frac{k_2}{s + \alpha_2} \quad \textbf{(11-21)}$$

For reasons that will become clear, we expand this function in the following special form. If we let $k_1' = k_1 - k_2$, we obtain

$$G(s) = \frac{k_0}{s} + \frac{k_1}{s + \alpha_1} + \frac{k_2}{s + \alpha_2} - \frac{k_2}{s + \alpha_1} \quad \textbf{(11-22)}$$

The important feature of this expansion is the last term. But regardless of its unique form, we recognize the step-response waveform as

$$g(t) = k_0 + k_1 e^{-\alpha_1 t} + k_2 (e^{-\alpha_2 t} - e^{-\alpha_2 t}) \qquad \text{(11-23)}$$

What is interesting about this form of the response is that the term multiplied by k_2 is zero at $t = 0$, and is also zero at $t = \infty$ since the circuit is stable $\alpha_1 > 0$ and $\alpha_2 > 0$. Consequently the initial and final values of the step-response waveform are

$$\text{IV} = g(0) = k_0 + k_1$$
$$\text{FV} = g(\infty) = k_0 \qquad \text{(11-24)}$$

It is clear that $k_1 = (\text{IV} - \text{FV})$. Hence the step response can always be written as

$$g(t) = \underbrace{\text{FV} + (\text{IV} - \text{FV}) e^{-\alpha_1 t}}_{\text{One-pole term}} + \underbrace{k_2 (e^{-\alpha_2 t} - e^{-\alpha_1 t})}_{\text{Two-pole term}} \qquad \text{(11-25)}$$

The interesting result here is that we have been able to express the step response as the sum of a one-pole term and a two-pole term. The one-pole term is exactly like the step response of a one-pole transfer function. A plot of this term is shown in Figure 11-12. The one-pole term carries the initial and final values, since the two-pole term is zero at $t = 0$ and $t = \infty$.

The two-pole term is the difference between two exponentials with the same amplitude but different time constants. Since $\alpha_2 > \alpha_1$, the term $k_2 e^{-\alpha_2 t}$ dies out more rapidly and the two-pole term approaches $k_2 e^{-\alpha_1 t}$ as t approaches infinity. These observations are reflected in the form of the double-pole term as shown in Figure 11-12.

To complete our task we must now determine the constants IV, FV, and k_2 from the transfer function. Using the same technique as for the single-pole response, Eq. 11-17, we have

$$\text{FV} = k_0 = sG(s) \Big|_{s=0} = H(s) \Big|_{s=0} = H(0)$$

$$\text{IV} = k_0 + k_1 = sG(s) \Big|_{s=\infty} = H(s) \Big|_{s=\infty} = H(\infty)$$

$$k_2 = (s + \alpha_2)G(s) \Big|_{s=-\alpha_2} = \frac{b_2\alpha_2^2 - b_1\alpha_2 + b_0}{\alpha_2(\alpha_2 - \alpha_1)} \qquad \text{(11-26)}$$

After some algebraic manipulation, k_2 can be written as

$$k_2 = \frac{\alpha_2\text{IV} + \alpha_1\text{FV} - b_1}{(\alpha_2 - \alpha_1)} \qquad \text{(11-27)}$$

Again we find that all of the data needed to construct the step response are readily available in the transfer function. The values of $H(s)$ at $s = 0$ and $s = \infty$ give us the final and initial values respectively. The location of the two real poles reveals the time constants of the two exponentials involved. Since $\alpha_2 > \alpha_1$, the pole closest to the origin in the s plane (α_1) decays less rapidly and tends to dominate the response as t becomes

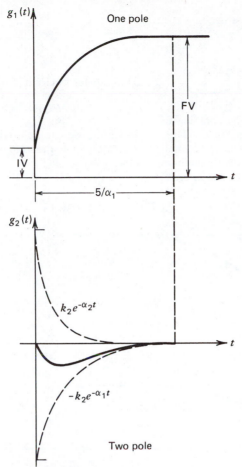

FIGURE 11-12
Components of the step response of a two-pole transfer function with real poles.

large. In other words, the time required for the response to be within 1 percent of its final value is determined by the pole *closest* to the origin, known as the **dominant** pole, and is approximately $5/\alpha_1$.

Example 11-6

Find and sketch the step response of circuit $C1$ in Figure 11-11.

The transfer functions can be obtained from voltage division as

$$H(s) = \frac{1/Cs}{R + Ls + 1/Cs} = \frac{1}{LCs^2 + RCs + 1}$$

By inspection of $H(s)$:

$$IV = H(\infty) = 0 \qquad FV = H(0) = 1$$
$$\text{and} \qquad b_1 = 0$$

If we assume that the transfer function has two real poles, then k_2 can be obtained from Eq. 11-27 as

$$k_2 = \frac{\alpha_2 IV + \alpha_1 FV - b_1}{(\alpha_2 - \alpha_1)} = \frac{\alpha_1}{(\alpha_2 - \alpha_1)}$$

Hence the step response is

$$g(t) = \underbrace{1 - e^{-\alpha_1 t}}_{\substack{\text{One-pole} \\ \text{term}}} + \underbrace{\frac{\alpha_1}{(\alpha_2 - \alpha_1)}(e^{-\alpha_2 t} - e^{-\alpha_1 t})}_{\text{Two-pole term}}$$

The manner in which these two terms combine to produce the step response is shown in Figure 11-13.

If $\alpha_2 \gg \alpha_1$, the two-pole term is relatively small in amplitude so that the one-pole term will dominate the response. The step response can then be approximated as

$$g(t) \simeq 1 - e^{-\alpha_1 t}$$

Hence

$$G(s) \simeq \frac{1}{s} - \frac{1}{s + \alpha_1} = \frac{\alpha_1}{s(s + \alpha_1)}$$

The transfer function is then

$$H(s) = sG(s) \simeq \frac{\alpha_1}{s + \alpha_1}$$

In words, we can approximate the transfer function with a one-pole function by ignoring the remote pole at $s = -\alpha_2$.

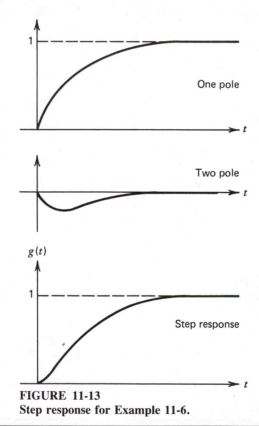

FIGURE 11-13
Step response for Example 11-6.

Example 11-7

Find the step response of circuit C2 in Figure 11-11 and sketch the result.

The transfer function was found in Chapter 10 (Example 10-14) as

$$H(s) = \frac{R_1 R_2 C_1 C_2 s^2 + (R_1 C_1 + R_2 C_2)s + 1}{R_1 R_2 C_1 C_2 s^2 + (R_1 C_1 + R_2 C_2 + R_1 C_2)s + 1}$$

By inspection of the transfer function,

$$IV = H(\infty) = \frac{R_1 R_2 C_1 C_2}{R_1 R_2 C_1 C_2} = 1$$

and

$$FV = H(0) = 1$$

Assuming that the transfer function has two real poles, we can write the step response as

$$g(t) = 1 + k_2 (e^{-\alpha_2 t} - e^{-\alpha_1 t})$$

In this case the one-pole term reduces to a step function and the only transient term is contributed by the two-pole term. The terms combine to produce the step response as illustrated in Figure 11-14.

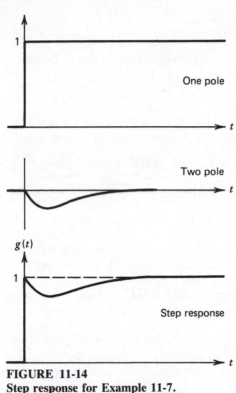

FIGURE 11-14
Step response for Example 11-7.

For Case B the two poles are complex and so we write the step-response transform by completing the square as

$$G(s) = \frac{H(s)}{s} = \frac{b_2 s^2 + b_1 s + b_0}{s(s^2 + a_1 s + a_0)} \qquad \textbf{(11-28)}$$
$$= \frac{b_2 s^2 + b_1 s + b_0}{s[(s + \alpha^2) + \beta^2]}$$

where

$$\alpha = \frac{a_1}{2}$$

$$\beta = \sqrt{a_0 - \frac{a_1^2}{4}}$$

To extract the step response from this transform, we expand it as

$$G(s) = \frac{k_0}{s} + \frac{k_1(s + \alpha)}{(s + \alpha)^2 + \beta^2} + \frac{k_2\beta}{(s + \alpha)^2 + \beta^2} \qquad \textbf{(11-29)}$$

The first term represents a step function, while the next two terms are the transforms of a damped cosine and a damped sine respectively. Hence $g(t)$ can be written as

$$g(t) = k_0 + k_1 e^{-\alpha t} \cos(\beta t) + k_2 e^{-\alpha t} \sin(\beta t) \qquad \textbf{(11-30)}$$

Again we can evaluate k_0 and k_1 from the initial and final conditions.

$$\text{FV} = g(\infty) = k_0 \qquad \textbf{(11-31)}$$
$$\text{IV} = g(0) = k_0 + k_1$$

Combining these two equations indicates that $k_1 = (\text{IV} - \text{FV})$. The step response for the complex-pole case takes the form

$$g(t) = \underbrace{\text{FV} + (\text{IV} - \text{FV})e^{-\alpha t} \cos(\beta t)}_{\text{Cosine term}} + \underbrace{k_2 e^{-\alpha t} \sin(\beta t)}_{\text{Sine term}} \qquad \textbf{(11-32)}$$

For complex poles the step response is the sum of a damped cosine and a damped sine term. The waveforms of these two components are shown in Figure 11-15. The cosine term carries the initial and final values since the sine term is zero at $t = 0$ and $t = \infty$. Although these two components combine in an intricate manner, they do preserve the main feature of the single-pole and double-pole decomposition used in Case A. One component meets the end conditions and the other is zero at both end points.

We now turn to the task of relating the amplitudes of the components to the transfer function. Following the same techniques as for the single-pole response, we obtain the same relation as Eq. 11-17.

$$\text{FV} = k_0 = sG(s)\Big|_{s=0} = H(s)\Big|_{s=0} = H(0)$$

$$\text{IV} = k_0 + k_1 = sG(s)\Big|_{s=\infty} = H(s)\Big|_{s=\infty} = H(\infty)$$

As we have seen before, the final and initial values are determined by the value of the transfer function at $s = 0$ and $s = \infty$ respectively. The value of k_2 can be found by evaluating $G(s)$ at $s = -\alpha$. By combining Eqs. 11-28 and 11-29 we obtain

$$\frac{b_2\alpha^2 - b_1\alpha + b_0}{-\alpha\beta^2} = \frac{k_0}{-\alpha} + \frac{k_2}{\beta}$$

After a modest amount of algebra, this reduces to

$$k_2 = \frac{b_1 - \alpha(\text{IV} + \text{FV})}{\beta} \qquad \textbf{(11-33)}$$

Again we see that all of the data needed to determine the step response are readily available in the transfer function. The value of $H(s)$ at the extremes of the s plane gives us the end points of $g(t)$ in the t domain. The real part of the complex poles produces the exponential damping. In fact the duration of the two sinusoidal terms is $5/\alpha$ as shown in Figure 11-15. Finally the imaginary part of the poles gives the frequency of the damped oscillation.

Example 11-8

Find the step response of the active circuit $C4$ in Figure 11-11.

The transfer function found in Chapter 10 (Example 10-17).

$$H(s) = \frac{K}{(RCs)^2 + (3 - K)RCs + 1}$$

By inspection of the transfer function,

$$\text{IV} = H(\infty) = 0 \qquad \text{FV} = H(0) = K$$
$$\text{and} \qquad b_1 = 0$$

For complex poles k_2 is found from Eq. 11-33 as

$$k_2 = -\frac{\alpha}{\beta} K$$

Hence the step response is

$$g(t) = K - Ke^{-\alpha t} \cos(\beta t) - \frac{\alpha}{\beta} K e^{-\alpha t} \sin(\beta t)$$

If $K \gg \alpha/\beta$, then the sine term has a relatively small amplitude and the step response is approximately

$$g(t) \simeq K(1 - e^{-\alpha t} \cos(\beta t))$$

Thus the step-response waveform approximates the cosine term in Figure 11-15 with IV = 0.

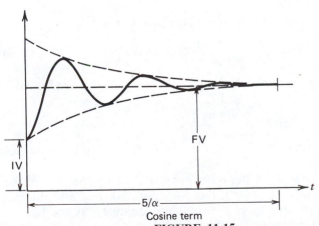

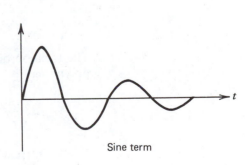

Sine term

FIGURE 11-15
Cosine and sine components of the step response of a two-pole transfer function with complex poles.

Example 11-9

Find the step response of the circuit $C3$ in Figure 11-11.

The transfer function was found in Chapter 10 (Example 10-16).

$$H(s) = \frac{RCs}{2LCs^2 + (RC + GL)s + 1}$$
$$= \frac{Rs/2L}{s^2 + (RC + GL)s/2LC + 1/2LC}$$

By inspection,

$$IV = H(\infty) = 0 \qquad FV = H(0) = 0$$
$$\text{and} \qquad b_1 = R/2L$$

In this case both IV and FV are zero. For complex poles this means that the cosine term in Eq. 11-32 is zero. The step response reduces to

$$g(t) = \frac{b_1}{\beta} e^{-\alpha t} \sin(\beta t)$$

This waveform contains only the sine term shown in Figure 11-15.

11-4 DAMPING AND NATURAL FREQUENCY

Our study of the step response of two-pole transfer functions involved two distinct paths. Case A assumed that the transfer function had two real, unequal poles, while Case B involved complex conjugate poles. The two cases have several similarities, although the forms of the step responses are strikingly different. For this reason, it is important to introduce some universal parameters that help us to determine which case prevails in a given situation.

The general form of the two-pole transfer function is

$$T(s) = \frac{b_2 s^2 + b_1 s + b_0}{s^2 + a_1 s + a_0} = H(s) \qquad \textbf{(11-34)}$$

The location of the poles is determined by the roots of the denominator polynomial.

$$s_1, s_2 = \frac{-a_1 \pm \sqrt{a_1^2 - 4a_0}}{2} \qquad \textbf{(11-35)}$$

The form (real or complex) of the roots is dependent on the sign of the quantity $D = a_1^2 - 4a_0$. If $D > 0$, the two roots are real and distinct (Case A). If $D < 0$, the two roots are complex conjugates (Case B).

To highlight this important distinction, we reintroduce two parameters first seen in Chapter 8 and write the denominator polynomial as

$$s^2 + a_1 s + a_0 = s^2 + 2\zeta\omega_0 s + \omega_0^2 \qquad \textbf{(11-36)}$$

The parameter ζ (zeta) is called the damping ratio, or simply the damping. The parameter ω_0 (omega zero) is called the undamped natural frequency, or simply the natural frequency. By equating the coefficients in Eq. 11-36 we have

$$2\,\zeta\omega_0 = a_1 \quad \text{and} \quad \omega_0^2 = a_0 \qquad \textbf{(11-37)}$$

Essentially all we have done so far is replace the two polynomial coefficients a_0 and a_1 by two new parameters called damping and natural frequency.

The virtue of this replacement is revealed by finding the roots of the polynomial in terms of these new parameters.

$$s_1, s_2 = \frac{-2\,\zeta\omega_0 \pm \sqrt{(2\,\zeta\omega_0)^2 - 4\omega_0^2}}{2}$$
$$= \omega_0\,(-\,\zeta \pm \sqrt{\zeta^2 - 1}) \qquad \textbf{(11-38)}$$

The purpose of this substitution is now clear. The quantity under the radical depends only on the damping ratio. If $\zeta > 1$, the quantity is positive and the two roots are real and distinct (Case A). If $1 > \zeta$, the same quantity is negative and the two roots are complex conjugates (Case B).

By introducing this terminology we have tied the distinction between real and complex poles to the concept of damping. When the poles are real and distinct ($\zeta > 1$), we say that the circuit is **overdamped.** When the poles are complex ($1 > \zeta$), the circuit is called **underdamped.** This terminology agrees with the form of the step response. When the circuit is overdamped, the step-response waveform is basically a sum of sluggish exponentials. When the circuit is underdamped, the step-response waveform consists of wiggly sinusoids. Figure 11-16 shows representative step responses for Case A and Case B. Somehow the terms overdamped and underdamped seem to capture the essence of the contrast.

Now $\zeta = 1$ is the boundary between Case A and Case B. Examination of Eq. 11-38 shows that in this event the radical vanishes, and the two roots are equal. This leads to a step response that is different than either Case A or B. The form of the response is not important here. What matters is that this special case marks the boundary between underdamping and

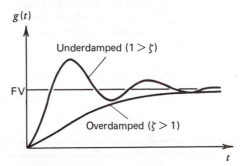

FIGURE 11-16
Comparison of underdamped and overdamped step responses.

overdamping. For this reason a circuit with $\zeta = 1$ is said to be **critically damped.**

The advantage of this terminology is that in simple circuits we can often relate the damping and natural frequencies to particular circuit elements. Sometimes these relationships are such that the damping and natural frequency can be adjusted independently. In particular we can often find the value of a parameter that leads to critical damping—the boundary between Case A and Case B.

Example 11-10

The transfer function for circuit $C1$ in Figure 11-11 is

$$H(s) = \frac{1}{LCs^2 + RCs + 1}$$

To identify the damping ratio and natural frequency, $H(s)$ must be written so that the coefficients of s^2 in the denominator is unity.

$$H(s) = \frac{1/LC}{s^2 + (R/L)s + 1/LC}$$

Comparing the denominator with the standard form $s^2 + 2\zeta\omega_0 s + \omega_0^2$ yields

$$2\zeta\omega_0 = \frac{R}{L} \quad \text{and} \quad \omega_0^2 = \frac{1}{LC}$$

The natural frequency depends on the product LC.

By adjusting this product we vary the natural frequency. Solving for the damping ratio,

$$\zeta = \frac{R}{2\omega_0 L}$$

The resistance R can be used to adjust the damping ratio since the natural frequency is determined by the LC product. Since ζ is proportional to R, the damping ratio can be anything from zero to infinity. The particular value of R that produces critical damping is called, curiously, the critical resistance. With $\zeta = 1$, we solve for the critical resistance as

$$R_{cr} = 2\omega_0 L$$

If the actual resistance is greater than this value, the circuit will be overdamped. If R is less than R_{cr}, the circuit will be underdamped.

Example 11-11

The transfer function for circuit $C4$ in Figure 11-11 is

$$H(s) = \frac{K}{(RCs)^2 + (3 - K)RCs + 1}$$

$$= \frac{K/(RC)^2}{s^2 + \frac{(3 - K)s}{RC} + \frac{1}{(RC)^2}}$$

Comparing the denominator with the standard form $s^2 + 2\zeta\omega_0 s + \omega_0^2$ yields

$$2\zeta\omega_0 = \frac{(3 - K)}{RC} \quad \text{and} \quad \omega_0^2 = \frac{1}{(RC)^2}$$

Clearly $\omega_0 = 1/RC$ and $\zeta = (3 - K)/2$. The product RC determines the natural frequency and the gain K determines the damping ratio. For $K = 1$ we get critical damping. Therefore for $1 > K$ the circuit is overdamped, and for $K > 1$ the circuit is underdamped. Moreover for $K > 3$ the damping ratio is negative. This is called negative damping, which is another name for instability. In other words, if the damping ratio is negative, the circuit is unstable.

Example 11-12

The two previous examples involved circuits that could be either underdamped or overdamped depending on the values of circuit parameters. This example shows that this latitude is not always available. The denominator polynomial of the transfer function in circuit C2 of Figure 11-11 is

$$R_1C_1R_2C_2s^2 + (R_1C_1 + R_2C_2 + R_1C_2)s + 1$$

With the leading coefficient made unity, the polynomial reads

$$s^2 + \frac{R_1C_1 + R_2C_2 + R_1C_2}{R_1C_1R_2C_2}s + \frac{1}{R_1C_1R_2C_2}$$

Comparison with the standard form $s^2 + 2\zeta\omega_0 s + \omega_0^2$ yields

$$2\zeta\omega_0 = \frac{R_1C_1 + R_2C_2 + R_1C_2}{R_1C_1R_2C_2}$$

and

$$\omega_0^2 = \frac{1}{R_1C_1R_2C_2}$$

About all that we can determine by inspection is that the natural frequency depends on two RC products. The algebraic process of determining parameter values for critical damping is not only formidable, it is impossible. This circuit is overdamped for all possible combinations of element values. To prove this, we combine the two foregoing equations to obtain

$$2\zeta\sqrt{R_1C_1R_2C_2} = R_1C_1 + R_2C_2 + R_1C_2$$

If R_1C_2 is subtracted from the right side of this equality, we obtain the inequality

$$2\zeta\sqrt{R_1C_1R_2C_2} > R_1C_1 + R_2C_2$$

or

$$2\zeta > \sqrt{\frac{R_1C_1}{R_2C_2}} + \sqrt{\frac{R_2C_2}{R_1C_1}}$$

The quantity on the right side of this inequality is of the form

$$f(x) = x + 1/x$$

It is easy to show that the minimum value of this function occurs at $x = 1$ and is $f(1) = 2$. Thus the quantity on the right side of the last inequality is never less than 2 for any values of the circuit parameters. Consequently we have $\zeta > 1$, and the circuit is always overdamped. As a corollary the circuit cannot produce complex poles.

We have used two sets of parameters to describe complex poles: (α, β) and (ζ, ω_0). These parameters can be related algebraically since

$$(s + \alpha)^2 + \beta^2 = s^2 + 2\alpha s + \alpha^2 + \beta^2$$
$$= s^2 + 2\zeta\omega_0 s + \omega_0^2$$

Equating the coefficients of like powers of s yields

$$\omega_0^2 = \alpha^2 + \beta^2 \qquad (11\text{-}39)$$
$$\zeta\omega_0 = \alpha$$

It is useful to represent these relationships geometrically as shown in Figure 11-17. The location of the complex pole can be determined using either set. The parameters α and β are akin to the rectangular coordinates of the poles. The parameter ω_0 is the radial distance from the origin to the pole. By geometry in Figure 11-17 we can write

$$\cos\theta = \zeta$$

Thus the quantities ζ and ω_0 are somewhat like the polar coordinates of the poles.

Example 11-13

A transfer function has a pair of complex poles at p_1, $p_2 = -1 \pm j$. Find the parameter α, the damped natural frequency β, the undamped natural frequency ω_0, and the damping ratio ζ.

From Eq. 11-39 or Figure 11-17, we immediately find that $\alpha = 1$, $\beta = 1$, and hence $\omega_0 = \sqrt{2}$ and $\zeta = \cos 45° = 0.707$.

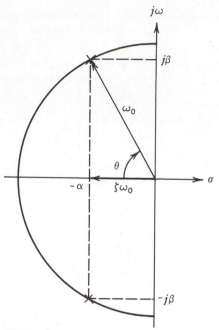

FIGURE 11-17
Geometric relationships between α, β, ζ, and ω_0.

We have seen that the distinction between Case A and Case B can be related to the concept of damping, and that damping can be tied to the values of circuit parameters. Example 11-12 demonstrated that some circuits cannot exhibit both underdamped and overdamped step responses. In fact the last three circuit examples illustrate (they do not prove) the following general principles.

1. *Example 11-12*. Circuits consisting of resistors and capacitors (*RC* circuits) or resistors and inductors (*RL* circuits) are always overdamped. Their poles must be real and must always lie on the negative real axis in the *s* plane.

2. *Example 11-10*. Circuits consisting of resistors, capacitors, and inductors (*RLC* circuits) can be either overdamped or underdamped. Their poles can be either real or complex and can lie anywhere in the left half of the *s* plane.

3. *Example 11-11*. Circuits consisting of resistors, capacitors, and gain elements (active *RC* circuits) can be overdamped, underdamped, or negatively damped (unstable). Their poles can lie anywhere in the *s* plane, including the right-half plane.

11-5   STEP-RESPONSE PARTIAL DESCRIPTORS

We have shown that a one-pole or two-pole transfer function contains all of the data needed to construct a complete description of its step-response waveform. Often we do not need a complete description but work with partial descriptors of the response. The step responses of devices, circuits, and systems are often specified in terms of these partial descriptors.

The step responses of many circuits and systems have waveforms similar to that shown in Figure 11-18. The response is initially driven toward its final value, which it overshoots, then undershoots, and eventually settles down on. The initial rise is often described in terms of **rise time** (T_R).

$$T_R = T_2 - T_1 \tag{11-40}$$

where T_1 is the time at which $g(t) = 0.1\, g(\infty)$ and T_2 is the time at which $g(t) = 0.9\, g(\infty)$. The constants 0.1 and 0.9 are the most common values, although other values sometimes are used. Rise time is an important descriptor of the speed of response of a system. Rise times range from fractions of nanoseconds for high-speed digital computers to several seconds for large electromechanical systems.

For some systems the time between $t = 0$ and $t = T_1$ is large compared with the rise time. In such situations an additional parameter called **delay time** T_D is often specified. Delay is defined as the time interval between $t = 0$ and the time at which $g(t) = 0.5\, g(\infty)$. In other words, it is the time

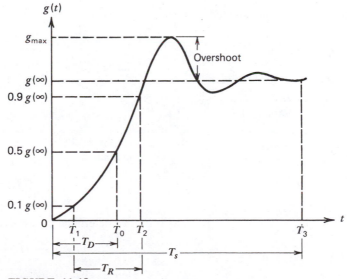

FIGURE 11-18
Representative step response showing rise time (T_R), delay time (T_D), settling time (T_S), and overshoot.

required for the response to get halfway to its final value. This is indicated as T_0 in Figure 11-18.

A distinguishing feature of many step responses is **overshoot.** Quantitatively the percent overshoot is defined as

$$P_0 = \text{Percent overshoot} = \frac{g_{max} - g(\infty)}{g(\infty)} \times 100\% \qquad (11\text{-}41)$$

Percent overshoot is not defined if $g(\infty) = 0$. For example, overshoot cannot be defined for the step responses found in Examples 11-3 and 11-9.

The **settling time** (T_S) is a measure of how long it takes for the response to "settle down." Mathematically T_S is the time interval between $t = 0$ and the time at which the response remains within a certain small percentage of its final value, T_3 in Figure 11-18. Typical values of this small percentage are 1, 3, and 5 percent.

The parameters rise time, delay time, percent overshoot, and settling time often describe step responses in sufficient detail for specification and design. The reader should remember that the definitions given here cannot be blindly applied to every type of step response. As noted overshoot is not defined if the final value is zero. In such cases rise time is defined as the time interval between $g(t) = 0.1\, g_{max}$ and $g(t) = 0.9\, g_{max}$; that is, the time between reaching 10 and 90 percent of the peak value of the response.

Example 11-14

Determine T_R, T_D, T_S, and percent overshoot for circuits with the transfer function

$$H(s) = \frac{b_0}{s + \alpha}$$

By using the methods of Sec. 11-2 we obtain

$$g(\infty) = \text{FV} = H(0) = b_0/\alpha$$
$$g(0) = \text{IV} = H(\infty) = 0$$

Hence

$$g(t) = \frac{b_0}{\alpha}(1 - e^{-\alpha t})$$

To determine T_R we write

$$g(T_1) = \frac{b_0}{\alpha}(1 - e^{-\alpha T_1}) = 0.1\, b_0/\alpha$$

Hence

$$e^{-\alpha T_1} = 0.9 \qquad \text{or} \qquad e^{\alpha T_1} = 10/9$$

and

$$T_1 = \frac{\ln 10 - \ln 9}{\alpha}$$

Similarly,

$$g(T_2) = \frac{b_0}{\alpha}(1 - e^{-\alpha T_2}) = 0.9\, b_0/\alpha$$

Hence

$$e^{-\alpha T_2} = 0.1 \qquad \text{or} \qquad e^{\alpha T_2} = 10$$

and

$$T_2 = \frac{\ln 10}{\alpha}$$

So finally,

$$T_R = T_2 - T_1 = \frac{\ln 9}{\alpha}$$
$$= 2.20/\alpha$$

To determine T_D we write

$$g(T_D) = \frac{b_0}{\alpha}(1 - e^{-\alpha T_D}) = 0.5\, b_0/\alpha$$

Hence

$$e^{-\alpha T_D} = 0.5 \qquad \text{or} \qquad e^{\alpha T_D} = 2$$

and

$$T_D = \frac{\ln 2}{\alpha} = 0.693/\alpha$$

The settling time for a 1 percent criterion is about five time constants as discussed in Chapter 6. Hence

$$T_S \cong 5/\alpha$$

All of the time partial descriptors are proportional to the reciprocal of the pole location or, equivalently, to the circuit time constant. Finally the percent overshoot is zero since the step response is an exponential rise that approaches its final value asymptotically with no overshoot.

Example 11-15

For the circuit in Figure 11-19 determine the maximum allowable capacitance for the circuit to have a rise time of at least 1 μs.

Using the results from Example 11-14, the requirement is

$$\frac{2.2}{\alpha} < 10^{-6}$$

The problem now is to find α or, equivalently, the circuit time constant. There are several ways to do this, but perhaps the simplest is to determine the time constant by finding the Thévenin resistance "seen" by the capacitor. Turning the voltage source in Figure 11-19 off and looking from the capacitor terminals, we see two 50-Ω resistors in parallel, or a Thévenin equivalent resistance of 25 Ω. Hence we can write

$$T_c = 25C = \frac{1}{\alpha}$$

and the requirement is that

$$2.2 \times 25C < 10^{-6}$$

or

$$C < 0.0182 \ \mu F$$

This is a fairly small capacitor. Thus even small capacitors can seriously affect a circuit's rise time.

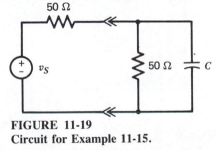

FIGURE 11-19
Circuit for Example 11-15.

Example 11-16

To determine the settling time (T_S) and percent overshoot (PO) of the step response of the two-pole transfer function

$$H(s) = \frac{b_0}{(s + \alpha)^2 + \beta^2}$$

We use the methods of Sec. 11-3 to write

$$g(0) = IV = H(\infty) = 0$$
$$g(\infty) = FV = H(0) = b_0/(\alpha^2 + \beta^2)$$

And since $b_2 = 0$, eq. 11-33 yields

$$k_2 = -\frac{\alpha}{\beta} FV$$

Hence the step response is

$$g(t) = FV[1 - e^{-\alpha t}(\cos \beta t + \frac{\alpha}{\beta} \sin \beta t)]$$

The settling time using a 1 percent criterion is

$$T_S \simeq 5/\alpha$$

The percent overshoot is found by first differentiating the step response:

$$\frac{dg}{dt} = \frac{(\alpha^2 + \beta^2)}{\beta} \, \text{FV} \, e^{-\alpha t} \sin \beta t$$

The first zero of dg/dt occurs at $t = \pi/\beta$ and this point is the maximum value of $g(t)$. Hence

$$g_{\max} = \text{FV}(1 + e^{-\alpha\pi/\beta})$$

The percent overshoot is then found as

$$\begin{aligned}
\text{PO} &= \frac{g_{\max} - g(\infty)}{g(\infty)} \times 100 \\
&= \frac{\text{FV}(1 + e^{-\alpha\pi/\beta}) - \text{FV}}{\text{FV}} \times 100 \\
&= 100 \, e^{-\alpha\pi/\beta}
\end{aligned}$$

It is informative to see how specifying bounds on T_S and PO leads to limitations on the location of the poles of $T(s)$. If we specify T_S as

$$B_1 \geq T_S = 5/\alpha$$

then

$$\alpha \geq 5/B_1$$

This leads to the s-plane pole restriction shown in Figure 11-20a. To meet a settling time specification the complex poles must lie to the left of the vertical line $\alpha = 5/B_1$. If the percent overshoot is specified as

$$B_2 \geq \text{PO} = 100 \, e^{-\alpha\pi/\beta}$$

then

$$\frac{\alpha}{\beta} \geq \frac{1}{\pi} \ln (100/B_2)$$

This bound translates into the pie-shaped region shown in Figure 11-20b. A combined T_S and PO specification then translates as shown in Figure 11-20c. Any pair of complex, conjugate poles lying in the unshaded region would meet both the settling time ($T_S < B_1$) and the overshoot (PO $< B_2$) requirement.

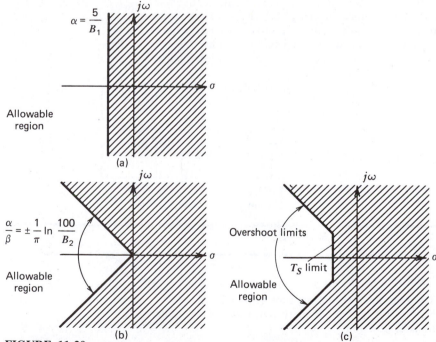

FIGURE 11-20
Allowable complex pole locations for specified settling time (T_S) and percent overshoot (PO). (*a*) Settling time restrictions of pole locations. (*b*) Overshoot restriction on pole locations. (*c*) Combined restrictions on pole locations.

Example 11-17

An effect known as propagation delay occurs in digital circuits. For circuits exhibiting this effect the output due to a step function input is essentially a step function delayed by an amount of time called the propagation delay (T_P). The circuit in Figure 11-21 illustrates this effect. The input step function $Au(t)$ drives an RC circuit whose output then drives a nonlinear device called a comparator. A comparator (see Sec. 4-8) is a differential input device such that

$$\text{If } v_P > v_N, \text{ then } v_{OUT} = A$$
$$\text{and If } v_P < v_N, \text{ then } v_{OUT} = 0$$

For the circuit in Figure 11-21 inverting input v_N is held fixed by the battery at

$$v_N = A/2$$

The noninverting input v_P is determined by the output of the RC circuit. For a step input $Au(t)$ then

$$v_P = A(1 - e^{-t/RC})$$

As long as $v_P < v_N$ the comparator output is zero. As soon as $v_P > v_N$ the output switches to $v_{OUT} = A$. This switch will occur when

$$v_P = v_N = \frac{A}{2} = A(1 - e^{-t/RC})$$

or

$$e^{-t/RC} = 1/2$$

And finally,

$$t = T_P = RC \ln 2$$

For this circuit the propagation delay equals the RC circuit delay time. The circuit output is then

$$v_{OUT} = A\, u(t - T_P)$$

Thus the input is a step function, and the output is as well, except that the output is delayed by an interval T_P. Propagation delays are cumulative. That is, if the output of this circuit served as the input to another identical circuit, then the second output would be $A\, u(t - 2T_P)$, a step function delayed by $2T_P$. Since a subsequent operation must wait for the step function to *propagate* through the system, the speed of operation of many devices is limited by their propagation delays. Often much emphasis is placed on reducing these delays. Consequently propagation delay is an important partial descriptor of the step response of digital circuits.

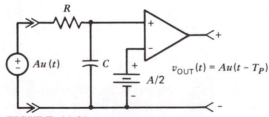

FIGURE 11-21
RC—comparator circuit illustrating propagation delay.

SUMMARY

- The impulse response, the step response, and the transfer function are all interrelated, and once one is found, the others can easily be determined.
- The general form of a one-pole step response is given by

$$g(t) = \text{FV} + (\text{IV} - \text{FV})e^{-\alpha t}$$

- The general form of a two-pole step response for two real, distinct poles is given by

$$G(s) = (b_2 s^2 + b_1 s + b_0)/(s^2 + a_2 s + a_0)$$

$$g(t) = \underbrace{\text{FV} + (\text{IV} - \text{FV})\,e^{-\alpha_1 t}}_{\text{One-pole term}} + \underbrace{k_2\,(e^{-\alpha_2 t} - e^{-\alpha_1 t})]}_{\text{Two-pole term}}$$

where k_2 is given by

$$k_2 = \frac{\alpha_2 \text{IV} + \alpha_1 \text{FV} - b_1}{(\alpha_2 - \alpha_1)}$$

- The general form of a two-pole step response for two complex, conjugate poles is given by

$$g(t) = \underbrace{\text{FV} + (\text{IV} - \text{FV})e^{-\alpha t}\cos(\beta t)}_{\text{Cosine term}} + \underbrace{k_2 e^{-\alpha t}\sin(\beta t)}_{\text{Sine term}}$$

where

$$k_2 = \frac{b_1 - \alpha(\text{IV} + \text{FV})}{\beta}$$

- Two-pole circuits can be classified using the parameters ζ, the damping ratio, and ω_0, the natural frequency. To determine these parameters, write the denominator polynomial as follows:

$$s^2 + a_1 s + a_0 = s^2 + 2\,\zeta\omega_0 s + \omega_0^2$$

- Table 11-1 summarizes the relationships between the various parameters for two-pole circuits.

ζ	Damping	Poles	Stability
> 1	Overdamped	Real, unequal	Stable
$= 1$	Critically damped	Real, equal	Stable
$0 < \zeta < 1$	Underdamped	Complex conjugate	Stable
$= 0$	Undamped	Pure imaginary	Marginal
< 0	Negatively damped	Positive real parts	Unstable

TABLE 11-1

● There are four partial descriptors used to highlight important aspects of a step response $g(t)$:

T_R, the rise time, the time required to go from 10 to 90 percent of its final value.

P_0, percent overshoot—

$$P_0 = \frac{g_{max} - g(\infty)}{g(\infty)} \times 100$$

T_S, settling time, the time required for a response to remain within a fixed percentage of its final value.

T_D, delay time, the time required for the response to reach half of its final value.

EN ROUTE OBJECTIVES
AND RELATED EXERCISES

11-1 IMPULSE AND STEP RESPONSE (SEC. 11-1)

Given any one of the following five quantities—
$g(t)$, $G(s)$, $h(t)$, $H(s)$, $T(s)$, determine any of the
other four.

Exercises

11-1-1   Using the transfer function $T(s)$ found in Exercise 10-3-2, find the circuit impulse response $h(t)$.

11-1-2   Using the transfer function $T(s)$ found in Exercise 10-3-8, find the circuit impulse response $h(t)$.

11-1-3   Using the transfer function $T(s)$ found in Exercise 10-3-12, find the circuit impulse response $h(t)$.

11-1-4   Listed in Table 11-2 are several functions of $f(t)$ and $F(s)$.

Column A	Column B
$u(t)$	$\dfrac{1}{s(s+1)}$
$r(t)$	$\dfrac{1}{s+1}$
e^{-t}	$\dfrac{1}{s^2+1}$
$1 - e^{-t}$	$\dfrac{s}{s+1}$
$\sin t$	$\dfrac{s}{s^2+1}$
$\cos t$	$\dfrac{1}{s^2}$
$1 - \cos t$	$\dfrac{1}{s(s^2+1)}$
	1
	$\dfrac{s^2}{s^2+1}$
	$\dfrac{1}{s}$

TABLE 11-2

     (1)   If $f(t)$ in column A is an impulse response, identify the corresponding transfer function in column B and find the step response $g(t)$.

     (2)   If $f(t)$ in column A is a step response, identify the corresponding transfer function in column B and find the impulse response $h(t)$.

11-1-5   Given the following step response $g(t)$, determine the corresponding transfer function $T(s)$.

     (a) $g_1(t) = 1 - e^{-at} \sin bt$

     (b) $g_2(t) = 1 - e^{-at} \cos bt$

     (c) $g_3(t) = 1 - te^{-at}$

     (d) $g_4(t) = 1 - e^{-at} - te^{-at}$

11-1-6   For each of the following transfer functions determine $H(s)$, $h(t)$, $G(s)$, and $g(t)$.

     (a) $T_1(s) = \dfrac{s}{s^2 + 5s + 6}$

     (b) $T_2(s) = \dfrac{s^2}{s^2 + 3s + 2}$

11-2 ONE-POLE STEP RESPONSE (SEC. 11-2)

    Given a circuit with a one-pole transfer function:

    (a)   Determine the circuit transfer function in terms of circuit parameters.

    (b)   Determine the IV, FV, and α in terms of circuit parameters.

    (c)   Write an expression for the step response $g(t)$ and sketch its waveform.

Exercises

The following set of exercises uses transfer functions obtained in Chapter 10. For each exercise determine the step response $g(t)$ in terms of circuit parameters and sketch its waveform.

11-2-1   Use $T(s)$ found in Exercise 10-3-1.

11-2-2   Use $T(s)$ found in Exercise 10-3-2.

11-2-3   Use $T(s)$ found in Exercise 10-3-3.

11-2-4   Use $T(s)$ found in Exercise 10-3-4.

11-2-5   Use $T(s)$ found in Exercise 10-3-5.

11-2-6   Use $T(s)$ found in Exercise 10-3-6.

11-3 TWO-POLE STEP RESPONSE (SEC. 11-3)

    Given a circuit with a two-pole transfer function:

    (a)   Determine the circuit transfer function in terms of circuit parameters.

(b)   Determine IV and FV in terms of circuit parameters.

(c)   For given numerical values of the circuit parameters determine the form of g(t) and sketch its waveform.

Exercises

The following set of exercises uses transfer functions obtained in Chapter 10. For each exercise determine $g(t)$ and sketch its waveform when all resistors are 1 kΩ, all capacitors are 1 μF, and all inductors are 1 H.

11-3-1   Use $T(s)$ found in Exercise 10-3-7.

11-3-2   Use $T(s)$ found in Exercise 10-3-9.

11-3-3   Use $T(s)$ found in Exercise 10-3-11.

11-3-4   Use $T(s)$ found in Exercise 10-3-13.

11-3-5   Use $T(s)$ found in Exercise 10-3-18.

11-3-6   Use $T(s)$ found in Exercise 10-3-19.

11-4 DAMPING AND NATURAL FREQUENCY (SECS. 11-4 and 11-5)

Given a circuit with a two-pole transfer function:

(a)   Derive relationships between the circuit parameters and the damping ratio and the undamped natural frequency.

(b)   Select circuit parameters to obtain a specified damping ratio and undamped natural frequency.

(c)   Calculate the circuit's step-response partial descriptors.

Exercises

The following set of exercises uses transfer functions obtained in Chapter 10. For each exercise first determine the relationships between each exercise and solve for the damping ratio and undamped natural frequency in terms of circuit parameters, then select values for the circuit parameters to obtain the indicated damping ratios and natural frequencies.

11-4-1   Use $T(s)$ found in Exercise 10-3-7 and obtain $\zeta = 0.5$ and $\omega_0 = 10^3$.

11-4-2   Use $T(s)$ found in Exercise 10-3-11 and obtain $\zeta = 0.5$ and $\omega_0 = 10^3$.

11-4-3   Use $T(s)$ found in Exercise 10-3-12 and obtain $\zeta = 1.0$ and $\omega_0 = 10^3$.

11-4-4   Use $T(s)$ found in Exercise 10-3-17 and obtain $\zeta = 0.5$ and $\omega_0 = 10^3$.

11-4-5   Use $T(s)$ found in Exercise 10-3-19 and obtain $\zeta = 1.5$ and $\omega_0 = 1.0$.

11-4-6   Use $T(s)$ found in Exercise 10-3-21 and obtain $\zeta = 0.0$ and $\omega_0 = 10^3$.

11-4-7   For the circuits in Figure E11-4-7 calculate T_R, T_S, T_D, and percent overshoot.

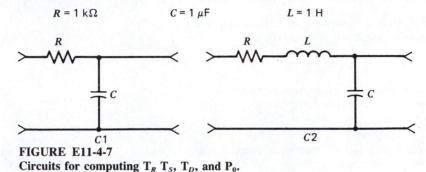

FIGURE E11-4-7
Circuits for computing T_R T_S, T_D, and P_0.

11-4-8   For the step response shown in Figure E11-4-8 compute T_R, T_S, T_D, and percent overshoot (PO).

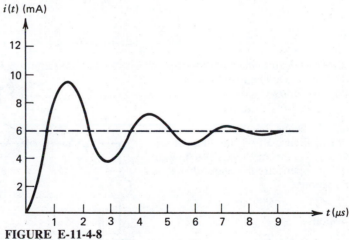

FIGURE E-11-4-8
Waveform for Exercise 11-4-8.

PROBLEMS

P11-1 (Analysis)
For a linear circuit we have seen that the step response and impulse response are related as

$$g(t) = \int_0^t h(t)dt \quad \text{or} \quad h(t) (=) \, dg/dt$$

Now if $f(t)$ is the response of a linear circuit due to a unit ramp input $r(t) = u(t)t$, show that

$$f(t) = \int_0^t g(t)dt \quad \text{and} \quad g(t) = \frac{df}{dt}$$

Use this result to find the unit ramp responses corresponding to the step responses given in Exercise 11-1-5.

P11-2 (Analysis)
A linear circuit has an impulse response

$$h(t) = e^{-t} \sin 2\pi t$$

Without performing any detailed mathematical calculations, sketch the response due to the following inputs:
(a) $u(t)$
(b) $u(t) - u(t - 5)$
(c) $100[u(t) - u(t - 0.01)]$

P11-3 (Analysis)
The circuit in Figure P11-3 contains a nonlinear element called an ideal diode D. The i-v characteristics of an ideal diode are as follows:

If $i > 0$, then $v = 0$ If $v \geq 0$, then $i = 0$

Roughly speaking, this means that if current "tends" to flow to the right in Figure P11-3, the ideal diode acts as a short circuit. If current "tends" to flow to the left, it acts as an open circuit. Show that this circuit has a step response

$$g(t) = (1 - e^{-t/RC}) \, u(t)$$

and an impulse response

$$h(t) = u(t)/RC$$

thus showing that $h(t) \neq dg/dt$ for a nonlinear circuit.

P11-4 (Analysis)
We have seen that the initial and final values of the step response can be found by evaluating the transfer function at $s = \infty$ and $s = 0$ respectively. If we apply these conditions directly to the circuit elements, we can often determine the initial and final values directly from the circuit without first finding the circuit transfer function. Since

$$Z_C = 1/Cs \quad \text{and} \quad Z_L = Ls$$

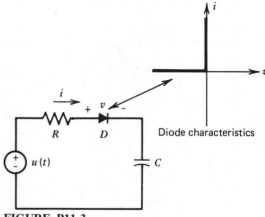

FIGURE P11-3
Nonlinear circuit for Problem 11-3.

for the initial value $s = \infty$ and

$$Z_C = 0 \text{ (short circuit)} \quad \text{and}$$
$$Z_L = \infty \text{ (open circuit)}$$

To determine the initial value, replace all capacitors by short circuits and all inductors by open circuits, and then analyze the resulting memoryless circuit. Conversely for the final value $s = 0$, and hence

$$Z_C = \infty \text{ (open circuit)} \quad \text{and}$$
$$Z_L = 0 \text{ (short circuit)}$$

We can, replace all capacitors by open circuits and all inductors by short circuits to determine the final value. Use this approach to determine the initial and final values of the step responses of the four circuits shown in Figure 11-11. Then check your results with the answers given in Examples 11-6 through 11-9.

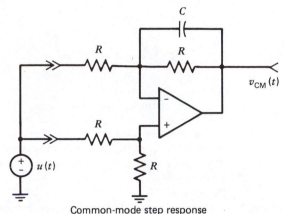

Common-mode step response

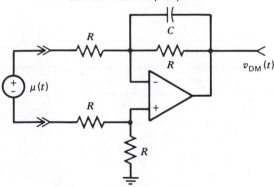

Difference-mode step response

FIGURE P11-5
Common- and difference-mode step-response circuits.

P11-5 (Analysis, Evaluation)
The step responses of circuits with two inputs and a single output are often specified in terms of a common-mode and a difference-mode response. The two responses can be obtained by applying the step input as shown in Figure P11-5. Find the common-mode and difference-mode step responses of this circuit and comment on any important differences and similarities.

P11-6 (Analysis)
The impulse response of a series RC circuit is $(1/RC)e^{-t/RC}$. Find the response due to a pulse input

$$x(t) = [u(t) - u(t - T)]/T$$

and show that the two responses are essentially the same if $T <<< RC$.

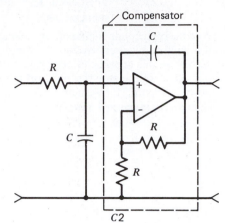

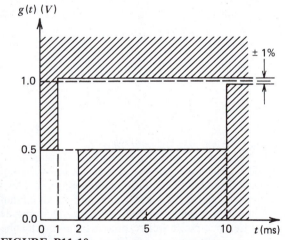

FIGURE P11-7
Re circuit and compensator.

P11-7 (Analysis, Evaluation)
The response time of the circuit $C1$ in Figure P11-7 has been found to be too slow. It is claimed that by adding the compensator shown in $C2$ the response time can be reduced to zero! Show that the claim is true, but point out any practical difficulties you see with this circuit.

P11-8 (Design)
Design circuits to realize the following step responses.
(a) $g(t) = e^{-100t}$
(b) $g(t) = 1 - e^{-100t}$
(c) $g(t) = 1 + e^{-100t}$

P11-9 (Design)
Design circuits to realize the following step responses.
(a) $g(t) = e^{-100t} - e^{-200t}$
(b) $g(t) = 1 - 2e^{-100t} + e^{-200t}$
(c) $g(t) = -e^{-100t} + 2e^{-200t}$

P11-10 (Design)
Design a one-pole circuit whose step response lies entirely within the unshaded region in Figure P11-10.

FIGURE P11-10
Design criteria for step-response problem.

P11-11   (Analysis, Design)

The general two-pole transfer function can be decomposed as

$$T(s) = b_2 T_2(s) + b_1 T_1(s) + b_0 T_0(s)$$

where for complex poles

$$T_2 = \frac{s^2}{(s + \alpha)^2 + \beta^2}$$

$$T_1 = \frac{s}{(s + \alpha)^2 + \beta^2}$$

$$T_0 = \frac{1}{(s + \alpha)^2 + \beta^2}$$

The transfer functions T_2, T_1, and T_0 are called the high-pass, bandpass, and low-pass canonic forms respectively. Figure P11-11 is a functional block diagram showing how each of the canonic forms can be combined to produce the step response of $T(s)$.

(a) Determine the individual step responses $g_2(t)$, $g_1(t)$, and $g_0(t)$.

(b) Show that the final value of $g(t)$ is determined by the signal path through $T_0(s)$ in Figure P11-11.

(c) Show that the initial value of $g(t)$ is determined by the signal path through $T_2(s)$ in Figure P11-11.

(d) Show that

$$g_1(t) = \frac{dg_0}{dt} \quad \text{and} \quad g_2(t) = \frac{dg_1}{dt}$$

(e) Construct an alternative functional block diagram using two differentiators and only the transfer function $T_0(s)$ that produces the step response $g(t)$.

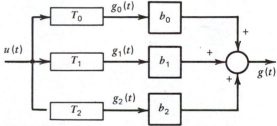

FIGURE P11-11
Functional block diagram for Problem 11-11.

P11-12   (Analysis)

The time-delay property of the Laplace transformation states that

If $\mathcal{L}\{f(t)\} = F(s)$, Then $\mathcal{L}\{f(t - T)\} = e^{-Ts} F(s)$

The step response of an ideal delay circuit is $u(t - T)$, where T is the delay time. Use the time-delay property to show that the transfer function of an ideal delay is $T(s) = e^{-T}s$.

Chapter 12
Sinusoidal Response

The vector diagram of sine waves gives the best insight into the mutual relationships of alternating currents and emf's.

Charles Steinmetz

In this chapter we study the response of linear circuits to the second premier test input signal—the sinusoidal waveform. The key concept here is the sinusoidal steady-state response, a condition that is achieved after the natural response has vanished and in which all voltages and currents in the circuit are sinusoidal. We find that this steady-state response can be derived from the circuit transfer function in a rather straightforward way. We also introduce a new way to represent the sinusoidal waveform called a phasor. The phasor representation of sinusoidal waveforms is particularly useful in the analysis of circuits in the steady state. It greatly simplifies the analysis process and leads to other important concepts, such as resonance and conjugate matching. The final section describes an important application of phasor circuit analysis.

12-1   THE SINUSOIDAL STEADY STATE

The unique properties of the sinusoidal signal make it a very common and useful waveform for testing and describing circuits and systems. The two key properties are that the sum of two sinusoids of the same frequency produces a sinusoid, and that a sinusoid may be integrated or differentiated any number of times and the result remains a sinusoid. No other periodic waveform has these special properties.

The upshot is that if a linear, stable circuit is driven by a sinusoid, then ultimately all currents and voltages in the circuit are sinusoids of the *same* frequency. This result is both theoretically important and of great practical significance. As a practical matter sinusoidal waveforms are easily generated and are relatively convenient to measure using standard laboratory instruments. As a consequence the performance of a communication system is often specified in terms of its sinusoidal response. Likewise the waveform used to generate and distribute electric power is almost always a sinusoid with a frequency of 60 Hz in the United States, and 50 Hz in most of the rest of the world.

To deal with the sinusoidal response mathematically, we first consider the case in which the circuit is described by a transfer function $H(s)$. Figure 12-1 shows the pattern of interaction. There is an input $x(t)$ or $X(s)$ and a resulting response $y(t)$ or $Y(s)$. In the s domain the connection between the responses is the transfer function

$$Y(s) = H(s) X(s) \qquad \text{(12-1)}$$

In the present case the input is a sinusoid, which we write in general form as

$$x(t) = X_o \cos (\omega t + \phi) \qquad \text{(12-2)}$$

which in turn can be expanded as

$$x(t) = X_o(\cos \omega t \cos \phi - \sin \omega t \sin \phi)$$

In this form we recognize $\cos \omega t$ and $\sin \omega t$ and can write the input transform as

$$X(s) = X_o \left[\frac{s}{s^2 + \omega^2} \cos \phi - \frac{\omega}{s^2 + \omega^2} \sin \phi \right]$$

$$= X_o \left[\frac{s \cos \phi - \omega \sin \phi}{s^2 + \omega^2} \right]$$

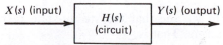

$X(s)$ (input) $\longrightarrow$ $\boxed{\begin{array}{c} H(s) \\ \text{(circuit)} \end{array}}$ $\longrightarrow$ $Y(s)$ (output)

FIGURE 12-1
Block diagram of the input, output, and transfer function.

This is the Laplace transform of the general sinusoidal waveform in Eq. 12-2. With this result we can express the response transform from Eq. 12-1 as

$$Y(s) = X_o \left[\frac{s \cos \phi - \omega \sin \phi}{(s - j\omega)(s + j\omega)} \right] H(s) \qquad (12\text{-}3)$$

In this form we see that the input signal has introduced two forced poles in the response, at $s = j\omega$ and at $s = -j\omega$. We now expand the response transform by partial fractions as

$$Y(s) = \underbrace{\frac{k}{s - j\omega} + \frac{k^*}{s + j\omega}}_{\text{Forced poles}} + \underbrace{\frac{k_1}{s - p_1} + \frac{k_2}{s - p_2} + \cdots + \frac{k_N}{s - p_N}}_{\text{Natural poles}}$$

where $p_1, p_2, \ldots p_N$ are the poles of the transfer function $H(s)$, and hence are natural poles and do not depend on the nature of the input. We now perform the inverse transformation and obtain the response waveform as

$$y(t) = \underbrace{k e^{j\omega t} + k^* e^{-j\omega t}}_{\text{Forced response}} + \underbrace{k_1 e^{p_1 t} + k_2 e^{p_2 t} + \cdots + k_N e^{p_N t}}_{\text{Natural response}}$$

This form gives a complete accounting of the poles of the response. There are response terms arising from the two forced poles, which are the forced response and N terms from the natural poles. The latter terms are the poles of the circuit descriptor $H(s)$.

At this point we introduce a key assumption or condition. If the circuit is stable, then all of the natural poles have negative real parts. Consequently all of the components of the natural response ultimately decay to zero, and eventually only the forced component of the response persists. It is important to realize that in electric circuits this process does not take a great deal of time. If we apply a sinusoidal input to a stable electronic circuit, then after a few hundred milliseconds at most, only the forced component of the response will be observable.

The persistent response is called the **sinusoidal steady-state response,** which is the terminology used in place of a more precise description— the forced component of the response of a stable, linear circuit due to a sinusoidal input. The term *steady state* is perhaps unfortunate since it implies a steady or constant response, when in fact the forced response is an ever-changing sinusoid. Nonetheless we will use the steady-state (SS) terminology, and we can now write it as

$$y_{SS}(t) = k e^{+j\omega t} + k^* e^{-j\omega t} \qquad (12\text{-}4)$$

In this form we see that the steady-state response is the sum of two conjugate terms. That is,

$$[k e^{+j\omega t}]^* = k^* e^{-j\omega t}$$

The sum of two conjugate terms is twice their real parts. For example, the conjugate of the complex number $a + jb$ is $a - jb$, and their sum is $a + jb + a - jb = 2a$. Hence we can express the steady-state response as

$$y_{SS}(t) = \text{Re}\{2ke^{j\omega t}\} \qquad (12\text{-}5)$$

where $\text{Re}\{\ \}$ stands for the real part.

To complete the description of the steady-state response we need to evaluate the constant $2k$, where k is the residue of the forced pole at $s = j\omega$. To evaluate this constant we use the cover-up algorithm in Eq. 12-3 to write

$$2k = 2X_o \left[\frac{s \cos \phi - \omega \sin \phi}{(s + j\omega)} \right] H(s) \Bigg|_{s=j\omega}$$

$$= 2X_o \left[\frac{j\omega \cos \phi - \omega \sin \phi}{j2\omega} \right] H(j\omega)$$

$$= X_o(\cos \phi + j \sin \phi)H(j\omega)$$

But by Euler's relationship

$$\cos \phi + j \sin \phi = e^{j\phi}$$

and defining $|H(j\omega)|$ as the magnitude and ϕ_H as the phase, we can write

$$H(j\omega) = |H(j\omega)|\ e^{j\phi_H}$$

We obtain $2k$ as

$$2k = X_o|H(j\omega)|\ e^{j(\phi + \phi_H)} \qquad (12\text{-}6)$$

When this result is substituted into Eq. 12-5 the result is

$$y_{SS}(t) = \text{Re}\{X_o|H(j\omega)|\ e^{j(\phi + \phi_H)}\ e^{j\omega t}\}$$
$$= X_o|H(j\omega)|\ \text{Re}\{e^{j(\omega t + \phi + \phi_H)}\}$$

But again using Euler's relationship

$$\text{Re}\{e^{jx}\} = \text{Re}\{\cos x + j \sin x\} = \cos x$$

and the steady-state response is therefore

$$y_{SS}(t) = X_o|H(j\omega)| \cos(\omega t + \phi + \phi_H) \qquad (12\text{-}7)$$

Thus *the steady-state component is a sinusoid of the same frequency as the input but with a different amplitude and phase angle.* The input-output relationships can be summarized in the statements:

Output amplitude = (Input amplitude) × (Magnitude of $H(j\omega)$)
Output phase = (Input phase) + (Angle of $H(j\omega)$)
Output frequency = Input frequency

Figure 12-2 presents these statements in an input-output format showing the effect of the circuit on the steady-state output. In the case of frequency there is no effect, since the input and output have the same frequency— but the amplitude and phase of the response are altered by the function $H(j\omega)$. The next two examples illustrate the application of these results.

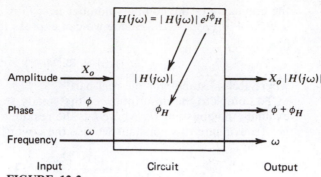

FIGURE 12-2
Signal transfer in the sinusoidal steady state.

Example 12-1

Assuming the input to the circuit in Figure 12-3 is $v_S(t) = V_o \cos (\omega t + \phi)$, determine the steady-state output.

To solve this problem we must first find the transfer function. By voltage division,

$$H(s) = \frac{1/Cs}{R + 1/Cs} = \frac{1}{RCs + 1}$$

Hence $H(j\omega)$ is

$$H(j\omega) = \frac{1}{1 + j\omega RC}$$

and so the magnitude is

$$|H(j\omega)| = \frac{1}{\sqrt{1 + (\omega RC)^2}}$$

while the phase shift is

$$\phi_H = \tan^{-1}(-\omega RC) = -\tan^{-1}(\omega RC)$$

Now applying the steady-state input-output rules we have

$$\text{Output amplitude} = V_0|H(j\omega)| = \frac{V_o}{\sqrt{1 + (\omega RC)^2}}$$

$$\text{Output phase} = \phi + \phi_H = \phi - \tan^{-1}(\omega RC)$$

$$\text{Output frequency} = \omega$$

The steady-state response is

$$v_{SS}(t) = \frac{V_o}{\sqrt{1 + (\omega RC)^2}} \cos [\omega t + \phi - \tan^{-1}(\omega RC)]$$

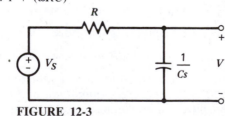

FIGURE 12-3
Circuit for Example 12-1.

Example 12-2

The circuit in Figure 12-4 has an input current of the form

$$i_S(t) = I_o \cos (\omega t + \phi)$$

Find the steady-state voltage appearing across the input. We begin our solution by noting that the input-output relationship is an impedance

$$V(s) = Z(s)I_S(s)$$

where

$$H(s) = Z(s) = Ls + R$$

Consequently,

$$H(j\omega) = Z(j\omega) = R + j\omega L$$

and

$$|H(j\omega)| = \sqrt{R^2 + (\omega L)^2}$$
$$\phi_H = \tan^{-1}(\omega L/R)$$

Applying the steady-state input-output rules:

Output amplitude $= I_o|H(j\omega)| = I_o \sqrt{R^2 + (\omega L)^2}$

Output phase $= \phi + \phi_H = \phi + \tan^{-1}(\omega L/R)$

Output frequency $= \omega$

The steady-state response is

$$v_{SS}(t) = I_o \sqrt{R^2 + (\omega L)^2}$$
$$\cos\left[\omega t + \phi + \tan^{-1}(\omega L/R)\right]$$

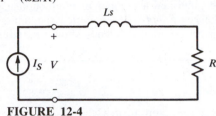

FIGURE 12-4
Circuit for Example 12-2.

12-2   PHASORS

The concept of a **phasor** has a long and fabled history, and its use in electrical engineering was well established long before the term was borrowed by *Star Trek*. Although the phasor concept sometimes "stuns" beginning students, there is absolutely no connection. The idea was originally put forth by the American engineer Charles Steinmetz at the International Electrical Congress in 1893. Steinmetz also popularized the phasor by demonstrating its many applications so that by the early twentieth century it was universally used in the study of ac circuits and systems. Thus the phasor has been with us for almost as long as electrical engineering has been a recognized discipline.

Simply put, a phasor is an alternative way to represent a sinusoid. More formally:

A phasor is a complex number representing the amplitude and phase of a sinusoidal waveform.

Thus if the sinusoidal signal $v(t)$ is written as

$$v(t) = A \cos(\omega t + \phi) \tag{12-8}$$

its phasor representation is

$$\mathbf{V} = Ae^{j\phi} \tag{12-9}$$

Alternatively, if the waveform is written as

$$v(t) = a \cos(\omega t) + b \sin(\omega t) \tag{12-10}$$

Historical Note

Charles Proteus Steinmetz (1865–1923) was born in Breslau, Germany, and educated in that country and Switzerland. He emigrated to the United States in 1889 and became associated with the General Electric Company in 1893, where he remained for the rest of his life. His major contributions to the early understanding of electrical engineering included the phenomenon of hysteresis, the concept of impedance, and the use of complex quantities (now called phasors) to describe ac devices and systems. Steinmetz had a wide range of interests and published a number of nontechnical papers. In a 1918 paper, *America's Energy Supply*, he computed the limits on the available energy supply and suggested that efficiency should be stressed because these limits ultimately would be reached.

---

then its phasor representation is

$$\mathbf{V} = a + jb \tag{12-11}$$

Obviously there is a relationship between these two representations. Expanding Eq. 12-9 by Euler's relationship yields

$$\mathbf{V} = A \cos (\phi) + j A \sin (\phi)$$

and comparing with Eq. 12-11 yields

$$a = A \cos (\phi)$$
$$b = A \sin (\phi) \tag{12-12}$$

or inversely

$$A = \sqrt{a^2 + b^2}$$
$$\phi = \tan^{-1}(b/a) \tag{12-13}$$

Equations 12-12 and 12-13 are familiar to us from our study of the sinusoid in Chapter 6. We know that there are two equivalent ways to represent a sinusoidal waveform. The **amplitude-phase** form leads to the *polar* form of the phasor, while the **Fourier coefficient** format leads to the *rectangular* form of the phasor. We can convert from one phasor form to the other using the simple right-triangle identities in Eqs. 12-12 or 12-13. All of this can be visualized by means of the graphical representation of a phasor shown in Figure 12-5.

There are two things about the phasor that require emphasis. First, phasors are indicated by boldface symbols to distinguish them from other signal representations, such as $v(t)$ and $V(s)$. Second, the phasor itself does not contain any information about the frequency of the sinusoid. Since all signals in the sinusoidal steady state are sinusoids of the same frequency, this information is not crucial. On the other hand, a frequency *must* be given or implied in any particular situation. Thus waveforms and phasors are simply two different and essentially equivalent ways to represent the eternal sinusoid. All of the data needed to construct the phasor

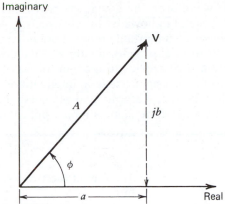

FIGURE 12-5
Graphical interpretation of the phasor representation.

are contained in the waveform, and conversely, all of the data (except frequency) required to define the waveform are contained in the phasor.

But why do we need yet another signal representation? The answer is insight. The phasor provides an insight into the relationship of waveforms in the sinusoidal steady state that is difficult to achieve in any other way. In sum, the justification for the phasor is found by studying its applications in circuit analysis. The first of these applications is the simple problem of finding the sum of two or more sinusoids of the same frequency.

Given two sinusoidal waveforms in Fourier coefficient form,

$$v_1(t) = a_1 \cos(\omega t) + b_1 \sin(\omega t)$$
$$v_2(t) = a_2 \cos(\omega t) + b_2 \sin(\omega t)$$

we know from our study in Chapter 6 that because the waveforms have the same frequency we can write their sum as

$$v_1(t) + v_2(t) = (a_1 + a_2) \cos(\omega t) + (b_1 + b_2) \sin(\omega t) \qquad \textbf{(12-14)}$$

Now the phasor representations of these two waveforms are

$$\mathbf{V}_1 = a_1 + jb_1$$
$$\mathbf{V}_2 = a_2 + jb_2$$

The sum of these two phasors is

$$\mathbf{V}_1 + \mathbf{V}_2 = (a_1 + a_2) + j(b_1 + b_2) \qquad \textbf{(12-15)}$$

The waveform corresponding to this phasor sum is none other than Eq. 12-14. Obviously this result could be extended to the sum of more than two sinusoids.

This discussion may seem somewhat pedantic. It may seem that the complex number representation (the phasor) simply provides a book-keeping system. The real part corresponds to the coefficient of the cosine term in the waveform, the imaginary part represents the sine term, and the symbol j simply allows us to tell which is which. This is, of course, true, but misses an important point. There is a one-to-one correspondence between a sinusoidal waveform sum and the corresponding phasor sum. Any constraint that applies to the sum of waveforms also applies to the sum of phasors. Among the several implications of this is that Kirchhoff's laws apply to both waveforms and phasors. But before we exploit this observation we illustrate some simple applications of phasors.

Example 12-3

Given the waveforms

$$v_1(t) = 10 \cos (10t - 45°)$$

and

$$v_2(t) = 5 \cos (10t + 30°)$$

construct their phasor representations and use these phasors to find the waveform

$$v_3(t) = v_1(t) + v_2(t)$$

Since the sinusoids have the same frequency, their phasors are

$$\mathbf{V}_1 = 10 \, e^{-j45°}$$
$$= 10 \cos (-45°) + j10 \sin (-45°)$$
$$= 7.07 - j7.07$$
$$\mathbf{V}_2 = 5 \, e^{j30°}$$
$$= 5 \cos (30°) + j5 \sin (30°) =$$
$$4.33 + j2.50$$

The sum of the two signals can be found in phasor form as

$$\mathbf{V}_3 = \mathbf{V}_1 + \mathbf{V}_2 = \underbrace{11.4 - j4.57}_{\text{Rectangular form}} = \underbrace{12.3 \, e^{-j21.8°}}_{\text{Polar form}}$$

Hence

$$v_2(t) = 12.3 \cos (10t - 21.8°)$$

Figure 12-6 shows the phasor diagram and indicates graphically how the summation takes place. This simple graphical picture helps us visualize the process. Finally note that the frequency plays no role in this calculation. We should point out that all three sinusoids do have the same frequency.

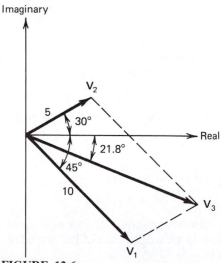

FIGURE 12-6
Phasor diagram for Example 12-3.

Example 12-4

Determine the waveforms corresponding to the following phasors and find the sum of the waveforms. (All three phasors have $\omega = 1000$.)

$$\mathbf{I}_A = 5 \, e^{j50°} \qquad \mathbf{I}_B = 5 \, e^{j170°} \qquad \mathbf{I}_C = 5 \, e^{-j70°}$$

All of the data are available immediately to write the waveform as

$$i_A(t) = 5 \cos (1000t + 50°)$$
$$i_B(t) = 5 \cos (1000t + 170°)$$
$$i_C(t) = 5 \cos (1000t - 70°)$$

The sum of these signals can be found in phasor form as

$$\mathbf{I}_A = 5 \cos (50°) + j5 \sin (50°) = 3.21 + j3.83$$

$$\mathbf{I}_B = 5 \cos (170°) + j5 \sin (170°) = -4.92 + j0.87$$
$$\mathbf{I}_C = 5 \cos (-70°) + j5 \sin (-70°) = 1.71 - j4.70$$

Hence

$$\begin{aligned} \mathbf{I}_A + \mathbf{I}_B + \mathbf{I}_C &= (3.21 - 4.92 + 1.71) \\ &\quad + j(3.83 + 0.87 - 4.70) \\ &= 0 + j0 \end{aligned}$$

The fact that the signals sum to zero is not at all obvious from examining their waveforms. However, the phasor diagram in Figure 12-7 makes the reason clear. We have three equal magnitude phasors displaced in phase by 120 degrees, so that the sum of any two is equal and opposite to the third.

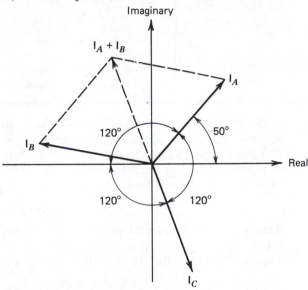

FIGURE 12-7
Phasor diagram for Example 12-4.

Example 12-5

Find the phasors for the waveforms

$$v_1(t) = \cos (\omega t)$$
$$v_2(t) = -\sin (\omega t)$$
$$v_3(t) = -\cos (\omega t)$$
$$v_4(t) = \sin (\omega t)$$

Using the identity $\cos (x + y) = \cos (x) \cos (y) - \sin (x) \sin (y)$ we write

$$\begin{aligned} v_2(t) &= \cos (\omega t + 90°) \\ &= \cos (\omega t) \cos (90°) - \sin (\omega t) \sin (90°) \\ &= -\sin (\omega t) \end{aligned}$$
$$\begin{aligned} v_3(t) &= \cos (\omega t + 180°) \\ &= \cos (\omega t) \cos (180°) - \sin (\omega t) \sin (180°) \\ &= -\cos (\omega t) \end{aligned}$$
$$\begin{aligned} v_4(t) &= \cos (\omega t + 270°) \\ &= \cos (\omega t) \cos (270°) - \sin (\omega t) \sin (270°) \\ &= \sin (\omega t) \end{aligned}$$

Hence

$$V_1 = 1\ e^{j0°} = \cos(0°) + j\sin(0°) = 1 + j0$$
$$V_2 = 1\ e^{j90°} = \cos(90°) + j\sin(90°) = 0 + j1$$
$$V_3 = 1\ e^{j180°} = \cos(180°) = j\sin(180°)$$
$$= -1 + j0$$

$$V_4 = 1\ e^{j270°} = \cos(270°) + j\sin(270°)$$
$$= 0 - j1$$

Figure 12-8 shows the phasor diagram indicating that the phasors are spaced at 90 degree increments and point in the four cardinal directions in the complex plane.

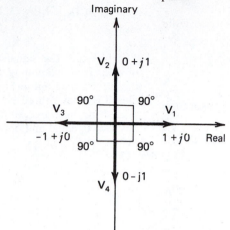

FIGURE 12-8
Phasor diagram for Example 12-5.

So far we have shown that a phasor is another way to describe a sinusoid that can be useful in such simple problems as finding signal sums. But there is much more to the phasor than this. The first indication of its usefulness is that in the sinusoidal steady state we can write the general input-output relationship of Eq. 12-1 in phasor form as

$$\mathbf{Y} = H(j\omega)\ \mathbf{X} \qquad (12\text{-}16)$$

That is, in the sinusoidal steady state the transfer function provides a relationship between the input phasor and the output phasor. To see that the relationship in Eq. 12-16 is true, we use the notation

$$\mathbf{Y} = Y_o e^{j\theta}$$
$$\mathbf{X} = X_o e^{j\phi}$$
$$H(j\omega) = |H(j\omega)|\ e^{j\phi_H}$$

Inserting these expressions into Eq. 12-16 yields

$$Y_o\ e^{j\theta} = |H(j\omega)|\ e^{j\phi_H} X_o\ e^{j\phi}$$

or

$$Y_o\ e^{j\theta} = |H(j\omega)| X_o\ e^{j\phi + \phi_H} \qquad (12\text{-}17)$$

Now two phasors can be equal only if they have the same amplitude and phase angle. Hence Eq. 12-17 tells us that:

$$\text{Output amplitude} = Y_o = |H(j\omega)| X_o$$
$$\text{Output phase} = \theta = \phi + \phi_H$$
$$\text{Output frequency} = \omega$$

But these are the same input-output rules that we derived in the previous section using Laplace transforms. Thus we can think of the input and the output as phasors, and the transfer function as the connection between the two. An example will illustrate this connection.

Example 12-6

Find the sinusoidal steady-state output for the circuit of Figure 12-9a.

To determine the steady-state output we first write the input phasor as

$$\mathbf{V}_S = 10\, e^{j0°} = 10 + j0$$

The circuit transfer function is found by voltage division as

$$T(s) = \frac{R}{Ls + R}$$

Since we are dealing with sinusoidal steady state, we can replace s by $j\omega$. Therefore, using the numerical values given in the figure, we find $T(j\omega)$ to be,

$$T(j1000) = \frac{1000}{1000 + j1000} = \frac{1}{1 + j1} = \frac{1 - j1}{2}$$

Now applying the input-output relationship in phasor form:

$$\mathbf{V}_o = T(j\omega)\, \mathbf{V}_S = \frac{1 - j1}{2}(10 + j0) = 5 - j5$$
$$= 7.07\, e^{-j45°}$$

Hence the sinusoidal steady-state output is

$$v_{SS}(t) = 7.07 \cos(1000t - 45°)$$

Figure 12-9 also shows the phasor diagram for the input and output. When the angle between input and output is negative (-45 degrees in this case) we say that the output **lags** the input. If the angle is positive, we say that the output **leads** the input.

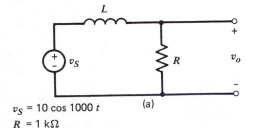

$v_S = 10 \cos 1000\,t$
$R = 1\,k\Omega$
$L = 1\,H$

FIGURE 12-9
Circuit and phasor diagram for Example 12-6. (*a*) Circuit. (*b*) Phasor diagram.

12-3 PHASOR CIRCUIT ANALYSIS

By phasor circuit analysis we mean the analysis of circuits in the sinusoidal steady state in which the signals are represented as phasors. We have seen that circuit analysis is based on an equilibrium established by constraints of two types: (1) connection constraints (Kirchhoff's laws) and (2) device constraints (element equations). To deal with phasor circuit analysis we must first see how these constraints are written in phasor form.

In the sinusoidal steady state the application of Kirchhoff's voltage law around a loop in the circuit would lead to an equation of the form

$$V_1 \cos(\omega t + \phi_1) + V_2 \cos(\omega t + \phi_2) + V_3 \cos(\omega t + \phi_3) = 0$$

But as we saw in the preceding section, there is a one-to-one correspondence between waveform sums and phasor sums. Hence it must be also true that

$$\mathbf{V}_1 + \mathbf{V}_2 + \mathbf{V}_3 = 0$$

Clearly this result would apply for any number of voltages and to sinusoidal currents in applying Kirchhoff's current law as well. In sum, the connection constraints are unchanged. Kirchhoff's laws apply to waveforms *and* phasors.

Turning now to the device constraints we first write down the element equations in the *s* domain in terms of impedances:

$$
\begin{aligned}
\text{Resistor} \quad & V_R(s) = Z_R I_R(s) = R I_R(s) \\
\text{Inductor} \quad & V_L(s) = Z_L I_L(s) = L s I_L(s) \\
\text{Capacitor} \quad & V_C(s) = Z_C I_C(s) = \frac{1}{Cs} I_C(s)
\end{aligned}
\tag{12-18}
$$

Now we know from Eq. 12-16 that the transfer function is the relationship between the input and output phasors. If we think of the element constraints in Eq. 12-18 as transfer functions in microscopic form, with the current as the input and the voltage as the output, then we can apply Eq. 12-16 here and write the element equations in phasor form as

$$
\begin{aligned}
\text{Resistor} \quad & \mathbf{V}_R = R\, \mathbf{I}_R \\
\text{Inductor} \quad & \mathbf{V}_L = j\omega L\, \mathbf{I}_L \\
\text{Capacitor} \quad & \mathbf{V}_C = \frac{1}{j\omega C}\, \mathbf{I}_C
\end{aligned}
\tag{12-19}
$$

In words, we find that the *i-v* relations in phasor form are simply the *s*-domain impedances with *s* replaced by *jω*, where *ω* is the frequency of the sinusoidal inputs.

Since the element constraints are relationships between a voltage and a current appearing at the same pair of terminals, we can just as easily think of the voltage as the input and the current as the response. If we had done so, we would have obtained

$$
\begin{aligned}
\text{Resistor} \quad & \mathbf{I}_R = \frac{1}{R}\, \mathbf{V}_R \\
\text{Inductor} \quad & \mathbf{I}_L = \frac{1}{j\omega L}\, \mathbf{V}_L \\
\text{Capacitor} \quad & \mathbf{I}_C = j\omega C\, \mathbf{V}_C
\end{aligned}
\tag{12-20}
$$

In this form we have the *s*-domain admittance with *s* again replaced by *jω*. In sum, in phasor circuit analysis the element constraints can be summarized in the form of impedances or admittances of the form

$$
\begin{aligned}
Z_R = R \qquad & \text{or} \qquad Y_R = \frac{1}{R} \\
Z_L = j\omega L \qquad & \text{or} \qquad Y_L = \frac{1}{j\omega L} \\
Z_C = \frac{1}{j\omega C} \qquad & \text{or} \qquad Y_C = j\omega C
\end{aligned}
$$

Which form we use depends on what is known and what we want to find.

There is an important distinction to be made here. Although impedances (or admittances) can be complex numbers, they are *not* phasors. This important distinction is not a new concept. We have seen that transforms and transfer functions can have the same form, but are really quite different things. One represents a signal and the other a circuit. And so it is here as well. Phasors represent signals and impedances describe circuit elements. These are quite different things, although both are complex numbers.

We see that the phasor circuit analysis constraints have a very familiar pattern. Kirchhoff's laws apply to phasors and the element constraints are algebraic relationships between phasors that have an Ohm's law-like form. This is the same pattern that we first saw with memoryless resistance circuits and then rediscovered in our study of *s*-domain circuit analysis in Chapter 10. Therefore phasor circuit analysis does not involve any new analysis techniques. In fact, all of the techniques (equivalence, voltage division, current division, reduction, Thévenin's theorem, superposition, node analysis, mesh analysis, etc.) that we have come to know as old friends apply here as well. The major difference is that the responses we find by analysis are phasors and not waveforms or transforms.

We can think of phasor circuit analysis in terms of the flow diagram in Figure 12-10. We begin in the time domain with a circuit that is operating in the sinusoidal steady state. We transform the circuit into the frequency domain, which means that all signals become phasors and the circuit elements are described by their impedances with *s* replaced by $j\omega$. We can then employ our lexicon of algebraic analysis techniques to solve for

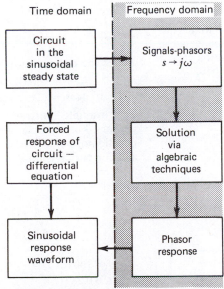

FIGURE 12-10
Pattern of circuit analysis using phasors.

the phasor response. This phasor response can then be taken back into the time domain, if need be, to obtain the response waveform. Figure 12-10 also points out that there is another route to this result. We could apply the methods discussed in Chapter 8 to find the circuit differential equation, and then determine the forced response by classical techniques. Generally the phasor approach is simpler. But more important, use of the phasor provides an insight into the sinusoidal steady-state response that is essential to the understanding of much of the terminology of electrical engineering.

Since phasor circuit analysis follows a familiar pattern, we need only illustrate the process. The next four examples serve this purpose.

Example 12-7

Determine $i(t)$, $i_C(t)$ and $i_R(t)$ for the circuit of Figure 12-11a.

From the data given we know that the circuit is operating in the sinusoidal steady state. Figure 12-11b shows the frequency-domain version of the circuit with the desired responses shown as phasors. To solve for these responses we will use circuit reduction. The resistor and capacitor are connected in parallel and can be replaced by an equivalent impedance Z_{EQ1}:

$$Z_{EQ1} = \frac{Z_R Z_C}{Z_R + Z_C} = \frac{(3000)(-j1000)}{3000 - j1000}$$

$$= \frac{-j3000}{3 - j1} = 300 - j900$$

This reduction is shown in Figure 12-11c. In this circuit the equivalent impedance Z_{EQ1} is connected in series with the inductor and the combination can be replaced by an equivalent input impedance Z_{EQ2}:

$$Z_{EQ2} = Z_{EQ1} + Z_L$$

$$= 300 - j900 + j500 = 300 - j400$$

This reduction is shown in Figure 12-11d. We now have an equivalent impedance at the input terminals and the current I is found as

$$I = \frac{V_S}{Z_{EQ2}} = \frac{100 + j0}{300 - j400} = \frac{1}{3 - j4}$$

$$= \frac{3 + j4}{25} = 0.2 \, e^{j53.1°}$$

We can write the corresponding current waveform as

$$i(t) = 0.2 \cos (2000t + 53.1°)$$

To find the other required current responses we observe from Figure 12-11b that the current I flows through the inductor and upon reaching the parallel combination of the resistor and capacitor divides into two components. Therefore by current division we have

$$I_R = \frac{Z_C}{Z_R + Z_C} I = \frac{-j1000}{3000 - j1000} 0.2 e^{j53.1°}$$

$$= 0.0632 \, e^{-j18.5°}$$

and

$$I_C = \frac{Z_R}{Z_R + Z_C} I = \frac{3000}{3000 - j1000} 0.2 e^{j53.1°}$$

$$= 0.190 \, e^{j71.5°}$$

Hence the two waveforms are

$$i_R(t) = 0.0632 \cos (2000t - 18.5°)$$
$$i_C(t) = 0.190 \cos (2000t + 71.5°)$$

This example illustrates some of the versatility of phasors. For example, we have demonstrated the application of parallel equivalence, series equivalence, and current division in phasor form.

(a)

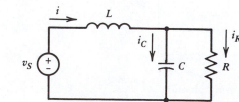

$v_S = 100 \cos 2000t$
$L = 0.25$ H
$C = 0.5 \, \mu F$
$R = 3 \, k\Omega$

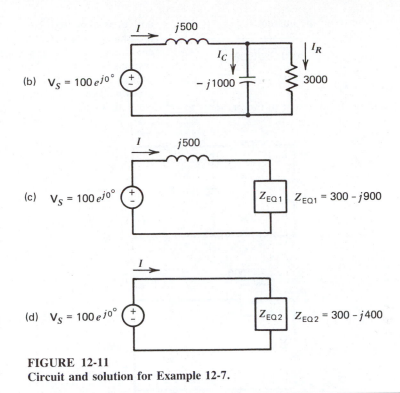

FIGURE 12-11
Circuit and solution for Example 12-7.

Example 12-8

Find the current I_R in the circuit of Figure 12-12a.

In this example we will use Thévenin's theorem to find the desired resistor current. Since this is the same circuit used in the previous example, we expect to obtain the same result. To find the Thévenin equivalent we first determine the open-circuit voltage at the interface by means of voltage division as

$$\mathbf{V}_{OC} = \mathbf{V}_T = \frac{Z_C}{Z_L + Z_C} \mathbf{V}_S = \frac{-j1000}{j500 - j1000}(100)$$
$$= \frac{-j1000}{-j500}100$$
$$= 200 + j0$$

This result points out an interesting aspect of the sinusoidal steady state. Here we have an open-circuit

voltage that is *larger* than the source voltage. Such a result cannot possibly be obtained with resistance circuits. To complete the Thévenin equivalent we need the short-circuit current. With a short circuit connected at the interface, the capacitor is ineffective, so that

$$\mathbf{I}_{SC} = \frac{\mathbf{V}_S}{Z_L} = \frac{100}{j500} = 0 - j0.2$$

Hence the Thévenin equivalent impedance is

$$Z_T = \frac{\mathbf{V}_{OC}}{\mathbf{I}_{SC}} = \frac{200}{-j0.2} = j1000$$

This Thévenin equivalent is shown in Figure 12-12b. From the Thévenin equivalent we can use series equivalence to determine the current I_R.

$$I_R = \frac{V_T}{R + Z_T} = \frac{200}{3000 + j1000}$$

$$= \frac{0.2}{3 + j1} = 0.0632\, e^{-j18.5°}$$

This is the same result obtained in the previous ex-

ample, as expected. Notice that we need not convert the phasor to a waveform to show that the answers are the same. We can simply compare the phasor responses obtained. This example illustrates further uses of phasors. We have shown the use of voltage division, Thévenin's theorem, and series equivalence methods in phasor circuit analysis.

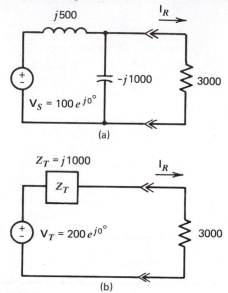

(a)

(b)

FIGURE 12-12
Circuit and its Thévenin equivalent for Example 12-8. (a) Given circuit. (b) Thévenin equivalent.

One of the important consequences of linearity is superposition. Superposition can be applied in phasor circuit analysis if we observe certain precautions. If all of the sources have the *same* frequency, then the superposition (summation) can be made in phasor form. That is, we can find the phasor response due to each source separately and obtain the response due to all sources by simply adding the individual phasors. However, if the sources have *different* frequencies, we cannot simply add the individual phasors to find the total response. Remember that the phasor contains no information about the frequency of the sinusoid it represents. We cannot add two phasors unless they have the same frequency. All is not lost since we can still use superposition and phasor analysis in the multiple-frequency case. However, we must use caution when finding the individual responses, taking due account of the changes in element impedances because of the different frequencies. But then the individual phasor response for each frequency must be converted into waveforms and superposition applied in the time domain. The multiple-frequency case gives rise to what is termed frequency response, a subject we discuss in the next chapter. In Example 12-9 (below) we treat the case in which the sources have the same frequency.

These examples illustrate most of the basic analysis methods in which we work directly with the circuit model. The methods parallel those we used in memoryless circuits and again in s-domain circuit analysis. The only complication is that we must manipulate complex numbers, and be willing to deal with the phasor abstraction. It takes a while to get used to the notion that a voltage can be expressed as $32 - j50.6$. But by definition a phasor is a complex number representing a sinusoidal waveform. The advantage of the phasor is that it simplifies the process of finding the sinusoidal steady-state response. With some practice we begin to think in terms of phasors without converting them into waveforms. That is, we actually think of the voltage as being $30 - j50.6$ rather than a sinusoidal waveform.

In more complicated circuits these direct methods of analysis could prove cumbersome and we would need to appeal to a general method such as node analysis or mesh analysis. The formulation of node-voltage or mesh-current equilibrium equations in phasor form follows exactly the

---

Example 12-9

Solve for the voltage $v_R(t)$ in the circuit of Figure 12-13a using superposition.

Figure 12-13a shows that the two voltage sources have the same frequency. Our objective is to determine the voltage $\mathbf{V}_R$. Figure 12-13b shows the phasor circuit with source 2 turned off (replaced by a short circuit). We can determine the contribution of source 1 to the desired response by voltage division, if we first combine R and L_2 into a single equivalent impedance:

$$Z_{EQ1} = \frac{Z_R Z_{L2}}{Z_R + Z_{L2}} = \frac{(500)(j4000)}{500 + j4000}$$

$$= \frac{j4000}{1 + j8} = 492 + j61.5$$

Then by voltage division,

$$\mathbf{V}_{R1} = \frac{Z_{EQ1}}{Z_{L1} + Z_{EQ1}} \mathbf{V}_{S1}$$

$$= \frac{492 + j61.5}{492 + j61.5 + j500} 100\, e^{-j30°}$$

$$= 21.0 - j63$$

Figure 12-13c shows the phasor circuit with source 1 off. The contribution by source 2 can be found by voltage division, if we first reduce the parallel combination of R and L_1 into an equivalent impedance.

$$Z_{EQ2} = \frac{Z_R Z_{L1}}{Z_R + Z_{L1}} = \frac{(500)(j500)}{500 + j500}$$

$$= \frac{j500}{1 + j1} = 250 + j250$$

Then by voltage division,

$$\mathbf{V}_{R2} = \frac{Z_{EQ2}}{Z_{EQ2} + Z_{L2}} \mathbf{V}_{S2}$$

$$= \frac{250 + j250}{250 + j250 + j4000} 200 e^{+j90°}$$

$$= 11 + j12.4$$

Since the two sources have the same frequency, we can now add the individual phasor responses to obtain the total response as

$$\mathbf{V}_R = \mathbf{V}_{R1} + \mathbf{V}_{R2} = (21.0 - j63) + (11.0 + j12.4)$$
$$= 32 - j50.6$$
$$= 59.9\, e^{-j57.7°}$$

Therefore the response waveform is

$$v_R(t) = 59.9 \cos(2000t - 57.7°)$$

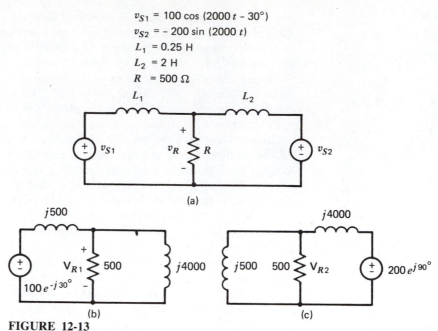

FIGURE 12-13
Circuit analysis using superposition for Example 12-9. (*a*) Given circuit. (*b*) Source no. 2 off. (*c*) Source no. 1 off.

same pattern that we have seen before. A typical node equation would be of the form

$$Y_1 V_1 - Y_2 V_2 - Y_3 V_3 = I_{S1}$$

where the phasors V_1, V_2, and V_3 are the node voltages, Y_1 is the sum of admittances connected to node ①, Y_2 is the admittance connected between nodes ① and ②, Y_3 is the admittance connected between nodes ① and ③, and finally, I_{S1} is the sum of the current source inputs connected to node ①. Similarly a typical mesh equation would be

$$Z_1 I_1 - Z_2 I_2 = Z_3 I_3 = V_{S1}$$

where I_1, I_2, and I_3 are the mesh currents, Z_1 is the sum of impedances in mesh 1, Z_2 is the impedance common to meshes 1 and 2, Z_3 is the impedance common to meshes 1 and 3, and V_{S1} is the sum of voltage source inputs in mesh 1.

Thus formulating a set of equilibrium equations in phasor form is a straightforward process involving concepts with which we are already familiar. Solving these equations for selected phasor responses can be accomplished by means of Cramer's rule, although this means that we would need to manipulate an array of complex numbers. In principle this is possible, but as a practical matter any circuit with more than two nodes or two meshes is almost always dealt with using a computer. The reason

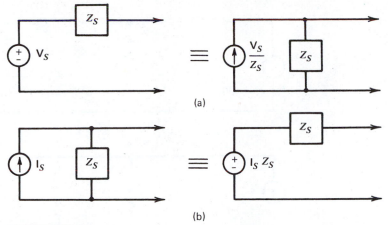

FIGURE 12-14
Phasor source transformations. (*a*) Voltage source to current source. (*b*) Current source to voltage source.

is quite simple. Most ordinary mortals simply cannot manipulate anything more than a two-by-two array of complex numbers without making mistakes.

Finally we note that the formulation of general equilibrium equations can be simplified if we use source transformations. Figure 12-14 shows the two transformations in phasor form. To convert from one type of source to the other, we either multiply or divide by the source impedance. Since the source impedance can be a complex quantity, the transformation may change the phase angle of the equivalent source. Our final example of phasor circuit analysis uses source transformation and node equations.

Example 12-10

The circuit in Figure 12-15*a* is the same one used in the previous example to illustrate the application of superposition. As before, our objective is to determine the voltage $\mathbf{V}_R$. In this example we observe that the two voltage sources are connected in series with impedances and therefore can be converted to equivalent current sources. In equation form the transformations are

$$\mathbf{I}_{S1} = \frac{\mathbf{V}_{S1}}{Z_{L1}} = \frac{100 \, e^{-j30°}}{j500} = -0.1 - j0.173$$

$$\mathbf{I}_{S2} = \frac{\mathbf{V}_{S2}}{Z_{L2}} = \frac{j200}{j4000} = 0.05 + j0$$

After the source transformations the resulting circuit in Figure 12-15*b* has only one nonreference node. The

node voltage equation at this node is

$$(Y_{L1} + Y_R + Y_{L2}) \mathbf{V}_R = \mathbf{I}_{S1} + \mathbf{I}_{S2}$$

or in numerical form,

$$\left[\frac{1}{j500} + \frac{1}{500} + \frac{1}{j4000} \right] \mathbf{V}_R = -0.05 - j0.173$$

After multiplying the equation by 4000, this becomes

$$(8 - j9) \mathbf{V}_R = -200 - j692$$

which yields

$$\mathbf{V}_R = 32 - j50.6$$

This is the same result obtained in the previous example using superposition.

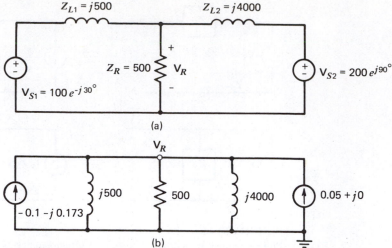

(a)

(b)

FIGURE 12-15
Circuit demonstrating source transformations for Example 12-10. (*a*) Given circuit.
(*b*) Circuit after source transformations.

12-4   IMPEDANCE AND ADMITTANCE

The concepts of impedance and admittance play key roles in the behavior of circuits in the sinusoidal steady state. The impedances of the three passive elements are

Resistor	$Z_R = R$
Inductor	$Z_L = j\omega L$
Capacitor	$Z_C = \dfrac{1}{j\omega C}$

The impedance of a resistor is purely real while the impedances of the memory elements are purely imaginary. When we combine these elements in a circuit to obtain an equivalent impedance, we find that it can be written as

$$Z = R + jX \qquad (12\text{-}21)$$

In this rectangular form we call the real part of the impedance R **resistance** and the imaginary part X **reactance.**

For example, the series RLC circuit in Figure 12-16 has an equivalent impedance of

$$Z = Z_R + Z_L + Z_C$$
$$= R + j\omega L + \frac{1}{j\omega C}$$
$$= R + j\left(\omega L - \frac{1}{\omega C}\right)$$

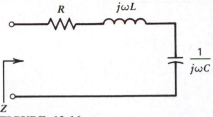

FIGURE 12-16
A series *RLC* circuit.

and we can identify the resistance and reactance as

$$R = R \quad \text{and} \quad X = \left(\omega L - \frac{1}{\omega C} \right) \tag{12-22}$$

In this case it is easy to see that the real part of the input impedance comes from the resistor and the reactive component from the inductor and capacitor. The real part is always positive but the reactive part can be either positive or negative. In fact we see that at one frequency the reactive part can be made zero. Specifically at the frequency

$$\omega_0 = \frac{1}{\sqrt{LC}} \tag{12-23}$$

the reactance vanishes and the input impedance is purely resistive. When this occurs, the circuit is said to be in **resonance** and the frequency at which it occurs is called the **resonant frequency.**

Admittance is the reciprocal of impedance ($Y \equiv 1/Z$), and the three passive elements have admittances that are

Resistor $Y_R = \dfrac{1}{R}$

Inductor $Y_L = \dfrac{1}{j\omega L}$

Capacitor $Y_C = j\omega C$

When we combine elements to produce an equivalent admittance, we find that it can be written as

$$Y = G + jB \tag{12-24}$$

In this form we call the real part of the admittance G **conductance** and the imaginary part B **susceptance.**

An example is the parallel *RLC* circuit in Figure 12-17. The equivalent input admittance of this circuit is

$$
\begin{aligned}
Y &= Y_R + Y_L + Y_C \\
&= \frac{1}{R} + \frac{1}{j\omega L} + j\omega C \\
&= \frac{1}{R} + j\left(\omega C - \frac{1}{\omega L} \right)
\end{aligned}
$$

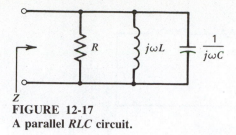

FIGURE 12-17
A parallel *RLC* circuit.

and we identify the conductance and susceptance as

$$G = \frac{1}{R} \quad \text{and} \quad B = \left(\omega C - \frac{1}{\omega L} \right) \tag{12-25}$$

The conductance is always positive and is contributed by the resistance. The susceptance comes from the inductor and capacitor and can be either positive or negative. In fact the susceptance vanishes at the frequency

$$\omega_0 = \frac{1}{\sqrt{LC}} \tag{12-26}$$

Thus the frequency at which *either* the reactance *or* the susceptance is zero is called the resonant frequency.

In these two simple cases it is easy to identify the resistance and reactance or conductance and susceptance with particular circuit elements. Our next example shows that the dependence can be a bit more complicated.

Example 12-11

Find the equivalent input impedance of the circuit of Figure 12-18.

The input impedance of the circuit can be found by first combining the resistor and capacitor into a single equivalent impedance.

$$Z_{RC} = \frac{Z_R Z_C}{Z_R + Z_C} = \frac{R/j\omega C}{R + 1/j\omega C} = \frac{R}{1 + j\omega RC}$$

This result can be put in the $R + jX$ form by multiplying and dividing by the conjugate of the denominator:

$$Z_{RC} = \frac{R}{(1 + j\omega RC)} \frac{(1 - j\omega RC)}{(1 - j\omega RC)}$$

$$= \frac{R}{1 + (\omega RC)^2} - j \frac{\omega R^2 C}{1 + (\omega RC)^2}$$

This impedance can now be combined with the inductor to produce the input impedance as

$$Z = Z_L + Z_{RC}$$

$$= \underbrace{\frac{R}{1 + (\omega RC)^2}}_{R} + j \underbrace{\left[\omega L - \frac{\omega R^2 C}{1 + (\omega RC)^2} \right]}_{X}$$

The input resistance is always positive but in this circuit depends on the resistor, the capacitor, and the frequency. The reactance depends on all three elements and the frequency, and can be either positive, negative, or zero. The resonant frequency can be found by setting the reactance to zero and solving for ω_0 as

$$\omega_0 L = \frac{\omega_0 R^2 C}{1 + (\omega_0 RC)^2}$$

or

$$\omega_0 = \sqrt{\frac{1}{LC} - \frac{1}{(RC)^2}}$$

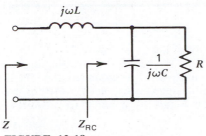

FIGURE 12-18
Circuit for Example 12-11.

The real and imaginary parts of an impedance or admittance generally depend on frequency. As a consequence the circuit will respond to different input frequencies in different ways. In particular, when the input frequency corresponds to a resonant frequency, one at which the imaginary part is zero, the circuit acts as if it were made up of pure resistances. For the reactances to cancel to produce resonance, the circuit must have at least one inductor and one capacitor. As a consequence passive RC or RL circuits cannot exhibit resonance.

In addition to the resonant frequencies, we are often interested in the circuit's response at zero and infinite frequencies. Table 12-1 contains a summary of the impedances and admittances of the circuit elements. First note that the impedance of the resistor does not depend on frequency. However, the other two elements have zero impedance or admittance at zero and infinite frequencies. Now since $\mathbf{V} = Z\mathbf{I}$, zero impedance means $\mathbf{V} = 0$ regardless of $\mathbf{I}$. But this condition describes a short circuit. Thus the inductor acts as a short circuit at zero frequency and the capacitor as a short circuit at infinite frequency.

Conversely, since $\mathbf{I} = Y\mathbf{V}$, zero admittance means the $\mathbf{I} = 0$ regardless of $\mathbf{V}$. But $\mathbf{I} = 0$ for any $\mathbf{V}$ describes an open circuit. Hence the capacitor acts as an open circuit at zero frequency and the inductor at infinite frequency. These observations are summarized in Table 12-2 (page 555). These results are very important because capacitors and inductors are often used in circuits to short-circuit or block either very high or very low frequencies. To study a circuit at zero or infinite frequency we can replace the capacitors and inductors by short or open circuits and analyze the remaining memoryless circuit. In this way we can begin to see how to obtain circuits that are frequency selective.

	At $\omega = 0$		At $\omega = \infty$	
Resistor	$Z_R = R$	$Y_R = G$	$Z_R = R$	$Y_R = G$
Inductor	$Z_L = 0$	$Y_L = \infty$	$Z_L = \infty$	$Y_L = 0$
Capacitor	$Z_C = \infty$	$Y_C = 0$	$Z_C = 0$	$Y_C = \infty$

TABLE 12-1

Example 12-12

Analyze the low and high frequency behavior of the OP AMP circuit of Figure 12-19a.

We see that the OP AMP circuit contains a capacitor. At zero frequency the capacitor acts as an open circuit and we obtain the equivalent circuit in Figure 12-19b. This is the familiar inverting amplifier configuration whose gain is

$$\text{Gain} = -\frac{R_2}{R_1} \quad \text{at } \omega = 0$$

At infinite (very high) frequency the capacitor acts as a short circuit and we obtain the equivalent circuit in Figure 12-19c. Here the output is short-circuited to the inverting input. Since the noninverting input is grounded, the inverting input voltage must be zero by the OP AMP constraint $V_N = V_P$. In other words, at infinite frequency the output will be zero, and thus

$$\text{Gain} = 0 \quad \text{at } \omega = \infty$$

From this it seems reasonable to suggest that the OP AMP circuit will amplify low frequencies and attenuate high frequencies. But in any case we can see that the circuit response will be frequency dependent.

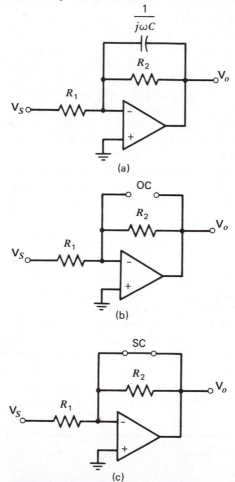

(a)

(b)

(c)

FIGURE 12-19
OP AMP circuit and its equivalents under certain assumptions for Example 12-12. (a) OP AMP circuit. (b) Equivalent circuit at $\omega = 0$. (c) Equivalent circuit at $\omega = \infty$.

	Behavior at $\omega = 0$	Behavior at $\omega = \infty$
Inductor	Short circuit	Open circuit
Capacitor	Open circuit	Short circuit

TABLE 12-2

12-5 POWER IN THE SINUSOIDAL STEADY STATE

To investigate power in the sinusoidal steady state we must return to the time domain since power, as we shall see, cannot be represented as a phasor. If we write the voltage and current at the input to a circuit as

$$v(t) = V_o \cos (\omega t + \phi_V)$$
$$i(t) = I_o \cos (\omega t + \phi_I) \qquad \text{(12-27)}$$

Then the power delivered at the input is

$$p(t) = [v(t)][i(t)]$$
$$= V_o I_o \cos (\omega t + \phi_V) \cos (\omega t + \phi_I)$$

Using the trigonometric identity,

$$2 \cos x \cos y = \cos (x - y) + \cos (x + y)$$

we can write the power as

$$p(t) = \underbrace{\frac{V_o I_o}{2} \cos (\phi_V - \phi_I)}_{\text{Constant term}} + \underbrace{\frac{V_o I_o}{2} \cos (2\omega t + \phi_V + \phi_I)}_{\text{Double-frequency sinusoid}} \qquad \text{(12-28)}$$

The instantaneous power in the sinusoidal steady state turns out to be the sum of two terms. The first term is a constant and the second is a sinusoid whose frequency is twice that of the current and voltage. In dealing with power we are usually interested in its average value. In Chapter 6 we studied the average value of various waveforms. For the sinusoid we found the average value to be zero because the positive area under the waveform is exactly canceled by the negative area. Consequently the double-frequency term in Eq. 12-28 makes zero contribution to the average power and therefore

$$P_{\text{AVE}} = \frac{V_o I_o}{2} \cos (\phi_V - \phi_I) \qquad \text{(12-29)}$$

The average power is proportional to the cosine of the difference between the voltage and current phase angles. It is convenient, as well as traditional, to represent this phase difference as the angle

$$\theta = \phi_V - \phi_I \qquad \text{(12-30)}$$

and then to write Eq. 12-29 as

$$P_{\text{AVE}} = \frac{V_o I_o}{2} \cos \theta \tag{12-31}$$

When written in this form, the term $\cos \theta$ is called the **power factor.**

It is also useful to relate the average power to the input impedance. We can express the phasors of the input current and voltage in Eq. 12-27 as

$$\mathbf{V} = V_o\, e^{j\phi_V} \quad \text{and} \quad \mathbf{I} = I_o\, e^{j\phi_I}$$

These two phasors are related by an input impedance, which we write in polar form as

$$Z = |Z|\, e^{j\phi_Z}$$

But since $\mathbf{V} = \mathbf{ZI}$ we can write

$$\frac{\mathbf{V}}{\mathbf{I}} = \frac{V_o\, e^{j\phi_V}}{I_o\, e^{j\phi_I}} = \frac{V_o}{I_o}\, e^{j(\phi_V - \phi_I)} = |Z|\, e^{j\phi_Z}$$

and we find that the difference between the voltage and current phase angles equals the angle of the impedance. In other words,

$$\theta = \phi_Z$$

But writing the impedance in rectangular form as

$$Z = R + jX$$

we see that the cosine of ϕ_Z, and hence the power factor, is

$$\cos \phi_Z = \cos \theta = \frac{R}{|Z|}$$

and Eq. 12-31 becomes

$$P_{\text{AVE}} = \frac{V_o I_o}{2} \frac{R}{|Z|}$$

But $V_o = |Z|\, I_o$ and so we obtain the result we seek as

$$P_{\text{AVE}} = \frac{I_o^2}{2} R \tag{12-32}$$

Equation 12-32 is another way to determine average power and it points out that average power is proportional to the input resistance. If we apply this idea to the individual circuit elements, we see that the resistor absorbs power, but that the inductor and capacitor cannot absorb power since they are purely reactive and have no resistance. This does not mean that the instantaneous power drawn by the reactive elements is zero since, as Eq. 12-28 points out, there is a double-frequency sinusoidal term. Thus the reactive elements extract power from the circuit on one-half of the cycle but promptly return exactly the same amount of power on the next half-cycle, thus drawing no power on the average.

A final comment is that in Chapter 6 we developed the concept of the rms value of a waveform as a measure of its ability to carry energy. Specifically we found that the peak value and rms value of the sinusoid are related as

$$I_{\text{RMS}} = \frac{I_{\text{PEAK}}}{\sqrt{2}}$$

Now I_o in Eq. 12-32 is a peak value, a fact that can be verified by looking back at Eq. 12-27. Hence yet another way to express average power is

$$P_{\text{AVE}} = I_{\text{RMS}}^2 \, R \qquad \text{(12-33)}$$

This form looks exactly like the expression we derived in Chapter 2 for the resistor. This is the reason that most ac measuring instruments are calibrated in rms value since it gives a direct indication of the average power carried by the signal.

Example 12-13

In Example 12-7 we found the input voltage, current, and impedance of the circuit in Figure 12-11 to be

$$\mathbf{V} = 100 \, e^{j0°} \quad \mathbf{I} = 0.2 \, e^{j53.1°} \quad \mathbf{Z} = 300 - j400$$

Calculate the average input power.

To find the average input power we can use the voltage, current, and power factor as

$$P_{\text{AVE}} = \frac{V_o I_o}{2} \cos \theta = \frac{(100)(0.2)}{2} \cos (53.1°) = 6 \text{ W}$$

Alternatively we can find the average power from the input current and resistance:

$$P_{\text{AVE}} = \frac{I_o^2}{2} R = \frac{(0.2)^2}{2} 300 = 6 \text{ W}$$

An important illustration of the use of average power is the maximum power transfer theorem. The general situation is shown in Figure 12-20a. We have a fixed-source circuit consisting of linear elements and sources, and an adjustable load circuit containing linear elements but not sources. The problem is to select the load that will extract the maximum average power from the fixed source. Our first step is to convert both circuits to their Thévenin equivalents as shown in Figure 12-20b. In terms of the parameters shown, the quantities $\mathbf{V}_T$, R_T, and X_T are fixed, and we are to select R_L and X_L so that maximum average power is transferred across the interface.

We can write the average power using Eq. 12-32 as

$$P_{\text{AVE}} = \frac{|\mathbf{I}|^2}{2} R_L$$

But the magnitude of the phasor current is

$$|\mathbf{I}| = \frac{|\mathbf{V}_T|}{|Z_T + Z_L|}$$

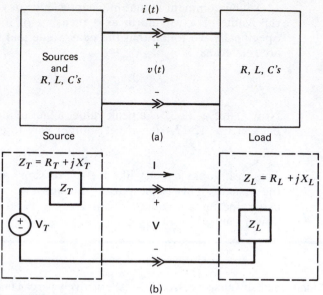

FIGURE 12-20
Maximum power transfer. (*a*) Source–load interface. (*b*) Equivalent phasor circuit.

Hence the power is

$$P_{\text{AVE}} = \frac{1/2 \, |\mathbf{V}_T|^2 \, R_L}{(R_T + R_L)^2 + (X_T + X_L)^2} \qquad \text{(12-34)}$$

Clearly we will minimize the denominator (thus maximize the power) if we set

$$X_L = - X_T$$

and we can do this because reactances can be both positive and negative. When this is done, Eq. 12-34 reduces to

$$P_{\text{AVE}} = \frac{1}{2} \frac{|\mathbf{V}_T|^2 \, R_L}{(R_T + R_L)^2}$$

But this is the same equation we encountered in Chapter 3 in dealing with maximum power transfer in resistance circuits. We know it is maximized if $R_L = R_T$. In sum, to obtain maximum power transfer we must set

$$X_L = - X_T \qquad \text{and} \qquad R_L = R_T$$

These conditions can be compactly expressed as

$$Z_L = Z_T{}^* \qquad \text{(12-35)}$$

That is, the load impedance must be the conjugate of the source impedance. For this reason the maximum power transfer condition is called **conjugate matching.**

Conjugate matching requires that we have a resonant circuit. The reactances cancel (the condition for resonance) and we then match the real parts. When the conjugate matching conditions are inserting into Eq. 12-34 we find that the maximum available power is

$$P_{\text{AVE MAX}} = \frac{|\mathbf{V}_T|^2}{8R_T} \qquad (12\text{-}36)$$

Example 12-13

For the circuit in Figure 12-21a determine the actual power delivered to the load and the maximum available power, and design a load to extract the maximum power from the source.

The phasor Thévenin equivalent is shown in Figure 12-21b. The maximum available power is

$$P_{\text{AVE MAX}} = \frac{|\mathbf{V}_T|^2}{8R_T} = \frac{5}{(8)(40)} = 15.6 \text{ mW}$$

To find the delivered power we first find the interface current as

$$\mathbf{I} = \frac{\mathbf{V}_T}{Z_T + Z_L} = \frac{\sqrt{5}\, e^{-j63.4°}}{40 - j80 + 200} = 8.84\, e^{-j45°} \text{ mA}$$

Hence the delivered power is

$$P_L = \frac{1}{2}|\mathbf{I}|^2 R_L$$

$$= 0.5(8.84 \times 10^{-3})^2(200) = 7.81 \text{ mW}$$

Since $Z_T = 40 - j80$ we need a load $Z_L = 40 + j80$ to extract the maximum available power. This means

$$R_L = 40 \qquad \text{and} \qquad X_L = 80$$

Since the required reactance is positive we can use an inductor,

$$\omega L = 80 \qquad \text{or} \qquad L = 80/10^6 = 80 \ \mu\text{H}$$

$v_S = 5 \cos 10^6 t$
$R = 200 \ \Omega$
$R_L = 200 \ \Omega$
$C = 0.01 \ \mu\text{F}$

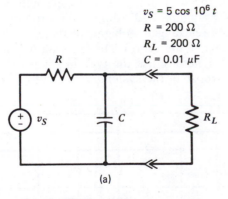

(a)

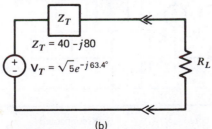

(b)

FIGURE 12-21
Circuits for Example 12-13. (a) Given circuit. (b) Phasor Thévenin equivalent.

12-6   THREE-PHASE POWER

One of the major applications of phasor analysis is in the study of ac electric power systems. Although the loads on power systems vary during any one day, these variations are extremely slow compared with the period of the sinusoid involved. Consequently the analysis of a power system can be carried out using steady-state concepts and the phasor. Historically speaking, it was the analysis of ac power equipment that led Steinmetz to propose the phasor as an analysis tool.

The predominant form for generating and distributing electric power is the three-phase system shown in Figure 12-22. The system has three lines (A, B, C), which are used to transmit power from the source to the loads. The figure also shows a fourth line, labeled N, for neutral. In a balanced system, the only type we consider, this fourth line carries no current or power, but does serve as a reference for defining voltages. The three-phase source in the illustration is modeled as three independent sources, although a three-phase generator is a single piece of equipment with three separate windings. Similarly the loads are modeled as three separate impedances, although a three-phase load may be housed within a single container.

The terminology Y-connected and Δ-connected refers to the two different ways the source and loads can be electrically arranged. The significance of the terminology is not especially apparent from examining Figure 12-22. Figure 12-23 shows the same electrical connections with the elements rearranged to highlight their "Y" and "Δ" natures (the Δ is upside down in the figure). The two diagrams are electrically equivalent but we will generally use the type in Figure 12-22 because it highlights the basic purpose of the system, which is to transmit power from source

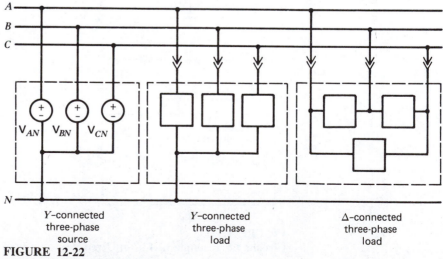

Y–connected
three-phase
source

Y–connected
three-phase
load

Δ–connected
three-phase
load

FIGURE 12-22
A three-phase power system.

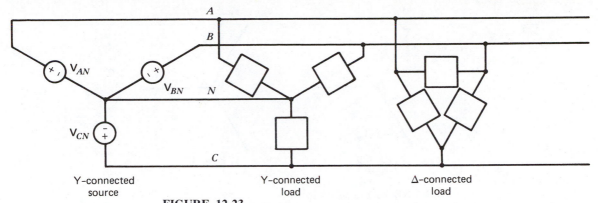

FIGURE 12-23
Three-phase power system rearranged to highlight the Y and Δ terminology.

to load interfaces via three lines. All the student need remember is that in a Y connection the three devices are connected from line to neutral, while devices in the Δ connection are wired from line to line. In almost all systems the source is Y connected while the loads can be either Y or Δ, although Δ is more common.

To identify voltages in the system we use a special double-subscript notation. The reason, other than tradition, is that there are at least six voltages to deal with; the three line-to-line voltages and the three line-to-neutral voltages. If we were to use the usual plus and minus reference marks to define all of these voltages, our circuit diagram would be cluttered and confusing. Consequently we use two subscripts to define the points across which the voltage is defined. For example, $\mathbf{V}_{AN}$ means the voltage of line *A* with respect to neutral (*N*), with the implied plus reference mark at *A* and the implied minus at *N*. The three line-to-neutral voltages, called the **phase voltages,** are identified as $\mathbf{V}_{AN}$, $\mathbf{V}_{BN}$, and $\mathbf{V}_{CN}$. The line-to-line voltages, called simply the **line voltages,** are identified as $\mathbf{V}_{AB}$, $\mathbf{V}_{BC}$, and $\mathbf{V}_{CA}$. In this notation it follows that $\mathbf{V}_{BN} = -\mathbf{V}_{NB}$, and consequently $\mathbf{V}_{AB} = \mathbf{V}_{AN} + \mathbf{V}_{NB} = \mathbf{V}_{AN} - \mathbf{V}_{BN}$.

A balanced three-phase source produces phase voltages that obey the following two constraints

$$|\mathbf{V}_{AN}| + |\mathbf{V}_{BN}| + |\mathbf{V}_{CN}| = V_P \tag{12-37}$$
$$\mathbf{V}_{AN} + \mathbf{V}_{BN} + \mathbf{V}_{CN} = 0$$

That is, the phase voltages all have the same amplitude (V_P), and they sum to zero. There are two possible arrangements that satisfy these constraints:

Positive sequence	Negative sequence	
$\mathbf{V}_{AN} = V_P \angle 0°$	$\mathbf{V}_{AN} = V_P \angle 0°$	
$\mathbf{V}_{BN} = V_P \angle -120°$	$\mathbf{V}_{BN} = V_P \angle -240°$	(12-38)
$\mathbf{V}_{CN} = V_P \angle -240°$	$\mathbf{V}_{CN} = V_P \angle -120°$	

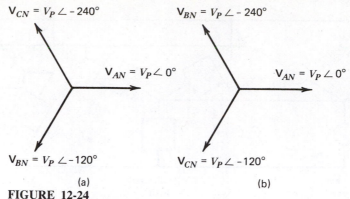

$\mathbf{V}_{CN} = V_P \angle -240°$ $\mathbf{V}_{BN} = V_P \angle -240°$

$\mathbf{V}_{AN} = V_P \angle 0°$ $\mathbf{V}_{AN} = V_P \angle 0°$

$\mathbf{V}_{BN} = V_P \angle -120°$ $\mathbf{V}_{CN} = V_P \angle -120°$

(a)   (b)

FIGURE 12-24
The two possible phase sequences. (*a*) Positive. (*b*) Negative.

These two cases are called the positive phase sequence and the negative phase sequence respectively, and Figure 12-24 shows the corresponding phasor diagrams. It is apparent that either case involves three phasors of equal length, which are all separated by a phase angle of 120 degrees, so that the sum of any two exactly cancels the third. It is also apparent that we can convert one phase sequence into the other by simply interchanging the labels on lines *B* and *C*. Thus there is no conceptual difference between the two sequences. In what follows we always use the positive phase sequence.

The reader is cautioned that although there is no conceptual difference, this is not the same as saying that the phase sequence is unimportant. It turns out that three-phase motors run in *one* direction when the positive sequence is applied, and in the *opposite* direction for the negative sequence. This can be a matter of considerable importance if the motor is driving, say, a conveyor belt. In practice, then, it is essential that there be no confusion about which is line *A, B, or C,* and what the phase sequence is.

The relationship between the line voltages and the phase voltages can be obtained by manipulating the phase voltage phasors:

$$\begin{aligned}
\mathbf{V}_{AB} &= \mathbf{V}_{AN} - \mathbf{V}_{BN} \\
&= V_P \angle 0° - V_P \angle -120° \\
&= V_P + j0 - (-0.5V_P - j\sqrt{3}\,V_P/2) \\
&= V_P(1.5 + j\sqrt{3}/2) \\
&= \sqrt{3}\,V_P \angle 30°
\end{aligned} \tag{12-39}$$

By a similar process the other line voltages are

$$\begin{aligned}
\mathbf{V}_{BC} &= \sqrt{3}\,V_P \angle -90° \\
\mathbf{V}_{CA} &= \sqrt{3}\,V_P \angle -210°
\end{aligned} \tag{12-40}$$

Figure 12-25 shows the phasor diagram of these results. The line voltages are all of equal amplitude and displaced by 120 degrees, and hence obey

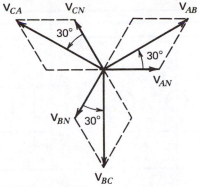

FIGURE 12-25
Phasor diagram showing the relationships
between phase voltages and line voltages.

the same constraints as the phase voltages. If we denote the amplitude of the line voltages as V_L, then

$$V_L = \sqrt{3}\, V_P \qquad (12\text{-}41)$$

Thus the line voltage amplitude is $\sqrt{3}$ greater than the phase voltage amplitude. This ratio appears in descriptions such as 120/208 V three phase. The 120 is the phase voltage and 208 is the line voltage.

The voltages are also normally expressed in terms of rms values rather than peak values. For sinusoids the relationship between peak and rms is $\sqrt{2}$, so that the equation for the average power can be written as

$$P = \frac{|V|}{\sqrt{2}}\frac{|I|}{\sqrt{2}}\cos\theta = V_{\text{RMS}}\, I_{\text{RMS}}\cos\theta \qquad (12\text{-}42)$$

Since we will make frequent use of this equation to determine power, it is more convenient to use the rms value. In dealing with three-phase power systems, phasor magnitudes always represent the rms value of a current or voltage rather than the peak value.

The phase reference for all voltages and currents is taken to be the line A phase voltage, that is, $\mathbf{V}_{AN} = V_P \angle 0°$. This practice can be seen in Figures 12-24 and 12-25, and will be used consistently here. If we report that a current is $14.7 \angle -137°$, this means that the amplitude of the current is 14.7 A rms and it is displaced from the line A phase voltage by -137 degrees.

We now turn to the analysis of a balanced three-phase circuit shown in Figure 12-26. The load is balanced; that is, the impedances in the legs of the Y-connected load are equal. Our goal is to determine the source–load interaction at the interface in terms of the three line currents $\mathbf{I}_A$, $\mathbf{I}_B$, and $\mathbf{I}_C$, and the total power delivered to the load.

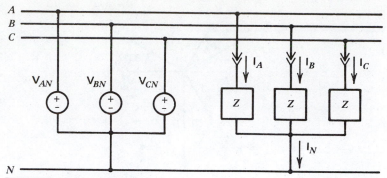

FIGURE 12-26
Y-connected source and load.

$$\mathbf{I}_A = \frac{\mathbf{V}_{AN}}{Z} = \frac{V_P \angle 0°}{|Z| \angle \theta} = \frac{V_P}{|Z|} \angle - \theta$$

$$\mathbf{I}_B = \frac{\mathbf{V}_{BN}}{Z} = \frac{V_P \angle - 120°}{|Z| \angle \theta} = \frac{V_P}{|Z|} \angle - 120° - \theta \qquad \textbf{(12-43)}$$

$$\mathbf{I}_C = \frac{\mathbf{V}_{CN}}{Z} = \frac{V_P \angle - 240°}{|Z| \angle \theta} = \frac{V_P}{|Z|} \angle - 240° - \theta$$

Figure 12-27 shows the phasor diagram relating the line currents and phase voltages. The three line currents are of equal amplitude and symmetrically disposed at 120-degree intervals. This means that $\mathbf{I}_A + \mathbf{I}_B + \mathbf{I}_C = 0$. If we apply KCL at the neutral connection of the load in Figure 12-26, we conclude that $\mathbf{I}_N = 0$. In a balanced three-phase system no

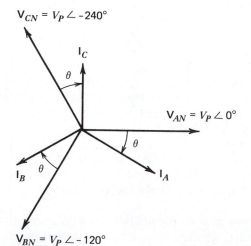

FIGURE 12-27
Phasor diagram relating line currents and phase voltages in a balanced three-phase system.

current flows in the neutral wire. The neutral wire could be replaced by any impedance, including infinity. In other words, the neutral wire can be disconnected without upsetting the system balance. In real systems the neutral wire may or may not be present. However, in working problems it is usually helpful to draw the neutral line since it serves as the reference for the phase voltages.

Example 12-14

For a balanced Y load with $Z = 10 + j10$, determine the line currents and total load power for a line voltage of 220 V_{RMS}.

Since we are given the line voltage we must first find the phase voltage:

$$V_P = \frac{V_L}{\sqrt{3}} = 127 \text{ V}$$

We can now write

$$I_A = \frac{127 \angle 0°}{10 + j10} = 8.98 \angle -45°$$

$$I_B = \frac{127 \angle -120°}{10 + j10} = 8.98 \angle -165°$$

$$I_C = \frac{127 \angle -240°}{10 + j10} = 8.98 \angle -285°$$

The power delivered to phase A is

$$\begin{aligned} P_A &= |V_{AN}| |I_{AN}| \cos \theta \\ &= (127)(8.98) \cos(-45°) \\ &= 806 \text{ W} \end{aligned}$$

But, since the system is balanced, each phase draws the same power and

$$P_T = 3P_A = 2.42 \text{ kW}$$

We now turn to the more common balanced Δ load shown in Figure 12-28. Our objective remains to determine the interface line currents and the total load power. But since the impedances here are connected from line to line, we must go through the intermediate step of first determining the phase currents I_1, I_2, and I_3 from the phase impedance and the line voltages.

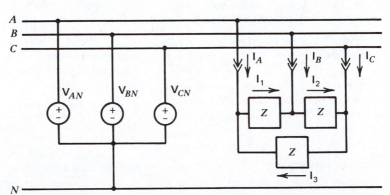

FIGURE 12-28
Y-connected source and a Δ-connected load.

$$\mathbf{I}_1 = \frac{\mathbf{V}_{AB}}{Z} = \frac{V_L \angle 30°}{Z} = \frac{0.866V_L + j0.5V_L}{Z}$$

$$\mathbf{I}_2 = \frac{\mathbf{V}_{BC}}{Z} = \frac{V_L \angle -90°}{Z} = \frac{-jV_L}{Z} \qquad \text{(12-44)}$$

$$\mathbf{I}_3 = \frac{\mathbf{V}_{CA}}{Z} = \frac{V_L \angle -210°}{Z} = \frac{-0.866V_L + j0.5V_L}{Z}$$

The phase currents are of equal amplitude (the system is balanced) and disposed at 120-degree intervals. Armed with these results, we now apply KCL at the connection points of the impedance to obtain the line currents.

$$\mathbf{I}_A = \mathbf{I}_1 - \mathbf{I}_3 = \frac{\sqrt{3}\,V_L}{Z} = \frac{\sqrt{3}\,V_L}{|Z|} \angle -\theta$$

$$\mathbf{I}_B = \mathbf{I}_2 - \mathbf{I}_1 = \frac{-0.866V_L - j1.5V_L}{Z} = \frac{\sqrt{3}\,V_L}{|Z|} \angle -120° - \theta \qquad \text{(12-45)}$$

$$\mathbf{I}_C = \mathbf{I}_3 - \mathbf{I}_2 = \frac{-0.866V_L + j1.5V_L}{Z} = \frac{\sqrt{3}\,V_L}{|Z|} \angle -240° - \theta$$

The power factor angle of the line currents is the angle of the impedance Z. This was true for the Y load (see Eq. 12-43) because the impedances were connected from line to neutral. But it is also true here even though the impedances are connected from line to line. The power factor of the load is determined by its impedance and it matters not whether the load is Y or Δ connected. Finally, by comparing Eqs. 12-44 and 12-45, we find that

$$I_L = \sqrt{3}\,I_P \qquad \text{(12-46)}$$

The line current amplitude in a Δ load is $\sqrt{3}$ greater than the phase current.

Example 12-15

Given a balanced Δ-connected load with $Z = 100 + j20$ and an rms line voltage of 3 kV, determine the line currents and the total power to the load.

We must begin by finding the phase currents.

$$\mathbf{I}_1 = \frac{\mathbf{V}_{AB}}{Z} = \frac{3000 \angle 30°}{100 + j20} = 29.4 \angle 18.7°$$
$$= 27.8 + j9.43$$

Since the system is balanced, we know that the other phase currents are of equal amplitude and displaced at 120-degree intervals. Hence we can simply write

$$\mathbf{I}_2 = 29.4 \angle 18.7° - 120° = -5.76 - j28.8$$
$$\mathbf{I}_3 = 29.4 \angle 18.7° - 240° = -22.1 + j19.4$$

We can find the line currents as

$$\mathbf{I}_A = \mathbf{I}_1 - \mathbf{I}_3 = 49.9 - j9.97 = 50.9 \angle -11.3°$$

Again we need not calculate the other line currents explicitly (except possibly as a check on our work)

since they must also have equal amplitudes and 120-degree displacements.

$$\mathbf{I}_B = 50.9 \angle -11.3° - 120° = 50.9 \angle -131.3°$$
$$\mathbf{I}_C = 50.9 \angle -11.3° - 240° = 50.9 \angle -251.3°$$

We now can determine the load power from the phase A line current and phase voltage. Since $V_L = \sqrt{3}\,V_P$, we write

$$P_A = V_P I_L \cos\theta$$
$$= \frac{3000}{\sqrt{3}}(50.9)\cos(-11.3°)$$
$$= 86.5 \text{ kW}$$

And the total load power is

$$P_T = 3P_A = 259 \text{ kW}$$

Actually we could have determined the power after the first step in our calculation since the phase current

I_1 and the line voltage V_{AB} allow us to determine the phase power as

$$P_1 = |V_{AB}||I_1| \cos \theta$$
$$= (3000)(29.4) \cos (30° - 18.7°)$$

$$= 86.5 \text{ kW}$$

which, of course, agrees with the result using the line current and phase voltage.

Three-phase loads are often described in terms of the power they draw at the interface, rather than their impedances. In such cases we still have the goal of determining the line currents. For a given total power P_T and power factor ($\cos \theta$), in a balanced system we can obtain the amplitude of the line current from the phase–power relationship as

$$I_L = \frac{P_T/3}{V_P \cos \theta} \tag{12-47}$$

This yields the amplitude of the line current but not its phase angle. The phase angle is provided by specifying that the power factor is *lagging* or *leading*. This means that the phase A line current is

$$I_A = I_L \angle \pm \theta \tag{12-48}$$

where the plus sign applies for leading power factors and the minus for lagging.

Example 12-16

A balanced three-phase source with an rms line voltage of 1500 V delivers 30 kW to a balanced Y load at a power factor of 0.8 lagging. Determine the line currents and phase impedances.

We first determine the phase voltage as

$$V_P = \frac{V_L}{\sqrt{3}} = 866 \text{ V}$$

and the magnitude of the line current from

$$I_L = \frac{30,000/3}{(866)(0.8)} = 14.4 \text{ A}$$

The power factor is given as lagging, and since cos (36.9°) = 0.8, we have the phase A line current:

$$I_A = 14.4 \angle -36.9°$$

The negative phase angle means that the phase A line current lags the corresponding phase voltage V_{AN}. The remaining line currents are obtained simply by subtracting increments of 120° from this result.

$$I_B = 14.4 \angle -156.9°$$
$$I_C = 14.4 \angle -276.9°$$

We can also determine the load impedance if need be. Since the load is Y connected

$$Z = \frac{V_{AN}}{I_A} = \frac{866 \angle 0°}{14.4 \angle -36.9°} = 60.1 \angle 36.9°$$

$$= 48.1 + j36.1$$

These examples illustrate common analysis methods used to determine the interface conditions in three-phase systems. An interesting interfacing problem arises when there is an impedance between the source and the load as illustrated in Figure 12-29. This impedance, called the wire impedance, represents the transmission system between the source and load.

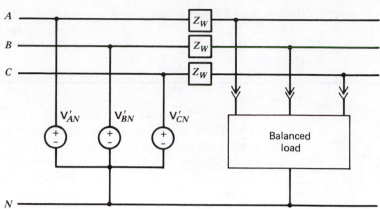

FIGURE 12-29
Three-phase system with wire impedances between source and load.

The constraint is that the source (power company) has agreed to deliver a given amount of power at a specified voltage level at the load (customer) interface. The problem is to determine what the source should produce to meet this interface constraint. Notice that this is not a maximum power-transfer situation. Here the load is fixed and the source is adjusted to meet conditions at the interface. In a maximum power-transfer situation, the source is fixed and the load is adjusted to extract maximum power. The two situations are completely different and "matching" has absolutely nothing to do with the power company's interfacing problem. If it could, the power company would reduce the wire impedances to zero. However, the best that can be done is to make the wire impedances all equal so that the system remains balanced.

Given the load constraint at the interface, we can determine the line currents as in Example 12-16. With these data and the specified voltage level, we work backward to the source to determine the required source voltages, that is, the phase voltages indicated as $\mathbf{V}'_{AN}$, $\mathbf{V}'_{BN}$, and $\mathbf{V}'_{CN}$ in Figure 12-29. For example, for phase A

$$\mathbf{V}'_{AN} = \mathbf{V}_{AN} + \mathbf{I}_A Z_W \qquad (12\text{-}49)$$

Once we have the source voltages, we can determine the total power the source must produce to supply both the interface load and any losses incurred in the transmission system. Our final example illustrates the process.

Example 12-17

Suppose that the load (30 kW at $V_L = 1500$ V) specified in Example 12-16 is connected to a source through wire impedances of $2 + j20$. Determine the required source voltage and the total power produced by the source.

In Example 12-16 we found that

$$\mathbf{V}_{AN} = 866 \angle 0° \qquad \text{and} \qquad \mathbf{I}_A = 14.4 \angle -36.9°$$

Hence the required phase A source voltage is

$V'_{AN} = 866 + j0 + (14.4 \angle -36.9°) (2 + j20)$
$= 1080 \angle 11.3°$

The angle between the phase voltage and line current at the source is $11.3° + 36.9° = 48.2°$ because of the intervening wire impedance. The total power that must be produced by the source is then

$P_T = 3P'_A = 3[(1080)(14.4) \cos (48.2)]$
$= 31.1 \text{ kW}$

The interface load requires 30 kW. The increase required of the source represents the losses in the transmission system wire resistance.

If we now look back at Eq. 12-47, we see that for a fixed total power and voltage level (the interface constraints), the line currents are inversely proportional to the power factor ($\cos \theta$). If the power factor is reduced, the line currents increase and greater losses are incurred in the wire impedances. For this reason power companies generally charge large industrial customers a premium rate for low-power-factor loads to reflect the true cost of meeting the interface constraint.

SUMMARY

- The sinusoidal steady-state response of a linear, stable circuit is the forced response due to a sinusoidal input.
- The sinusoidal steady-state response of a linear circuit with a transfer function $H(s)$ and an input $x(t) = X_o \cos(\omega t + \phi)$ is

$$y_F(t) = X_o |H(j\omega)| \cos[\omega t + \phi + \angle H(j\omega)]$$

That is,

$$\text{Response amplitude} = X_o|H(j\omega)|$$
$$\text{Response phase} = \phi + \angle H(j\omega)$$
$$\text{Response frequency} = \omega$$

- A phasor is a complex number representing a sinusoidal waveform. The waveform

$$x(t) = X_o \cos(\omega t + \phi)$$

is represented in phasor form as

$$\mathbf{X} = X_o e^{j\phi} = X_o \cos\phi + jX_o \sin\phi = a + jb$$

The magnitude and angle of the phasor correspond to the amplitude and phase angle of the sinusoid. The real and imaginary parts correspond to the Fourier coefficients of the sinusoid.

- Phasors provide an easy way to combine sinusoidal waveforms. Two phasors are equal if, and only if, they have the same amplitude and phase angle or the same real and imaginary parts. In the sinusoidal steady state, phasors obey Kirchhoff's laws.
- In the sinusoidal steady state, the element i-v relationships are written in terms of phasors as
Resistor

$$\mathbf{V}_R = Z_R \mathbf{I}_R \qquad Z_R = R \qquad Y_R = \frac{1}{R}$$

Inductor

$$\mathbf{V}_L = Z_L \mathbf{I}_L \qquad Z_L = j\omega L \qquad Y_L = \frac{1}{j\omega L}$$

Capacitor

$$\mathbf{V}_C = Z_C \mathbf{I}_C \qquad Z_C = \frac{1}{j\omega C} \qquad Y_C = j\omega C$$

- Using phasors and element impedances (or admittances), the sinusoidal steady-state response of circuits can be determined in a manner analogous to memoryless resistance circuits. Phasor circuit analysis techniques include circuit reduction, the Thévenin-Norton theorem, superposition, node analysis, and mesh analysis.

- In the sinusoidal steady state the equivalent impedance at a pair of terminals can be written as

$$Z_{EQ} = R(\omega) + jX(\omega)$$

where $R(\omega)$ is called the resistance and $X(\omega)$ is called the reactance. Likewise an equivalent admittance can be written as

$$Y_{EQ} = G(\omega) + jB(\omega)$$

where $G(\omega)$ is called the conductance and $B(\omega)$ the susceptance.

- The nature of Z_{EQ} (or Y_{EQ}) at $\omega = 0$ can be determined directly from the circuit by replacing capacitors by open circuits and inductors by short circuits. Conversely the nature of Z_{EQ} at $\omega = \infty$ can be determined from the circuit by replacing capacitors by short circuits and inductors by open circuits.

- The frequencies at which an impedance or admittance is purely real are called resonant frequencies. That is, if

$$X(\omega_0) = 0 \quad \text{or} \quad B(\omega_0) = 0$$

then ω_0 is called a resonant frequency.

- In the sinusoidal steady state the average power delivered at an interface by the signals

$$v(t) = V_o \cos (\omega t + \phi_V) \quad \text{and} \quad i(t) = I_o \cos (\omega t + \phi_I)$$

is

$$P_{AVE} = \frac{V_o I_o}{2} \cos (\phi_V - \phi_I)$$

- In the sinusoidal steady state the maximum average power will be delivered by a fixed source to an adjustable load if the source and load impedances are conjugates.

- Three-phase systems are commonly used to generate and distribute electric power. Three-phase sources are generally Y connected, and loads are either Y connected or Δ connected.

- In a balanced three-phase system the individual line or phase voltages are of equal magnitude and displaced in increments of 120 degrees. The same condition applies to line or phase currents.

- The total power delivered by a balanced three-phase source to a balanced three-phase load is

$$P_T = 3 \, V_P \, I_L \cos \theta$$

where V_P is the phase voltage, I_L the line current, and $\cos \theta$ the power factor.

EN ROUTE OBJECTIVES
AND RELATED EXERCISES

12-1 PHASORS (SECS. 12-1 and 12-2)

Given two or more sinusoids of the same frequency, convert the waveforms or their sum into phasor form. Conversely, given two or more phasors representing sinusoids of the same frequency, convert the phasors or their sum into waveforms.

Exercises

12-1-1   Convert the following waveforms into phasor form and find $v_1(t) + v_2(t)$ and $i_1(t) + i_2(t)$.
  (a) $v_1(t) = 100 \cos (\omega t - 90°)$
  (b) $v_2(t) = 200 \sin (\omega t) + 50 \cos (\omega t)$
  (c) $i_1(t) = 3 \cos (\omega t) + 4 \sin (\omega t)$
  (d) $i_2(t) = 5 \cos (\omega t - 233°)$

12-1-2   Convert the following phasors into sinusoidal waveforms or the form $a \cos (\omega t) + b \sin (\omega t)$.
  (a) $\mathbf{V}_1 = 10 \, e^{j30°}$
  (b) $\mathbf{V}_2 = 60 \, e^{-j270°}$
  (c) $\mathbf{I}_1 = 5 \, e^{j180°}$
  (d) $\mathbf{I}_2 = 15 \, e^{j70°}$

12-1-3   Convert the following phasors into waveforms and find $v_1(t) + v_2(t)$ and $i_1(t) + i_2(t)$.
  (a) $\mathbf{V}_1 = \dfrac{10 + j10}{2 - j3}$
  (b) $\mathbf{V}_2 = [3 - j8][5 \, e^{-j60°}]$
  (c) $\mathbf{I}_1 = \dfrac{10}{1 + j3}$
  (d) $\mathbf{I}_2 = \dfrac{1 + j3}{1 - j3}$

12-1-4   Given the waveforms
$$v_1(t) = 50 \cos (\omega t - 45°)$$
and
$$v_2(t) = 25 \sin (\omega t)$$
determine the sinusoidal waveform $v_3(t)$ such that $v_1(t) + v_2(t) + v_3(t) = 0$.

12-1-5   The following constraints apply to three phasors:
$$\mathbf{I}_1 + \mathbf{I}_2 - \mathbf{I}_3 = 0$$
and
$$2\mathbf{I}_1 + 3\mathbf{I}_2 + \mathbf{I}_3 = 10 \, e^{j30°}$$

If $i_1(t) = 12 \cos (\omega t + 90°)$, what are $i_2(t)$
and $i_3(t)$?

12-2 *PHASOR CIRCUIT ANALYSIS (SEC. 12-3)*

*Given a linear circuit with one or more sinu-
soidal inputs of the same frequency:*

(a) *Convert the circuit into phasor form.*

(b) *Solve for selected response phasors using circuit reduction, Thévenin's or Norton's theorem, superposition, mesh analysis, or node analysis.*

(c) *Convert the response phasors into response waveforms.*

Exercises

12-2-1 Find the indicated sinusoidal steady-state
response waveforms in each of the circuits
in Figure E12-2-1.

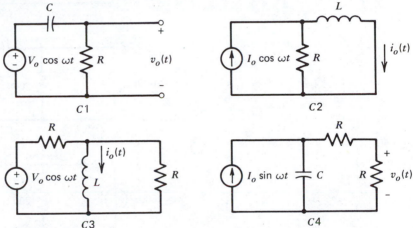

FIGURE E-12-2-1
**Circuits for finding sinusoidal steady-state responses
for Exercise 12-2-1.**

12-2-2 Find the indicated sinusoidal steady-state responses $v_o(t)$ and $i_o(t)$ in each of the circuits in Figure E12-2-2.

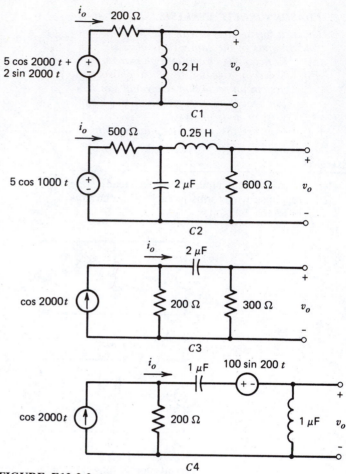

FIGURE E12-2-2
Circuits for Exercises 12-2-2.

12-2-3   Find the phasor response **I** in each of the
circuits in Figure E12-2-3.

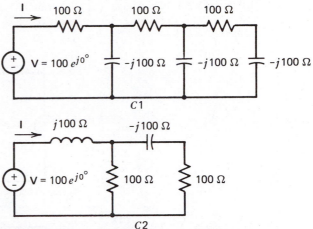

FIGURE 12-2-3
**Circuits for finding the phasor response for Exercise
12-2-3.**

12-2-4   Show that the sinusoidal steady-state output
of the circuit in Figure E12-2-4 will be zero
if $Z_2 Z_3 = Z_1 Z_4$.

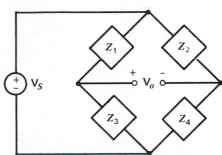

FIGURE E12-2-4
A Wheatstone bridge.

12-2-5 Find the Thévenin equivalent at the indicated interface for each of the circuits in Figure E12-2-5. Then use the Thévenin equivalent to determine the phasor responses **V** and **I**.

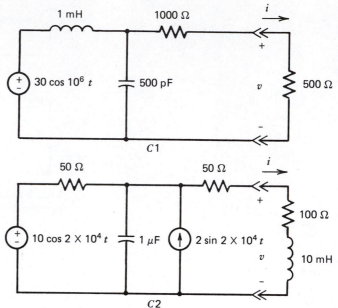

FIGURE 12-2-5
Circuits for Exercise 12-2-5.

12-3 IMPEDANCE (SEC. 12-4)

Given a linear circuit:

(a) *Determine the sinusoidal steady-state impedance (or admittance) at a specified pair of terminals and express the result as*

$$Z_{EQ} = R(\omega) + jX(\omega)$$
or
$$Y_{EQ} = G(\omega) + jB(\omega)$$

(b) *Determine the value of the impedance at selected frequencies including $\omega = 0$ and $\omega = \infty$.*

(c) *Determine the resonant frequencies (if any).*

Exercises

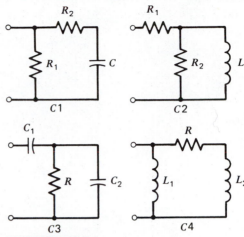

FIGURE E12-3-1
Circuits for Exercise 12-3-1.

12-3-1 Find the equivalent impedance at the indicated terminals for each circuit in Figure E12-3-1 and express the result in the form $R(\omega) + jX(\omega)$. Determine the values of $R(0)$, $R(\infty)$, $X(0)$, and $X(\infty)$. Do these values agree with those obtained directly from the circuit with $\omega = 0$ and $\omega = \infty$? Do any of the circuits have a resonant frequency?

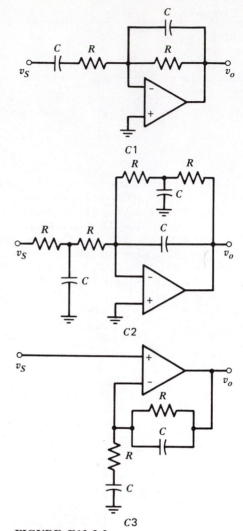

FIGURE E12-3-2
OP AMP circuits.

12-3-2 Determine the gain of each of the OP AMP circuits in Figure E12-3-2 at $\omega = 0$ and $\omega = \infty$.

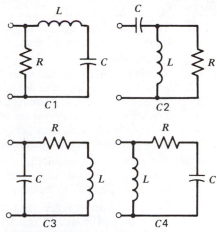

FIGURE E12-3-3
Circuits for Exercise 12-3-3.

12-3-3 Determine the equivalent impedance of each of the circuits in Figure E12-3-3 and express the result in the form $R(\omega) + jX(\omega)$. Determine the resonant frequency ω_0 of each circuit and the value of $R(\omega_0)$.

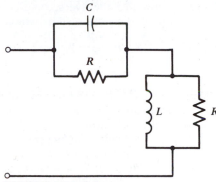

FIGURE E12-3-4
Circuit for duality Exercise 12-3-4.

12-3-4 For the circuit in Figure E12-3-4 show that if $GL = RC$, then $Z_{EQ} = R$ at all frequencies.

12-3-5 Show that if $Z_{EQ} = R + jX$, then

$$Y_{EQ} = \frac{R}{\sqrt{R^2 + X^2}} - j\frac{X}{\sqrt{R^2 + X^2}}$$

12-4 POWER (SEC. 12-5)

Given a linear circuit in the sinusoidal steady state:

(a) *Determine the average power delivered at a specified interface.*

(b) *Determine the maximum available average power at a specified interface.*

(c) *Determine the load impedance required to draw the maximum available average power at a specified interface.*

Exercises

12-4-1 For each of the circuits in Exercise 12-2-3 determine the average power delivered by the source.

12-4-2 For each circuit in Exercise 12-2-5 determine the average power delivered at the indicated interface. Determine the maximum available average power at the interface. Determine the load impedance required to draw the maximum available average power.

12-4-3 For each of the circuits in Exercise 12-2-2 determine the maximum available average power at the indicated output terminals.

12-4-4 For the circuit in Figure E12-4-4 determine the values of R and C that will cause the maximum average power to be delivered to the load circuit.

12-4-5 Show that the average power delivered at an interface can also be written as

$$P_{\text{AVE}} = \text{Re}\{\frac{\mathbf{VI}^*}{2}\}$$

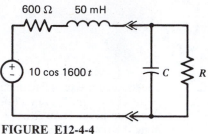

FIGURE E12-4-4
Circuit for Exercise 12-4-4.

12-5 THREE-PHASE POWER (SEC. 12-6)

Given a balanced three-phase system in the sinusoidal steady state:

(a)   Determine the line currents and the total average power given the line or phase voltage and the load impedances.

(b)   Determine the line currents and load impedances given the total average power, the power factor, and the line or phase voltage.

(c)   Determine the generator voltage required to deliver a specified average power at a fixed voltage to a specified load interface.

Exercises

12-5-1 A balanced Y-connected three-phase source and Y-connected load with a per-phase impedance of $10 + j5$ operates with a line voltage of 208 V_{RMS}. Determine the line currents and the total average power delivered to the load.

12-5-2 A balanced Y-connected three-phase source with $V_L = 4000$ V_{RMS} is connected to a balanced Δ-connected load with phase impedances of $100 \angle 30°$. Determine the line currents, the total average power delivered to the load, and the power factor.

12-5-3 A balanced Y-connected three-phase source with a phase voltage of 240 V_{RMS} is connected to a balanced Δ-connected load with phase impedances of $5 - j15$ and a balanced Y-connected load with phase impedances $3 + j4$. Determine the line current, the total average power delivered to the loads, and the overall power factor.

12-5-4 A balanced Y-connected three-phase source delivers a total of 10 kW to a balanced Y-connected load. The line voltage is 440 V_{RMS} and the power factor is 0.75 lagging. Determine the line currents and the phase impedances.

12-5-5 Repeat Exercise 12-5-4 for a balanced Δ-connected load.

12-5-6 The load interface conditions given in Exercise 12-5-4 are delivered by a balanced Y-connected source through balanced wire impedances of $1 + j5$. Determine the source line voltage required and the total power delivered by the source.

12-5-7 A balanced three-phase load absorbs 10 kW at a line voltage of 1 kV with a leading power factor of 0.8. A balanced Y-connected three-phase source delivers the power through balanced wire impedances of $2 + j50$. Determine the source line voltages and total power.

PROBLEMS

P12-1 (Analysis)
Normal ac voltmeters measure only the amplitude of a sinusoid and do not indicate phase angle. By appropriate measurement they can be used to determine both amplitude and phase. Figure P12-1 shows a relay coil with unknown resistance and inductance. With the circuit operating in the sinusoidal steady state at 1 kHz, the following voltmeter measurements are made.

$$|\mathbf{V}_o| = 10 \text{ V} \qquad |\mathbf{V}_1| = 4 \text{ V} \qquad |\mathbf{V}_2| = 8 \text{ V}$$

Determine R and L, and construct a phasor diagram of the three voltages.

P12-2 (Analysis)
The unit output method studied in Chapter 3 can be applied to ladder circuits in the sinusoidal steady state. We assume that the output is a unit phasor $(1 \angle 0°)$, and then by alternate application of the element impedance constraints and Kirchhoff's laws work our way backward to determine the input phasor. We then obtain the transfer function as

$$T(j\omega) = \frac{\text{Output phasor}}{\text{Input phasor}} = \frac{1 \angle 0°}{\text{Input phasor}}$$

Apply this method to the circuit in Figure P12-2 and show that $\angle T(j\omega) = -90$ degrees when $\omega = 1/(\sqrt{5}\ RC)$ and $\angle T(j\omega) = -180$ degrees when $\omega = \sqrt{6}\ /RC$.

P12-3 (Analysis)
The circuit in Figure P12-3 is operating in the sinusoidal steady state at its resonant frequency. Show that the output amplitude is greater than the input amplitude if $L > R^2C$.

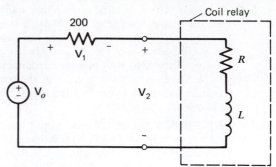

FIGURE P12-1
Coil relay problem.

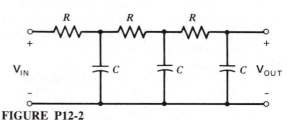

FIGURE P12-2
Circuit for appling unit output method using phasors.

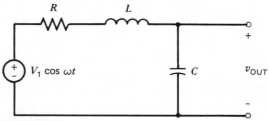

FIGURE P12-3
A resonant *RLC* circuit.

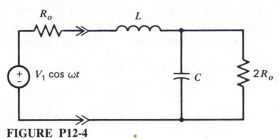

FIGURE P12-4
An impedance matching problem.

P12-4 (Analysis)
Show that the source–load interface in Figure P12-4 is matched if

$$R_oC = \frac{1}{2\omega} \quad \text{and} \quad \frac{R_o}{L} = \omega$$

Then determine the circuit resonant frequency.

P12-5 (Design, Evaluation)
A single-phase load draws 3 kW from a 60-Hz source with a voltage of 220 V_{RMS}. The load power factor is 0.8 lagging. We want to connect a capacitor in parallel with this load so that the combination draws 3 kW with a power factor of unity. Determine the capacitance required. Then explain why this might be a desirable step.

P12-6 (Analysis)
The RL circuit in Figure P12-6 is operating in the sinusoidal steady state. From our previous studies we know that the circuit can be described by its time constant $T_C = L/R$. This problem demonstrates that it can also be described by its "half-power" frequency $\omega_c = R/L$.
(a) Show that at zero frequency the average power delivered to the resistance is

$$P_o = V_1^2/2R$$

(b) Show that the average power delivered at the "half-power" frequency is $P_o/2$.

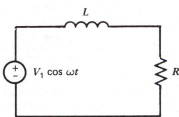

FIGURE P12-6
Circuit for determining the half-power frequency.

P12-7 (Analysis)
A balanced Y-connected three-phase source with a phase voltage of 120 V_{RMS} is connected to an unbalanced Y load. The phase impedances of the load are $Z_A = 100\ \Omega$, $Z_B = 100\ \Omega$, and $Z_C = j100\ \Omega$.
(a) With the neutral wire connected, determine the line currents and the total power delivered to the load.
(b) Repeat part *a* with the neutral wire disconnected.

P12-8   (Analysis)
Figure P12-8 shows a balanced Y-connected load and a balanced Δ-connected load. Show that the two loads are equivalent by showing that the two loads draw the same line currents and average power from a balanced three-phase source.

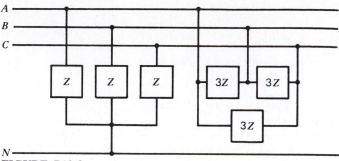

FIGURE P12-8
Circuit for showing equivalence of Y and Δ loads.

P12-9   (Analysis)
Figure P12-9 shows a balanced Δ-connected source with phase B connected to ground. This system is sometimes used in aircraft where the structure serves as the ground return. The system saves weight since only two wires are required to transmit three-phase power. For $V_L = 120\ V_{\mathrm{RMS}}$.
(a)   Determine the line currents (including I_B) and the total power delivered to the load.
(b)   Determine the voltage of nodes Ⓐ and Ⓒ with respect to ground.

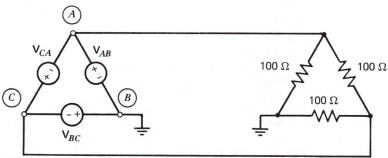

FIGURE P12-9
Aircraft power distribution problem.

Chapter 13
Frequency Response

The advantage of the straight-line approximation is, of course, that it reduces the complete characteristic to a sum of elementary characteristics.

H. W. Bode

In the previous chapter we studied the steady-state response due to a sinusoidal input. In this chapter we extend the concept of sinusoidal response to the case where the input signal contains many different frequencies. The concept is introduced using the Fourier series representation of a periodic waveform, and then extended to the general notion of a signal spectrum. This leads to an interpretation of signal processing in terms of frequency response. The transfer function continues to play a central role in our study. Its values on the j axis in the s plane relate the poles and zeros to the frequency response of the circuit. One-pole transfer functions are treated first, and these concepts are then extended to transfer functions with several real poles and zeros using Bode plots. Two-pole transfer functions are discussed in terms of their Bode plots for the low-pass, bandpass, and high-pass cases. The final section deals with the design of filters to achieve highly selective frequency responses.

13-1   FOURIER ANALYSIS

Up to this point we have studied circuits whose input is a step function or a sinusoid. In reality circuits are driven by very complex waveforms whose characteristics are to some degree unpredictable. However, it often turns out that real signals can be described in terms of a set of sinusoids at different frequencies. Note the difference here. In Chapter 12 we studied the response due to sinusoids at a single frequency. In this chapter we are concerned with the case in which the input signal can be represented by a set of sinusoids at different frequencies. This leads us to investigate the frequency content of signals and consequently the frequency response of circuits. We begin our study with periodic waveforms.

The spectral analysis of periodic waveforms has its origin in the works of the French mathematician-scientist Jean Baptiste Joseph Fourier (1768–1830). Fourier analysis is a mathematical process that serves as a bridge between the time domain (waveforms) and the frequency domain (spectra). The Fourier series applies to most periodic waveforms. To review, a periodic waveform is one that is endlessly repetitive. More precisely, a waveform $v(t)$ is said to be periodic if

$$v(t) = v(t + T_0) \qquad \textbf{(13-1)}$$

where T_0 is called the period, and is the smallest nonzero interval for which Eq. 13-1 is true. The concept of frequency is then related to the period as

$$\omega_0 = 2\pi f_0 = 2\pi/T_0 \qquad \textbf{(13-2)}$$

The sinusoid is the premier periodic waveform. Some other periodic waveforms are shown in Figure 13-1.

Fourier analysis shows that a reasonably well-behaved,[1] periodic waveform can be expressed as an infinite series of sinusoids with harmonically related frequencies. If $v(t)$ is periodic, with period T_0, and is reasonably well-behaved, then

$$
\begin{aligned}
v(t) = a_0 + \\
a_1 \cos(\omega_0 t) + b_1 \sin(\omega_0 t) + \\
a_2 \cos(2\omega_0 t) + b_2 \sin(2\omega_0 t) + \\
a_3 \cos(3\omega_0 t) + b_3 \sin(3\omega_0 t) + \\
a_4 \cos(4\omega_0 t) + b_4 \sin(4\omega_0 t) + \\
a_5 \cos(5\omega_0 t) + b_5 \sin(5\omega_0 t) + \ldots
\end{aligned}
\qquad \textbf{(13-3)}
$$

or more compactly,

$$v(t) = a_0 + \sum_{n=1}^{\infty} \left[a_n \cos(n\omega_0 t) + b_n \sin(n\omega_0 t) \right] \qquad \textbf{(13-4)}$$

---

[1]Reasonably well-behaved means that $v(t)$ is single-valued, that the integral of $|v(t)|$ over a period is finite, and that $v(t)$ has a finite number of discontinuities in any one period. These conditions, called the Dirichlet conditions, are met by real signals. They are not necessarily met by the models we make of real signals.

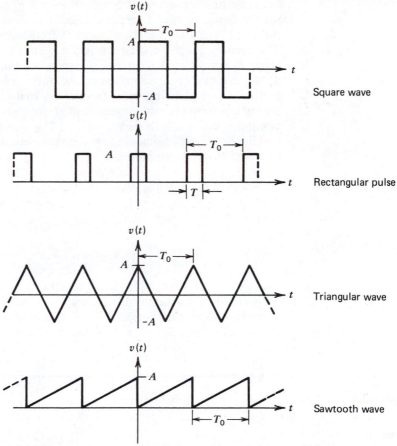

FIGURE 13-1
Some periodic waveforms.

where the fundamental frequency is $\omega_0 = 2\pi/T_0$, and a_0, a_n, and b_n are constants.

This infinite sum representation of a periodic waveform is called a **Fourier series.** The constants in the series are the Fourier coefficients of each of the discrete frequencies contained in the signal. An enumeration, equation, or graph that gives the coefficients for each frequency is the frequency domain description, or **spectrum,** of the waveform. For a general periodic waveform an enumeration of the spectrum would be

a_0	a_1, b_1	a_2, b_2	$\ldots$	$a_n, b_n \ldots$
at	at	at		at
$\omega = 0$	$\omega = \omega_0$	$\omega = 2\omega_0$		$\omega = n\omega_0$

The constant a_0 is the zero-frequency or dc component of the waveform, while a_1 and b_1 are the Fourier coefficients of the component at the fundamental frequency ω_0. Similarly a_2 and b_2 are the Fourier coefficients for the component at $2\omega_0$, the second harmonic. In general a_n and b_n are the coefficients of the component at frequency $n\omega_0$, the nth harmonic. The zero-frequency component is an a coefficient since $a_0 \cos (0) = a_0$ while $b_0 \sin (0) = 0$. This zero-frequency component is often called the dc (for direct current) component. All of the other frequencies are integer multiples of the fundamental frequency, that is, harmonics. Collectively the harmonics are sometimes called the ac (for alternating current) components.

The Fourier coefficients for a given periodic waveform $v(t)$ can be obtained from the equations

$$a_0 = \frac{1}{T_0} \int_0^{T_0} v(t)\, dt$$

$$a_n = \frac{2}{T_0} \int_0^{T_0} v(t) \cos (2\pi n t/T_0)\, dt \qquad n = 1, 2, 3, \ldots \qquad \textbf{(13-5)}$$

$$b_n = \frac{2}{T_0} \int_0^{T_0} v(t) \sin (2\pi n t/T_0)\, dt \qquad n = 1, 2, 3, \ldots$$

The integration involved in these equations is shown to extend from 0 to T_0. Actually the integration can span any convenient interval as long as it spans exactly one period of $v(t)$. The next example illustrates the application of these equations to obtain the Fourier coefficients of a periodic waveform.

Example 13-1

Determine the Fourier coefficients for the square wave shown in Figure 13-1.

Since the integration in Eq. 13-5 extends over one period we need a mathematical expression for the waveform over this time span.

$$v(t) = \begin{cases} A & 0 \le t \le T_0/2 \\ -A & T_0/2 < t \le T_0 \end{cases}$$

Now applying Eqs. 13-5,

$$a_0 = \frac{1}{T_0} \int_0^{T_0/2} A\, dt + \frac{1}{T_0} \int_{T_0/2}^{T_0} (-A)\, dt$$

$$= \frac{A}{T_0} \left(\frac{T_0}{2} - 0 - T_0 + \frac{T_0}{2} \right)$$

$$= 0$$

$$a_n = \frac{2}{T_0} \int_0^{T_0/2} A \cos (2\pi n t/T_0)\, dt +$$

$$\frac{2}{T_0} \int_{T_0/2}^{T_0} (-A) \cos (2\pi n t/T_0)\, dt$$

$$= \frac{2A}{T_0} \left[\frac{\sin (2\pi n t/T_0)}{2\pi n/T_0} \right]_0^{T_0/2}$$

$$- \frac{2A}{T_0} \left[\frac{\sin (2\pi n t/T_0)}{2\pi n/T_0} \right]_{T_0/2}^{T_0}$$

$$= \frac{A}{n\pi} \left[\sin n\pi - \sin 0 \right.$$

$$\left. - \sin 2n\pi + \sin n\pi \right]$$

$$= 0$$

$$b_n = \frac{2}{T_0} \int_0^{T_0/2} A \sin (2\pi n t/T_0)\, dt$$

$$+ \frac{2}{T_0} \int_{T_0/2}^{T_0} (-A) \sin (2\pi n t/T_0)\, dt$$

$$= \frac{2A}{T_0} \left[-\frac{\cos (2\pi n t/T_0)}{2\pi n/T_0} \right]_0^{T_0/2}$$

$$- \frac{2A}{T_0} \left[-\frac{\cos (2\pi n t/T_0)}{2\pi n/T_0} \right]_{T_0/2}^{T_0}$$

$$= \frac{A}{n\pi} \left[-\cos n\pi + \cos 0 + \cos 2n\pi - \cos n\pi \right]$$

$$= \frac{2A}{n\pi} (1 - \cos n\pi)$$

Note that $a_0 = 0$ and $a_n = 0$ for all n, so that only the b coefficients need be evaluated.

$$b_1 = \frac{2A}{\pi}(1 - \cos \pi) = \frac{4A}{\pi}$$

$$b_2 = \frac{2A}{2\pi}(1 - \cos 2\pi) = 0$$

$$b_3 = \frac{2A}{3\pi}(1 - \cos 3\pi) = \frac{4A}{3\pi}$$

$$b_4 = \frac{2A}{4\pi}(1 - \cos 4\pi) = 0$$

From this it is clear that only the odd values of n are nonzero, and the Fourier series of the square wave can be written as

$$v(t) = \frac{4A}{\pi}\left(\underbrace{\sin \omega_0 t}_{\text{Fundamental}} + \underbrace{\frac{1}{3}\sin 3\omega_0 t}_{\text{Third harmonic}}\right.$$

$$\left. + \underbrace{\frac{1}{5}\sin 5\omega_0 t}_{\text{Fifth harmonic}} + \cdots\right)$$

Figure 13-2 shows how the individual harmonics combine to produce a square wave with A = 2.

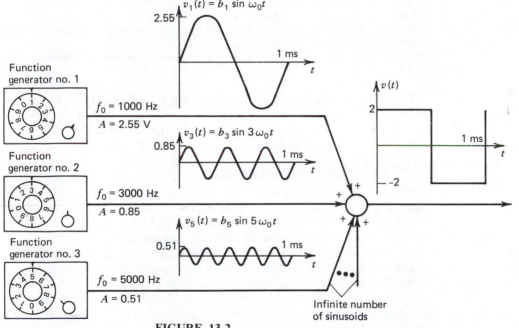

FIGURE 13-2
Construction of a square-wave from sinusoids.

The previous example illustrates the use of Eqs. 13-5 to derive the Fourier coefficients of a given periodic waveform. Fortunately it is not necessary to go through this process for every periodic waveform. Extensive tables of Fourier series expansions are available in a number of sources.[2] For our purposes the abbreviated list in Figure 13-3 will suffice. For each waveform graphically portrayed, the values of a_0, a_n, and b_n are presented in both equation and graphical form.

[2]For example see Spiegel, Murray D., *Mathematical Handbook*, Schaum's Outline Series, McGraw-Hill Book Company, New York, 1968.

Historical Note

The terms dc and ac have deep historical roots in electrical engineering. In the last quarter of the nineteenth century there were two rival systems for supplying electric power. The earlier dc system was championed by Thomas Edison. The ac system was developed somewhat later by Nikola Tesla and George Westinghouse. The competition between these two systems in the late 1880s has come to be known as the "war of the currents." Edison attacked the high-voltage ac system as a public danger. He sponsored studies to show that high-voltage alternating current provided a humane way to execute criminals. The New York Legislature adopted electrocution, and in fact one legislator suggested that "to Westinghouse" be used instead of "to electrocute." Westinghouse opposed the idea of electrocution since it tended to discredit his system. The "war" was ultimately settled when the Westinghouse Electric Company won the contract to install an ac electric power system for the Niagara Falls project. After the Niagara project the future belonged to alternating current. Through a series of mergers the Edison Electric Company eventually became the General Electric Company of today.

The process of calculating or comparing Fourier coefficients can be simplified if the waveform has certain symmetries. First of all, the constant a_0 is none other than the average value of the waveform as defined in Chapter 6 (compare definition of a_0 with Eq. 6-29.) Thus if the waveform has equal area above and below the $t = 0$ axis, it has zero average value. The square wave and triangular wave in Figure 13-3 are examples of periodic waveforms with zero average value, and hence $a_0 = 0$ in both cases. If the waveform is defined such that $v(-t) = -v(t)$, then it is said to have odd symmetry and the Fourier series will contain only sine terms (b_n). The square wave in Figure 13-3 is an example of this form of symmetry. Likewise if the waveform is define so that $v(-t) = v(t)$, then it is said to have even symmetry and only the cosine terms (a_n) will appear in the Fourier series. The triangular wave in Figure 13-3 is an example of a waveform with this type of symmetry. Finally, if the waveform has the property that $v(t) = -v(t \pm T_0/2)$, it is said to have half-wave symmetry, and only the odd harmonics will appear in the Fourier series. Both the square wave and the triangular wave are examples of signals with half-wave symmetry.

Example 13-2

Apply the results in Figure 13-3 to determine the Fourier series of a triangular wave.

As noted, the waveform has zero average value and even symmetry and hence $a_0 = 0$ and $b_n = 0$ as noted in the illustration. Thus we can write

$$a_1 = \frac{8A}{\pi^2} \sin^2 \pi/2 = \frac{8A}{\pi^2}$$

$$a_2 = \frac{8A}{(2\pi)^2} \sin^2 \pi = 0$$

$$a_3 = \frac{8A}{(3\pi)^2} \sin^2 3\pi/2 = \frac{8A}{(3\pi)^2}$$

$$a_4 = \frac{8A}{(4\pi)^2} \sin^2 2\pi = 0$$

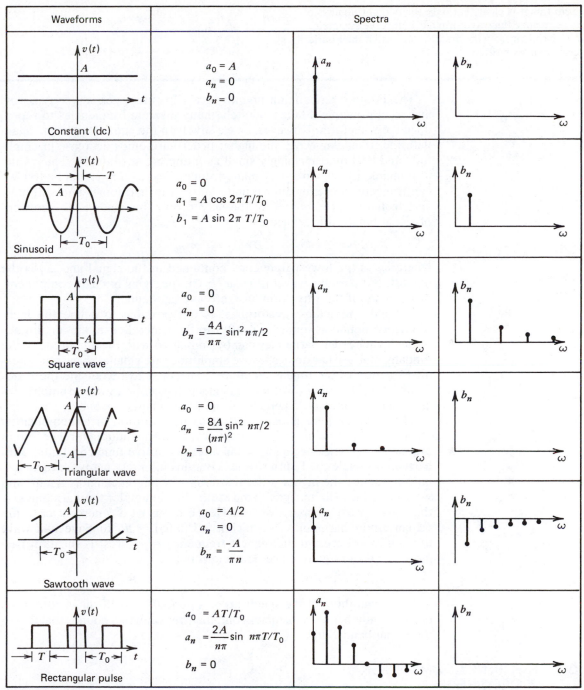

Waveforms	Spectra

Constant (dc)
$$a_0 = A$$
$$a_n = 0$$
$$b_n = 0$$

Sinusoid
$$a_0 = 0$$
$$a_1 = A \cos 2\pi T/T_0$$
$$b_1 = A \sin 2\pi T/T_0$$

Square wave
$$a_0 = 0$$
$$a_n = 0$$
$$b_n = \frac{4A}{n\pi} \sin^2 n\pi/2$$

Triangular wave
$$a_0 = 0$$
$$a_n = \frac{8A}{(n\pi)^2} \sin^2 n\pi/2$$
$$b_n = 0$$

Sawtooth wave
$$a_0 = A/2$$
$$a_n = 0$$
$$b_n = \frac{-A}{\pi n}$$

Rectangular pulse
$$a_0 = AT/T_0$$
$$a_n = \frac{2A}{n\pi} \sin \ n\pi T/T_0$$
$$b_n = 0$$

FIGURE 13-3
Fourier series of some periodic waveforms.

591

From this it is clear that only the odd harmonics are present since the waveform has half-wave symmetry. It is also clear from the first few terms that the Fourier series can be written as

$$v(t) = \frac{8A}{\pi^2}\left(\cos \omega_0 t + \frac{1}{9}\cos 3\omega_0 t + \frac{1}{25}\cos 5\omega_0 t + \cdots\right)$$

The Fourier series is an infinite series for most periodic waveforms and so in principle it takes infinitely many discrete frequencies to reproduce the waveform. However, as we saw with the square wave and again with the triangular wave, the higher order harmonics decrease in amplitude and become vanishingly small as n approaches infinity. Thus while in principle it takes infinitely many frequencies, as a practical matter we can truncate the series after a finite number of harmonics and still have an adequate representation of the waveform. This leads to the concept of signal **bandwidth** defined as

$$BW = \omega_H - \omega_L \qquad (13\text{-}6)$$

where ω_L is the lowest frequency contained in the signal and ω_H is the highest. For periodic waveforms, $\omega_L = 0$ if the signal has a dc component, or $\omega_L = \omega_0$, if the waveform has zero average value.

For most periodic waveforms it is not possible to define the high-frequency bound ω_H quite so neatly. In the general case, the amplitudes of the higher order harmonics can be compared with the amplitude of the fundamental. All harmonics whose amplitudes are smaller than some specified fraction of the fundamental are considered to be outside the bandwidth interval. In other words, we neglect all harmonics whose amplitudes are smaller than, say, 5 percent of the fundamental.

The situation here is quite similar to the duration of an exponential waveform. In principle the exponential waveform endures forever, but in practice it becomes negligibly small after about five time constants. Similarly in principle the Fourier series contains infinitely many frequencies, but in practice the higher order harmonics become negligibly small, and we can define a finite signal bandwidth. For example, the harmonics of the square wave decrease as $1/n$. Hence by using a 5 percent criterion we can ignore harmonics for which $n > 1/(0.05) = 20$. But since only odd harmonics are present the highest frequency of interest is the 19th harmonic, and the square wave bandwidth is

$$BW = \omega_H - \omega_L = 19\omega_0 - \omega_0 = 18\omega_0$$

Thus we can think of the spectrum of a periodic waveform as consisting of a finite number of frequencies contained between two bounds that define the signal bandwidth.

13-2 FREQUENCY-DOMAIN SIGNAL PROCESSING

We have introduced the Fourier series representation of periodic waveforms to illustrate the fact that signals can be described in terms of their

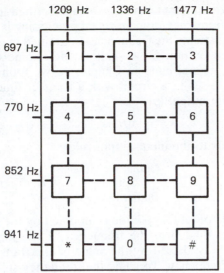

FIGURE 13-4
Push-button telephone dialing frequencies.

frequency content, that is, in terms of a frequency spectrum. The notion of the frequency content of signals is an important one, and it applies to situations in which the waveforms are not periodic. For example, the push-button telephone dial system shown in Figure 13-4 represents each digit as a mixture of two sinusoidal frequencies. When the 3 button is depressed, the instrument transmits a mixture of sinusoids at 697 and 1477 Hz. If the 5 button is depressed, the dial signal is a mixture of 770 and 1336 Hz. The frequencies are not harmonic, which accounts for the atonal sounds produced by the push-button dial, and means that the signals are not periodic. However, the important point is that the push-button dial system produces signals whose spectrum lies in the range from 697 to 1477 Hz. In other words, we can characterize the signal in terms of its frequency content without knowing the specific number being dialed.

Periodic waveforms and the push-button dial system signals are examples of signals that contain well-defined discrete frequencies. Other types of signals contain a continuum of frequencies. For example, most of the spectral content of human speech is concentrated in the range from about 100 Hz to roughly 5 kHz. This is a much greater range than is required for intelligibility. The spectrum of the telephone voice signal is limited to the range from about 300 to 3300 Hz. This is not to say that the human voice signal does not contain frequencies outside this range, or that the human ear cannot detect frequencies below 300 Hz or above 3300 Hz. It only means that most listeners can extract the data from the signal (that is, understand the speech) even though the signal's spectrum is limited to this band of frequencies. The important notion here is that the data content of the signal is relatively easy to describe in terms of frequencies, and that the spectrum involves a continuum of frequencies.

We have seen that signals can be described in terms of their frequency content. Frequency-domain signal processing is performed when a circuit selectively affects the frequencies contained in the input signal to produce the desired output. How shall we determine the frequency selectivity properties of a circuit? In Chapter 12 we found that for an input signal $A \cos(\omega t + \phi)$ to a circuit with a transfer function $H(s)$, the sinusoidal steady-state output is

$$y_{SS} = A|H(j\omega)| \cos(\omega t + \phi + \angle H(j\omega)) \tag{13-7}$$

which leads to the input-output rules:

$$\text{Output amplitude} = A|H(j\omega)|$$
$$\text{Output phase} = \phi + \angle H(j\omega)$$
$$\text{Output frequency} = \omega$$

In sum, the output amplitude is modified by the magnitude of the transfer function evaluated at the frequency of the input. Likewise the output phase angle is altered by the angle of the transfer function evaluated at the same frequency. To study the frequency selectivity characteristics of a circuit we need to look at its transfer function for all frequencies of interest.

The effect of the circuit on the frequency response is represented by the functions $|H(j\omega)|$ and $\angle H(j\omega)$, called the gain and the phase respectively. The gain and phase functions are the frequency-dependent relationships that are used to describe the **frequency response** of the circuit. In effect then, if we evaluate a transfer function $H(s)$ at $s = j\omega$, then the magnitude of $H(j\omega)$ is the circuit gain as a function of frequency and the angle is its phase shift. The gain and phase functions of a circuit can be expressed mathematically or graphically. When presented graphically the result is called a frequency response plot, such as the example in Figure 13-5.

A system of terminology based on the gain of the circuit is used to describe its frequency response. For example, Figure 13-5 shows that the gain is essentially constant if the input frequency is below ω_c. The range of frequencies over which this occurs is called a **transmission band** or

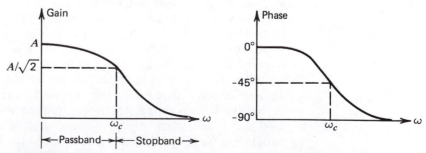

FIGURE 13-5
Typical frequency response plots.

passband. Conversely, for frequencies above ω_c, the gain falls off so that the output signals in this range are smaller than those in the passband. The range of frequencies over which this occurs is called an **attenuation band** or **stopband.**

The frequency associated with the boundary between a transmission band and an adjacent attenuation band is called the **cutoff frequency** (ω_c in Figure 13-5). As Figure 13-5 shows, the transition from the passband to the stopband is gradual, and therefore the precise location of the cutoff frequency is a matter of definition. The most widely used definition is that cutoff occurs when the gain has decreased by a factor of $1/\sqrt{2} = 0.707$ from the maximum gain in the passband. Again this definition is arbitrary since there is no sharp boundary between a transmission band

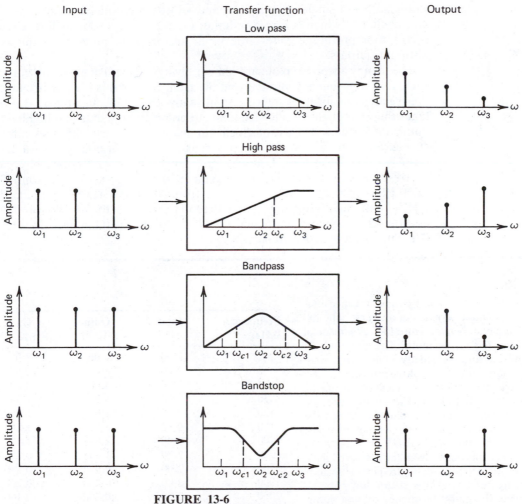

FIGURE 13-6
Four prototype filter gain characteristics.

and the adjacent attenuation band. However, the definition is motivated by the fact that the power carried by a waveform (current or voltage) is proportional to the square of its amplitude. If the amplitude is reduced by a factor of $1/\sqrt{2}$, then the power carried is reduced by a factor of one-half. For this reason the cutoff frequency is also called the **half-power frequency.**

Additional frequency response terminology is based on the four prototype gain characteristics shown in Figure 13-6. A **low-pass** gain has single transmission band extending from zero frequency (dc) to some cutoff frequency ω_c. Conversely a **high-pass** characteristic has a single transmission band extending from some cutoff frequency ω_c to infinite frequency. A **bandpass** gain has a single transmission band with two cutoff frequencies ω_{c1}, ω_{c2}, neither of which is zero or infinite. Finally the **bandstop** gain has a single attenuation band with two cutoff frequencies ω_{c1}, ω_{c2}, neither of which is zero or infinite. Figure 13-6 also shows how each of the four prototypes alters the input spectrum to produce distinctly different outputs.

Frequency response plots are almost always made using a logarithmic scale for the frequency variable and the gain. In part this is done because the frequency ranges of interest span many orders of magnitude. But a logarithmic frequency scale also tends to compress the data and highlight dominant features. The use of a logarithmic frequency scale involves some special terminology. Any frequency range whose end points have a 2:1 ratio is called an **octave.** Likewise any range whose end points have a 10:1 ratio is called a **decade.** For example, the frequency range from 10 to 20 Hz is one octave, as is the range from 20 to 40 MHz. The standard UHF (ultra high-frequency) band extends from 0.3 to 3 GHz, or one decade. The common audio range (20 Hz to 20 kHz) is then three decades (20:200, 200:2000, 2000:20,000).

Example 13-3

This example illustrates the experimental measurement of the gain characteristics of a circuit. The experimental setup is shown in Figure 13-7. A sinusoidal source is connected at the input. The amplitude of the input sinusoid and the steady-state output sinusoid are measured by the voltmeters shown. By changing the frequency of the input and recording the input and output amplitudes in the steady state, we obtain the measured data in Table 13-1. Since the transfer function is a ratio of the output over the input, it follows that the gain is A_0/A. The gain in the last column is then calculated from the measured data. A log-log plot of the gain-versus-frequency scale is shown in the figure. The gain is clearly low pass, so we can estimate the cutoff frequency by first noting that the maximum passband gain is 2.00. Cutoff oc-

Frequency (kHz)	Input $A(V)$	Output $A_0(V)$	Gain A_0/A
1	2.00	4.00	2.00
2	2.00	4.00	2.00
5	2.00	4.00	2.00
10	2.00	3.98	1.99
20	2.00	3.92	1.96
50	2.00	3.72	1.86
100	2.00	2.82	1.41
200	2.00	1.79	0.895
500	2.00	0.970	0.486
1000	2.00	0.400	0.200
2000	2.00	0.200	0.100

TABLE 13-1

curs when the gain reduced to $2/\sqrt{2} \sim 1.41$. Either the tabular data or the gain plot shows that this occurs at about 100 kHz. The logarithmic scales highlight the passband and stopband regions. The total fre-

quency range of the data spans three decades (1 kHz to 1 MHz) plus one octave (1 to 2 MHz). In summary, the circuit is a low-pass filter with a passband gain of 2.00 and a cutoff frequency of 100 kHz.

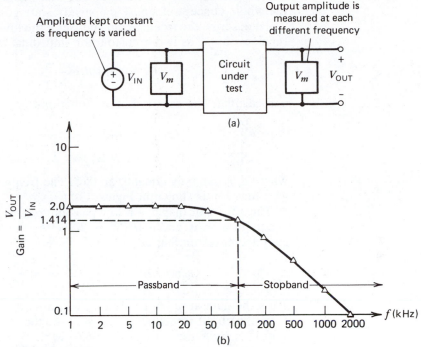

FIGURE 13-7
**Experimental measurement of frequency response. (*a*) Experimental setup.
(*b*) Experimental data.**

13-3   ONE-POLE TRANSFER FUNCTIONS

Since the transfer function is the key to understanding the frequency response of a circuit, it should not be surprising to learn that poles and zeros are important in this regard. In this section we treat the one-pole transfer function. Transfer functions with several poles and zeros can be studied using one-pole transfer functions as building blocks. These methods, called Bode plots, are studied in the following section.

We begin our study by dividing the general one-pole transfer function into a low-pass and a high-pass prototype.

$$H(s) = \frac{b_1 s + b_0}{s + \alpha} = \underbrace{\frac{b_1 s}{s + \alpha}}_{\text{High pass}} + \underbrace{\frac{b_0}{s + \alpha}}_{\text{Low pass}} \tag{13-8}$$

The constants b_1, b_0, and α must be real numbers. The first two of these constants can be negative (in OP AMP circuits, for example) but α must be positive so that the pole (located at $s = -\alpha$) is in the left-half plane. This requirement is needed to ensure that the circuit is stable, otherwise the whole concept of sinusoidal steady state goes out the window.

To describe the frequency response of the low-pass prototype, we first replace s by $j\omega$ as is our custom for sinusoidal steady-state problems,

$$H(j\omega) = \frac{b_0}{j\omega + \alpha} \tag{13-9}$$

and calculate the gain and phase functions as

$$\text{Gain} = |H(j\omega)| = \frac{|b_0|}{\sqrt{\omega^2 + \alpha^2}}$$
$$\text{Phase} = \angle\, H(j\omega) = \angle\, b_0 - \tan^{-1}(\omega/\alpha) \tag{13-10}$$

where $\angle b_0$ can only equal $0°$ or $180°$. The frequency response plot of the gain function is shown in Figure 13-8.

The gain plot displays a low-pass characteristic. The maximum gain occurs at $\omega = 0$, and is $|H(0)| = |b_0|/\alpha$. As the frequency increases, the gain decreases until at $\omega = \alpha$,

$$\text{Gain} = |H(j\omega)| = \frac{|b_0|}{\sqrt{\alpha^2 + \alpha^2}} = \frac{|b_0|}{\sqrt{2}\,\alpha} = \frac{|H(0)|}{\sqrt{2}} \tag{13-11}$$

Thus the gain at $\omega = \alpha$ is reduced from $|H(0)|$ by a factor of $1/\sqrt{2}$. In sum, the cutoff frequency is $\omega_c = \alpha$. At the cutoff frequency the phase shift is $\angle\, b_0 - \tan^{-1}(\alpha/\alpha) = \angle\, b_0 - 45°$.

The low- and high-frequency gain asymptotes shown in Figure 13-8 are important. For $\omega \ll \alpha$ the gain becomes $|H(j\omega)| \to |b_0|/\alpha = |H(0)|$. For $\omega \gg \alpha$ the gain becomes $|H(j\omega)| \to |b_0|/\omega$. Clearly these two asymptotes intersect at $\omega = \alpha$. In other words, the low- and high-frequency asymptotes intersect at the cutoff frequency. The low-frequency asymptote is the horizontal line in Figure 13-8, while the high-frequency asymptote is the diagonal line. Their intersection forms a "corner," and since this occurs at the cutoff frequency, it is sometimes called the **corner frequency.**

The actual gain curve shown in Figure 13-8 does not differ greatly from the two asymptotes except around the cutoff frequency. At cutoff the difference is about 30 percent. One octave on either side of cutoff the difference is only about 10 percent. One decade on either side the difference is about 0.5 percent. Thus the two straight-line asymptotes are fairly good approximations to the gain response. Note that this approximation was made visible by using a log-log plot.

To construct the straight-line approximation to the gain response we use the low-frequency asymptote (a horizontal line whose ordinate equals $|H(0)|$ out to the cutoff, or corner frequency. Above cutoff we use the high-frequency asymptote, a line of the form $|b_0|/\omega$. In the high-frequency region the gain must decrease by a factor of 10 whenever the frequency increases by 10. On a log-log plot a straight line whose ordinate decreases

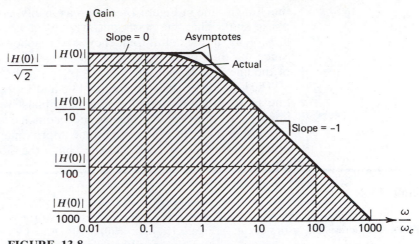

FIGURE 13-8
Gain response of a one-pole low-pass prototype.

by 10 whenever the abscissa increases by 10 is said to have a slope of
-1.

The semilog plot of the phase shift of a one-pole low-pass prototype is shown in Figure 13-9. The phase shift at low frequency is $\angle\, b_0$. At the cutoff frequency the phase is $\angle\, b_0 - 45°$, and at high frequencies it is $\angle\, b_0 - 90°$. Note that the $\angle\, b_0$ is zero if b_0 is positive and is $-180°$ if b_0 is negative. The bulk of the phase swing occurs in the range from $\omega_c/10$ to $10\omega_c$, that is, one decade on either side of the cutoff frequency. In fact,

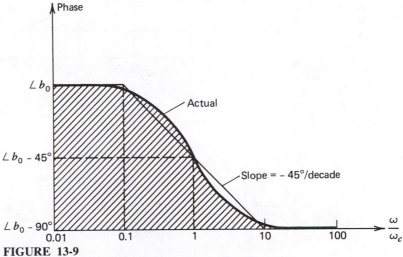

FIGURE 13-9
Phase response of a one-pole low-pass prototype.

the straight line in Figure 13-9 is a reasonably good approximation to the phase shift. This line starts out at $\angle b_0$ at $\omega = \omega_c/10$, passes through $\angle b_0 - 45°$ at the cutoff frequency, and reaches $\angle b_0 - 90°$ at $\omega = 10\omega_c$. Since the total phase swing is $-90°$ over a two-decade range, the slope of the line is $-45°/\text{decade}$.

The logarithmic scales in Figures 13-8 and 13-9 allow us to make straight-line approximations to both the gain and phase responses. These approximations are quite useful when we need a quick evaluation of the frequency reponse of a circuit. The straight-line approximations are not sufficient if precise results are needed, but they have the advantage that results can

Example 13-4

Construct the straight-line approximations to the gain and phase characteristics of the circuit in Figure 13-10a.

The voltage transfer function for the RC circuit can be found using voltage division as

$$H(s) = \frac{V_2}{V_1} = \frac{Z_2}{Z_2 + Z_1} = \frac{1/Cs}{1/Cs + R} = \frac{1}{RCs + 1}$$

Therefore,

$$H(j\omega) = \frac{1}{j\omega RC + 1}$$

By inspection $H(0) = 1$, and hence $|H(0)| = 1$ and $\angle H(0) = 0°$. Also by inspection $\omega_c = \alpha = 1/RC$. With these three quantities we construct the straight-line approximations to the gain and phase responses shown in Figure 13-4.

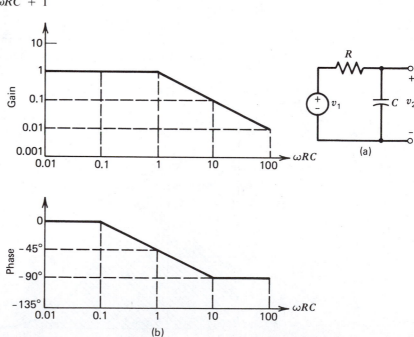

(a)

(b)

FIGURE 13-10
Circuit and frequency response for Example 13-4. (*a*) Circuit. (*b*) Frequency response.

be obtained quickly using graphical methods. Often a quick graphical estimate is all we need.

To construct the straight-line approximations for a one-pole prototype we need only two numbers, $H(0)$ and α. The quantity $H(0)$ gives us both the low-frequency gain $|H(0)|$ and the low-frequency phase shift since $\angle b_0 = \angle H(0)$. The location of the pole then gives us the cutoff frequency $\omega_c = \alpha$. The required quantities $H(0)$ and α can usually be determined by inspection of the transfer function $H(s)$. Examples 13-4 and 13-5 illustrate this.

Example 13-5

Construct the straight-line approximation to gain and phase characteristics of the circuit in Figure 13-11a.

The voltage transfer function of the inverting OP AMP circuit shown is

$$H(s) = \frac{V_2}{V_1} = \frac{-Z_2}{Z_1}$$

$$= -\frac{1/(Cs + 1/R_2)}{R_1} = -\frac{R_2}{R_1}\frac{1}{(R_2Cs + 1)}$$

so that

$$H(j\omega) = -\frac{R_1}{R_1}\frac{1}{(j\omega R_2C + 1)}$$

By inspection $H(0) = -R_2/R_1$, and hence $|H(0)| = R_2/R_1$ and $\angle H(0) = -180°$. Also by inspection $\omega_c = 1/R_2C$. The straight-line approximations to the gain and phase response in Figure 13-11 are constructed from these results.

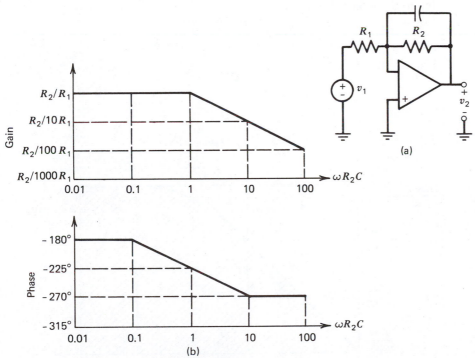

(a)

(b)

FIGURE 13-11
Circuit and frequency response for Example 13-5. (a) Circuit. (b) Frequency response.

If we turn our attention to the one-pole high-pass prototype

$$H(s) = \frac{b_1 s}{s + \alpha} \tag{13-12}$$

we see that it differs from the low-pass case by the introduction of a zero at $s = 0$. Once again we replace s by $j\omega$ and obtain the gain and phase functions as

$$\text{Gain} = |H(j\omega)| = \frac{|b_1|\omega}{\sqrt{\omega^2 + \alpha^2}}$$

$$\text{Phase} = \angle\, H(j\omega) = \angle\, b_1 + 90° - \tan^{-1}(\omega/\alpha) \tag{13-13}$$

The frequency response plot of the gain is shown in Figure 13-12.

The gain-versus-frequency plot displays a high-pass characteristic. The maximum gain occurs as $\omega \to \infty$ and is $|H(\infty)| = |b_1|$. As the frequency decreases, the gain decreases until at $\omega = \alpha$,

$$\text{Gain} = |H(j\omega)| = \frac{|b_1|\alpha}{\sqrt{\alpha^2 + \alpha^2}} = \frac{|b_1|}{\sqrt{2}} = \frac{|H(\infty)|}{\sqrt{2}} \tag{13-14}$$

which shows that the cutoff frequency is $\omega_c = \alpha$. At the cutoff frequency the phase shift is $\angle\, b_1 + 90° - \tan^{-1}(\alpha/\alpha) = \angle\, b_1 + 45°$.

The low- and high-frequency gain asymptotes can be used to approximate the gain response.

For $\omega \gg \alpha$

$$|H(j\omega)| \to |b_1| \qquad \text{(high frequency)}$$

For $\omega \ll \alpha$

$$|H(j\omega)| \to |b_1|\frac{\omega}{\alpha} \qquad \text{(low frequency)}$$

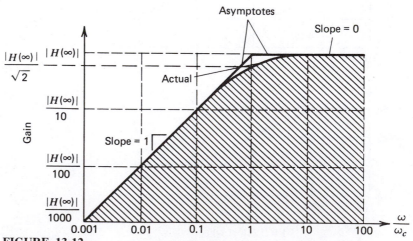

FIGURE 13-12
Gain response of a one-pole high-pass prototype.

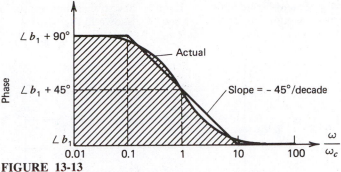

FIGURE 13-13
Phase response of a one-pole high-pass prototype.

The high-frequency asymptote is the horizontal line in Figure 13-12. The low-frequency asymptote is the diagonal line, with slope $= +1$ since the gain increases by a factor of 10 whenever ω increases by 10. The two asymptotes intersect at the cutoff or corner frequency. At the corner frequency the difference between the actual gain and the asymptotes is about 30 percent. An octave above or below the corner the differences are about 10 percent. As an engineering approximation the two asymptotes are a reasonably good description of the gain characteristic.

The semilog plot of the phase shift of the one-pole high-pass function is shown in Figure 13-13. The phase shift at high frequency is $\angle b_1$. It is $\angle b_1 + 45°$ at the corner frequency and approaches $\angle b_1 + 90°$ at low frequency. The bulk of the phase swing (90°) occurs in the two-decade range on either side of the corner frequency. The phase shift can be approximated by the straight line shown in the figure. Note that $\angle b_1$ will be 0° if b_1 is positive and $-180°$ if b_1 is negative.

Like that of the low-pass prototype, the frequency response of a high-pass prototype can also be approximated by straight lines. To construct the approximation we need two numbers, $H(\infty)$ and α. The location of the pole identifies the cutoff frequency $\omega_c = \alpha$. The quantity $H(\infty)$ gives us both the high-frequency gain $|H(\infty)|$ and the high-frequency phase shift since $\angle b_1 = \angle H(\infty)$. These quantities can usually be determined by inspection of the transfer function as illustrated in the next two examples.

Example 13-6

Construct the straight-line approximations to the gain and phase characteristics of the circuit in Figure 13-14a.

The voltage transfer function for the *RL* circuit in Figure 13-14 can be found from voltage division as

$$H(s) = \frac{V_2}{V_1} = \frac{Z_2}{Z_2 + Z_1} = \frac{Ls}{Ls + R}$$

Therefore,

$$H(j\omega) = \frac{j\omega L}{j\omega L + R}$$

By inspection $H(\infty) = 1$, and hence $|H(\infty)| = 1$ and $\angle H(\infty) = 0°$. Also by inspection the corner frequency is $\omega_c = \alpha = R/L$. With these quantities we construct the straight-line approximations to the gain and phase shown in the figure.

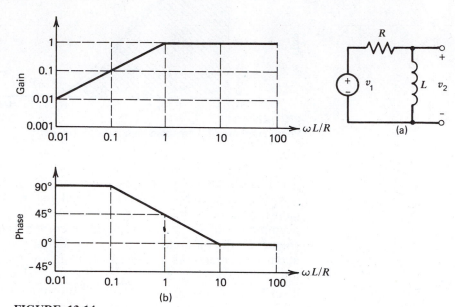

FIGURE 13-14
Circuit and frequency response for Example 13-6. (*a*) Circuit. (*b*) Frequency response.

Example 13-7

Construct the straight-line approximations to the gain and phase characteristics of the circuit in Figure 13-15*a*.

The voltage transfer function of the inverting OP AMP circuit in Figure 13-15 is

$$H(s) = \frac{V_2}{V_1} = \frac{-Z_2}{Z_1} = \frac{-R_2}{R_1 + 1/Cs} = \frac{-R_2Cs}{R_1Cs + 1}$$

and

$$H(j\omega) = \frac{-R_2j\omega C}{R_1j\omega C + 1}$$

By inspection $H(\infty) = -R_2/R_1$, and hence $|H(\infty)| = R_2/R_1$ and $\angle H(\infty) = -180°$. Also by inspection the corner frequency is $\omega_c = \alpha = 1/R_1C$. The straight-line approximations to the gain and phase shift shown in the figure are then constructed using these results.

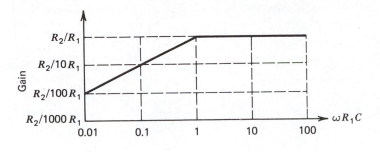

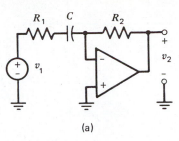

(a)

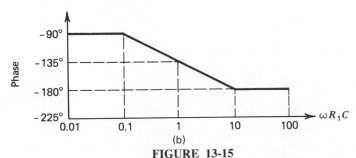

(b)

FIGURE 13-15

Circuit and frequency response for Example 13-7. (*a*) Circuit. (*b*) Frequency response.

13-4 BODE PLOTS

Frequency response plots that use a logarithmic scale for both the gain and frequency axes are called **Bode plots.** These log-log plots highlight dominant features, reveal certain symmetries, and open the way to new insights into the relationships between poles and zeros and circuit frequency response. Bode plots involve simple straight-line approximations to the gain and phase response that are particularly useful when the poles and zeros all lie on the negative real axis in the *s* plane. The name honors H. W. Bode, an American engineer, who first extensively used logarithmic plots to study the frequency response properties of network functions.

To illustrate the process consider the transfer function:

$$H(s) = K \frac{(s + \alpha_1)(s + \alpha_2)}{(s + \alpha_3)(s + \alpha_4)} \tag{13-15}$$

This function has two poles and two zeros, which for now we assume all lie on the *negative real axis*. $H(s = j\omega)$ can be written in the form

$$H(j\omega) = K \frac{\alpha_1\alpha_2}{\alpha_3\alpha_4} \frac{(1 + j\omega/\alpha_1)(1 + j\omega/\alpha_2)}{(1 + j\omega/\alpha_3)(1 + j\omega/\alpha_4)} \tag{13-16}$$

Each of the factors in this equation is of the form $(1 + j\omega/\alpha)$. Using the notation

$$\text{Magnitude} = M = |1 + j\omega/\alpha| = \sqrt{1 + (\omega/\alpha)^2}$$
$$\text{Angle} = \theta = \angle\ (1 + j\omega/\alpha) = \tan^{-1}\ (\omega/\alpha) \qquad \textbf{(13-17)}$$

Equation 13-16 can be written as

$$H(j\omega) = H(0)\ \frac{(M_1\ e^{j\theta_1})(M_2\ e^{j\theta_2})}{(M_3\ e^{j\theta_3})(M_4\ e^{j\theta_4})} \qquad \textbf{(13-18)}$$

where

$$H(0) = K\ \frac{\alpha_1\alpha_2}{\alpha_3\alpha_4}$$

In this notation the gain and phase functions are

$$\text{Gain} = |H(j\omega)| = |H(0)|\ \frac{M_1 M_2}{M_3 M_4}$$
$$\text{Phase} = \angle\ H(j\omega) = \angle\ H(0) + (\theta_1 + \theta_2) - (\theta_3 + \theta_4) \qquad \textbf{(13-19)}$$

This result is clearly general and is not limited to the special case of two poles and two zeros considered here. In words, the gain is equal to $|H(0)|$ times a quotient that is the product of the zero magnitudes divided by the product of the pole magnitudes. The phase angle equals $\angle\ H(0)$ plus the sum of the zero angles minus the sum of the pole angles.

In general then, each pole and zero makes its own contribution to the gain and phase of the transfer function. Because of the form of the equations for the magnitude and angle, zeros tend to increase the gain and phase angle while poles tend to decrease gain and phase.

The interesting feature of the phase is that the contribution of each pole and zero can be found by simply summing the contribution of each individual factor. It would indeed be nice if the same process applied to the gain. But this is what happens when we plot gain using a logarithmic scale. In effect a log scale implies taking the logarithm of the gain. Hence the gain in Eq. 13-19 can be written as

$$\log |H(j\omega)| = \log |H(0)| + \overset{\text{Zeros}}{\log(M_1)} + \log(M_2) - \overset{\text{Poles}}{\log(M_3)} - \log(M_4)$$

In other words, with a logarithmic scale the gain equals the zero frequency gain plus the contribution of each zero minus the contribution of each pole.

We are now ready for the *coup de grace*. Each of the individual magnitudes is of the form $\sqrt{1 + (\omega/\alpha)^2}$. Therefore the log-log plot of the contribution of a pole or a zero to the overall gain takes the form shown in Figure 13-16. The low-frequency asymptote of the magnitude is one (flat) out to the individual corner frequency defined by the location of the pole or zero $(\omega = \alpha)$. Thereafter the high-frequency asymptote is a straight line with a slope of $+1$ for a zero and -1 for a pole. To construct a

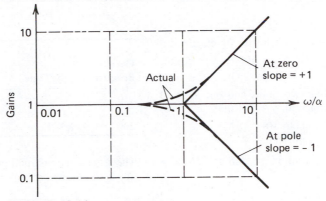

FIGURE 13-16
Gain contribution of a real pole or zero.

straight-line approximation to the gain, we draw a horizontal line, whose ordinate is $|H(0)|$, out to the first corner frequency. If the first corner is a zero, then the gain slope changes to $+1$. If it is a pole, the slope is -1. By repeating this process at each corner frequency (zero or pole) we obtain what is called a **corner** or **Bode** plot—an approximation to the overall gain.

The same approach can be used to obtain an approximation to the phase curve. Figure 13-17 shows the phase shift contribution of a pole or a zero. The only difference is that the slope changes occur one decade below and above the corner frequency. The slope changes are $+45°$/decade for a zero and $-45°$/decade for a pole.

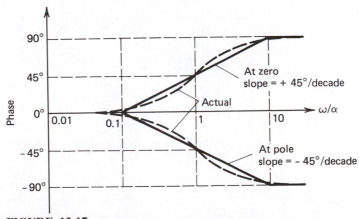

FIGURE 13-17
Phase contribution of a real pole or zero.

Example 13-8

Construct a Bode plot of the transfer function

$$H(s) = \frac{4000(s + 10)}{(s + 40)(s + 200)}$$

The function $H(s)$ has two poles and one zero. If we let $H(s = j\omega)$ and $\omega = 0$, we find that

$$H(0) = \frac{(4000)(10)}{(40)(200)} = 5$$

The low-frequency gain and phase shift are $|H(0)| = 5$ and $\angle H(0) = 0°$

Corner Frequency	Type	Gain Slope Change	Phase Slope Change
10	Zero	+1	+45°/decade
40	Pole	−1	−45°/decade
200	Pole	−1	−45°/decade

The graphical construction of the gain and phase curve approximations are shown in Figure 13-18. The gain

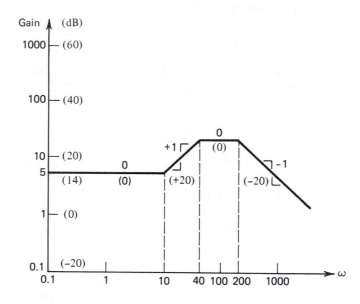

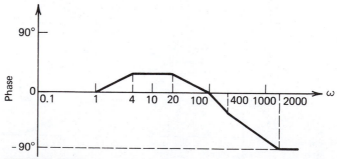

FIGURE 13-18
Frequency response for Example 13-8.

is 5 out to the first corner frequency caused by the zero at $s = -10$. The slope is $+1$ until the next corner frequency due to the pole at $s = -40$, at which point the net slope is zero. This zero slope line continues to the last corner due to the pole at $s = -200$. There-

after the gain falls off with a slope of -1. Construction of the phase curve is similar except that the breaks in the curve occur one decade below the corner frequencies and the slope changes are $\pm 45°$/decade.

Example 13-9

The ideal OP AMP model introduced in Chapter 4 assumes that the device has an infinite gain and an infinite bandwidth. A more realistic model is shown in Figure 13-19a. The controlled source in this model has a low-pass frequency dependence with

$$\text{Gain} = \mu$$
$$\text{Bandwidth} = 1/T$$

A Bode plot of this open-loop characteristic is shown in Figure 13-19b. The area within the passband is called the gain–bandwidth product (G-BW). For the open-loop characteristic

$$\text{G-BW} = \mu/T \qquad \text{(open loop)}$$

To determine the closed-loop transfer function we write device and connection equations as

$$V_o(s) = \frac{\mu}{Ts + 1}(V_P - V_N)$$

$$\text{(OP AMP constraint)}$$

$$V_P = V_S \text{ (Input connection)}$$
$$V_N = V_o \text{ (Feedback connection)}$$

Substituting the connection conditions into the device constraint yields

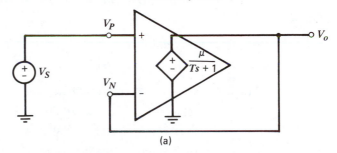

(a)

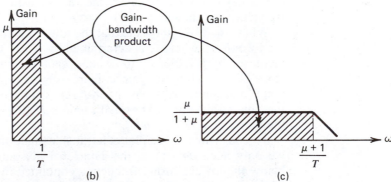

(b)   (c)

FIGURE 13-19
Gain–bandwidth product of a nonideal OP AMP circuit. (a) Nonideal OP AMP circuit. (b) Open loop. (c) Closed loop.

$$V_o(s) = \frac{\mu}{Ts + 1}(V_S - V_o)$$

which produces the closed-loop transfer function as

$$\frac{V_o}{V_S} = H(s) = \frac{\mu}{1 + \mu}\left[\frac{1}{Ts/(\mu + 1) + 1}\right]$$

A Bode plot of the closed-loop transfer function is shown in Figure 13-19c. The relevant parameters of this transfer function are

$$\text{Gain} = \frac{\mu}{1 + \mu}$$
$$\text{Bandwidth} = (1 + \mu)/T$$

The gain–bandwidth product for the closed-loop case is then

$$\text{G-BW} = \mu/T \qquad \text{(Closed loop)}$$

which is the same as the open-loop case. In other words, the gain–bandwidth product is an invariant parameter that is not changed by feedback. Feedback merely redistributes the open-loop gain–bandwidth product over a wider frequency range but at a lower gain. Thus the gain–bandwidth product of the OP AMP is a better descriptor of its limitations or capabilities than is open-loop gain alone.

The process to find a Bode diagram as outlined begins by determining the low-frequency behavior of $H(s)$, that is, $|H(0)|$ and $\angle H(0)$, and then introducing slope changes at the corner frequencies with increasing ω. This process must be modified if $H(s)$ has zeros at $s = 0$, since $H(0) = 0$.

If $H(s)$ has a zero of order n at $s = 0$, then it can be written as

$$H(s) = s^n H_1(s) \qquad \textbf{(13-20)}$$

where $H_1(s)$ is finite at the origin, that is, $H_1(0) \neq 0$. Replacing s by $j\omega$,

$$H(j\omega) = (j)^n(\omega)^n H_1(j\omega) \qquad \textbf{(13-21)}$$

The low-frequency asymptotic behavior of $H(s)$ is then

$$\text{Gain} = |H(j\omega)| = \omega^n |H_1(0)|$$
$$\text{Phase} = \angle H(j\omega) = \angle H_1(0) + \angle j^n = \angle H_1(0) + n90° \qquad \textbf{(13-22)}$$

where by low frequency we do not mean zero (we cannot get there on a log scale) but a frequency range that is well below the lowest corner frequency contained in $H_1(s)$.

What Eq. 13-22 shows is that the low-frequency asymptote is a line of the form $K\omega^n$, where K is determined by $H_1(0)$. On a log-log plot this line increases by a factor of 10^n whenever ω increases by 10^n; therefore its slope is $+n$. Thus the gain is increasing at low frequency with a slope that is determined by the number of zeros at the origin. At low frequencies the phase shift approaches $n90°$, unless $H_1(0)$ is negative, in which case the phase is $-180° + n90°$.

To construct a Bode plot of a transfer function with zeros at the origin, we only need determine the gain level at some frequency that is well below the lowest corner frequency contained in $H_1(0)$. Above this level-setting frequency, the gain increases with a slope of $+n$ until we encounter the first corner frequency of $H_1(s)$. Thereafter the process is the same as before. Poles introduce slope changes of $+1$ and zeros of -1.

Example 13-10

Construct a Bode plot of the transfer function

$$H(s) = \frac{400s}{(s + 40)(s + 200)}$$

$H(s)$ has a single ($n = 1$) zero at the origin. Hence

$$H_1(s) = \frac{400}{(s + 40)(s + 200)}$$

By inspection $H_1(0) = 1/20$, and hence $|H_1(0)| = 1/20$ and $\angle H_1(0) = 0°$. The low-frequency asymptotes of $H(s)$ are

$$|H(j\omega)| = \omega/20 \quad \text{and} \quad \angle H(j\omega) = 90°$$

$\omega = 1$ is a convenient level-setting frequency since it is more than a decade below the lowest corner frequency due to the pole at $s = -40$. At this frequency the asymptotic gain is $1/20 = 0.05$. Figure 13-20 shows the Bode plots for this $H(s)$. The low-frequency gain is 0.05 at $\omega = 1$. The gain increases with a slope of $+1$ until we encounter the corner due to the pole at $s = -40$, at which point the net slope becomes zero. The slope remains zero until the corner due to the second pole; thereafter it falls off with a slope of -1. The phase plot is constructed in the usual way except that the low-frequency angle is 90°.

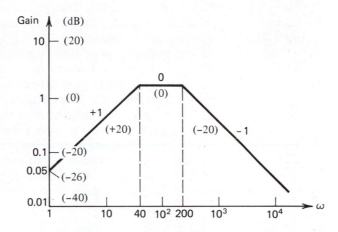

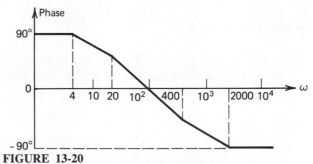

FIGURE 13-20
Frequency response for Example 13-10.

As we have seen, Bode plots involve a log-log plot of the gain versus frequency. For reasons that are heavily woven into the fibers of electrical engineering tradition the logarithm of gain is given a special unit called the **decibel (dB).** The name honors the American inventor Alexander Graham Bell and the term is defined as

$$\text{GAIN}_{dB} = 20 \log_{10}(\text{GAIN})$$

We have dealt with gain as a ratio of voltages or currents. Using the decibel definition, a gain of unity corresponds to 0 dB, while a gain ratio of 10 is 20 dB. In general a gain of 10^N is equivalent to $20N$ dB. Other important values of gain are $\sqrt{2}$ and $1/\sqrt{2}$, which correspond to $+3$ dB and -3 dB, respectively. Recall that our definition of cutoff is the frequency at which the gain has decreased by a factor of $1/\sqrt{2}$ from the passband gain. In decibel notation this means that the gain has decreased by -3 dB from the passband gain. Thus the terms *cutoff, half-power, corner,* and *3-dB down* frequency all mean the same thing and are used interchangeably by electrical engineers.

The slopes of the asymptotes in Bode plots are often expressed in decibels per decade. A slope of -1 is equivalent to -20 dB per decade, while a slope of $+2$ is equivalent to $+40$ dB per decade. This means that there is no change in how a Bode plot looks whether it is plotted in decibels or as a "straight" ratio. The only difference is the scale of the vertical axis of the gain plot and the manner in which the slopes are described. Figures 13-18 and 13-20 show in parentheses how a Bode gain plot would look if the gain was expressed in decibels. Although the decibel notation is very pervasive in the literature of electrical engineering, we will continue to use gain as a ratio of a voltage or a current.

13-5   TWO-POLE TRANSFER FUNCTIONS

To continue our study of frequency response we now consider transfer functions that have two poles. The general form of such a function is

$$H(s) = \frac{b_2 s^2 + b_1 s + b_0}{s^2 + 2\zeta\omega_0 s + \omega_0^2} \qquad \text{(13-23)}$$

Following the pattern used in the one-pole case, we consider prototype functions defined by setting combinations of coefficients in the numerator to zero. Specifically $b_2 = b_1 = 0$ defines the low-pass prototype. Setting $b_2 = b_0 = 0$ yields a bandpass prototype, while $b_1 = b_0 = 0$ is a high-pass prototype.

We begin our two-pole study with the bandpass prototype. Setting b_2 and b_0 to zero in Eq. 13-23 and replacing s by $j\omega$ yields

$$H(j\omega) = \frac{b_1 j\omega}{-\omega^2 + 2\zeta\omega_0 j\omega + \omega_0^2} \qquad \text{(13-24)}$$

As we might expect, the damping and natural frequency play key roles in the gain and phase response of this function. To illustrate this let us find the low- and high-frequency gain asymptotes.

For $\omega << \omega_0$ (low frequency)

$$|H(j\omega)| \rightarrow \frac{|b_1|\omega}{\omega_0^2} \qquad \text{(13-25)}$$

For $\omega >> \omega_0$ (high frequency)

$$|H(j\omega)| \rightarrow \frac{|b_1|}{\omega}$$

The low-frequency asymptote is directly proportional to frequency (slope = +1 on a log-log plot), while the high-frequency asymptote is inversely proportional to frequency (slope = −1 on a log-log plot). Clearly these two asymptotes intersect when $\omega = \omega_0$, as shown in Figure 13-21. The ordinate, or gain, at the intersection is $|b_1|/\omega_0$.

The two asymptotes suggest that the transfer function displays a band-pass characteristic centered about the natural frequency ω_0. To examine the gain response in more detail, we must consider the effect of circuit damping as well. To do this we write Eq. 13-24 as

$$H(j\omega) = \frac{b_1/\omega_0}{2\,\zeta + j\left(\dfrac{\omega}{\omega_0} - \dfrac{\omega_0}{\omega}\right)} \qquad \text{(13-26)}$$

In this special form the only part of $H(j\omega)$ that varies with frequency is the imaginary part in the denominator. Clearly the maximum gain will

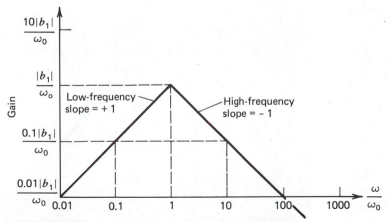

FIGURE 13-21
Gain asymptotes for a two-pole bandpass prototype.

occur when the imaginary part vanishes, that is, when $\omega = \omega_0$. The maximum or **midband** gain is

$$|H(j\omega)| = \frac{|b_1|/\omega_0}{2\,\zeta} \qquad (13\text{-}27)$$

The quantity in the numerator of this expression is the gain at which the low- and high-frequency asymptotes intersect. It follows that the maximum gain will lie above the asymptotes if $\zeta < 0.5$, and below the asymptotes when $\zeta > 0.5$.

Figure 13-22 compares the actual gain of a two-pole bandpass prototype with the asymptotic gain for several values of damping. As predicted, the actual gain curve lies above the asymptotes for $\zeta < 0.5$. Moreover the gain curve displays a sharp narrow peak. For $\zeta > 0.5$ the gain curve flattens out and lies below the asymptotes. When $\zeta = 0.5$ the actual gain is generally rather close to the asymptotes.

In any event it is clear that the function has a bandpass gain characteristic. There are two cutoff frequencies in this case, one on either side of the center frequency ω_0. It turns out that these two frequencies occur when the imaginary part in the denominator of Eq. 13-26 equals $\pm 2\zeta$. When this occurs the gain calculated from Eq. 13-26 is

$$|H(j\omega)| = \frac{|b_1|/\omega_0}{|2\,\zeta \pm j2\zeta|} = \frac{|H(j\omega_0)|}{\sqrt{2}} \qquad (13\text{-}28)$$

It is apparent that for this condition the gain is reduced by a factor of $1/\sqrt{2}$ from its maximum or midband value at $\omega = \omega_0$.

To determine the two cutoff frequencies we set the imaginary part in the denominator of Eq. 13-26 equal to $\pm 2\zeta$.

$$\frac{\omega}{\omega_0} - \frac{\omega_0}{\omega} = \pm 2\zeta \qquad (13\text{-}29)$$

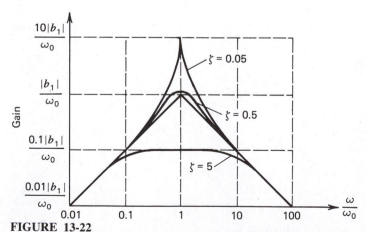

FIGURE 13-22
Gain of a two-pole bandpass prototype showing variations with ζ.

which yields the quadratic equation

$$\omega^2 \mp 2\zeta\omega_0\omega - \omega_0^2 = 0 \tag{13-30}$$

whose roots are

$$\omega = \omega_0 \left(\pm \zeta \pm \sqrt{1 + \zeta^2} \right) \tag{13-31}$$

Only the two positive roots have physical significance, and hence

$$\omega_{c1} = \omega_0 (- \zeta + \sqrt{1 + \zeta^2})$$
$$\omega_{c2} = \omega_0 (+ \zeta + \sqrt{1 + \zeta^2}) \tag{13-32}$$

Note that $\omega_{c1} < \omega_0$ is the lower cutoff frequency, while $\omega_{c2} > \omega_0$ is the upper cutoff frequency. Also by multiplying Eqs. 13-32 together we obtain

$$\omega_0^2 = \omega_{c2}\omega_{c1} \tag{13-33}$$

which reveals that the center frequency ω_0 is the geometric mean of the two cutoff frequencies. These two cutoff frequencies define the passband of the gain characteristic.

Often we are interested in the width of the passband, called the band-width (BW). Subtracting Eqs. 13-32 yields

$$BW = \omega_{c2} - \omega_{c1} = 2\zeta\omega_0 \tag{13-34}$$

The bandwidth is proportional to the damping ratio. In fact, if $\zeta < 0.5$, then BW $< \omega_0$, and if $\zeta > 0.5$, BW $> \omega_0$. Thus $\zeta = 0.5$ is the boundary between two extreme cases. The **narrowband** case ($\zeta < 0.5$) involves a sharp peak in the gain response and bandwidths that are small compared with the circuit natural frequency. In the **wideband** case ($\zeta > 0.5$) the gain response is relatively flat over a frequency range that is often much greater than the circuit natural frequency. These two contrasting cases both are used in practice. The narrowband response is highly selective, passing only a very restricted range of frequencies, often less than one octave. In contrast the wideband response usually encompasses several decades within its passband.

The two extreme cases also show up in the phase response. Using Eq. 13-26 we write

$$\angle H(j\omega) = \angle b_1 - \tan^{-1} \left[\frac{(\omega/\omega_0) - (\omega_0/\omega)}{2\zeta} \right] \tag{13-35}$$

The asymptotic values of the phase angle are
$\omega << \omega_0$ (low frequency)

$$\angle H(j\omega) = \angle b_1 + 90°$$

$\omega = \omega_0$ (midband)

$$\angle H(j\omega) = \angle b_1 \tag{13-36}$$

$\omega >> \omega_0$ (high frequency)

$$\angle H(j\omega) = \angle b_1 - 90°$$

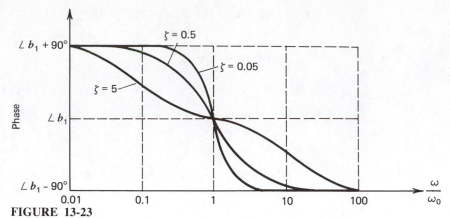

FIGURE 13-23
Phase of a two-pole bandpass prototype.

Graphs of the phase response for several values of damping are shown in Figure 13-23. The total phase swing is 180°. The transition is very abrupt for the narrowband case and more majestic for the wideband situation. The phase shift is the same at the center frequency in either case, and in either case is relatively linear in the passband.

If we look back to Figure 13-22 we note that ζ has a lot to do with the bandwidth of the network. The bandwidth becomes extremely narrow for $\zeta < 0.5$. Yet simply giving the circuit's bandwidth does not reveal the sharpness of its peak. For example, a bandwidth of 100 Hz may be small or narrowband at 1 MHz but flat or broadband at a 60-Hz center frequency. A new parameter is used to measure the sharpness or selectivity of the peak. This parameter, called the **quality factor Q**, is defined as

$$Q \equiv \frac{\text{Center frequency}}{\text{Bandwidth}} \tag{13-37}$$

Hence in our example a 100-Hz bandwidth at 1 MHz equates to a Q of 10,000 while at 60 Hz it equates only to a Q of 0.6. The crossover between what Q is considered narrowband and what Q is considered broadband is somewhat arbitrary but is often taken as 1.

We can rewrite the bandpass prototype Eq. 13-23 to include the definition of Q as follows:

$$H(s) = \frac{b_1 s}{s^2 + (\omega_0/Q)s + \omega_0^2} \tag{13-38}$$

Rewritten in the foregoing form the relation between Q and ζ is immediately evident:

$$Q = \frac{1}{2\zeta} \tag{13-39}$$

As expected small ζ means large Q and a high degree of selectivity while large ζ means just the opposite.

Example 13-11

The transfer function of the circuit in Figure 13-24 is

$$H(s) = \frac{V_2}{V_1} = \frac{Ls/(LCs^2 + 1)}{R + Ls/(LCs^2 + 1)}$$

$$= \frac{s/RC}{s^2 + s/RC + 1/LC}$$

which is the form of a two-pole bandpass prototype. By inspection of $H(s)$,

$$\omega_0{}^2 = \frac{1}{LC}$$

$$2\zeta\omega_0 = BW = \frac{1}{RC}$$

$$|H(j\omega_0)| = 1$$

Thus the center frequency is determined by the LC product, the bandwidth can be set by adjusting R, and the midband gain is always one regardless of these parameters. This circuit could be used to produce either a narrow or a wideband response since the center frequency and bandwidth can be adjusted independently. An example of a narrowband case is the response required in the intermediate-frequency (IF) amplifiers in commercial AM receivers. For such circuits $f_0 = 455$ kHz and BW = 10 kHz. This implies a two-pole response with

$$\zeta = \frac{BW}{2f_0} = \frac{10}{2 \times 455} = 0.0110$$

Thus the circuit must be very lightly damped, or equivalently, have a very narrowband frequency response. Solving for the circuit parameters required to produce this response

$$LC = 1/\omega_0{}^2 = 1/(2\pi 455 \times 10^3)^2 = 1.22 \times 10^{-13}$$
$$RC = 1/BW = 1/(2\pi \times 10^4) = 0.159 \times 10^{-4}$$

There are two constraints and three parameters. Selecting $L = 1$ mH from experience then yields $C = 122$ pF, and $R = 130$ kΩ. The resulting gain response is shown in Figure 13-24. Note the very narrowband response required in this application. The Q for this circuit is calculated to be

$$Q = \omega_0/BW = 1/2\zeta = 45.5$$

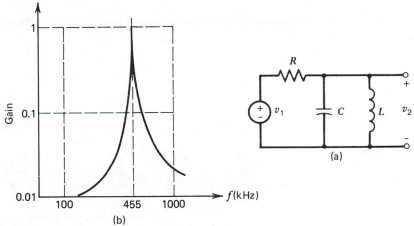

FIGURE 13-24
A tuned circuit (a) Circuit. (b) Frequency response.

Example 13-12

The transfer function of the OP AMP circuit in Figure 13-25a is

$$H(s) = \frac{-R_2C_1s}{(R_1C_1s + 1)(R_2C_2s + 1)}$$

which is of the bandpass form. This circuit cannot produce a narrowband response since the poles are always real. The circuit can be used to produce very wideband responses since the pole locations can be adjusted independently. When the poles are widely separated it is useful to write $H(s)$ as

$$H(s) = -\underbrace{\frac{R_1C_1s}{(R_1C_1s + 1)}}_{\text{High pass}}\underbrace{\frac{1}{(R_2C_2s + 1)}}_{\text{Low pass}}\underbrace{\frac{R_2}{R_1}}_{\text{Gain}}$$

that is, to partition $H(s)$ into a one-pole high-pass prototype, a one-pole low-pass prototype, and a gain term, see Figure 13-25b. The low-frequency rolloff is provided by the high-pass term and the high-frequency rolloff by the low-pass term. Since both of these terms have passband gains of one, the midband gain of the composite is determined by the gain term. For widely separated poles ($R_1C_1 > 10R_2C_2$) the cutoff frequency and midband gains are approximately

$$\omega_{c1} = 1/R_1C_1,$$
$$\omega_{c2} = 1/R_2C_2,$$
$$|H(j\omega_0)| = R_2/R_1$$

As an example consider an audio amplifier with cutoff frequencies of 20 Hz and 20 kHz and a midband gain of 10. Note that these cutoff frequencies imply

$$f_0 = \sqrt{f_{c1}f_{c2}} = \sqrt{4 \times 10^5} = 632 \text{ Hz}$$

and

$$\zeta = \text{BW}/2f_0 = (20{,}000 - 20)/(2 \times 632) = 15.8$$

Thus the circuit is heavily damped or, equivalently, has a broadband frequency response. Solving for the circuit parameters required

$$R_1C_1 = 1/(40\pi)$$
$$R_2C_2 = 1/(40000\pi)$$
$$R_2/R_1 = 10$$

There are three constraints and four parameters. If $R_1 = 10\text{k}\Omega$, then $R_2 = 100 \text{ k}\Omega$, $C_2 = 0.796 \text{ }\mu\text{F}$, and $C_2 = 79.6 \text{ pF}$. The resulting gain is shown in Figure 13-25c. Note that the gain is essentially flat over a three-decade range.

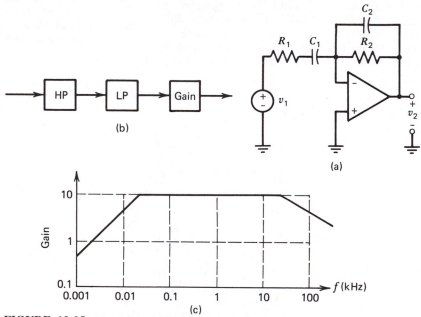

FIGURE 13-25
Wide-band OP AMP circuit. (a) Circuit. (b) Prototypes. (c) Frequency response.

The frequency response of a two-pole low-pass prototype can be studied by setting $b_2 = b_1 = 0$ in Eq. 13-23 and replacing s by $j\omega$.

$$H(j\omega) = \frac{b_0}{-\omega^2 + 2\zeta\omega_0 j\omega + \omega_0^2} \qquad \textbf{(13-40)}$$

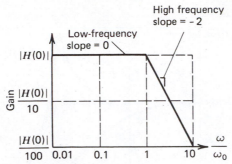

FIGURE 13-26
Gain asymptotes of a two-pole low-pass prototype.

The low- and high-frequency gain asymptotes of this function are
For $\omega \ll \omega_0$ (low frequency)

$$|H(j\omega)| \rightarrow \frac{|b_0|}{\omega_0^2} = |H(0)| \tag{13-41}$$

For $\omega \gg \omega_0$ (high frequency)

$$|H(j\omega)| \rightarrow \frac{|b_0|}{\omega^2}$$

The low-frequency asymptote is flat at a level that equals the zero frequency or dc gain. The high-frequency asymptote is inversely proportional to the square of the frequency (slope $= -2$ on a log-log scale). These two asymptotes intersect at $\omega = \omega_0$, as shown in Figure 13-26. The two asymptotes suggest a low-pass filter characteristic with a passband gain of $|H(0)|$ and a slope or rolloff of -2 in the stopband.

The influence of the damping ratio on the gain can be illustrated by evaluating the gain at the natural or corner frequency.

$$|H(j\omega_0)| = \frac{|b_0|/\omega_0^2}{2\zeta} = \frac{|H(0)|}{2\zeta} \tag{13-42}$$

Thus the gain at the natural frequency will be above the low-frequency asymptote for $\zeta < 0.5$ and below the asymptote for $\zeta > 0.5$.

Figure 13-27 shows the gain of a two-pole low-pass function for several values of damping. As suggested, the gain curves lie above the asymptotes for $\zeta < 0.5$ and below the asymptotes for $\zeta > 0.5$. For $\zeta = 0.5$ the gain curve is generally fairly close to the asymptotes.

The distinctive feature of the two-pole case is that lightly damped circuits exhibit a pronounced peak in the gain response near the natural frequency. Overdamped circuits do not exhibit this peak. In fact is it fairly easy to show that the peak in the gain response occurs at

$$\omega_P = \omega_0 \sqrt{1 - 2\zeta^2} \tag{13-43}$$

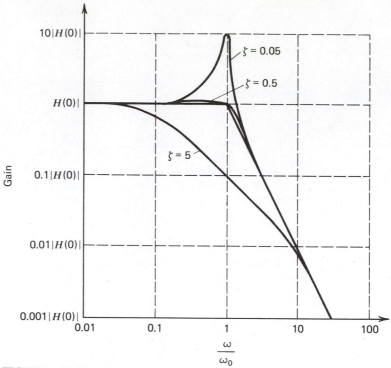

FIGURE 13-27
Gain of a two-pole low-pass transfer function.

The quantity under the radical is positive when $\zeta < 1/\sqrt{2}$, and hence the gain does not exhibit a peak unless the damping ratio is below this bound.

The cutoff frequency of the two-pole low-pass function is a rather complicated function of the damping ratio. What is clear, however, is that if $\zeta = 1/\sqrt{2}$, then from Eq. 13-42 the gain at $\omega = \omega_0$ is

$$|H(j\omega_0)| = \frac{|H(0)|}{\sqrt{2}} \qquad (13\text{-}44)$$

In other words, the cutoff frequency is ω_0 when $\zeta = 1/\sqrt{2}$. Figure 13-28 compares the gain in this case with the one-pole low-pass prototype. The important difference is the sharper stopband rolloff of the two-pole prototype. The cause of this additional rolloff is, of course, the additional pole that causes the high-frequency asymptote to fall off as $1/\omega^2$, rather than $1/\omega$ as in the one-pole case. The general principle is that the gain of an n-pole low-pass function falls off as $1/\omega^n$ at high frequency. Thus the more poles we have in a low-pass transfer function, the greater the attenuation will be in the stopband. This is an extremely important concept, as we shall see in the design of filters in the next section.

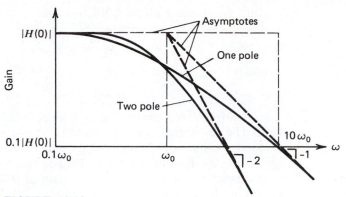

FIGURE 13-28
Comparison of a one-pole and a two-pole low-pass prototype.

Example 13-13

The transfer function of the active RC circuit in Figure 13-29 was found in Example 10-17 as

$$H(s) = \frac{V_2}{V_1} = \frac{K}{(RCs)^2 + (3 - K)RCs + 1}$$

By inspection $\omega_0 = 1/RC$, $\zeta = (3 - K)/2$, and $|H(0)| = K$. The natural frequency is determined by the RC product, while the damping ratio is independently adjusted by fixing the device gain K. The gain at $\omega = \omega_0$ is $|H(j\omega_0)| = K/(3 - K)$. To design the circuit so that the cutoff frequency is ω_0, we set $\zeta = 1/\sqrt{2}$, hence $(3 - K) = \sqrt{2}$, or $K = 1.29$.

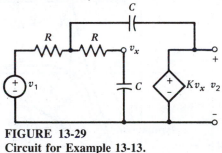

FIGURE 13-29
Circuit for Example 13-13.

To complete our discussion of two-pole prototypes we briefly treat the high-pass case. By setting $b_1 = b_0 = 0$ in Eq. 13-23 and setting $s = j\omega$, we obtain

$$H(j\omega) = \frac{-b_2\omega^2}{-\omega^2 + 2\zeta\omega_0 j\omega + \omega_0^2} \tag{13-45}$$

The high-pass situation is actually the mirror image of the low-pass case and so our discussion need not be detailed. The low- and high-frequency

asymptotes of the high pass function are
For $\omega \ll \omega_0$ (low frequency)

$$|H(j\omega)| \rightarrow |b_2|\omega^2/\omega_0^2 \qquad \text{(13-46)}$$

For $\omega \gg \omega_0$ (high frequency)

$$|H(j\omega)| \rightarrow |b_2| = |H(\infty)|$$

The high-frequency gain is flat at a level equal to the infinite frequency gain. The low-frequency asymptote is proportional to the square of the frequency (slope $= +2$ on a log-log scale). These two asymptotes intersect at $\omega = \omega_0$, as shown in Figure 13-30. The two asymptotes indicate a high-pass characteristic with a passband gain of $|H(\infty)|$ and a slope or rolloff of $+2$ in the stopband.

The influence of the damping ratio can be illustrated by determining the gain at the corner frequency ω_0.

$$|H(j\omega_0)| = \frac{|b_2|}{2\zeta} = \frac{|H(\infty)|}{2\zeta} \qquad \text{(13-47)}$$

Thus the gain at the corner frequency will be above the high-frequency asymptote for $\zeta < 0.5$ and below for $\zeta > 0.5$. Figure 13-30 shows the gain of the two-pole high-pass prototype for several values of damping. As indicated, the gain curve lies above the asymptotes for $\zeta < 0.5$ and below

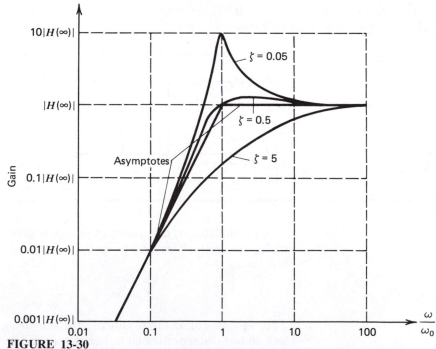

FIGURE 13-30
Gain of a two-pole high-pass transfer function.

the asymptotes for $\zeta > 0.5$. For $\zeta = 0.5$ the gain curve is generally fairly close to the asymptotes.

Again the distinctive feature of the two-pole case is that lightly damped circuits exhibit a pronounced peak in the gain response near the corner frequency. Overdamped circuits do not exhibit this peak. Thus as we have now seen in the bandpass, low-pass, and high-pass cases, circuits whose step response would be a lightly damped sinusoid have pronounced peaks or resonances in their frequency response.

Example 13-14

The transfer function of the circuit in Figure 13-31 can be found by voltage division as

$$H(s) = \frac{V_2}{V_1} = \frac{Ls}{Ls + R + 1/Cs}$$

$$= \frac{s^2}{s^2 + sR/L + 1/LC}$$

By inspection this is a high-pass characteristic with $|H(\infty)| = 1$ and

$$\omega_0^2 = 1/LC \qquad 2\zeta\omega_0 = R/L$$

Thus the corner frequency is determined by the LC product and the damping can be adjusted by selecting R. For example, for $\omega_0 = 10^6$ and $\zeta = 0.5$ we have the two constraints

$$LC = 10^{-12} \qquad \text{and} \qquad \frac{R}{L} = 10^6$$

Thus selecting $C = 0.001 \ \mu\text{F}$, we obtain $L = 1$ mH and $R = 1000 \ \Omega$.

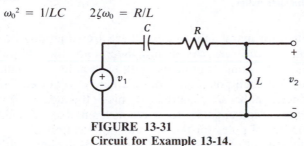

FIGURE 13-31
Circuit for Example 13-14.

13-6 FILTERS

One of the main applications of frequency response techniques is in the design and analysis of filters. Simply stated, a filter is a frequency-sensitive circuit. A filter can pass signals with high or low frequencies or a selected band of frequencies at the exclusion of all others. In Figure 13-32 the transfer functions of four ideal filter prototypes are presented. Contrast these ideal filter characteristics with the more realistic responses sketched in Figure 13-6.

Achieving the sharp corners or rolloffs at the corner frequencies has been the goal of several generations of design engineers. Consider the characteristics of the low-pass filters of Figure 13-33. To attain the low-pass filtering action a simple, single RC voltage divider would suffice. It has only one pole and it would roll off at a slope of -1. While this filter

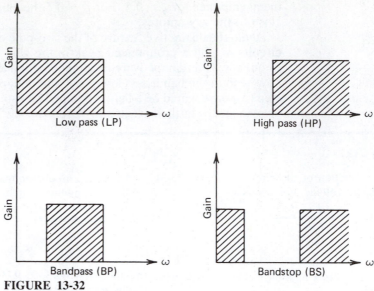

FIGURE 13-32
Ideal filter characteristics.

is good enough for many applications, there are situations where a more "ideal" filter is necessary. As seen in Sec. 13-5, the rolloff is directly proportional to the number of poles, with each pole increasing the slope by -1. Since for each pole a memory element is required, and since an infinite number of poles are required to achieve the ideal case, it becomes the task of the design engineer to arrive at a suitable trade-off as to the optimum number of poles for the intended application.

Before the advent of OP AMPs the task of designing multipole filters was considered by many to be an art to be undertaken only while wearing garlands of garlic. To illustrate this point, consider the task of trying to design a simple two-pole RC filter with a cutoff frequency ω_c. Let us find the transfer functions of the circuits in Figure 13-34.

The transfer function of the single-pole circuit can be written down immediately as

$$H(s) = \frac{1}{RCs + 1} \tag{13-48}$$

The cutoff frequency is readily found as $1/RC$ and it is a simple task to design such a filter for whatever cutoff frequency is desired. It would be nice to assume that if we wanted the same cutoff for the double-pole filter of Figure 13-34*b* we could pick both R's and both C's to be the same, and that filter would have a cutoff of $1/RC$ and a rolloff at -2. Unfortunately it is not that simple. After some effort the transfer function of the double-pole circuit is found to be

$$H(s) = \frac{1}{R_1 R_2 C_1 C_2 s^2 + (R_1 C_1 + R_1 C_2 + R_2 C_2)\, s + 1} \tag{13-49}$$

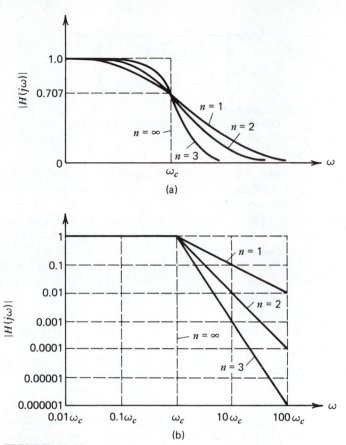

FIGURE 13-33
Real filters. (*a*) **Linear graph.** (*b*) **Bode plot.**

It can be readily seen that if we pick $R_1 = R_2 = R$ and $C_1 = C_2 = C$, the poles will not be at $\omega_c = 1/RC$, but rather at

$$p_1, p_2 = \frac{1}{2RC}(-3 \pm \sqrt{5})$$

While the final slope will be -2, the rolloff will be -1 between pole p_1

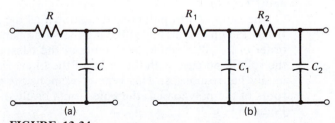

FIGURE 13-34
Low-pass filters. (*a*) **Single pole.** (*b*) **Double pole.**

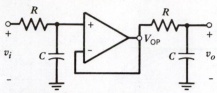

FIGURE 13-35
OP AMP providing isolation to a two-pole
RC low-pass filter.

and pole p_2. This usually is not desired. It takes little imagination to see the complexity that would arise to design the 10- or more-pole filter required in many applications.

OP AMPs have done much to alleviate the task of designing filters. Consider the filter of Figure 13-35. This is nothing more than the two-pole filter of Figure 13-34b with an OP AMP follower placed between the two RC networks. Since an ideal OP AMP draws no current at its input, there is no loading of the first RC filter by the second. The output of the OP AMP is

$$V_{op}(s) = \frac{1}{RCs + 1} V_1(s) \tag{13-50}$$

and the transfer function of the entire circuit is

$$H(s) = \frac{1}{(RCs + 1)} \times \frac{1}{(RCs + 1)} \tag{13-51}$$

This is indeed the result we wanted. The follower has allowed us to greatly simplify our design. While it is not the most effective method, we can easily design an n-pole filter by simply fabricating n identical RC circuits and inserting an OP AMP follower between each one and its neighbor.

Even this approach has some problems; as we cascade more and more stages, the value of our transfer function at the cutoff frequency is reduced by 0.707^n where n is the order of the filter. The result is that our bandwidth is reduced for every stage added. More efficient filter design techniques employ the use of OP AMP circuits with transfer functions that have their poles skillfully located to attain certain desired results. The location of the poles is a function of the polynomial in the denominator of the transfer function. Two such famous polynomials were developed by *Butterworth* and *Chebyshev*.

The Butterworth polynomial is known as the **maximally flat** response. It has the characteristic of having the same bandwidth regardless of the order of the filter while the flatness of the passband and the steepness of the rolloff increase with the order of the filter. Figure 13-33a shows maximally flat responses. This type of response is obtained by locating the roots of the polynomial, the poles, on a circle in the s plane as shown in Figure 13-36.

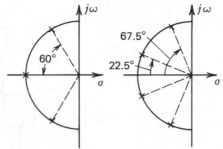

FIGURE 13-36
Location of poles of third- and fourth-order Butterworth filters.

The polynomials that give rise to these poles known as the Butterworth polynomial are given as

$$H(s) = \frac{1}{(s/\omega_c)^2 + 2A(s/\omega_c) + 1}$$ **(13-52)**

where ω_c is the desired cutoff frequency and $2A$ is the appropriate Butterworth coefficient obtained from Table 13-2.

A prototype design of this kind of filter using OP AMPs is given in Figure 13-37. The filter in Figure 13-37a will realize a single-pole response, while the filter in Figure 13-37b will realize a double-pole response. Higher order filters are easily generated by cascading appropriate numbers of single- or double-pole prototypes, all with identical R's and C's but with a feedback gain set by R_1 and R_2 determined from the appropriate Butterworth coefficient.

The transfer function of the double-pole low-pass prototype is found to be (see Example 10-17):

$$H(s) = K\,\frac{(1/RC)^2}{s^2 + \left(\dfrac{(3-K)}{RC}\right)s + \left(\dfrac{1}{RC}\right)^2}$$ **(13-53)**

Order	Butterworth Polynomial
1	$(s + 1)$
2	$(s^2 + 1.41\,s + 1)$
3	$(s + 1)(s^2 + s + 1)$
4	$(s^2 + 0.765\,s + 1)(s^2 + 1.85\,s + 1)$
5	$(s + 1)(s^2 + 0.618\,s + 1)(s^2 + 1.62\,s + 1)$
6	$(s^2 + 0.518\,s + 1)(s^2 + 1.41\,s + 1)(s^2 + 1.93\,s + 1)$

TABLE 13-2

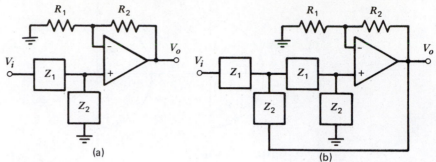

(a)

(b)

FIGURE 13-37
Single - and double-pole Butterworth prototypes. (*a*) Single pole. (*b*) Double pole. $Z_1 = R$ for low pass, C for high pass. $Z_2 = C$ for high pass, R for low pass. $\omega_c = 1/RC$ in either case.

Example 13-15

Design a fourth-order Butterworth low-pass filter with an ω_c of 100 rad/second.

For a fourth-order filter we recognize that we will need to cascade two second-order prototypes. The polynomial for a fourth-order system is found from Table 13-2 to be

$$(s^2 + 0.765\ s + 1)(s^2 + 1.85\ s + 1)$$

We are now ready to design our filter. First of all we will select our R's and C's so that our cutoff frequency is 100 rad/second. A suitable combination would be $R = 10\ \text{k}\Omega$, $C = 1\ \mu\text{F}$. We must now select the gain K of each of our two OP AMP to match the required Butterworth polynomials. We must first put our transfer function into proper form:

$$H(s) = K\ \frac{1}{(RCs)^2 + (3 - K)\ (RCs) + 1}$$

If we compare the coefficients of the s^2 term of Eq. 13-52 with ours we see that $\omega_c = 1/RC$. Now we

compare the s term of our transfer function with the appropriate s term of the desired Butterworth polynomial, that is,

$$(3 - K)RC = \frac{2A}{\omega_c} = 2ARC$$

where $2A$ is the Butterworth coefficient. This results in two equations and two unknowns:

$$3 - K_1 = (0.765)$$

and

$$3 - K_2 = (1.85)$$

This yields $K_1 = 2.24$ and $K_2 = 1.15$.

Using the noninverting OP AMP relation we can select R_1 and R_2 to match the gains we want:

$$K_1 = \frac{R_1 + R_2}{R_1} = 2.24$$

and

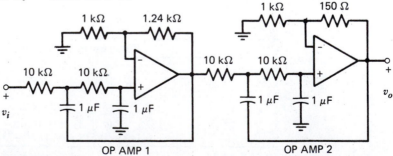

FIGURE 13-38
Four-pole Butterworth low-pass filter.

$$K_2 = \frac{R_1 + R_2}{R_1} = 1.15$$

If we pick R_1 to equal 1 kΩ in both cases we find that

for OP AMP$_1$, R_2 = 1.24 kΩ, and for OP AMP$_2$, R_2 = 150 Ω. Thus our final design is shown in Figure 13-38.

While the Butterworth maximally flat response is used extensively with satisfactory results, it may not be the optimum solution. An alternative design centers around a different set of polynomials developed by Chebyshev. Instead of locating the poles on a circle of constant radius, the poles of Chebyshev polynomials are located on an ellipse and result in a steeper rolloff near the cutoff frequency. Of course there is a price to pay, and that is that there is a ripple in the passband. Usually this ripple is specified in decibels (dB), which often is ±1 dB or sometimes even ±3 dB. Figure 13-39 shows the location of the poles for a second- and fourth-order Chebyshev filter, and how they compare with an equivalent Butterworth filter. Figure 13-40 shows a more revealing comparison between Chebyshev and Butterworth. The major advantage of Chebyshev is that the rolloff in the transition region is considerably steeper than with a comparable Butterworth. The price is, of course, the ripple. For this

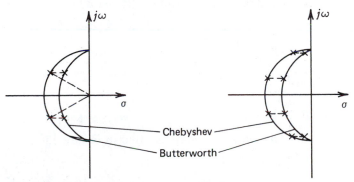

FIGURE 13-39
Comparison between Butterworth and Chebyshev poles.

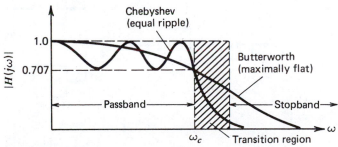

FIGURE 13-40
Comparison of same-order Butterworth and Chebyshev filter characteristics.

reason the Chebyshev function is often called the **equal ripple** or **stagger tuned** function. It should be noted that once out of the transition range, both filters roll off at a slope of $-n$, reflecting their order.

Up to now we have been discussing varied and sundry ways to design single- and multiple-pole high- and low-pass filters. We have avoided discussing bandpass and bandstop filters. The reason is that once the design of high- and low-pass filters is understood, it is a simple transition to combine those designs into either a bandpass or a bandstop filter. Figure 13-41 shows how high- and low-pass filters can be combined to achieve the desired bandpass or bandstop characteristics.

In designing the filters of Figure 13-41 a tacit but important assumption has been made. It has been assumed that the cascaded or otherwise combined filters do not load each other. In general, if active filters are used, the assumption is usually justified. However, if passive filters are used in the design, the results are far less predictable. If we assume that loading is not a problem then we see that cascading a high-pass and a low-pass filter will produce a bandpass filter (Figure 13-41a) provided we select the cutoff frequency of the high-pass filter, ω_{cHPF}, to be smaller than the cutoff frequency of the low-pass filter, ω_{cLPF}. In fact if the two cutoff frequencies are separated by more than one decade, the two cutoff frequencies will be approximately equal to the cutoff frequencies of the bandpass filter desired. That is, if

$$\omega_{cLPF} > 10\omega_{cHPF}$$

then

$$\omega_{cHPF} \simeq \omega_{c1BPF} = \omega_{c1}$$

and

$$\omega_{cLPF} \simeq \omega_{c2BPF} = \omega_{c2}$$

This design is demonstrated in Figure 13-42.

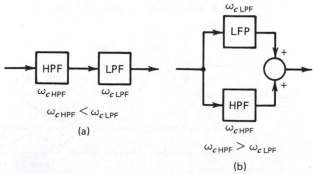

FIGURE 13-41
Bandpass and bandstop filters. (*a*) Bandpass filter. (*b*) Bandstop filter.

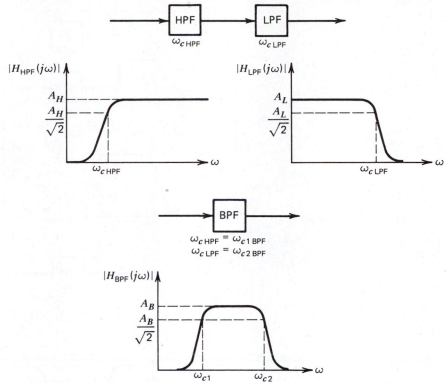

FIGURE 13-42
Bandpass filter design.

Designing a bandstop filter is only a bit more challenging than a band-pass filter (Figure 13-41b). Here the input signal is split so as to take two paths. The top path allows signals up to the cutoff frequency of the low-pass filter to pass, that is, up to ω_{cLPF}, while the lower path permits signals from the cutoff frequency of the high-pass filter on up to pass, that is, from ω_{cHPF}. If ω_{cLPF} is less than ω_{cHPF}, then the band between these two frequencies is a stopband. This design is shown in Figure 13-43.

Once again, if the two critical frequencies are separated by more than one decade, then the two cutoff frequencies are approximately the cutoff frequencies of the bandstop filter. In this case, if

$$\omega_{cHPF} > 10\omega_{cLPF}$$

then

$$\omega_{cLPF} \simeq \omega_{c1BSF} = \omega_{c1}$$

and

$$\omega_{cHPF} \simeq \omega_{c2BSF} = \omega_{c2}$$

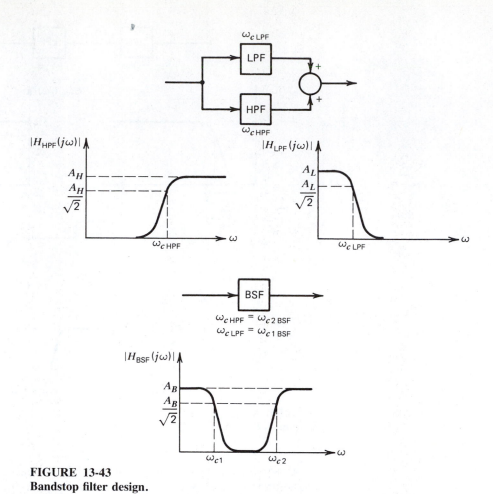

FIGURE 13-43
Bandstop filter design.

Example 13-16

Design, using one-pole filters, an audio filter that will pass all frequencies between 20 Hz and 20 kHz. The bandpass gain will be variable from 0.1 to 10.

There are several approaches to this problem, but probably the easiest will be to use OP AMP sections to realize each of the blocks, Figure 13-44a. Since we are to design a bandpass filter the approach of Figure 13-42 is applicable. The low-pass filter is readily designed with $\omega_{cLPF} = 20 \times 10^3 \times 2\pi$ rad/sec, as shown in Figure 13-44b. The high-pass filter is de-

signed using $\omega_{cHPF} = 20 \times 2\pi$ rad/sec, as shown in Figure 13-44c. The variable gain is simply added via a cascaded inverter as shown in Figure 13-44d. Since OP AMP circuits can be cascaded without loading each other, the resulting transfer function of the cascade is simply the product of the three circuit transfer functions. Clearly the design will work but is not the most efficient. A more efficient design is shown in Figure 13-44e, where the variable gain is included within the high-pass filter circuit.

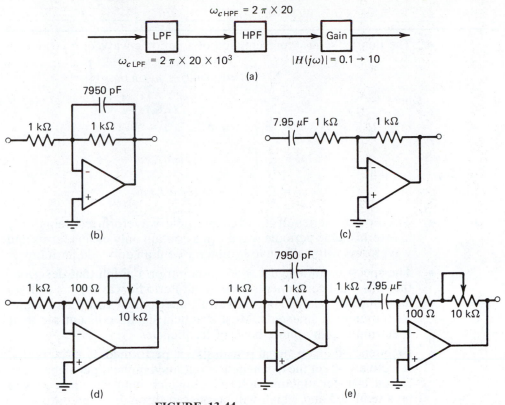

FIGURE 13-44
Design solutions for Example 13-16. (*a*) Design approach. (*b*) Low-pass filter with $\omega_{cLPF} = 2\pi \times 20 \times 10^3$ rad/second. (*c*) High-pass filter with $\omega_{cHPF} = 2\pi \times 20$ rad/second. (*d*) Variable gain inverter from 0.1 to $\simeq$ 10. (*e*) More efficient design.

SUMMARY

- The Fourier series representation of a periodic waveform $v(t)$ is

$$v(t) = a_0 + \sum_{n=1}^{\infty} \left[a_n \cos(n\omega_0 t) + b_n \sin(n\omega_0 t) \right]$$

  where
$$\omega_0 = 2\pi/T_0$$

$$a_0 = \frac{1}{T_0} \int_0^{T_0} v(t)\, dt$$

$$a_n = \frac{2}{T_0} \int_0^{T_0} v(t) \cos(2\pi nt/T_0)\, dt$$

$$b_n = \frac{2}{T_0} \int_0^{T_0} v(t) \sin(2\pi nt/T_0)\, dt$$

- The Fourier series of an even periodic waveform contains only sine terms. Odd periodic waveforms contain only cosine terms and waveforms with half-wave symmetry contain only odd harmonics.

- The spectrum of a signal is an equation or a graph that describes the sinusoidal components in a signal. Periodic waveforms contain harmonically related sinusoids while aperiodic waveforms contain a continuum of sinusoids. Most practical signals concentrate their spectrum within a finite band of frequencies.

- Frequency-domain signal processing is performed by circuits that selectively affect the frequencies contained in the input signal. The circuit transfer function determines which input frequencies will be attenuated and which will be passed through to the output.

- For a sinusoidal input $A \cos(\omega t + \phi)$ to a circuit with a transfer function $H(s)$, the sinusoidal steady-state output is

$$y_{SS} = A|H(j\omega)| \cos[\omega t + \phi + \angle\, H(j\omega)]$$

  The functions $|H(j\omega)|$ and $\angle\, H(j\omega)$ are called the circuit gain and phase respectively.

- The four prototype circuit gain characteristics are low pass, high pass, bandpass, and bandstop. The frequency associated with the boundary between a transmission band and the adjacent stopband is called the cutoff frequency.

- The major features of the gain of the one-pole low-pass and high-pass prototypes

$$H_{LP} = \frac{b_0}{s + \alpha} \qquad H_{HP} = \frac{b_1 s}{s + \alpha}$$

  are described by the low-frequency and high-frequency asymptotes:
  Low frequency $\omega \ll \alpha$

$$H_{LP} \to b_0/\alpha \qquad H_{HP} \to b_1\omega/\alpha$$

High frequency $\omega >> \alpha$

$$H_{LP} \to b_0/\omega \qquad H_{HP} \to b_1$$

These two asymptotes intersect at the cutoff frequency $\omega = \alpha$.

- Bode plots are simple straight-line approximations to the gain and phase characteristics of transfer functions. Bode plots are particularly useful if the critical frequencies (poles and zeros) are all located on the negative real axis in the s plane. Each critical frequency introduces a gain slope change of ± 1 and a phase slope of $\pm 45°$/decade, where the plus sign applies to zeros and the minus sign to poles.

- The major features of the two-pole low-pass, bandpass, and high-pass prototypes

$$H_{LP} = \frac{b_0}{s^2 + 2\zeta\omega_0 s + \omega_0^2} \qquad H_{BP} = \frac{b_1 s}{s^2 + 2\zeta\omega_0 s + \omega_0^2}$$
$$H_{HP} = \frac{b_2 s^2}{s^2 + 2\zeta\omega_0 s + \omega_0^2}$$

are described by the low- and high-frequency asymptotes
Low frequency $\omega << \omega_0$

$$H_{LP} \to b_0/\omega_0^2 \qquad H_{BP} \to b_1\omega/\omega_0^2 \qquad H_{HP} \to b_2\omega^2/\omega_0^2$$

High frequency $\omega >> \omega_0$

$$H_{LP} \to b_0/\omega^2 \qquad H_{BP} \to b_1/\omega \qquad H_{HP} \to b_2$$

These asymptotes intersect at $\omega = \omega_0$.

- The bandwidth of the two-pole bandpass prototype is

$$BW = 2\ \zeta\omega_0$$

For the wideband case $\zeta > 0.5$ and $BW > \omega_0$. For the narrowband case $\zeta < 0.5$ and $BW < \omega_0$.

- The quality factor of a bandpass response defined as

$$Q = \frac{\text{Center frequency}}{\text{Bandwidth}}$$

is a measure of the selectivity of the circuit. For a two-pole bandpass prototype $Q = 1/(2\zeta)$. The boundary between narrowband and wideband response occurs for $Q = 1$.

- Filters are frequency-selective circuits designed to achieve one of the four prototype responses; low pass, bandpass, high pass, or bandstop. The design of multiple-pole filters is greatly simplified by using OP AMP circuits. Individual OP AMP circuits are designed to realize one or two poles and the overall multipole transfer function obtained by cascading the individual circuits.

EN ROUTE OBJECTIVES
AND RELATED EXERCISES

13-1 *FOURIER SERIES (SEC. 13-1)*

Given an equation or graph of a periodic waveform:

(a) *Determine the Fourier series representation of the waveform.*

(b) *Plot the amplitude spectrum of the waveform.*

(c) *Determine the signal bandwidth using a specified criterion.*

Exercises

In each of the following exercises an equation or graph is given that defines a periodic waveform.

(a) Determine the Fourier coefficients a_o, a_n, and b_n.

(b) Plot the amplitude spectrum of the signal.

(c) Determine the signal bandwidth using the criterion that harmonics, whose amplitude is less than 5 percent of the fundamental, are negligible.

13-1-1   Use the periodic waveform shown in Figure E13-1-1.

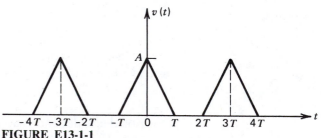

FIGURE E13-1-1
Waveform for Exercise 13-1-1.

13-1-2   Use the periodic waveform shown in Figure E13-1-2.

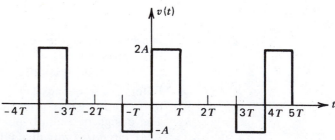

FIGURE E13-1-2
Waveform for Exercise 13-1-2.

13-1-3 Use the periodic waveform defined as $v(t) = |A \sin (2\pi t/T_0)|$.

13-1-4 Use the periodic waveform defined as $v(t) = |A \cos (2\pi t/T_0)|$.

13-1-5 Use the periodic waveform defined in Figure E13-1-5.

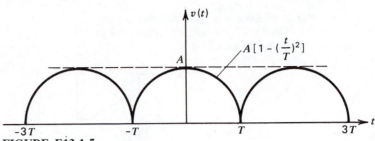

FIGURE E13-1-5
Waveform for Exercise 13-1-5.

13-1-6 Use the periodic waveform defined in Figure E13-1-6.

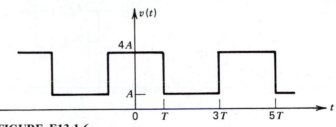

FIGURE E13-1-6
Waveform for Exercise 13-1-6.

13-2 *ONE-POLE FREQUENCY RESPONSE (SECS. 13-2 to 13-4)*

Given a circuit with a one-pole low-pass or high-pass transfer function:

(a)   Determine the quantities H(s), ω_c, |H(0)|, |H(jω_c)|, |H(∞)|, $\angle$ H(0), $\angle$ H(jω_c), and $\angle$ H(∞) in terms of circuit parameters.

(b)   Construct the straight-line approximation to the gain and phase characteristics and sketch the actual curves.

Exercises

The following set of exercises uses transfer functions obtained in Chapter 10. For each exercise determine the straight-line approximations to the gain and phase characteristics in terms of circuit parameters and sketch the actual curves.

13-2-1 Use the transfer function found in Exercise 10-3-1.

13-2-2 Use the transfer function found in Exercise 10-3-2.

13-2-3 Use the transfer function found in Exercise 10-3-3.

13-2-4 Use the transfer function found in Exercise 10-3-4.

The following set of exercises uses different combinations of the circuits $C1$ through $C8$ connected in cascade as shown in Figure E13-2-1.

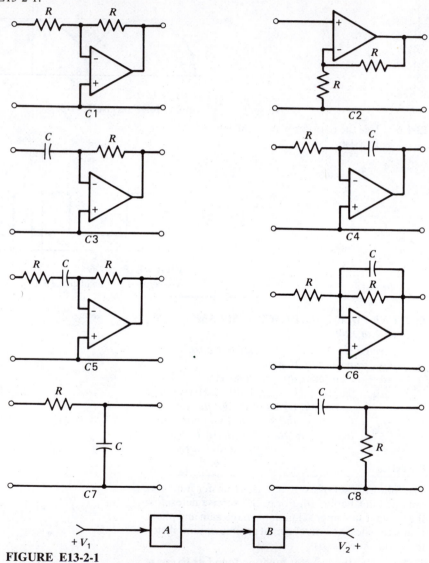

FIGURE E13-2-1
OP AMP and other circuits for Exercises 13-2-5 through 13-2-12.

For each exercise connect the indicated circuits in cascade and solve for the transfer function of the combination. Then determine the straight-line approximations to the gain and phase characteristics and construct the frequency response plots in terms of circuit parameters.

Exercise	Circuit A is	Circuit B is	$T(s)$ is
13-2-5	C1	C6	V_2/V_1
13-2-6	C2	C8	V_2/V_1
13-2-7	C3	C6	V_2/V_1
13-2-8	C4	C8	V_2/V_1
13-2-9	C5	C2	V_2/V_1
13-2-10	C6	C2	V_2/V_1
13-2-11	C7	C1	V_2/V_1
13-2-12	C8	C1	V_2/V_1

13-2-13 Show that the block diagram of Figure E13-2-13 is a low-pass filter with $\omega_c = K$.

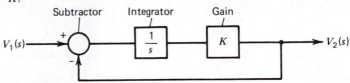

FIGURE E13-2-13
Block diagram of a low-pass filter.

13-2-14 Using only an integrator, a subtractor, and a gain block, devise a block diagram similar to Figure E13-2-13 for a high-pass filter.

13-2-15 Realize the filter of Figure E13-2-13 using standard OP AMP building blocks. Assume $\omega_c = 100$ rad/second.

13-3 BODE PLOTS (SEC. 13-4)

Given a transfer function with real poles and zeros, construct the straight-line approximations to the gain and phase characteristics. Conversely, given the straight-line approximations to the gain or phase characteristics, determine the corresponding transfer function.

Exercises

13-3-1 Construct the Bode plots (gain and phase) for the following transfer functions.

(a) $H(s) = \dfrac{1000}{(s + 10)(s + 50)}$

(b) $H(s) = \dfrac{100s}{(s + 10)(s + 50)}$

(c) $H(s) = \dfrac{10^5 s}{(s + 10)(s + 50)(s + 1000)}$

(d) $H(s) = \dfrac{1000s(s + 100)}{(s + 10)(s + 50)(s + 1000)}$

13-3-2　Repeat Exercise 13-3-1 for the following transfer functions.

(a) $H(s) = \dfrac{10s}{(s + 2)}$

(b) $H(s) = \dfrac{100s}{(s + 2)^2}$

(c) $H(s) = \dfrac{10s^2}{(s + 2)^2}$

(d) $H(s) = \dfrac{4000}{(s + 2)^2(s + 2000)}$

13-3-3　Determine the transfer function corresponding to the Bode plot in Figure E13-3-3.

13-3-4　Repeat Exercise 13-3-3 for the Bode plot in Figure E13-3-4.

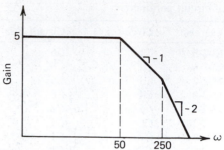

FIGURE E13-3-3
Gain characteristic for Exercise 13-3-3.

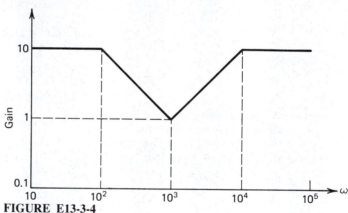

FIGURE E13-3-4
Gain characteristic for Exercise 13-3-4.

13-3-5 Repeat Exercise 13-3-3 for the Bode plot in Figure E13-3-5.

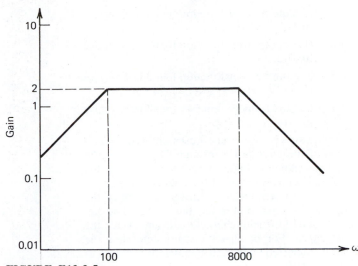

FIGURE E13-3-5
Gain characteristic for Exercise 13-3-5.

13-4 *TWO-POLE FREQUENCY RESPONSE (SECS. 13-2, 13-4, and 13-5)*

Given a circuit with a two-pole low-pass, band-pass, or high-pass transfer function:

(a)   Determine $|H(0)|$, $|H(j\omega_0)|$, and $|H(\infty)|$ in terms of circuit parameters.

(b)   Construct the straight-line approximation to the gain characteristic and sketch the actual curve for specified numerical values of the circuit parameters.

(c)   Determine the bandwidth and cutoff frequencies of bandpass circuits.

Exercises

The following set of exercises uses transfer functions obtained in Chapter 10. For each exercise construct the straight-line approximations to the gain characteristic and sketch the actual curve when all resistors are 1 kΩ, all capacitors are 1 μF, and all inductors are 1 H. For bandpass circuits determine the circuit bandwidth and cutoff frequencies using these numerical values.

13-4-1 Use the transfer function found in Exercise 10-3-7.

13-4-2 Use the transfer function found in Exercise 10-3-8.

13-4-3 Use the transfer function found in Exercise 10-3-9.

13-4-4 Use the transfer function found in Exercise 10-3-11.

13-4-5 Use the transfer function found in Exercise 10-3-12.

13-4-6 Use the transfer function found in Exercise 10-3-17.

13-4-7 Use the transfer function found in Exercise 10-3-18.

The following set of exercises uses different combinations of the circuits $C5$ through $C8$ connected in cascade as shown in Figure E13-2-1. For each of the following exercises, connect the indicated circuits in cascade and determine the transfer function of the combination. Then determine the straight-line approximation to the gain characteristic and sketch the actual curve. For bandpass circuits determine the circuit center frequency and bandwidth in terms of the product RC.

Exercise	Circuit A is	Circuit B is	$T(s)$ is
13-4-8	$C5$	$C6$	V_2/V_1
13-4-9	$C5$	$C8$	V_2/V_1
13-4-10	$C6$	$C6$	V_2/V_1
13-4-11	$C6$	$C8$	V_2/V_1
13-4-12	$C7$	$C5$	V_2/V_1
13-4-13	$C7$	$C7$	V_2/V_1
13-4-14	$C8$	$C7$	V_2/V_1
13-4-15	$C8$	$C8$	V_2/V_1

13-5 ACTIVE AND PASSIVE FILTER DESIGN (SEC. 13-6)

Design an active or passive circuit with up to four poles to realize a given filter specification.

Exercises

13-5-1 Design a two-stage RC-OP AMP circuit to realize the transfer function whose Bode plot is shown in Exercise 13-3-3.

13-5-2 Repeat Exercise 13-5-1 for the Bode plot given in Exercise 13-3-4.

13-5-3 Repeat Exercise 13-5-1 for the Bode plot given in Exercise 13-3-5.

13-5-4 Design a two-stage RC-OP AMP bandpass filter with a passband gain of 5 and a center frequency of $\omega_0 = 1000$, and a bandwidth $= 2000$.

13-5-5   Design a single-stage *RC*-OP AMP bandpass filter with a center frequency of $\omega_0 = 1000$ and a bandwidth $= 200$.

13-5-6   Design a two-stage *RC*-OP AMP circuit to realize a third-order Butterworth low-pass filter with $f_c = 100$ Hz.

13-5-7   A circuit is required to realize the high-pass transfer function

$$H(s) = \frac{1000s}{s + 100}$$

The following components are available:

Resistors	Capacitors	Inductors
100 Ω, $0.05 each	10 pF, $0.20	1 mH, $0.50
1 kΩ	100 pF, 0.20	10 mH, 0.50
10 kΩ	1000 pF, 0.20	100 mH, 0.50
100 kΩ	0.01 μF, 0.30	1 H, 1.00
330 kΩ	1.0 μF, 0.30	10 H, 2.00
1 MΩ	10 μF, 0.50	100 H, 10.00
	47 μF, 1.00	

741 OP AMPs, $0.60 each.
Power supply (± 15 V), $15.00

Design a minimum cost circuit.

PROBLEMS

P13-1 (Design)
Your small company is designing an inexpensive radio to receive the AM broadcast band. You are the design engineer for the intermediate-frequency (IF) stage. The total radio will sell for under $10.00, so your portion of the design should cost under $2.00. You are to design a fixed bandpass filter with the following specifications ±10 percent: center frequency (f_0)—455 kHz; bandwidth (BW)—8 kHz; selectivity (Q) ≥ 50; gain (K) at f_0 = 2. In estimating your cost use the following schedule:

Your time—$60/hour; estimate that 1000 radios will be built.

Standard values only:
Resistors	— ±5 percent (carbon),	0.05 each
	— ±10 percent (carbon),	0.02 each
Trim resistors	— ±1 percent (wirewound),	0.25 each
Capacitors	— ±5 percent (tubular),	0.45 each
	— ±10 percent (disk),	0.10 each
	— ±10 percent (electrolytic),	0.50 each
Inductors	— ±5 percent (air wound),	0.60 each
OP AMPs	—741's GBW = 1 MHz,	0.65 each

Power is already available and paid for, as is space on a printed circuit board.

P13-2 (Analysis)
A filter known as a quasi-low-pass filter is shown in Figure P13-2.[1] Find its transfer function and sketch its Bode gain diagram.

P13-3 (Evaluation)
A certain signal generator designed to operate up to 1 MHz can be modeled as shown in Figure P13-3a. The output of the generator must be filtered to keep spurious responses over 10 MHz from exiting the generator. Several designs are proposed in Figure

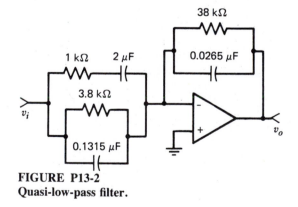

FIGURE P13-2
Quasi-low-pass filter.

[1]B. Welling, "Analysis and Design of Active Filters Using Operational Amplifiers," Application Note AN-438, p. 7, Motorola, Phoenix, Ariz. November 1968.

P13-3*b*, *c*, *d*. As design engineer for this task, which would you select and why? (*Hint:* Consider the effects of loading.)

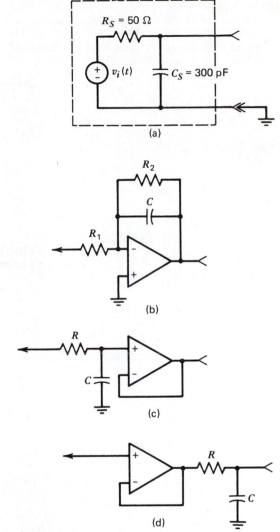

FIGURE P13-3
High-pass filters.

P13-4 (Analysis)

The circuit shown in Figure P13-4 produces a pair of zeros that are located on the imaginary axis and whose location depends on the gain K of the noninverting amplifier. Show that this is so.

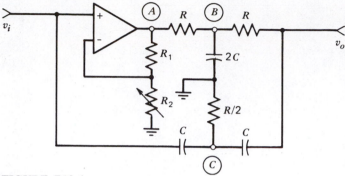

FIGURE P13-4
Circuit with zero on imaginary axis.

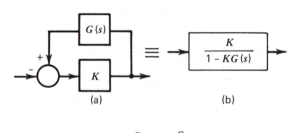

P13-5 (Analysis, Design)

Refer to the block diagram of a typical feedback system shown in Figure P13-5a. The transfer function of the feedback block is usually referred to as **feedback gain, $G(s)$.** An equivalent block diagram is shown in Figure P13-5b. We can readily see that the poles of the system are found by setting $1 - KG(s) = 0$. Consider the *Wien-bridge oscillator* shown in Figure P13-5c. Find its feedback transfer function and select K, the gain of the amplifier, to achieve oscillation (poles on imaginary axis).

FIGURE P13-5
Feedback circuit—a Wien-bridge oscillator.

P13-6 (Analysis)

All-pass networks are often used to alter a signal's phase while leaving its amplitude unchanged. Such an application occurs in steering the beams in phased-array antennas. Consider the circuit in Figure P13-6. Show that it indeed is a realization of a first-order all-pass filter, that is, show $|H(\omega)|$ = a constant and $\angle H(\omega)$ = function of ω.

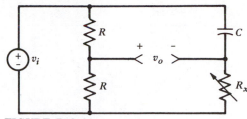

FIGURE P13-6
An all-pass (AP) network.

P13-7 (Analysis)

Some crystals, such as quartz, can be made to deform when a potential is applied across them. Such materials are known as piezoelectric and are used in applications where high-Q responses are needed. Figure P13-7b shows a circuit model of a crystal. The R, L, and C are the electrical equivalents of the miniature mechanical system the crystal more accurately resembles. C', however, is the electrostatic capacitance between the electrodes with the crystal as the dielectric. For the values given, determine the circuit's resonant frequency and Q.

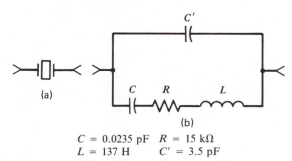

$$C = 0.0235 \text{ pF} \quad R = 15 \text{ k}\Omega$$
$$L = 137 \text{ H} \quad C' = 3.5 \text{ pF}$$

FIGURE P13-7
A piezoelectric crystal and equivalent circuit.

P13-8 (Design)

Touch-Tone telephones use a frequency multiplexing scheme to determine which key is depressed. Figure P13-8 shows the Touch-Tone frequencies. Whenever a key is touched, two audio frequencies are added together and transmitted—one from the high group, another from the low group. These multiple signals are decoded at a central office. Assuming each individual frequency to be pure (approximately 0 BW), design a block diagram for the decoding system. Then design a highly selective two-pole filter for separating the 1336-Hz signal from the others. How much attenuation is provided at 1209 Hz by your filter?

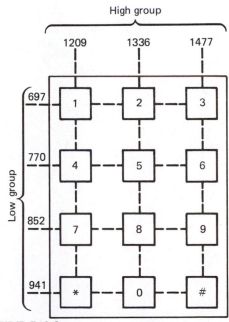

FIGURE P13-8
Touch-Tone ® keyboard and frequency groups.

Appendix A
Block Objectives and Related Problems

One must learn by doing the thing: for though you think it, you have no certainty until you try.

Sophocles

This appendix contains problems that support the block objectives given at the beginning of each of the three parts of this text. Each part has a set of three block objectives defined at the analysis, design, and evaluation levels. The en route objectives given in each chapter are designed to lead to the ability to meet the block objectives. The problems in this appendix are in turn designed to test the ability to meet one or more of the block objectives.

BLOCK I
MEMORYLESS CIRCUITS
(Chapters 1–5)

BLOCK OBJECTIVES

Analysis

Given a memoryless circuit with prescribed input signals, determine prescribed output signals or input-output relationships using any analysis technique.

Design

Devise a memoryless passive or active circuit or modify an existing circuit to obtain a specified output signal for given input signals or to implement a given input-output relationship.

Evaluation

Given two or more memoryless circuits that perform the same signal-processing function, select the best circuit based on given criteria such as performance, cost, parts count, power dissipation, and simplicity.

PROBLEMS

B1-1 (Analysis, Design, Evaluation)

Circuit $C1$ in Figure A-B1-1 is linear so its input-output relationship is of the form

$$v_o = K_2v_2 + K_1v_1$$

(a) (Analysis) Determine the constants K_1 and K_2 in terms of circuit resistances.

(b) (Design) For $R_1 = 1 \text{ k}\Omega$ and $R_2 = 2 \text{ k}\Omega$, select the values of the remaining circuit resistances to achieve

$$v_0 = 10(v_2 - v_1)$$

(c) (Evaluation) Show that the circuit $C2$ in Figure A-B1-1 meets the design requirement in (b). Compare these two designs on the basis of the number of devices required and the loading of the two input signal sources.

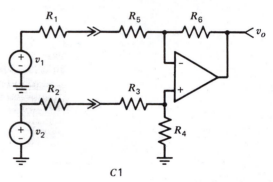

$C1$

$$R_3 = R_6 = 10 \text{ k}\Omega$$
$$R_4 = R_5 = 90 \text{ k}\Omega$$

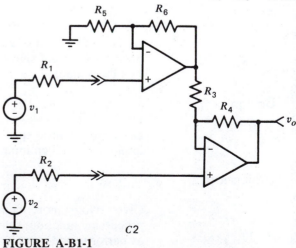

$C2$

FIGURE A-B1-1
Two linear summers.

B1-2 (Analysis, Design, Evaluation)
Circuit $C1$ in Figure A-B1-2 contains a photoresistor whose resistance varies inversely with the intensity of incident light. In complete darkness its resistance is 10 kΩ. In bright sunlight its resistance is 1 kΩ.

(a) (Analysis) At any one light intensity the circuit is linear so its input-output relationship is of the form $V_0 = Kv_1$. Determine the constant K in terms of circuit resistances.

(b) (Design) For $v_1 = +15$ V select the values of R and R_F so that $v_o = -10$ V in bright sunlight and $v_o = +10$ V in complete darkness.

(c) (Evaluation) Show that circuit $C2$ in Figure A-B1-2 meets the design requirement given in (b). Compare these two designs on the basis of the number of devices required and the total power dissipated.

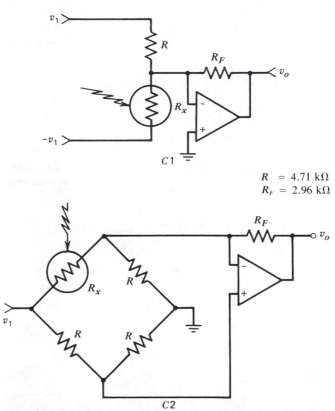

$R = 4.71$ kΩ
$R_F = 2.96$ kΩ

FIGURE A-B1-2
Two circuits for outputting a voltage proportional to temperature.

B1-3 (Analysis, Design)

The source circuit in Figure A-B1-3 is to be connected to R_L.

(a) Select R_L so that maximum power is delivered to the load.

(b) Select R_L so that maximum voltage is delivered to the load.

(c) Select R_L so that the load current is 10 mA.

(d) For $R_L = 1$ kΩ design an interface circuit so that the load current is 5 mA.

(e) Repeat (d) for a load current of 10 mA.

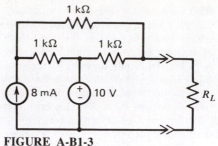

FIGURE A-B1-3
Maximum power transfer problem.

B1-4 (Analysis, Design)

The source circuit in Figure A-B1-4 contains an adjustable potentiometer. The load resistance is fixed. An interface circuit is to be designed so that the power delivered to the load varies between 0 and 20 mW as the potentiometer is adjusted over its full range.

(a) (Analysis) Show that the interface circuit must contain an amplifier with a voltage gain of at least 4.

(b) (Design) Design an interface circuit to meet the objective given above.

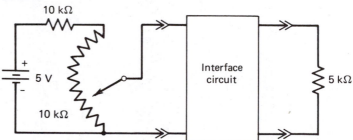

FIGURE A-B1-4
A potentiometer interfacing problem.

B1-5   (Design, Evaluation)

A system contains a sensor that requires an excitation voltage of 10 V ± 0.1 V. The input resistance of the sensor varies from 1 kΩ to 1 MΩ. The system contains a power supply that provides +15, −15, and +5 V. Figure A-B1-5 shows two integrated circuit packages that are on the approved parts list for the system and must be used in this design.

(a)   (Design) Design at least two interface circuits using only the two IC packages in the figure. You may use any number of packages but the best design uses the fewest total package count. There are at least a half dozen two-package designs.

(b)   (Evaluation) Compare your two designs in terms of the power dissipated in the R-$2R$ package. Two-package designs range from 5 to 100 mW.

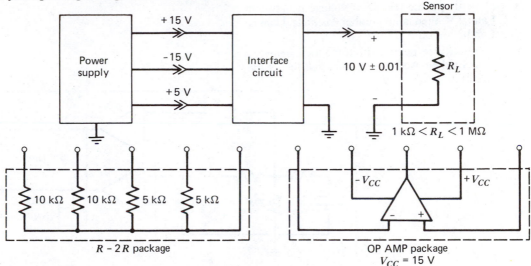

FIGURE A-B1-5
Interface design problem using OP AMP/R-$2R$ ICs.

B1-6 (Analysis)

Figure A-B1-6 shows a simplified version of a system called a logic analyzer. When the probe is connected to a circuit test point, the analyzer detects whether the voltage (relative to ground) is greater than 2 V (logic high), less than 1 V (logic low), or between 1 and 2 V (an ambiguous case indicating a circuit fault). In this circuit the output of either comparator is 5 V if $v_P > v_N$ and 0 V if $v_P < v_N$. The NAND gate is a digital device whose output is 5 V if, and only if, both of its inputs are zero.

(a) Show that if the test point voltage is greater than 2 V, then lamp number 1 is on and lamps 2 and 3 are off.

(b) Show that if the test point voltage is less than 1 V, then lamp number 2 is on and lamps 1 and 3 are off.

(c) Show that if the test point voltage is between 1 and 2 V, then lamp number 3 is on and lamps 1 and 2 are off.

(d) Identify three points within the logic analyzer that could be used to perform a self-test of the analyzer.

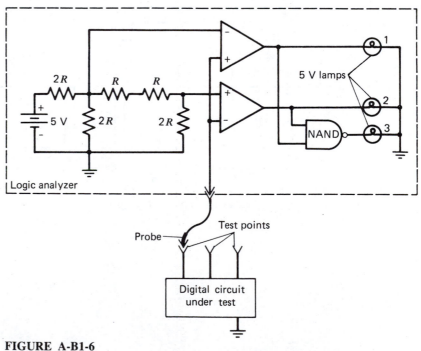

FIGURE A-B1-6
A logic analyzer.

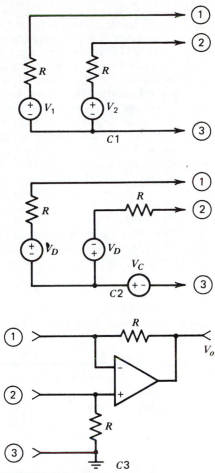

B1-7 (Analysis)

Circuit $C1$ in Figure A-B1-7 shows a three-terminal source configuration. Circuit $C2$ is another three-terminal source where

$$V_C = \frac{V_1 + V_2}{2} = \text{Common mode voltage}$$

$$V_D = \frac{V_1 - V_2}{2} = \text{Differential mode voltage}$$

(a) Show that $C2$ is equivalent to $C1$ by showing that the open-circuit voltage and short-circuit current "seen" between any two pairs of terminals in $C2$ are the same as those "seen" in $C1$.

(b) Now connect the output terminals of $C2$ to the input terminals of the OP AMP circuit $C3$ and show that $V_o = -2\,V_D$. That is, show that the circuit responds only to the differential mode input and rejects the common mode. The use of superposition is suggested.

FIGURE A-B1-7

Common mode and differential mode analysis problem.

B1-8 (Analysis)

Figure A-B1-8 shows an ideal voltage source in parallel with an adjustable potentiometer. This problem investigates the effect of adjusting the potentiometer on the Thévenin equivalent circuit seen at the interface.

(a) Determine the Thévenin equivalent at the indicated interface in terms of k, V_o, and R.

(b) Show that $R_{TH} = 0$ when either $k = 0$ or $k = 1$. Explain this result physically in terms of the position of the movable arm of the potentiometer.

(c) Show that the maximum power available at the interface is infinity for $k = 0$ and zero for $k = 1$. Explain this result physically in terms of the position of the movable arm of the potentiometer.

(d) Show that the short-circuit current is $V_o/(kR)$ for $k \neq 1$, but is physically difficult to determine $k \equiv 1$.

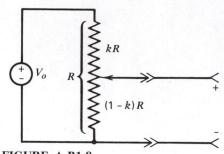

FIGURE A-B1-8
Thévenin equivalent of a variable voltage divider circuit.

B1-9 (Analysis)

Circuit $C1$ in Figure A-B1-9 shows a constant voltage source (V) in series with a source resistance (R) and connected at an interface to a load resistance (R_L). The Maximum Power Transfer Theorem indicates that the maximum power will be delivered to the load when the two resistors are matched $(R_L = R)$, and that the maximum power is

$$P_{\text{MAX}} = \frac{V^2}{4R}$$

(a) Circuit $C2$ shows the same source connected to a battery whose voltage is $V/2$. Show that the power delivered to the battery is equal to P_{MAX}.

(b) Since the battery clearly does not "match" the source resistance, can we conclude that circuit $C2$ disproves the Maximum Power Transfer Theorem? Before answering review the statement of the theorem in Chapter 3, noting carefully what it says and does not say.

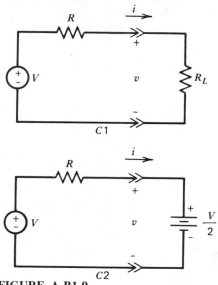

FIGURE A-B1-9
Maximum power transfer to a voltage source.

BLOCK II
CIRCUITS WITH MEMORY
(Chapters 6–10)

BLOCK OBJECTIVES

Analysis

Given a linear circuit containing not more than two memory elements with a prescribed input signal waveform, determine prescribed output signals or input-output relationships using either time-domain or s-domain techniques.

Design

Devise an active or passive circuit containing not more than two memory elements or modify an existing circuit to obtain a specified output signal for a given input signal or to implement a given input-output relationship.

Evaluation

Given two or more circuits with memory that perform the same signal processing function, select the best circuit based on given criteria such as performance, cost, parts count, power dissipation, and simplicity.

PROBLEMS

B2-1 (Analysis, Design)

The RC circuit in Figure A-B2-1 is driven by the exponential voltage shown.

(a) (Analysis) Determine $v_o(t)$ if there is no initial voltage on the capacitor. Identify the forced response, the natural response, the zero-state response, and the zero-input response.

(b) (Design) Determine the value of capacitance that is required to cause the response to be of the form Kte^{-100t}.

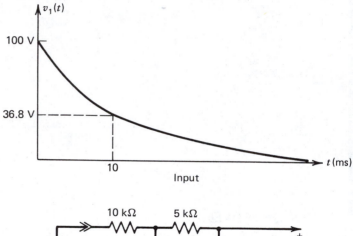

Input

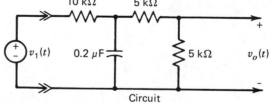

Circuit

FIGURE A-B2-1
Single-memory circuit and response.

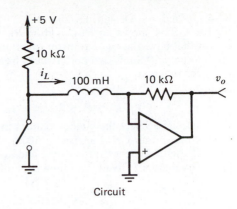

Circuit

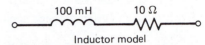

Inductor model

FIGURE A-B2-2
Analysis of "ideal" and "real" inductor in single-memory circuit.

B2-2 (Analysis)
The switch in Figure A-B2-2 has been open for a long time and then is closed at $t = 0$.
(a) Show that $i_L(t) = 0.5$ mA and $v_o(t) = -5$ V for all $t > 0$. That is, show that the circuit will "remember" the conditions that existed at $t = 0$ when the switch was closed.
(b) Now replace the 100-mH inductor by the model shown in the figure. Again let the switch be open for a long time and close at $t = 0$. Determine $i_L(t)$ and $v_o(t)$ for $t > 0$. About how long does it take for the circuit to "forget" the conditions that existed at $t = 0$?

B2-3 (Analysis)
The circuit in Figure A-B2-3 is driven by an input $v_1(t) = 10\ u(t)$. For $v_C(0) = 5$ V and $i_L(0) = 0$A:
(a) Determine the zero-state response and zero-input response.
(b) What initial condition would cause the natural response to be zero?
(c) What input would cause the forced response to be zero?

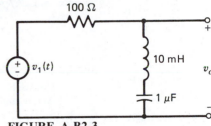

FIGURE A-B2-3
RLC **Circuit for analysis.**

B2-4   (Analysis, Design, Evaluation)
Figure A-B2-4 shows an *RLC* and an *RC*-OP AMP
circuit that have the same transfer function.
(a)   (Analysis) Show that both circuits have a transfer function

$$T(s) = \frac{V_2}{V_1} = \pm \frac{\omega_o^2}{s^2 + 2\zeta\omega_o s + \omega_o^2}$$

(b)   (Design) Select the element values in both circuits so that $\zeta = 1$ and $\omega_0 = 100$.
(c)   (Design) Repeat (b) for $\zeta = 1$ and $\omega_0 = 10^8$.
(d)   (Evaluation) Given the following criteria:

$$L \leqslant 1 \text{ H}$$
$$C \leqslant 10 \ \mu \text{ F}$$
$$\text{OP AMP bandwidth} \leqslant 10 \text{ MHz}$$

select the best design for each application.

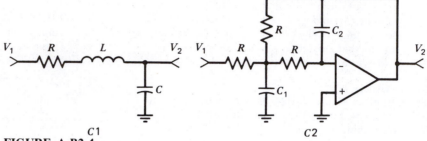

*C*1

FIGURE A-B2-4
Active and passive equivalent double-memory circuits.

B2-5   (Analysis, Design, Evaluation)
For circuits *C*1 and *C*2 in Figure A-B2-5:
(a)   (Analysis) Determine $T(s) = V_2/V_1$. Construct
a pole-zero diagram for each circuit transfer
function. Relate the pole and zero locations to
specific circuit elements.
(b)   (Design) Use a cascade connection of two of
these circuits to obtain the transfer functions

$$T_1(s) = \frac{s^2}{(s + 100)(s + 400)}$$
$$T_2(s) = \frac{(s + 200) s}{(s + 100)(s + 400)}$$
$$T_3(s) = \frac{(s + 50)(s + 200)}{(s + 100)(s + 400)}$$

(c)   (Evaluation) Show that these circuits *cannot*
be used to obtain the transfer function below
and explain why, in terms of poles and zeros.

$$T(s) = - \frac{s (s + 500)}{(s + 100)(s + 400)}$$

Then derive a rule that restricts the locations of the poles and zeros of $T(s)$ to ensure that it can be obtained using these circuits.

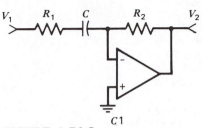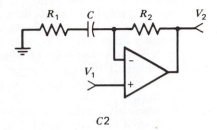

FIGURE A-B2-5
Two single-memory active circuits.

B2-6 (Analysis, Design)
The circuit in Figure A-B2-6 is a two-input cascade connection of three standard OP AMP building blocks.

(a) (Analysis) Find the s-domain input-output relationship in the from

$$V_o(s) = H(s)[V_1(s) + V_2(s)]$$

(b) (Analysis) Now remove the V_2 input source and make a feedback connection from the output to the former V_2 input, thus forcing the condition $V_2 = V_o$. Use the result found in (a) to determine the closed-loop transfer function $T(s) = V_o(s)/V_1(s)$

(c) (Analysis) Construct a pole-zero diagram of $T(s)$ and show that the closed-loop circuit is marginally stable.

(d) (Design) Select values of R and C so that the circuit will oscillate with a period of 10 ms.

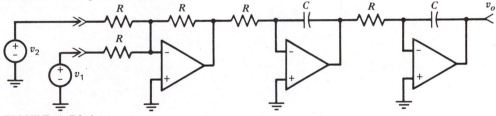

FIGURE A-B2-6
Cascade connection of three OP AMP buiding blocks.

B2-7 (Analysis, Evaluation)
The New Products Division of the RonA1 Corporation (founded by two well-known authors) has announced the availability of an IC package called the Universal Single-Pole Transfer Function Synthesizer (USPTS). Figure A-B2-7 shows the first page of the data sheet on this new device. Unfortunately the second page, which describes how to interconnect the circuit pins to obtain each transfer function, has been lost in the mail.

(a)   (Analysis) Reconstruct the missing second page by showing how to interconnect the circuit pins to obtain each transfer function listed. All connections are short circuits from pin to pin, pin to ground, or pin to power. No external components other than the power supply are required. (*Hint:* The output is always pin 7.)

(b)   (Evaluation) Explain why the transfer functions are independent of the load provided $R_L > 2 \ \text{k}\Omega$.

(c)   (Evaluation) Explain why the *RC* products are controlled to $\pm$ 0.1 percent whereas the OP AMP gain is controlled to within only a factor of 10.

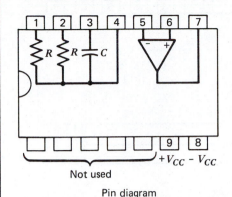

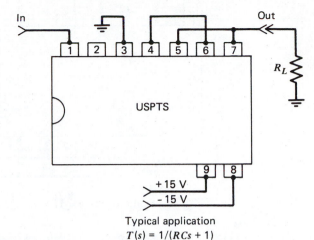

Typical application
$T(s) = 1/(RCs + 1)$

PERFORMANCE CHARACTERISTICS

1. OP AMP

For $V_{cc} = 15$ V output voltage swing $= \pm 14$ V
output current (MAX) $= 20$ mA
gain (MIN) $= 50$ V/mV
gain (MAX) $= 500$ V/mV

2. For $R_L > 2$ kΩ the available transfer functions are:

All Pole Type

$$\frac{1}{RCs + 1}, \frac{2}{RCs + 2}, \frac{1}{RCs + 2}, -\frac{1}{RCs + 1}, -\frac{1}{RCs}, -\frac{2}{RCs}$$

All Zero Type

$$RCs + 1, \frac{RCs + 2}{2}, RCs + 2, -(RCs + 1), -RCs, -\frac{RCs}{2}$$

Pole-Zero Type

$$\frac{RCs}{RCs + 1}, \frac{RCs}{RCs + 2}, \frac{RCs + 1}{RCs + 2}, \frac{RCs + 1}{RCs}, \frac{RCs + 2}{RCs}, \frac{RCs + 2}{RCs + 1}$$

3. To obtain different pole/zero locations order from standard models or order custom models in lots of 1000 or more.

Model	RC(ms)	Tolerance
USPTS— .1	0.1	± 1%
USPTS— .3	0.3	± 1%
USPTS— 1.0	1.0	± 1%
USPTS— 3.0	3.0	± 1%
USPTS—10.0	10.0	± 1%

USPTS DATA SHEET (PAGE 1)

FIGURE A-B2-7
(Ronal data sheet figures).

665

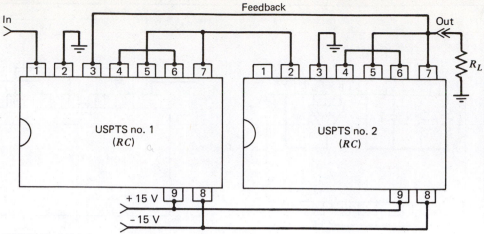

FIGURE A-B2-8
USPTS Application Note no. 1.

B2-8 (Analysis)

The RonA1 Corporation has issued Application Note No. 1 on the USPTS integrated circuit described in Problem B2-7. The note indicates that by interconnecting two circuits as shown in Figure A-B2-8 a two-pole transfer function with

$$\omega_0 = \sqrt{2}/RC$$

can be obtained.

(a) Verify the claim and show that the poles are complex.

(b) The Application Note further claims that by interchanging the connections at pins 2 and 3 on both circuits a two-pole transfer function with

$$\omega_0 = \sqrt{2}/RC$$

can be obtained. Verify this claim and show that the poles are complex.

B2-9 (Analysis)

Figure A-B2-9 shows two standard OP AMP circuits, an inverter and an inverting integrator. The two circuits have the same input and their outputs drive a comparator. The comparator in this circuit has an output $+V$ if $v_P > v_N$ and an output $-V$ if $v_P < v_N$. Throughout this problem we always assume that there is no initial voltage on the capacitor at $t = 0$.

(a) Show that if an input $+V$ is applied at $t = 0$, then the output is also $+V$ until $t = RC$ and is $-V$ thereafter.

(b) Show that if an input $-V$ is applied at $t = 0$, then the output is also $-V$ until $t = RC$ and is $+V$ thereafter.

(c) Now make a feedback connection from the output to the input, thus forcing the condition input = output. Assume that the initial comparator output is $+V$. Show that the circuit will oscillate so that the comparator output is a square wave with a period of $4RC$ and the integrator output is a triangular wave with the same period.

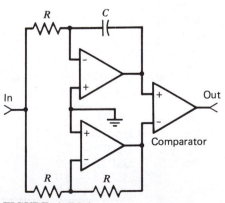

FIGURE A-B2-9
Linear and nonlinear OP AMP circuit.

666

BLOCK III
APPLICATIONS
(Chapters 11–13)

Block Objectives

Analysis

Given a linear circuit containing one or more memory elements, determine its step or impulse response including its damping and natural frequency, its sinusoidal response including the average power delivered at a prescribed interface, and its frequency response including a sketch of its Bode plot.

Design

Devise an active or passive circuit containing one or more memory elements if necessary or modify an existing circuit to obtain a specified step, sinusoidal, or frequency response.

Evaluation

Given two or more circuits that perform the same signal-processing function, select the best circuit based on given criteria such as performance, cost, parts count, power dissipation, and simplicity.

PROBLEMS

B3-1 (Analysis, Design)
A linear circuit is driven by an input $v_1(t) = \cos(2000t)$. The zero-state response due to this input is

$$v_o(t) = -e^{-1000t} + e^{-4000t} + 1.5 \sin(2000t)$$

(a) (Analysis) Determine the circuit transfer function.

(b) (Analysis) Determine the steady-state component of the output for an input $v_1(t) = \cos(4000t)$ and construct a phasor diagram showing the input and output.

(c) (Analysis) Determine and sketch the circuit step response.

(d) (Analysis) Construct the Bode plot of the circuit gain as a function of frequency.

(e) (Design) Design a circuit that would have all of these responses.

B3-2 (Analysis, Design, Evaluation)
Two double-pole bandpass filters are being considered for an application that requires a bandwidth of $1000/2\pi$ Hz. The step responses of the two filters are

$$g_1(t) = e^{-10t} \sin 100t$$
$$g_2(t) = e^{-10t} - e^{-1000t}$$

(a) Which filter best meets the requirement?

(b) Design a filter to meet this requirement.

B3-3 (Analysis, Evaluation)
Three single-pole filters are being considered for use in a low-speed digital application to eliminate high-frequency noise. The filter requirements are given in Table A-B3-3.

The Bode plots of the three candidate filters are shown in Figure A-B3-3.

(a) Which filter best meets the requirements?

(b) Design a filter that meets this requirement.

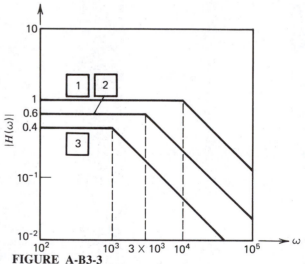

FIGURE A-B3-3
Three single-pole filter Bode diagrams.

Characteristic	Requirement
Steady-state output for 5 V-d-c input	2 to 7 V
Rise time for step input	Less than 1 ms
Gain at $f = 5$ kHz	Less than 10 percent of the maximum passband gain

TABLE A-B3-3

B3-4   (Analysis)

The RonA1 Corporation has issued Application Note No. 2 on the USPTS integrated circuit described in Problem B2-7. The note indicates that by interconnecting three USPTS circuits as shown in Figure A-B3-4 we obtain a third-order Butterworth low-pass filter with $\omega_o = 1/RC$. Verify this claim.

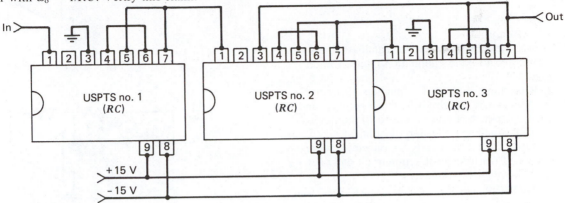

FIGURE A-B3-4
USPTS Application Note no. 2.

B3-5   (Analysis)

Filter characteristics are sometimes specified in terms of insertion ratio, which is defined as the quotient of the average power delivered to a load without and with the filter inserted.

(a)   Circuit $C1$ in Figure A-B3-5 shows a sinusoidal source directly connected to a load. Determine the average power delivered to the load in the sinusoidal steady state.

(b)   Circuit $C2$ in Figure A-B3-5 shows the same source and load with an inductor inserted between them. Determine the average power delivered to the load in the sinusoidal steady state with the inductor inserted.

(c)   Insertion ratio is defined as the ratio of power found in (a) to the power found in (b) above. Determine the insertion ratio of this circuit.

(d)   Show that the insertion ratio is one at zero frequency and explain this result physically in terms of the inductor impedance. Show that the insertion ratio is infinite at infinite frequency and explain this result physically.

(e)   Insertion loss in decibels is defined as
Insertion loss (dB) = 10 log (insertion ratio)
At what frequency is the insertion loss equal to 3 dB?

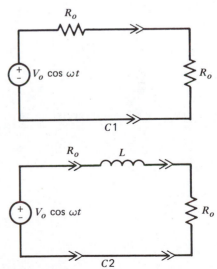

FIGURE A-B3-5
Insertion ratio problem.

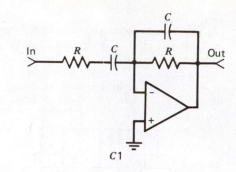

B3-6 (Analysis)

This problem demonstrates that bandpass filters can be converted to band elimination filters. Circuit $C1$ in Figure A-B3-6 is an RC-OP AMP bandpass filter.

(a) Determine the transfer function of circuit $C1$.

(b) Now insert circuit $C1$ between points A and B ($C1$ input to A and output to B) in circuit $C2$. Determine the transfer function of $C2$. Your analysis will be greatly simplified if you make use of the transfer function of $C1$ and view $C2$ as a weighted summer.

(c) Construct a pole-zero map of the transfer function found in (b). Sketch the gain response of the circuit noting the high- and low-frequency asymptotes and the location of the "notch."

B3-7 (Design)

Design a circuit whose gain response lies entirely within the unshaded region in Figure A-B3-7.

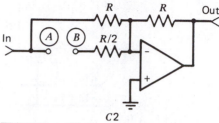

FIGURE A-B3-6
Bandpass–band elimination filter problem.

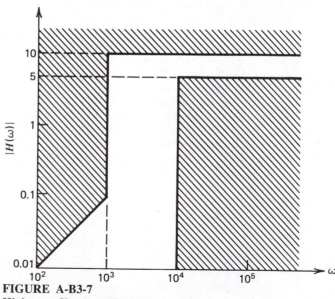

FIGURE A-B3-7
High-pass filter specifications.

B3-8   (Analysis)

We have noted several times that the two premier input test signals are the step function and the sinusoid. Both the step response and the frequency response of a circuit can be determined from the transfer function. It follows that we can infer something about the frequency response of a circuit by examining its step response. Figure A-B3-8 shows step responses of some two-pole circuits. Without doing any detailed calculations (actually you cannot as there are no scales on the figure):

(a)   Which circuits will have bandpass frequency responses?

(b)   Which circuits will have low-pass frequency responses?

(c)   Classify the circuits as broadband or narrowband based on what you can see in the step responses.

(d)   Now sketch a Bode plot (gain only) of each circuit roughly to scale indicating the high- and low-frequency asymptotes and any important peaks.

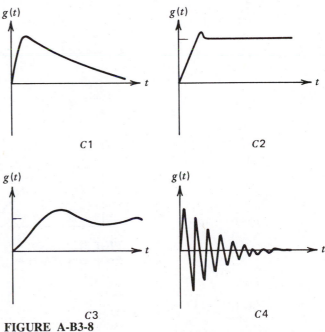

FIGURE A-B3-8
Step responses of various two-pole circuits.

B3-9   (Analysis)

This problem is the converse of Problem B3-8, which you should read before starting. In this problem we want to infer something about the step response of a circuit by examining its frequency response. Figure A-B3-9 shows the gain response of some double-pole circuits. Without doing any detailed calculations:

(a)   Which circuit(s) have step responses with IV = 0 and IV ≠ 0?

(b)   Which circuit(s) have step responses with FV = 0 and FV ≠ 0?

(c)   Which circuit(s) have step responses with lightly damped oscillations?

(d)   Now roughly sketch the step response of each circuit indicating the initial and final values and any important "ringing."

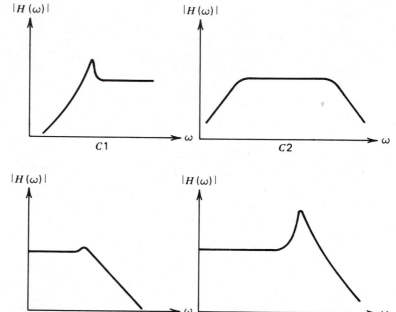

FIGURE A-B3-9

Frequency responses of various two-pole circuits.

Appendix B
Operational Amplifiers

OP AMP Characteristics

As an amplifier so connected can perform the mathematical operations of arithmetic and calculus on the voltages applied to its input, it is hereafter termed an 'operational amplifier.'

John R. Ragazzini

The integrated circuit OP AMP can be viewed as a circuit or as a device. The former viewpoint is necessary for the design of OP AMP circuits, while the device perspective is appropriate for the analysis and design of signal-processing circuits using OP AMPs. The dominant feature of the OP AMP device is high gain, but there are secondary charcteristics that limit the range of applications of any single device type. There is no single "best" OP AMP, only trade-offs to be made depending on the requirements of each application. This appendix discusses some of the important characteristics that limit the performance of OP AMPs.

673

OPERATIONAL AMPLIFIERS

A typical integrated circuit operational amplifier consists of a dozen or so transistors, roughly the same number of resistors, and perhaps one or two capacitors. These individual components are fabricated and interconnected on a tiny piece of silicon, and then packaged as a complete amplifier with only certain terminals brought out to the outside world. The finished package may contain as many as four individual amplifiers as illustrated in Figure B-1. Basically this package contains four independent amplifiers except that they share the same input power pins ($+V_{CC}$ and $-V_{CC}$) and offset control pins.

OP AMP characteristics are specified in many ways, but some of the important parameters are input offset, voltage gain, unity gain bandwidth, slew rate, and maximum output current.

Offset refers to the fact that the amplifier output is not generally zero when the input is zero. Offset is specified in terms of an input voltage (usually in millivolts) that must be applied to reduce the output to zero. The input offset current (usually in picoamperes) is the input current that exists when the output has been reduced to zero. Offset is important in low-level applications where the signal output may be small compared with the no signal output, that is, the offset. The offset can be reduced

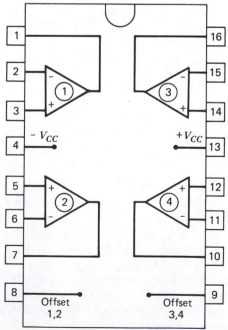

FIGURE B-1
Quad OP AMP integrated circuit.

No.	OP AMP Type	Input Offset Voltage (mV)	Input Offset Current (pA)	Voltage Gain (V/mV)	Unity Gain Bandwidth (MHz)	Slew Rate (V/μs)	Maximum Output Current (mA)
1	High power	5	200	75	1	3	500
2	Wideband	3	200	15	30	30	50
3	High slew rate	5	100	4	50	400	50
4	Precision input	0.05	2	500	0.4	0.06	1
5	General purpose	2	50	100	1	3	10

TABLE B-1
Representative Values of OP AMP Parameters

to zero by applying dc signals to the offset control pins. However, the offset parameters vary somewhat with temperature and from unit to unit.

The amplifier voltage gain is the parameter discussed in Chapter 4. It is usually specified in volts per millivolt. Thus a gain of 50 V/mV means $\mu = 50{,}000$. If the OP AMP is connected as a voltage follower, then the frequency range over which it exhibits a closed-loop gain of one is called the unity gain bandwidth (usually in megahertz), or equivalently, the gain–bandwidth product. Slew rate refers to the maximum allowable rate of change of the output voltage (usually in volts per microsecond). If the maximum slew rate is exceeded, the output becomes a ramp regardless of the input. Thus slewing is a nonlinear effect that must be avoided in linear applications. Finally, the maximum output current (usually in milliamperes) is given for a specified load.

Table B-1 lists some representative values of these parameters for different classes of amplifiers. The table is not intended to be definitive,

---

Historical Note

The first integrated circuit OP AMP to gain wide acceptance was the UA709. It required external discrete components to stabilize its operation. The UA709 was a very simple and versatile general-purpose OP AMP, but had rather poor latch-up characteristics. Latch-up refers to a condition that can occur when the output is driven into saturation. The output may "hang up" in this state even though the input is subsequently reduced to zero. The only way to remove the latch-up is to return the device to its zero state by completely deenergizing the power supply.

The next major advance in general-purpose OP AMP design was the UA741. The required stabilizing components were fabricated internally, which greatly simplifies the application of the device, but makes it somewhat less versatile than the UA709. The UA741 is fully protected against output short circuits, and does not latch up. The UA747 is a dual version of the UA741. A broad range of specialized OP AMPs was subsequently developed. Their characteristics include low input offsets, high voltage or current outputs, and high slew rates.

but only illustrative of the characteristics of several types of OP AMPs. As can be seen from the table, some of these parameters can be optimized for certain applications, but generally only at the expense of one or more of the other characteristics. For example, a high-power OP AMP can deliver a large output current but has relatively large offset parameters. Conversely a precision input amplifier has low offset and high gain, but at the price of bandwidth, slew rate, and output current. As the name suggests, the general-purpose amplifier strikes a middle ground and is intended to serve in a variety of applications.

Appendix C
Passive Components

Resistors
Potentiometers
Capacitors
Inductors

The circuit representation of systems for analysis and design is commonplace in electrical engineering, making it essential that the foundations and limitations of the circuit concept be well understood.

M. E. Van Valkenburg

Until about 1950 circuit theory was based on three passive components: resistors, capacitors, and inductors. With the development of solid-state electronics the number of basic building blocks rapidly proliferated, ultimately leading to the integrated circuit. In spite of the many advantages of ICs, there remains a considerable usage of the traditional components in both new systems and vintage designs. Thus the three traditional components, and the models we use to represent them, remain as foundational concepts in circuit theory. Some understanding of their physical characteristics and limitations is essential to the application of circuit theory in practical situations.

RESISTORS

Resistors can be fabricated from a wide variety of materials. Perhaps the simplest type to understand is a length of wire. Since a piece of wire has nonzero resistance, increasing its length increases its resistance. Certainly one can wrap a piece of wire of known resistance per unit length around an insulator with as many turns as necessary to attain the desired resistance. A toaster, lamp, or a wirewound resistor uses this technique. However, most common resistors are known as composition resistors, which means simply a conductive epoxy molded into a cylindrical shape and impregnated with graphite or some other highly conductive material, the percentage of which is varied to attain the desired resistance. Other resistors are made by vacuum depositing a thickness of resistive metal, or silk screening a conductive paste on an insulating substance. Such "thin"- or "thick"-film techniques are used with IC chips to produce networks called "hybrid circuits." In these types of resistors the resistivity of the material is held constant but either its thickness, its length, or its width is varied (see Figure C-1a).

Two parameters determine a resistor's value: its physical dimensions and its resistivity. Mathematically stated, the value of resistance of a bar of resistive material of length l, area A (width w × thickness t), and resistivity ρ is

$$R = \rho \, (l/A) \; \Omega \qquad\qquad (C-1)$$

The most common resistors are hot-molded composition fixed resistors. Figure C-1b shows a cutaway view of such a resistor.

The resistors come in standard values of resistance. Their physical size depends on their power-handling capability or wattage rating. The larger the physical size, the larger is the power-handling capability. Composition resistors come in ratings of 1/8, 1/4, 1/2, 1, and 2 W. Table C-1 lists the approximate physical size of each wattage rating.

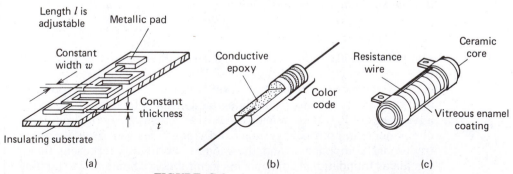

FIGURE C-1
Various resistors. (*a*) Thin or thick film. (*b*) Composition. (*c*) Wirewound.

| Wattage | Approximate Size (inch) | |
	Diameter	Length
1/8	0.062	0.145
1/4	0.090	0.250
1/2	0.140	0.375
1*	0.225	0.562
2*	0.312	0.688

TABLE C-1
Wattage Ratings and Physical Sizes of Composition Resistors
* A special heat sinking clamp can increase the wattage rating of 1-W resistor to 3 W and the 2-W resistor to 4 W.

Most common resistors do not have a numerical resistance rating printed on them, but rather have a color code. An exception is precision resistors, those with an accuracy or **tolerance** of better than ± 1 percent, and even these sometimes employ a color code. Shown in Figure C-2 is the standard color code for common composition resistors.

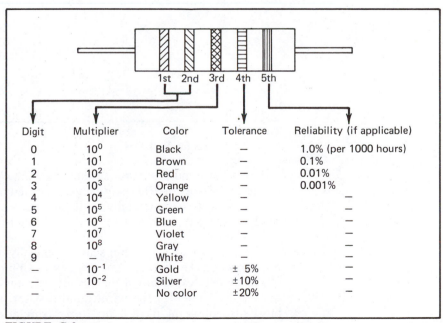

Digit	Multiplier	Color	Tolerance	Reliability (if applicable)
0	10^0	Black	—	1.0% (per 1000 hours)
1	10^1	Brown	—	0.1%
2	10^2	Red	—	0.01%
3	10^3	Orange	—	0.001%
4	10^4	Yellow	—	—
5	10^5	Green	—	—
6	10^6	Blue	—	—
7	10^7	Violet	—	—
8	10^8	Gray	—	—
9	—	White	—	—
—	10^{-1}	Gold	± 5%	—
—	10^{-2}	Silver	±10%	—
—	—	No color	±20%	—

FIGURE C-2
Resistor color code.

Example C-1

A resistor has the following colored bands and is the given size. What are its resistance, tolerance, and wattage rating?

The first band is the first digit, red = 2; the second band is the second digit, red = 2; the third band is the multiplier, red = 10^2; so that, we have 22×10^2 = 2.2 kΩ; the fourth band is the tolerance, gold ± 5 percent. The resistance can be 2.2 kΩ $\pm$ 5 percent. From its size the resistor is rated at 2 W.

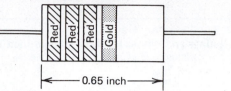

FIGURE C-3
Example for reading a resistor color code.

Resistors come in standard or preferred values depending on their tolerance. Table C-2 shows these values. Some precision resistors use the first three bands for digits, and the fourth band then becomes the multiplier. However, this type of resistor is not common.

Tolerance	Values					
$\pm 5\%$	10	15	22	33	47	68
	11	16	24	36	51	75
	12	18	27	39	56	82
	13	20	30	43	62	91
$\pm 10\%$	10	15	22	33	47	68
	12	18	27	39	56	82
$\pm 20\%$	10	15	22	33	47	68

TABLE C-2
Preferred Values for Resistors

For wirewound resistors, cermet resistors, high-power resistors, and precision resistors the values usually are printed on the devices themselves.

Resistors are commercially available over a very wide range of values. Table C-3 lists typical resistor types with their range, best tolerance, and relative cost.

Most resistors follow Ohm's law, especially if their wattage ratings are not exceeded. However, some resistors exhibit parasitic effects that are not negligible, especially if the circuit operates at higher frequencies. Of note is the model for the wirewound resistor, shown in Figure C-4, the most susceptible of all resistors to parasitic effects.

Wirewound resistors have a significant series inductance. Typical values for a 1-kΩ wirewound resistor are $L = 200\ \mu H$ and $C = 20$ pF. This results in a resonant frequency $f_R = 1/2\ \pi\sqrt{LC} = 2.5$ MHz. To be safe

Resistor Type	Range	Best Tolerance	Relative Cost
Composition	1 Ω–22 MΩ	±5%	Very low
Metal film	10 Ω–1 MΩ	±1/2%	Medium
Wirewound	1 Ω–270 kΩ	±1/2%	High
Thin film	25 Ω–100 kΩ	±0.01%	Medium
Thick film	10 Ω–1 MΩ	±2% → ±5%	Low
Diffused (on ICs)	20 Ω–50 kΩ	±10%	Low (as part of IC)

TABLE C-3
Some Characteristics of Resistors

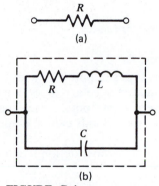

(a)

(b)

FIGURE C-4
Ideal and parasitic model of a resistor. (*a*)
Ideal. (*b*) **Extended model.**

we would limit our use to $0.1 f_R$ or 250 kHz, since beyond that frequency the impedance of the resistor could change significantly.

POTENTIOMETERS

Potentiometers are three-terminal devices. Two of the terminals are connected to the ends of a resistive element while the third is movable and can make contact with the resistive element at some intermediate point. By connecting the ends of the potentiometer to a fixed source, a variable percentage of the source voltage can be obtained between the third contact or wiper and the referenced end. In this most common application the potentiometer functions as a variable voltage divider.

Early potentiometers were only two-terminal devices with one end left unconnected. When used in this fashion the potentiometer functions as a variable resistor or rheostat.[1]

[1]George Little patented a variable resistor or rheostat in 1871. It was lost to history when the first potentiometer was built and used.

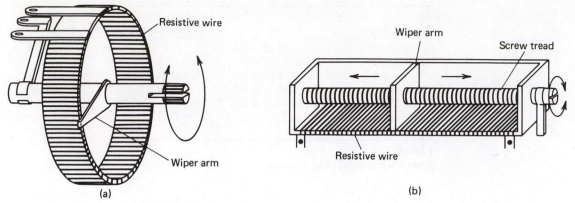

FIGURE C-5
Potentiometers. (*a*) Rotary. (*b*) Linear.

There are numerous types of potentiometer construction in use today. The two most common are the rotary potentiometer for frequent use and the linear potentiometers (trim "pots" being the most common) used for infrequent adjustments. Both kinds can be made from composition, film, or be wirewound.

Two of the most common construction techniques are shown in Figure C-5.

Some characteristics of potentiometers are contained in Table C-4.

Potentiometer Type	Range	Tolerance	Cost
Composition	100 Ω–5 MΩ	± 10%	Low
Film (metal)	50 Ω–10 kΩ	± 2 1/2%	Medium
Wirewound*	10 Ω–100 kΩ	± 2 1/2%	High

TABLE C-4
Potentiometer Characteristics
* Wirewound potentiometers are capable of handling large amounts of power.

CAPACITORS

Capacitors employ the physical property of *action at a distance*. Whenever charges are separated, a coulomb force of attraction or repulsion exists between them. This force takes the following form:

$$F = \frac{q_1 q_2}{4\,\pi\epsilon_o r^2} \tag{C-2}$$

where q_1 and q_2 are the value of the separated charges in coulombs, ϵ_o is the permittivity of free space (8.85×10^{-12} N/m), and r is the separation between the charges. From this equation it should be apparent that physical geometry will play a part in our determination of capacitance.

The simplest vehicle for collecting and separating charge is a pair of parallel plates. Since two separated parallel plates with charge on them will exhibit a force according to Eq. C-2, it is often convenient to measure this force in terms of an electric field $\mathscr{E}$ equal to the force per unit charge, that is,

$$\mathscr{E} = F/q \tag{C-3}$$

From Gauss' law (which states that the electric flux passing through any closed surface is equal to the total charge enclosed by that surface) one can relate the electric field to the geometry of the parallel plates, that is,

$$\mathscr{E} = q/\epsilon A \tag{C-4}$$

where A is the area of the parallel paltes and ϵ is the permittivity of the dielective medium. Furthermore, for a pair of parallel plates, the voltage can be found to be

$$v = \mathscr{E}d \tag{C-5}$$

where d is the separation between the two plates. Substituting Eq. C-4 into C-5, we obtain

$$v = \frac{qd}{\epsilon A} \tag{C-6}$$

If we call $\epsilon A/d$ the capacitance of the capacitor C, we have the familiar expression

$$q = Cv \tag{C-7}$$

where

$$C = \epsilon A/d \tag{C-8}$$

We see that the capacitance of a parallel plate capacitor is related to the area and the separation of the plates (the physical geometry) and a parameter ϵ or permittivity of the medium separating the plates. In other words, the capacitor (much as the resistor is) is affected by two things—its physical geometry and a parameter related to its medium.

Real capacitors employ either a geometric or dielctric medium variation, or both, to obtain a fixed value of capacitance.

Very small values of capacitance, usually in the picofarad range, are available as disk capacitors and these are usually made of a fixed insulator separating two disk-shaped metal plates. The entire capacitor is then epoxy coated for mechanical strength and isolation. An example is shown in Figure C-6.

Larger capacitors, from a few hundred picofarads to up to a microfarad or so, are made by rolling two strips of metal foil and separating them with a Mylar or other dielectric strip. The foil strips are connected on each rolled end. An example is shown in Figure C-7.

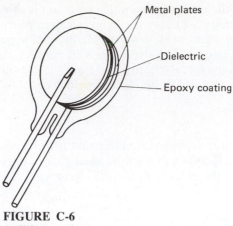

FIGURE C-6
A disk capacitor.

The largest capacitors of up to 1000 μF are made by forming an oxide coating electrochemically on the surface of a metal plate, usually aluminum. These capacitors are of a structure somewhat similar to that of the foil capacitors but they achieve higher capacitance because of the higher ϵ of the dielectric and the thinness of the oxide. It is important to note that these capacitors, called electrolytics, are polarized and must be connected correctly to function properly. An example is shown in Figure C-8.

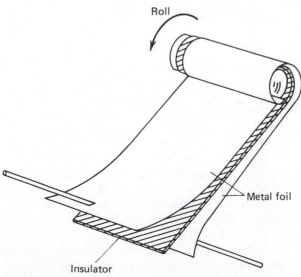

FIGURE C-7
A tubular capacitor.

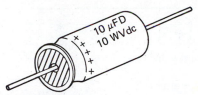

FIGURE C-8
An electrolytic capacitor.

Capacitors are relatively ideal devices, that is, their model fits their physical behavior quite closely. However, in certain restricted applications a more accurate equivalent circuit of a capacitor may be needed. Such an equivalent circuit is shown in Figure C-9.

The most troublesome of all the parasitic elements is R_P, the shunt resistance, R_P ranges from 10^{12} Ω in Teflon and polystyrene capacitors to a low of 10^7 Ω for some types of electrolytics. R_S, the series resistance, and L, the series inductance, are both very small and usually are ignored.

A limitation of all capacitors is the breakdown voltage. This is the voltage at which the dielectric medium becomes conductive and supports an arc. To ensure proper usage, capacitors are marked with at least two parameters, their capacitance and their breakdown voltage. Also, if the capacitor is an electrolytic, the polarity is plainly marked. Finally, many capacitors carry a tolerance factor. If the tolerance factor is missing, it can be assumed to be ± 20 percent as with the resistor. It is important to note that many real capacitors, for example, electrolytics, are polarized, that is, there is a positive terminal and a negative terminal. If this type of capacitor is placed in the circuit improperly, the capacitor will "leak" or may be destroyed. Many different materials are used to make capacitors, but they usually fall into one of the three categories mentioned. Table C-5 lists the range of capacitances, tolerances, and working voltages (WV dc) for capacitors in each of the three categories. Not all capacitors

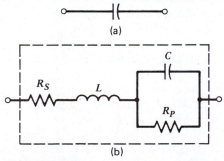

FIGURE C-9
Ideal and parasitic model of a capacitor.
(*a*) **Ideal.** (*b*) **Extended model.**

①.0, 1.2, 1.3, ①.5, 1.8, 2.0, ②.2, 2.4, 2.5, 2.7, 3.0, ③.3, 3.5, 3.9, 4.0, ④.7, 5.0, 5.1, 5.6, 6.0, 6.5, ⑥.8, 7.5, 8.0, 8.2, 9.1, 9.3

The most common values are circled.

Disk (ceramic, tantalum)
   Capacitance range: 3 pF → 0.05 μF
   Tolerances: ±10% → 80% − 20%
   Working voltages: 3 → 100 WV dc

Tubular (paper, Mylar, tantalum)
   Capacitance range: 0.005 μF → 2 μF (70 μF tantalum)
   Tolerances: ±5%, ±10%, ±20%
   Working voltages: 100 → 1600 WV dc

Electrolytic (aluminum, tantalum)
   Capacitance range: (0.0047 μF tantalum) 1 μF → 0.3 F
   Tolerances: ±10% (tantalum), −10% + 75%
   Working voltages: 3 → 450 WV dc

TABLE C-5
Standard Values for Capacitors

in each category are available in every possible combination of capacitance, working voltage, and tolerance.

INDUCTORS

Much like the capacitor, inductors exhibit the *action-at-a-distance* phenomenon. A compass placed near a wire carrying current would be deflected to align with a magnetic force field surrounding the wire. The amount of force exerted by each magnetic pole is called the magnetic field density B. This magnetic field density B is proportional to several key parameters, including the amount of current the wire is carrying and the distance from the wire. Mathematically,

$$B \propto \mu i / r^2 \qquad \textbf{(C-9)}$$

where μ is the magnetic permeability of the medium surrounding the conductor.

   In dealing with inductors it is often more convenient to talk about a quantity called magnetic flux, ϕ. This new parameter is related to B in the following way:

$$\phi = \oint B \cos \theta \, ds \qquad \textbf{(C-10)}$$

where s is the surface of integration and θ is the angle between the surface of integration and the field density B. This magnetic flux can be said to be represented by lines of flux surrounding the current-carrying wire.

Now in inductors there generally are many current-carrying wires, each creating its own line of flux. If these lines of flux are all in the same direction, as would be the case for an inductor, then these lines of flux add. Furthermore if, as in a typical inductor, each conductor carries the same current, the total flux can be represented by a simple relation known as flux linkages, λ.

$$\lambda = N\phi \qquad \text{(C-11)}$$

where N is the number of current-carrying conductors (or turns in an inductor).

Faraday's law now can be used to relate this basic parameter to a signal variable (voltage), that is,

$$v = d\lambda/dt \qquad \text{(C-12)}$$

or to the other signal variable (current) as

$$\lambda = Li \qquad \text{(C-13)}$$

where L is a constant of proportionality called the **self-inductance.** It should be pointed out that a current carrying conductor can produce flux linkages in another nearby conductor (and generation a current in it by transformer action). In this latter case Eq. C-13 is more correctly written as

$$\lambda_1 = M_{12}i_2 \qquad \text{(C-14)}$$

where the flux linkages λ_1 about conductor 1 are generated by i_2 in conductor 2. The constant of proportionality between conductors 1 and 2 M_{12} is called the **mutual inductance.** Both L and M are measured in henrys.

Inductors are not as ubiquitous as either capacitors and resistors, primarily because of their physical size and weight. Nevertheless many circuits employ inductors, either as simple inductors or as transformers. A

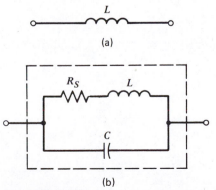

(a)

(b)

FIGURE C-10
Ideal and parasitic model of an inductor.
(*a*) **Ideal.** (*b*) **Extended model.**

1.0,   1.2,   1.5,   1.8,   2.2,   2.7,   3.3,   3.9,   4.7,   5.6,   6.8,   8.2

Inductance range: 0.1 μH $\rightarrow$ 180 mH
Tolerance: $\pm 5\%$, $\pm 10\%$, $\pm 20\%$

TABLE C-6
Standard Values for Inductors

practical inductor is the least ideal of the three devices (R, L, C) used. Its inherent resistance, as a result of the large number of wires used in most inductors, often cannot be ignored. Furthermore a parasitic capacitance also exists. In general a real inductor is often modeled as an ideal inductor in series with an ideal resistance with a capacitor in parallel across the series R_S and L as shown in Figure C-10. Typical values for R_S are from less than 1 Ω to several hundred ohms.

Real inductors take many shapes and most are rated in henrys, usually microhenrys or millihenrys. Occasionally inductors are listed in ohms, reflecting their impedance Z at the specific frequency in which they are designed to operate. Table C-6 lists the range, tolerance, and available standard values for inductors.

Appendix D
Initial and Final Value Theorems

The Initial Value Theorem
The Final Value Theorem

I am the alpha and the omega,
the beginning and the ending . . .

Revelations 1-8.
King James Version

It is often useful to be able to determine the initial and final values of a response $v(t)$ directly from its transform $V(s)$. This can be accomplished by relating the initial and final values of $v(t)$ to the values of $sV(s)$ at $s = \infty$ and $s = 0$ respectively.

THE INITIAL VALUE THEOREM

The initial value theorem states that if $v(t)$ and its first derivative are Laplace transformable, then

$$\text{IV} = \underset{t \to 0}{\text{Limit}}\, v(t) = \underset{s \to \infty}{\text{Limit}}\, sV(s) \qquad \textbf{(D-1)}$$

To prove this theorem we let s approach infinity in the equation for the Laplace transform of the derivative of $v(t)$.

$$\underset{s \to \infty}{\text{Limit}} \int_0^\infty \left[\frac{dv}{dt}\right] e^{-st} dt = \underset{s \to \infty}{\text{Limit}}\, [sV(s) - v(0)] \qquad \textbf{(D-2)}$$

Since s and t are two independent variables we can interchange the limiting and integration process in Eq. D-2. But since

$$\underset{s \to \infty}{\text{Limit}}\, e^{-st} = 0$$

the intergral vanishes so that the right side of Eq. D-2 becomes

$$\underset{s \to \infty}{\text{Limit}}\, [sV(s) - v(0)] = 0$$

or

$$\underset{s \to \infty}{\text{Limit}}\, sV(s) = v(0) = \underset{t \to 0}{\text{Limit}}\, v(t) \qquad \textbf{(D-3)}$$

as required.

THE FINAL VALUE THEOREM

The final value theorem states that

$$\text{FV} = \underset{t \to \infty}{\text{Limit}}\, v(t) = \underset{s \to 0}{\text{Limit}}\, sV(s) \qquad \textbf{(D-4)}$$

To prove this result we again apply limits to the Laplace transform of the derivative of $v(t)$.

$$\underset{s \to 0}{\text{Limit}} \int_0^\infty \left[\frac{dv}{dt}\right] e^{-st} dt = \underset{s \to 0}{\text{Limit}}\, [sV(s) - v(0)] \qquad \textbf{(D-5)}$$

But since s and t are independent variables we can interchange the limiting and integration process on the left side of Eq. D-5. Since $e^0 = 1$ the integral becomes

$$\int_0^\infty \left[\frac{dv}{dt}\right] \underset{s \to 0}{\text{Limit}}\, e^{-st} dt = \underset{t \to \infty}{\text{Limit}} \int_0^t \frac{dv}{dt} dt$$
$$= \underset{t \to \infty}{\text{Limit}}\, [v(t) - v(0)] \qquad \textbf{(D-6)}$$

Substituting this result into the left side of Eq. D-5 yields

$$\text{Limit } v(t) = \text{Limit } sV(s) \qquad \text{(D-7)}$$
$$\scriptstyle t\to\infty \qquad\qquad\qquad s\to 0$$

as required.

The final value theorem has certain restrictions. Basically the requirement is that a final value of $v(t)$ must exist. Since we determine the final value from $sV(s)$, the equivalent restriction is that all of the poles of $sV(s)$ must have negative real parts. That is, $sV(s)$ cannot have poles in the right half of the s plane *or* on the j axis. Note carefully the nature of this restriction. The limit of $sV(s)$ as s approaches infinity may very well exist and produce a number. But that number is not the final value of $v(t)$ unless $sV(s)$ is absolutely stable, that is, has no right half plane or j axis poles.

Example D-1

Determine the initial and final values of the waveforms corresponding to the transforms

$$V_1(s) = \frac{1}{s + 1}$$

$$V_2(s) = \frac{1}{(s + 1)^2}$$

$$V_2(s) = \frac{s}{s^2 + 1}$$

Applying the initial and final value theorems to $V_1(s)$ we obtain

$$\text{IV} = \text{Limit}_{s\to\infty} sV_1(s) = \text{Limit}_{s\to\infty} \frac{s}{s + 1} = 1$$

$$\text{FV} = \text{Limit}_{s\to 0} sV_1(s) = 0$$

Since we recognize $v_1(t) = e^{-t}$, we see that the results are valid. For $V_2(s)$ the initial and final value theorems yield

$$\text{IV} = \text{Limit}_{s\to\infty} sV_2(s) = \text{Limit}_{s\to\infty} \frac{s}{(s + 1)^2} = 0$$

$$\text{FV} = \text{Limit}_{s\to 0} sV_2(s) = \text{Limit}_{s\to 0} \frac{s}{(s + 1)^2} = 0$$

Since we recognize that $v_2(t) = te^{-t}$, we see that the results are valid. For $V_3(s)$ the initial and final value theorems yield

$$\text{IV} = \text{Limit}_{s\to\infty} sV_3(s) = \text{Limit}_{s\to\infty} \frac{s^2}{s^2 + 1} = 1$$

$$\text{FV} = \text{Limit}_{s\to 0} sV_3(s) = 0$$

We recognize that $v_3(t) = \cos t$. The initial value above is correct but the final value is not. The reason is that $sV_3(s)$ has poles on the j axis so that the final value theorem does not apply. Thus the limit of $sV(s)$ may exist but it is not the final value of $v(t)$ unless the restrictions on the locations of the poles of $sV(s)$ are satisfied.

Appendix E
Complex Numbers

Definitions
Arithmetic Operations
Exercises

There can be little mathematical terrain left over which holds the promise that [complex] function theory held for us.

H. W. Bode

The theory of complex variables is a basic part of applied mathematics. In addition the theory finds applications in almost all fields of engineering. The use of complex numbers in electrical engineering was introduced by Charles Steinmetz in 1894. Thus complex numbers have been with us for almost as long as electrical engineering has existed as a recognized discipline. This appendix reviews the basic definitions and arithmetic operations. These matters, though rudimentary, must be second nature to anyone claiming the title of electrical engineer. Exercises are provided to verify your mastery of these basic skills.

DEFINITIONS

A complex number consists of two components, one real and the other imaginary. A complex number expressed in rectangular form is written as

$$z = x + jy \qquad \text{(E-1)}$$

where x is called the real part of z and y (not jy) the imaginary part. This distinction is sometimes written as

$$x = \text{Re}\,\{z\} \qquad \text{and} \qquad y = \text{Im}\,\{z\} \qquad \text{(E-2)}$$

where Re stands for the **real part of** and Im means that **imaginary part of**. The symbol $j = \sqrt{-1}$ is an imaginary number that serves as a flag to identify the imaginary part of z. The symbol j is used here since the symbol i commonly used by mathematicians stands for current in an electrical engineering context.

Figure E-1 shows the graphical representation of the complex number $z = x + jy$. From the figure it is clear that we can write

$$x = |z|\cos\theta \qquad \text{and} \qquad y = |z|\sin\theta \qquad \text{(E-3)}$$

or conversely,

$$|z| = \sqrt{x^2 + y^2} \qquad \text{and} \qquad \theta = \tan^{-1}(y/x) \qquad \text{(E-4)}$$

These relationships provide an alternative representation of the complex number z

$$z = |z| \angle \theta \qquad \text{(E-5)}$$

called the polar form of z. The quantity $|z|$ is called the magnitude or modulus and θ the angle or argument of z.

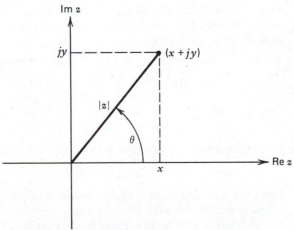

FIGURE E-1
Graphical representation of complex numbers.

The relationship between the rectangular and polar forms can also be expressed using Euler's identity:

$$e^{j\theta} = \cos\theta + j\sin\theta \tag{E-6}$$

Using this identity and Eq. E-3 we can write the rectangular form in Eq. E-1 as

$$\begin{aligned}
z &= |z|\cos\theta + j|z|\sin\theta \\
&= |z|(\cos\theta + j\sin\theta) \\
&= |z|\,e^{j\theta}
\end{aligned} \tag{E-7}$$

which is called the exponential form of the complex number. The exponential representation is written in terms of the magnitude and angle of z, as is the polar form, but is more easily converted to rectangular form using Euler's identity. Thus there are three possible representations: rectangular, polar, and exponential. When manipulating complex numbers we generally use the rectangular and exponential forms. The polar form is used mostly as a shorthand notation of the exponential form.

Associated with every complex number is another complex number called the conjugate and defined as

$$z^* = x - jy \tag{E-8}$$

The conjugate is obtained by simply reversing the sign of the imaginary part. The exponential form of the conjugate is

$$z^* = |z|\,e^{-j\theta} \tag{E-9}$$

which points out that the conjugate also can be obtained by simply reversing the sign of the angle of z.

ARITHMETIC OPERATIONS

The rules for mathematical operations make frequent use of the conversion from one form to another, and the fact that $j^2 = -1$. For addition or subtraction we write:

$$\begin{aligned}
\text{If } z_1 &= x_1 + jy_1 \quad \text{and} \quad z_2 = x_2 + jy_2, \\
\text{Then } z_1 + z_2 &= (x_1 + jy_1) + (x_2 + jy_2) \\
&= (x_1 + x_2) + j(y_1 + y_2) \\
\text{and } z_1 - z_2 &= (x_1 + jy_1) - (x_2 + jy_2) \\
&= (x_1 - x_2) + j(y_1 - y_2)
\end{aligned} \tag{E-10}$$

Note that when the addition rule is applied to conjugates we obtain

$$z + z^* = 2x = 2\,\text{Re}\{z\} \quad \text{and} \quad z - z^* = j2y = j2\,\text{Im}\{z\} \tag{E-11}$$

Thus the addition of conjugates produces twice the real part, while subtraction produces twice the imaginary part of z.

Addition or subtraction must be carried out using the rectangular form of the two numbers. However, multiplication and division can be accomplished in either form. For multiplication in rectangular form we write:

$$\text{If } z_1 = x_1 + jy_1 \quad \text{and} \quad z_2 = x_2 + jy_2,$$
$$\begin{aligned} \text{Then } z_1 z_2 &= (x_1 + jy_1)(x_2 + jy_2) \\ &= (x_1 x_2 + j^2 y_1 y_2) + j(x_1 y_2 + x_2 y_1) \\ &= (x_1 x_2 - y_1 y_2) + j(x_1 y_2 + x_2 y_1) \end{aligned} \tag{E-12}$$

For multiplication in exponential form, the rule is as follows:

$$\text{If } z_1 = |z| \, e^{j\theta_1} \quad \text{and} \quad z_2 = |z_2| \, e^{j\theta_2},$$
$$\begin{aligned} \text{Then } z_1 z_2 &= [|z_1| \, e^{j\theta_1}][|z_2| \, e^{j\theta_2}] \\ &= (|z_1| \, |z_2|)[e^{j\theta_1} \, e^{j\theta_2}] \\ &= |z_1| \, |z_2| \, e^{j(\theta_1 + \theta_2)} \end{aligned} \tag{E-13}$$

Multiplication is normally easier to carry out in exponential form, but both methods should be understood. If the multiplication rule is applied to conjugates, then

$$\begin{aligned} zz^* &= [|z| \, e^{j\theta}][|z| \, e^{-j\theta}] \\ &= |z|^2 \end{aligned} \tag{E-14}$$

Thus multiplying z by its conjugate produces the square of the magnitude of z.

The division rule in rectangular form reads:

$$\text{If } z_1 = x_1 + jy_i \quad \text{and} \quad z_2 = x_2 + jy_2,$$
$$\text{Then } \frac{z_1}{z_2} = \frac{x_1 + jy_1}{x_2 + jy_2}$$

This quotient is manipulated by multiplying the numerator and denominator by the conjugate of the denominator

$$\frac{z_1}{z_2} = \frac{x_1 + jy_1}{x_2 + jy_2} \cdot \frac{x_2 - jy_2}{x_2 - jy_2}$$

Applying the multiplication rule from Eq. E-12 to the numerator and the conjugate multiplication rule from Eq. E-14 to the denominator we obtain

$$\frac{z_1}{z_2} = \frac{(x_1 x_2 + y_1 y_2) + j(x_2 y_1 - x_1 y_2)}{x_2^2 + y_2^2} \tag{E-15}$$

The division rule in exponential form is written as follows:

$$\text{If } z_1 = |z_1| e^{j\theta_1} \quad \text{and} \quad z_2 = |z_2| \, e^{j\theta_2},$$
$$\begin{aligned} \text{Then } \frac{z_1}{z_2} &= \frac{|z_1|}{|z_2|} \frac{e^{j\theta_1}}{|z_2| \, e^{j\theta_2}} \\ &= \frac{|z_1|}{|z_2|} \, e^{j(\theta_1 - \theta_2)} \end{aligned}$$

As with multiplication, division is more easily carried out in exponential form although, again, both methods should be understood.

Example E-1

Find the sum, difference, product, and quotient of the complex numbers

$$z_1 = 1 - j2$$
$$z_2 = 2 + j3$$

Applying the rules described in the forgoing:

$$z_1 + z_2 = (1 - j2) + (2 + j3)$$
$$= (1 + 2) + j(-2 + 3)$$
$$= 3 + j1$$
$$z_1 - z_2 = (1 - j2) - (2 + j3)$$
$$= (1 - 2) + j(-2 - 3)$$
$$= -1 - j5$$
$$z_1 z_2 = (1 - j2)(2 + j3)$$
$$= (2 + 6) + j(3 - 4)$$
$$= 8 - j1$$
$$\frac{z_1}{z_2} = \frac{1 - j2}{2 + j3} \times \frac{2 - j3}{2 - j3} = \frac{(2 - 6) + j(-3 - 4)}{2^2 + 3^2}$$
$$= \frac{-4}{13} - j\frac{7}{13}$$

The product and quotient can also be found by converting z_1 and z_2 to exponential form

$$z_1 = \sqrt{1^2 + 2^2} \angle \tan^{-1}(-2/1) = \sqrt{5}\, e^{-j63.4°}$$
$$z_2 = \sqrt{2^2 + 3^2} \angle \tan^{-1}(3/2) \quad = \sqrt{13}\, e^{j56.3°}$$

Hence

$$z_1 z_2 = \sqrt{5}\, e^{-j63.4°} \sqrt{13}\, e^{j56.3°}$$
$$= \sqrt{65}\, e^{-j7.1°} = 8 - j1$$
$$\frac{z_2}{z_1} = \frac{\sqrt{5}\, e^{-j63.4°}}{\sqrt{13}\, e^{j56.3°}} = \frac{\sqrt{5}}{\sqrt{13}} e^{-j120°}$$
$$= -\frac{4}{13} - j\frac{7}{13}$$

which agrees with the results using the rectangular form.

Example E-2

Manipulation of complex numbers may involve a sequence of operations. For example. If

$$A = 1 + j2 \quad B = 3 - j4 \quad C = 0 + j2$$

evaluate $D = (A \times B + 1)/C$:

$$D = \frac{(1 + j2)(3 - j4) + 1}{j2}$$
$$= \frac{(3 + 8) + j(6 - 4) + 1}{j2}$$
$$= \frac{12 + j2}{j2} \times \frac{-j2}{-j2}$$
$$= \frac{4 - j24}{4}$$
$$= 1 - j6$$

Example E-3

In electrical engineering we often encounter the problem of evaluating the magnitude and angle of a function such as

$$H(\omega) = \frac{j\omega}{10 + j\omega}$$

for specific values of ω such as $\omega = 1$, 10, and 100. One way to deal with this situation is to treat the numerator and denominator as separate complex numbers. We first find the magnitude and angle of these numbers separately, and then apply the division rule. Treating the numerator first,

$$N(\omega) = 0 + j\omega$$

so that

$$|N(\omega)| = \sqrt{0^2 + \omega^2} = |\omega|$$
$$\angle N(\omega) = \angle \tan^{-1}(\omega/0) = \angle \tan^{-1}(\infty) = 90°$$

Note that the angle of the numerator is 90° regardless of the value of ω. Similarly for the denominator,

$$|D(\omega)| = \sqrt{10^2 + \omega^2} = \sqrt{100 + \omega^2}$$
$$\angle D(\omega) = \angle\tan^{-1}(\omega/10)$$

Now applying the division rule we obtain

$$|H(\omega)| = \frac{|N(\omega)|}{|D(\omega)|} = \frac{|\omega|}{\sqrt{100 + \omega^2}}$$
$$\angle H(\omega) = \angle N(\omega) - \angle D(\omega)$$
$$= \angle 90° - \angle\tan^{-1}(\omega/10)$$

We can now evaluate $H(\omega)$ for the values of ω given.

$$\omega = 1 \quad |H(1)| = \frac{1}{\sqrt{100 + 1}} = 0.0995$$
$$\angle H(1) = \angle 90° - \angle\tan^{-1}(1/10) = 84.3°$$

$\omega = 10$

$$|H(10)| = \frac{10}{\sqrt{100 + 100}} = 0.707$$
$$\angle H(10) = \angle 90° - \angle\tan^{-1}(10/10) = 45°$$

$\omega = 100$

$$|H(100)| = \frac{100}{\sqrt{100 + 10^4}} = 0.995$$
$$\angle H(100) = \angle 90° - \angle\tan^{-1}(100/10) = 5.71°$$

By using these values and a few others we can construct a sketch of $|H(\omega)|$ and $\angle H(\omega)$ as shown in Figure E-2.

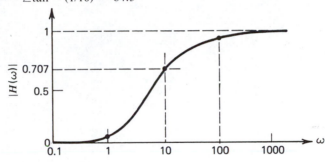

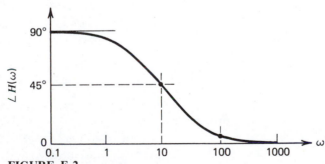

FIGURE E-2
Magnitude and phase plot of a complex function.

Exercises

E-1
Given
$A = 3 + j4 \qquad B = 5 - j7 \qquad C = -2 + j3$
$D = 5 \angle -30°$
Evaluate

(a) $A \times B$
(b) $C + D$
(c) BC/D
(d) $A^* + C \times A$
(e) $B + A \times D$

E-2

Find the magnitude and angle of the following functions at the indicated values of

(a) $\quad H(\omega) = \dfrac{10}{1 + j\omega} \qquad \omega = 0.1, 1, 10$

(b) $\quad H(\omega) = \dfrac{10 + j\omega}{j\omega} \qquad \omega = 1, 10, 100$

(c) $\quad H(\omega) = \dfrac{10 + j\omega}{1 + j\omega} \qquad \omega = .1, 1, 10, 100$

Appendix F
Answers to Selected Problems

En Route Objective Exercises and Chapter Problems
Block Objectives Problems
Complex Numbers Exercises

All wish to know, but none to pay the fee.

Juvenal

The nature of circuit analysis, and indeed of all engineering courses, is that learning is facilitated through the routine practice of solving problems. This text is organized to permit the student to grow from solving straightforward exercises based on a single idea or concept to more complex, broader based problems requiring the integration of several ideas. In terms of Bloome's taxonomy the en route objectives are essentially behaviorly motivated, that is, their mastery can be easily stated and measured. The student should attempt several exercises for each objective studied. Once routinely successful, the student proceeds to the next en route objective. These objectives test mastery at the *knowledge, comprehension,* and *application* level of the taxonomy. Subsequently the student proceeds to the problems at the end of each chapter. These problems encompass several objectives contained in the chapter and their solution requires mastery in *analysis, synthesis,* and *evaluation*—skills higher in the taxonomy. Ultimately, as the various blocks of the text are completed, the student can attempt problems that integrate ideas over several chapters and many en route objectives. These block problems represent the ultimate goal for the text. The block problems are contained in Appendix A, along with the block objectives.

CHAPTER 1

Exercises

1-2-1
(a)   4 MHz
(b)   16 ns
(c)   0.47 μF
(d)   100 kΩ
(e)   5 mH

1-2-2
(a)   five gigaradians per second
(b)   two joules
(c)   six and two-hundredths nanovolts
(d)   ten decibels
(e)   one hundred two femtoamperes

CHAPTER 2

Exercises

2-1-1
10^{11} electrons

2-1-3
1200 J

2-1-4
4.32 cents

2-1-5
0.125 mm/second

2-2-1
(a)   2.5 mA, 12.5 mW
(c) -0.6 V

2-2-2
C-1   10 mA, 10 V, 100 mW
C-2   5 A, 0 V, 0 W
C-3   1 A, 100 V, 100 W

2-2-3
$i = v/R$, so it will always intercept the i axis at (0,0) and the slope will vary as $1/R$.

2-2-7
(a) $i_{out}(t) = 0$ A
(b) $i_{out}(t) = 5$ A
(c) $i_{out}(t) = -e^{-t}$ A

2-3-2

C-1 (b)   1 and 2 are in series, 4 and 5 are in series.
C-2 (b)   1, 2, and 3 are in parallel.
C-3 (b)   2 and 3 are in parallel, 4 and 5 are in parallel—nothing is in series.
C-4 (b)   2 and 4 are in parallel, 5 and 6 are in parallel—nothing is in series.

2-3-3

C-1 (b)   1 and 2 are in series, 4 and 5 are in series.
C-2 (b)   1 and 2 are in parallel, 3 and 4 are in parallel.
C-3 (b)   1 and 2 are in parallel, 3 and 5 are in series.
C-4 (b)   1 and 4 are in parallel, 3 and 5 are in parallel—nothing is in series.

2-3-4

C-1 (b)   4 and 5 are in series.
C-2 (b)   1 and 3 are in parallel.
C-3 (b)   4 and 5 are in parallel, 1 and 3 are in series.
C-4 (b)   5 and 6 are in parallel, 1 and 4 are in series.

2-3-6

(a)   4 nodes
(b)   2 and 4 are in series, 3 and 5 are in series—nothing is in parallel.

2-3-8

(a)   3 nodes (node B and ground are now the same node).
(b)   3 and 5 are in series, 1 and 2 are in parallel.

2-4-1

C-1 $v_x = 5$ V
C-2 $v_x = -12$ V

2-4-2

C-1 $i_x = 1.4$ mA
C-2 $i_x = -1.89$ A

2-4-3

$v_1 = 5$ V $i_1 = -1/2$ A
$v_2 = 5$ V $i_2 = 1/2$ A
$v_3 = 0$ V $i_3 = 0$ A
$\qquad\quad i_4 = 1$ A
$\qquad\quad i_5 = -1/2$ A
$\qquad\quad i_6 = -1/2$ A

2-4-5

$i_1 = 50$ mA $p = 1$ W
$i_2 = 50$ mA
$i_3 = 25$ mA
$i_4 = 25$ mA
$i_5 = 25$ mA
$i_6 = 100$ mA

2-5-1

C-1   10 Ω
C-2   20 Ω
C-3   30 Ω
C-4   40 Ω
C-5   232.8 Ω

2-5-4

C-1   1 mΩ ±5 percent
C-2   10 Ω
C-3   110 Ω
C-4   5.5 Ω ±5 percent

2-5-5

Pins (10–11) 20 Ω, (11–12) 237.6 Ω, (2–3) 20 Ω, (1–2) 237.6 Ω, (1–13) 239.4 Ω, (1–6) ∞ Ω, (5–7) 139.2 Ω, (4–5) 6069 Ω, (1–7) 119.4 Ω, (8–13) 20 Ω, (12–13) 0 Ω

2-6-1

C-1 $i_x = \dfrac{R_1 I_s}{R_1 + R_2}$

C-2 $v_x = \dfrac{R_2 V_s}{R_1 + R_2 + R_3}$

C-3 $i_x = \dfrac{-R_1 I_s}{R_1 + R_2}$

C-4 $v_x = \dfrac{-R_4 V_s}{R_1 + R_2 + R_3 + R_4}$

2-6-2

C-1   0.6 V
C-2   0.34 V

2-6-3

C-1   2.5 A
C-2   0.91 V
C-3   0.714 V
C-4   0.4 A

2-6-5

Switch ON $v_{\text{OUT}} = 7.5$ V
Switch OFF $v_{\text{OUT}} = 3$ mV

2-6-6

$R_s = 50$ mΩ

2-6-7

$R_V = 1010$ Ω

2-6-8

Fuse will not blow.

2-6-10

50 percent

2-6-11

$$R_1 = \frac{R_x R_z}{R_x + R_y + R_z}$$

$$R_2 = \frac{R_x R_y}{R_x + R_y + R_z}$$

$$R_3 = \frac{R_y R_z}{R_x + R_y + R_z}$$

and

$$R_x = \frac{1}{R_3}(R_1 R_2 + R_2 R_3 + R_3 R_4)$$

$$R_y = \frac{1}{R_2}(R_1 R_2 + R_2 R_3 + R_3 R_4)$$

$$R_z = \frac{1}{R_1}(R_1 R_2 + R_2 R_3 + R_3 R_4)$$

2-7-1

C-1 $v_1 = -20$ V, $v_2 = 5$ V
C-2 $i_x = 2$ A
C-3 $i_1 = 0.32$ A, $i_3 = 0.06$ A
C-4 $v_x = 2.5$ V

2-7-2

C-1 $V_x = -4$ V
C-2 $V_x = 50$ V

Problems

2-1
$n(0) = 1.1 \times 10^{18}/\text{cm}^3$
$I(5\,L_n) = -6.72\ \mu\text{A}$

2-3
$\Delta I_B = 1.25\ \mu\text{A}$

2-5
$R/4$

2-7
$T_{\text{HOT}} = 2242\text{C}$

2-9
1106 W

CHAPTER 3

Exercises

3-1-1

C-1 $\quad i_o = \dfrac{I_S R_1}{(R_1 + R_2)}$

C-2 $\quad v_o = \dfrac{V_S R_2 R_3}{R_1 R_2 + R_1 R_3 + R_2 R_3}$

C-3 $\quad v_o = \dfrac{[R_3(R_2 + R_4) - R_4(R_1 + R_3)]I_S}{R_1 + R_2 + R_3 + R_4}$

C-4 $\quad v_o = \dfrac{R_1 R_3 I_S}{R_1 + R_2 + R_3}$

C-5 $\quad i_o = 0$

C-6 $\quad v_o = \dfrac{R_2 R_4}{R_1 R_2 + R_1 R_3 + R_1 R_4 + R_2 R_3 + R_2 R_4}$

3-1-3

C-1 $\quad v_o = 0.56$ V

C-2 $\quad i_o = 60$ mA

3-1-4

$V_S = 144$ V

3-2-1

C-1 $\quad V_o = \dfrac{3\,V_S}{8} + \dfrac{I_S R}{2}$

C-2 $\quad V_o = \dfrac{3}{8}V_1 + \dfrac{3}{8}V_2$

C-3 $\quad V_o = 2\,V_2/3$

C-4 $\quad V_o = I_1 R/3 - R I_2$

3-2-3

$V_o = \dfrac{1}{6}(2V_1 + 3V_2 + V_3)$

3-2-5

$R_L = 160\ \Omega$

3-3-2

	Open	Closed
C-1	$V_T = 0$	$V_T = \dfrac{R_2 V_S}{R_1 + R_2}$
	$R_T = R_2$	$R_T = \dfrac{R_1 R_2}{R_1 + R_2}$
C-2	$V_T = \dfrac{R_2 V_S}{2R_1 + R_2}$	$V_T = 0$
	$R_T = \dfrac{R_1 R_2}{2R_1 + R_2}$	$R_T = 0$
C-3	$V_T = V_S$	$V_T = \dfrac{R_2 V_S}{R_1 + R_2}$
	$R_T = R_1$	$R_T = \dfrac{R_1 R_2}{R_1 + R_2}$
C-4	$V_T = V_2$	$V_T = V_2 + \dfrac{R_2(V_1 - V_2)}{R_1 + R_2}$
	$R_T = R_2$	$R_T = \dfrac{R_1 R_2}{R_1 + R_2}$

3-3-3
C-1 $V_T = 3 V_S/8 + I_S R/2$ $R_T = R$
C-2 $V_T = 3 V_S/4$
$R_T = R$
C-3 $V_T = 2 V_2/3$
$R_T = 2R/3$
C-4 $V_T = I_1 R/3 - RI_2$
$R_T = 4R/3$

3-3-5
$V_T = V_o = V_1/8 + V_2/4 + V_3/2$
$R_T = R$

3-3-6
$V_T = 21$ V, $R_T = 18 \ \Omega$

3-3-7
$i_d \sim 1.0$ mA, $v_d \sim 0.7$ V

3-3-8
$V_S = 40$ V, $R_S = 10$ kΩ

3-4-2
C-1 $P = 2.5$ W
C-2 $P = 2.5$ mW
C-3 $P = 290$ mW
C-4 $P = 41.67$ mW
C-5 $P = 780$ mW
C-6 $P = 125$ W

3-4-4
(a) $R_L = 5 \ \Omega$
(b) $R_L = 10 \ \Omega$
(c) $R_L = 20/3 \ \Omega$
(d) $R_L = 20/3 \ \Omega$
(e) $R_L = 6 \ \Omega$
(f) $R_L = 10 \ \Omega$

3-5-1
(a) $R_L = 12.5 \ \Omega$
(b) $R_L = 50 \ \Omega$
(c) $R_L = \infty \ \Omega$
(d)   Impossible with passive circuit.
(e) $R_L = 0 \ \Omega$
(f) $R_L = \infty \ \Omega$
(g) $R_L = 131 \ \Omega$ or $19 \ \Omega$
(h) $50 \ \Omega$
(i)   No value of R_L exists that will work.

3-5-2
(a) $R_S \simeq 3 \ \text{k}\Omega$
(b) $R_S = 8.65 \ \text{k}\Omega$

3-5-4
(a) $R_2 = 16.67 \ \Omega$
(b) $R_2 = 25 \ \Omega$

Problems

3-1
$1.33 \ \text{mA} > i_L > 0.57 \ \text{mA}, \ 0 < R_L < 10 \ \text{k}\Omega$

3-2
$V_{AB} = 1.0 \ \text{V}, I_A = 2.5 \ \text{mA}; V_{CB} = 2.5 \ \text{V}, I_C = 22$
mA

3-4
$R_{\text{IN}} = 1.105 \ \text{k}\Omega, R_{\text{OUT}} = 99.507 \ \text{k}\Omega$

3-6
Interface is simply resistor R_L between top connectors, wire between bottom wires; $R_L = 123 \ \Omega$.

3-8
$R_{\text{THEVENIN}} = 500 \ \Omega, V_{\text{THEVENIN}} = 50 \ \text{V} \ P = 85 \ \text{W}$

CHAPTER 4

Exercises

4-1-1

C-1 (a) $\quad \dfrac{v_2}{v_S} = \dfrac{\mu R_L R_I}{(R_L + R_o)(R_S + R_I)}$

C-2 (a) $\quad \dfrac{v_2}{v_S} = \dfrac{R_L\, g\, R_I R_o}{(R_S + R_I)(R_L + R_o)}$

(b) Same as C-2 (a).

4-1-3

C-1 $\quad \dfrac{v_2}{v_S} = \dfrac{-\beta R_C}{R_S + R_e (1 + \beta)}$

C-2 $\quad \dfrac{v_2}{v_S} = \dfrac{(1 + \beta)R_x}{R_S}$

4-1-4

$v_2 = \beta (k_1 v_{S1} + k_2 v_{S2})$

4-1-6

$V_{AB}/I_{TEST} = r_0$

4-1-8

C-1 $\quad R_T = \dfrac{[R_S + R_e(1 + \beta)]R_e}{\beta(R_e + R_S)}$

C-2 $\quad R_T = R_e$

4-2-2

$R_B = 18.6\text{ k}\Omega,\ R_C = 1\text{ k}\Omega$

4-2-3

$R_B = 500\ \Omega$

4-2-4

$V_S = 10$ V, $V_o = 1.89$ V
$V_S = 5$ V, $V_o = 1.43$ V
$V_S = 1$ V, $V_o = 0.1$ V

4-3-1

C-1 (a) $v_1 = 5/3$ V, $P = 2.78$ mW
(b) 5-V source
C-2 (a) $v_1 = 5$ V, $P = 25$ mW
(b) OP AMP power supplies
C-3 (a) $v_1 = 10$ V, $P = 100$ mW
(b) OP AMP power supplies
C-4 (a) $v_1 = 15$ V, $P = 225$ mW
(b) OP AMP power supplies

4-3-4

C-1 $v_o = 2v_{S2} - 2v_{S1}$

C-2 $v_o = 2v_{S2} - 2v_{S1}$

4-3-7

$v_o = 33v_1 - 0.5 + v_2$

4-3-10

Case 1		Case 2
$0 < t \le 1$	$V_o = -3.75$ V	Same as Case 1
$1 < t \le 2$	$V_o = -1.88$ V	scaled by factor of 2.
$2 < t \le 3$	$V_o = -3.13$ V	
$3 < t \le 4$	$V_o = -0.63$ V	
$4 < t \le 5$	$V_o = -1.25$ V	

4-4-2

For minus gains use a standard inverter, that is, $v_o/v_1 = -R_F/R_1$. Connect pin 5 to $+V_{CC}$, pin 2 to $-V_{CC}$, pin 1 to ground, and pin 6 to pin D to achieve the gains indicated and the specific connections that follow.

-2 gain $=$ pin C to v_1; pin B to pin 3
-1 gain $=$ pin A to v_1; pin B to pin 3
$-\frac{1}{2}$ gain $=$ pin B to v_1; pin C to pin 3
-3 gain $=$ pin B to pin C to v_1; pin A to pin 3

For plus gains use a standard noninverter, that is, $v_o/v_1 = (R_1 + R_2)/R_2$. Connect pin 5 to $+V_{CC}$, pin 2 to $-V_{CC}$, pin 1 to v_1, pin D to pin 6 to achieve the gains indicated and the specific connections that follow.

$+3$ gain $=$ pin C to pin 4; pin B to pin 3
$+2$ gain $=$ pin A to pin 4; pin B to pin 3
$+1$ gain $=$ do not connect pin D to pin 6, simply connect pin 6 to pin 3.
$+\frac{1}{2}$ gain $=$ cannot be accomplished

4-4-3 See Figure.

4-4-5

Two inverters in cascade. Strain gage is used as the feedback resistor for the first inverter; all other resistors are 1 kΩ.

4-4-7

Use a diode as feedback element in a standard inverting OP AMP configuration. If $i_D = i_F = 10^{-14} e^{V_o/0.026}$, then $v_o = 0.026 \ln(-10^{14} v_S/R)$.

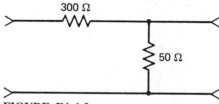

FIGURE F4-4-3
Possible design solution for Exercise 4-4-3.

4-5-1

For V_H Output

C-1	$v_S < 0$
C-2	$v_S > \dfrac{R_2 V_o}{R_1 + R_2}$
C-3	$v_S > -2$ V (first stage) then $v_S < -2$ V
C-4	When $v_S > V_o$, $V_o = V_H$
C-5	Never have V_H

For V_L Output

C-1	$v_S > 0$
C-2	$v_S < \dfrac{R_2 V_o}{R_1 + R_2}$
C-3	$v_S < -2$ V
C-4	$v_S < V_o$
C-5	Always V_L

4-5-3   See Figure.

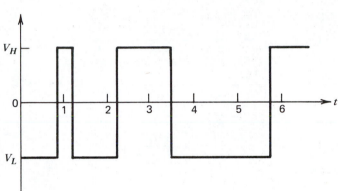

FIGURE F4-5-3
Output of comparator.

4-5-5

With input to "$+$" and "$-$" grounded, the comparator's output will be low until $t = 6$, then it becomes high (V_H) thus detecting when $v(t)$ goes positive.

Problems

4-1

$I_C/I_B = \beta^2 + 2\beta$

4-2

Circuit A is a NOR gate.
Circuit B is a NAND gate.

4-3

$\dfrac{\Delta v_o}{\Delta v_2} = -9.9$. If $v_1 = 3$ V, $v_o = 210.9$ V.

4-5

$h_{ie} \simeq 2100\ \Omega$

$h_{re} = 9.99 \times 10^{-4}$

$h_{fe} = 100$

$h_{oe} = 5.04 \times 10^{-5}\ \mho$

4-6

$v_0 = \beta R_C(v_2 - v_1)/R_B$

CHAPTER 5

Exercises

5-1-1

C-1 $V_x = 5$ V

C-2 $V_x = 30$ V

5-1-4

C-1 $V_o = \dfrac{\mu G_1 G_3}{(G_3 + G_5)(G_1 + G_2 + G_4 - \mu G_2)}$

C-2 $V_E = (1 + \beta)R_E R_1 I_S/[(1 + \beta)R_E + R_1 + R_2]$

C-3 $V_o = \left[\dfrac{V_S}{R_2} + \dfrac{rV_S}{(R_1 + r)R_3}\right]\dfrac{1}{R_{EQ}}$

where $\dfrac{1}{R_{EQ}} = \dfrac{1}{R_2} + \dfrac{1}{R_3} + \dfrac{1}{R_4}$

C-4 $V_A = \dfrac{R_4 R_5}{R_4 + R_5}\left[\dfrac{V_S}{R_5} + \dfrac{\beta \mu R_2 V_S}{R_1 + R_2}\right]$

5-1-5

C-1

$$V_B = \frac{\begin{vmatrix} 2 & 1 & 0 & -1 \\ -1 & 0 & 1 & 0 \\ 0 & 0 & 3 & -1 \\ -1 & -1 & -1 & 2 \end{vmatrix}}{\begin{vmatrix} 2 & -1 & 0 & -1 \\ -1 & 3 & 1 & 0 \\ 0 & -1 & 3 & -1 \\ -1 & 0 & -1 & 2 \end{vmatrix}}$$

$$V_C = \frac{\begin{vmatrix} 2 & -1 & 1 & -1 \\ -1 & 3 & 0 & 0 \\ 0 & -1 & 0 & -1 \\ -1 & 0 & -1 & 2 \end{vmatrix}}{\begin{vmatrix} 2 & -1 & 0 & -1 \\ -1 & 3 & 1 & 0 \\ 0 & -1 & 3 & -1 \\ -1 & 0 & -1 & 2 \end{vmatrix}}$$

C-2 $V_x = 0$ V

5-1-6

C-1 $V_o = -V_S R_{EQ}/R_1$ where

$$\frac{1}{R_{EQ}} = \frac{2R_3}{R_4 R_1} + \frac{2}{R_4} + \frac{R_3}{R_4{}^2}$$

C-2 $V_o = \dfrac{V_S R_4 R_5}{R_1(R_3 R_5 - R_4 R_2)}$

C-3 $V_o = -V_S \dfrac{R_2 R_3 R_5}{R_1(R_5 R_2 + R_3 R_{EQ})}$

where $\dfrac{1}{R_{EQ}} = \dfrac{1}{R_2} + \dfrac{1}{R_4} + \dfrac{1}{R_5}$

C-4 $V_o = \dfrac{V_S R_{EQ}}{R_1}$

where $\dfrac{1}{R_{EQ}} = \dfrac{1}{R_1} + \dfrac{1}{R_2} + \dfrac{1}{R_4}$

C-5 $V_o = \dfrac{V_S R_{EQ}}{R_3 K}$

where $\dfrac{1}{R_{EQ}} =$

$$\frac{2(R_3 + R_4)}{R_4 R_3} + \frac{(R_3 + R_4)}{R_4{}^2} + \frac{1}{R_3}$$

and $K = \dfrac{R_1 R_2}{R_2(R_1 + R_2)} - \dfrac{R_{EQ}}{R_4}$

C-6 $V_o = 0$

C-7 $V_o = -V_S$

C-8 $V_o = V_S \left(\dfrac{R_{EQ2}}{R_4} - \dfrac{R_{EQ1}}{R_1} \right) \bigg/ \left(\dfrac{R_{EQ1}}{R_2} - \dfrac{R_{EQ2}}{R_5} \right)$

where $\dfrac{1}{R_{EQ1}} = \dfrac{1}{R_1} + \dfrac{1}{R_2} + \dfrac{1}{R_3}$

and $\dfrac{1}{R_{EQ2}} = \dfrac{1}{R_4} + \dfrac{1}{R_5} + \dfrac{1}{R_6}$

C-9 $V_o = \dfrac{V_S R_{EQ}}{R_1}$

where $\dfrac{1}{R_{EQ}} = \dfrac{R_3}{R_2 R_8} + \dfrac{R_5}{R_6 R_7} - \dfrac{R_5 R_3}{R_2 R_6 R_4}$

5-1-9

$V_T = 3$ V, $R_T = 5/6\ \Omega$

For P_{MAX}, $R_L = 5/6\ \Omega$ and $P_{MAX} = 2.7$ W

5-1-10

$$V_o = V_I \left[\frac{1}{g_m r_D} + K_1 - \frac{2K_1}{g_m r_D} \right]^{-1}$$

5-2-2

(a) Use circuit reduction, then simple voltage divider equations can be used.

(b) Superposition. Need to know the R_T of the circuit so $R_L = R_T$.

(c) Observation. $V_x = V_1 - V_2$.

(d) Node analysis. Many devices in parallel, one equation will give V_x.

(e) Node analysis, with respect to node at center of circuit, then KCL. $I_{R2} = I_2 - I_1$.

(f) Circuit reduction [see (a)].

(g) Node analysis with respect to bottom right corner. This answer gives v_x and $v_x^2/R_x = P$.

(h) Thévenin circuit, because to "load" the circuit $R_L \gg R_T$.

(i) Node equations to find V at upper center node then $V/R_L = I$ and ask: Is $I > 1/4$ A?

(j) Circuit reduction [see (a)].

(k) R_{EQ} is needed, however, the Δ configuration is hard to work. A conversion to Y's could be done but this is messy and who remembers the formula, so it is probably best to use loop equations and add up the consumed power.

(l) Observation. $I_x = \dfrac{V_1 - V_2}{R}$.

Problems

5-1

$$i_{B1} = i_{B2} = \frac{V_{CC} - V_T}{(\beta + 2)R_{C1} + R_B} \simeq \frac{V_{CC} - V_T}{\beta R_{C1}}$$

$$v_{CE2} = V_{CC} - \frac{R_{C2}\,\beta[V_{CC} - V_T]}{(\beta + 2)R_{C1} + R_B} \simeq$$

$$V_{CC} - \frac{R_{C2}\,[V_{CC} - V_T]}{R_{C1}}$$

$$v_{CE1} = V_T + \frac{R_B\,[V_{CC} - V_T]}{(\beta + 2)R_{C1} + R_B} \simeq$$

$$V_T + \frac{R_B\,[V_{CC} - V_T]}{\beta R_{C1}}$$

5-2

$$I_1 = -V_2/R_o, \quad I_2 = V_1/R_o$$

CHAPTER 6

Exercises

6-1-2

(a) $v_A(t) = 0$

(b) $v_B(t) = 5\ u(t) + 5\ u(t-1)$

(c) $v_C(t) = 5\ u(t) - 10\ u(t-1)$

(d) $v_D(t) = 5\ u(t) + 10\ u(t-1)$

6-1-3

(a) (1) $5t,\ 0 < t < \infty$ (1) $5\ \delta\ (t)$
     0, otherwise

(b) (2) $-5t,\ 0 < t < \infty$ (2) $-5\ \delta\ (t)$
     0, otherwise

(c) (3) $5(t-1)\ u(t-1)$ (3) $5\ \delta\ (t-1)$

(d) (4) $-10(t-1)\ u(t-1)$ (4) $-10\ \delta\ (t-1)$

6-1-5

(a) $v_1(t) = 10\ u(t)\ e^{-2t}$ $T_C = 1/2$ second

(b) $v_2(t) = 10\ u(t)\ e^{-t/2}$ $T_C = 2$ seconds

(c) $v_3(t) = -10\ u(t)\ e^{-20t}$ $T_C = 50$ ms

(d) $v_4(t) = -10\ u(t)\ e^{-t/20}$ $T_C = 20$ seconds

6-1-7

(a) (1) $-5(1 - e^{-2t})\mu(t)$

     (2) $-20\ (1 - e^{-t/2})\mu(t)$

     (3) $\frac{1}{2}\ (1 - e^{-20t})\mu(t)$

     (4) $200\ (1 - e^{-t/20})\ \mu(t)$

(b) (1) $-20\ e^{-2t}\ u(t)$

     (2) $-5\ e^{-t/2}\ u(t)$

     (3) $200\ e^{-20t}\ u(t)$

     (4) $(1/2)\ e^{-t/20}\ u(t)$

6-1-10

$a = 14.14 \cos 45° = 10$

$b = 14.14 \sin 45° = 10$

6-1-12

$v_3(t) = 2.36 \cos (2000\pi t - 206.6°)$

     $f = 1000$ Hz $A = 22.36$

     $T_0 = 1$ ms $\phi = 206.6°$

6-1-14

	f	T_0	a	b
(a)	2 kHz	0.5 ms	-20	0
(b)	2 kHz	0.5 ms	0	$+20$
(c)	0.0025 Hz	400 seconds	21.21	21.21
(d)	1 kHz	1 ms	42.42	-42.42

6-1-17

$v(t) = a \cos (\omega t) + b \sin (\omega t)$

$$\frac{dv(t)}{dt} = -a\omega \sin (\omega t) + b \omega \cos (\omega t)$$

$$\frac{d^2v(t)}{dt^2} = -a\omega^2 \cos (\omega t) - b \omega^2 \sin (\omega t)$$

Let $k^2 v(t) = \omega^2 v(t)$

$\qquad = a\omega^2 \cos (\omega t) + b\omega^2 \sin (\omega t)$; therefore,

$$\frac{d^2v(t)}{dt^2} + k^2 v(t) = 0$$

6-2-2

(a) $u(t - 1) + u(t - 2) + u(t - 3) - 3u(t - 4)$

(b) $2[u(t) - u(t - 1)] - [u(t - 2) - u(t - 3)]$
$+ 2[u(t - 4) - u(t - 5)]$

(c) Each cycle needs an equation that would be added. The first cycle's equation is
$-0.5 + (3/2) (t + 1) [u(t + 1) - u(t - 1)] -$
$3(t - 1)[u(t - 1) - u(t - 2)]$

(d) $-1.5 + 3.5 \cos (2\pi ft - 180°)$

6-2-6

(a) $\qquad \dfrac{dv_1(t)}{dt} = u(t - 1)$

(b) $\qquad \dfrac{dv_2(t)}{dt} = -2 u(t - 1)$

(c) $\qquad \dfrac{dv_3(t)}{dt} = u(t) - 2 u(t - 2)$

(d) $\qquad \dfrac{dv_4(t)}{dt} = u(t) - 2 u(t - 1) + u(t - 2)$

6-2-7

To find max set $\dfrac{dv(t)}{dt} = 0$; solve for t.

Max occurs at $t = \left(\dfrac{T_1 T_2}{T_1 - T_2}\right) \ln\,(T_1/T_2)$.

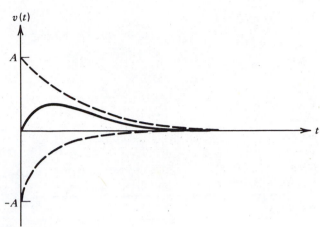

FIGURE F 6-2-7
Waveform for Exercise 6-2-7.

6-2-10
(a) $v_A(t) = 2.5r(t)[u(t) - u(t - 1)] -$
$2.5r(t - 3)[u(t - 3) - u(t - 5)]$
(b) $v_B(t) = 5[r(t) - r(t - 1)] +$
$5[r(t - 3) - r(t - 5)] + 5r(t - 6)$

6-3-1
(a) $V_P = V_{PP} = 10$
(b) $V_P = 10\,\sqrt{2}$, $V_{PP} = 20\,\sqrt{2}$,
$V_{\text{AVE}} = 0$, $V_{\text{RMS}} = 10$
(c) $V_P = 10$, $V_{PP} = 20$, $V_{\text{AVE}} = 0$, $V_{\text{RMS}} =$
$10/\sqrt{2}$
(d) $V_P = 100$, $V_{PP} = 200$

6-3-2
(a) $V_P = V_{PP} = 20$
(b) $V_P = V_{PP} = 0.048$
(c) $V_P = V_{PP} = 8.87$
(d) $V_P = V_{PP} = 6.97$

6-3-4
$V_P = V_{PP} = 3$
$V_{AVE} = 6/5$
$V_{RMS} = \sqrt{14/5}$

6-3-5
$V_P = 2$
$V_{PP} = 3$
$V_{AVE} = 5/6$
$V_{RMS} = 3/\sqrt{6}$

6-3-8
$V_{PP} = 79.98$ V

6-3-9
$P_{TOTAL} = 3.5$ W

Problems

6-2

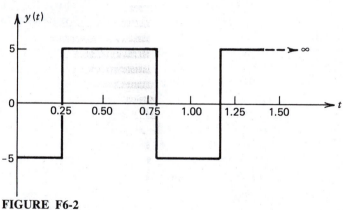

FIGURE F6-2
Output of comparator.

6-3

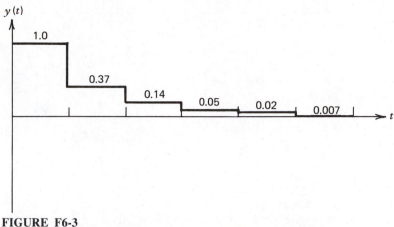

FIGURE F6-3
Output of sample-hold.

6-6

$v_o(t) = \cos 2\pi 10^6 t + \dfrac{m}{2} [\cos 2\pi (10^6 + 10^3)t$

$+ \cos 2\pi (10^6 - 10^3)t]$

Frequencies present—1001 kHz, 1000 kHz, and 999 kHz.

6-7

(a) $v_F(t) = \displaystyle\sum_{n-0}^{\infty} A \sin \dfrac{2\pi t}{T} \left\{ \left[u(t - nT) - u\left(t - nT - \dfrac{T}{2} \right) \right] - \left[u\left(t - nT - \dfrac{T}{2} \right) 2 - u\left(t - nT - T \right) \right] \right\}$

(b) $v_H(t) = \displaystyle\sum_{n=0}^{\infty} A \sin \dfrac{2\pi t}{T} \left[u(t - nT) - u\left(t - nT - \dfrac{T}{2} \right) \right]$

CHAPTER 7

Exercises

7-1-1

(a) $i_C = 10 \ \mu A$

(b) $i_C = 500 \cos 10t \ \mu A$

(c) $i_C = -40 \ e^{-2t} \ \mu A$

(d) $i_C = 100[100 \ e^{-t} \sin 100t - e^{-t} \cos 100 \ t] \ \mu A$

(e) $i_C = 50[u(t) - u(t - 2)] \ \mu A$

(f) $i_C = (-5000/\sqrt{2})(\cos 100t + \sin 100t) \ \mu A$

7-1-3

(a) $v_L = 4$ mV

(b) $v_L = -0.01 \sin 10t$ V

(c) $v_L = e^{-0.5t}$ μV

(d) $v_L = (-10\,e^{-0.5t} \cos 100t - 2000\,e^{-0.5t} \sin 100t)$ μV

(e) $v_L = 50\sqrt{2}\,(\cos 100t - \sin 100t)$

(f) $v_L = 0$

7-1-4

$i_1 = (5/2)$ μA, $\quad i_2 = (1/2)$ μA

7-1-5

$p = 25t/2$ μW, $\quad w = 25t^2/4$ μJ

7-1-8

C-1 $\quad C_{EQ} = 3.5$ μF

C-2 $\quad L_{EQ} = 1065.45$ μH

C-3 $\quad$ 11 μH in series with 1 μF in series with parallel combination of 1 μF and 10 μH.

7-1-10

A 1-μF capacitor is in the box.

7-1-12

(a) 2.3497 mH

(b) 2.35 mH can be fitted with a 10 percent 2.2-mH "standard" inductor. The issue of cost suggests use of a single inductor with some "tolerance." The choice must be tempered with the tolerance required in the circuit.

7-2-2

C-1 $\quad v_o = \left(\dfrac{-R}{L_1}\right) \int v_1\, dt - \left(\dfrac{R}{L_2}\right) \int v_2\, dt$

C-2 $\quad v_o = v_s(t) + \left(\dfrac{+L}{R}\right) \dfrac{dv_s(t)}{dt}$

C-3 $\quad v_o = \left(\dfrac{-L}{2R}\right) \dfrac{dv_1(t)}{dt} + \dfrac{v_1(t)}{2}$

7-2-4

C-1 $\quad 2x + 5\dfrac{dy}{dx} - \dfrac{d^2y}{dx^2} = y$

C-2 $\quad 5x - 100\dfrac{dy}{dx} - 5\dfrac{d^2y}{dx^2} = y$

C-3 $\quad \left(\dfrac{-1}{\omega_0^2}\right) \dfrac{d^2y}{dx^2} = y$

7-2-6

C-1

No. 1 $v_o = (1/250\pi)[\cos 500\pi t - 1]$

No. 2 $0 \leq t < 5 \qquad v_o = 0$

$\quad\quad\quad 5 \leq t < 10 \qquad v_o = (t - 5)$

$\quad\quad\quad 10 \leq t < 15 \qquad v_o = (t - 20)/2$

$\quad\quad\quad 15 \leq t < 20 \qquad v_o = -2.5 - (t - 15)^2/10$

No. 3 $0 \leq t < 1 \qquad v_o = 0.005t^2$

$\quad\quad\quad 1 \leq t < 2 \qquad v_o = -.005t^2 + 0.02t - 0.01$

C-2

No. 1 $v_o = 0.1\ \pi \cos 500\pi t$

No. 2 $0 \leq t < 5 \qquad v_o = 0$

$\quad\quad\quad 5 \leq t < 10 \qquad v_o = 0$

$\quad\quad\quad 10 \leq t < 15 \qquad v_o = 0$

$\quad\quad\quad 15 \leq t < 20 \qquad v_o = -2 \times 10^{-5}$

No. 3 $0 \leq t < 1 \qquad v_o = 10^{-6}$

$\quad\quad\quad 1 \leq t < 2 \qquad v_o = -10^{-6}$

C-3

No. 1 $v_o = -1000\ \pi \cos 500\ \pi t$

No. 2 $0 \leq t < 5 \qquad v_o = 0$

$\quad\quad\quad 5 \leq t < 10 \qquad v_o = 0$

$\quad\quad\quad 10 \leq t < 15 \qquad v_o = 0$

$\quad\quad\quad 15 \leq t < 20 \qquad v_o = 0.2$

No. 3 $0 \leq t < 1 \qquad v_o = -10^{-3}$

$\quad\quad\quad 1 \leq t < 2 \qquad v_o = 10^{-3}$

C-4   Input signals equal output signals.

7-3-1

$v_2(t) = 5 \cos 10t$

7-3-2

$i_2(t) \simeq 0.05 \cos 10t$

$v_2(t) = 5 \cos 10t$

7-3-4

$i = \dfrac{1}{L_T} \int v_S(t)dt \qquad \text{where} \qquad L_T = (L_1 + L_2$

$+ L_3 - 2\ M{\bullet}{\bullet} - 2\ M{\blacktriangle}{\blacktriangle} + 2\ M{\blacksquare}{\blacksquare})$

7-3-5

(a) $L = 22.94\ \mu H$

(b) $1\ H = 8300$ turns

$\quad\quad 10\ mH = 830$ turns

$\quad\quad 100\ \mu H = 83$ turns

(c)   0.00181-inch-diameter wire

7-3-7

(a) $i_1 = -200(e^{-10}t - 1), v_2 = 10\ e^{-10t}, i_2 = 0$

(b) $v_2 = 1 - e^{-5t}, i_2 = 0$

7-3-9

$k_m = 0, v_2(t) = 0$

$k_m = 0.5, v_2(t) = 14{,}300 \sin 10t$

7-4-1

$n = 1/13$

7-4-2

$n = 1/2.7$

7-4-3

$v_2 = 5.79\ V_{RMS}$

$v_3 = 45.83\ V_{RMS}$

7-4-4

$n = 1/2.27$

7-4-6

$i_2 = 20 \sin 10t,\ v_{OUT} = 0.2 \cos 10t$

Problems

7-1

$d = 3.19 \times 10^{-6}$ m, $C = 1\ \mu F$, $\mathscr{E} = 15.7 \times 10^6$ V/m Yes, it will break down.

7-3

(a) $i(t) = -e^{-t}$

(b) $v_C = -0.368$ V, $w_C = 0.0677$ J

(c) $i_{L1} = 0.368$ A, $w_{L1} = 0.0677$ J

7-4

$v_L(t) = -3\ e^{-3t}$ V

CHAPTER 8

Exercises

8-1-1

(a) $x = 5\ e^{-10t}$

(b) $x = 10 - 20\ e^{-10,000t}$

(c) $x = 1 - e^{10t}$

(d) $x = -e^{-10,000t} - 1$

(e) $x = -e^{-5t} + \cos 5t + \sin 5t$

(f) $x = 0.1 \cos 100t + 10 \sin 100t + 2.9\ e^{-10,000t}$

(g) $x = -0.915 \cos 100t + 3.145 \sin 100t + 0.915\ e^{-100t}$

8-1-2

C-1 $T = RC/2$ C-2 $T = 4RC$ C-3 $T = 4RC/3$

C-4 $T = L/2R$ C-5 $T = 4L/R$ C-6 $T = 5L/3R$

C-7 $T = 20$ ms   C-8 $T = 830\ \mu s$ C-9 $T = 100$ ms

C-10 $T = 100$ ns   C-11 $T = 0.366$ ms   C-12 $T = 0.6$ ms

8-1-4

C-1 $v(t) = [V_oR_2/(R_1 + R_2)]\ e^{-t/T}$

C-2 $i(t) = [V_o/R_1]\ e^{-t/T}$

C-3 $v(t) = V_o + [(V_oR_2/(R_1 + R_2) - V_o)]\ e^{-t/T}$

C-4 $i(t) = V_o/R_1$

8-1-6

C-1 $v(t) = V_o(1 - e^{-t/T})$

C-2 $i(t) = (V_o/R_2)(1 - e^{-t/T})$

C-3 $v(t) = V_o - 2V_o\, e^{-t/T}$

C-4 $i(t) = (V_o/R) - (2V_o/R)\, e^{-t/T}$

C-5 $v(t) = V_o(1 - e^{-t/T})$

C-6 $i(t) = (V_o/R)(1 - e^{-t/T})$

8-1-8

C-1 $v_C = 8(1 - e^{-10,000t})\, u(t)$

C-2 $i_L = 6.67(1 - e^{-500t})\, u(t) \times 10^{-3}$

C-3 $v_o = 3.125(1 - e^{-2500t})$

C-4 $i_o = 0.2777\, e^{-0.833t}$

8-1-11

C-1 $v_o = \dfrac{AR}{L^2\omega^2 + R^2}[R\cos\omega t + \omega L\sin\omega t]$

v_o lags v_S.

C-2 $v_o = \dfrac{A}{\sqrt{1 + (\omega RC)^2}}\cos[\omega t - \tan^{-1}(\omega RC)]$

v_o lags v_S.

C-3 $v_o = \dfrac{A/2}{\sqrt{1 + (\omega RC/2)^2}}\cos[\omega t - \tan^{-1}(\omega RC/2)]$

8-2-1

(a) $x = -5e^{-5t} + 5e^{-2t},\ t \geq 0$

(b) $x = 2t\, e^{-2t},\ t \geq 0$

(c) $x = e^{-t}\cos(2t - 90°),\ t \geq 0$

(d) $x = 10 - 0.625\, e^{-10t} + 5.625\, e^{-t},\ t \geq 0$

(e) $x = 4 + 50t\, e^{-5t} + 5e^{-5t},\ t \geq 0$

(f) $x = 10 + 9.72\, e^{-t}\cos(3t - 59.03°)$

8-2-3

C-1 $R = 2(2L/C)^{1/2}$

C-2 $R = (L/2C)^{1/2}$

C-3 $R = (L/2C)^{1/2}$

C-4 $R = 2(5L/3C)^{1/2}$

8-2-5

C-1 $v = 10 + 16.7\, e^{-7500t}\cos(5587t - 233.3°)$ V

C-2 $v_C = 10 + 8.37\, e^{-7500t}\cos(5587t - 53.3°)$ V

C-3 $i_L = 6.67 + 9\, e^{-3250t}\cos(3625t - 221.8°)$ mA

C-4 Same as C-3

8-2-7

C-1 $i_L = 350 + 352\, e^{-587.5t}\cos(5000t - 173.3°)$ μA

C-2 $v_C = 15 + e^{-19995t} - 16\, e^{-1255t}$ V

C-3 $v_C = 30 + 1.16\, e^{-18660t} - 16.16\, e^{-1340t}$ V

C-4 $i_L = 0.350 - 6.39\, e^{-587.5t}\cos(5000t - 186.7°)$ mA

8-2-9

C-1 $V_S = v_o + 2RC\dfrac{dv_o}{dt} + \dfrac{R^2C^2d^2v_o}{dt^2}$ In using quadratic equation, roots are always critically damped. C-2 and C-3 have same form as C-1.

8-2-11

$\mu < 3$

Problems

8-1

Use RC series circuit with voltage taken across capacitor. We want RC to equal 1.112 seconds. A suitable pair would be $C = 1\ \mu\text{F}$, $R = 1.1\ \text{M}\Omega$. Choose ± 5 percent tolerance components for cost.

8-3

$$i_S(t)R = \dfrac{R^2C^2d^2v_o(t)}{dt^2} + 3RC\,\dfrac{dv_o(t)}{dt} + v_o(t)$$

8-4

$i_L(t) = 30\,e^{-300t} - 10\,e^{-400t}$ A

8-5

$v_L(t) = e^{-5t}\,(1549\sin 1291t)$ V

8-6

Without the diode the sudden opening of the contacts causes an arc since the current in the inductor does not like to change abruptly. The diode does not conduct as long as the contacts are closed. When the contacts open, the current in the inductor now has a path through the diode. The resistor R_2 allows the "stored" energy to dissipate.

CHAPTER 9

Exercises

9-1-1

(a) $5/s$

(b) $-5/s$

(c) $(5/s)\,e^{-s}$

(d) $(-10/s)\,e^{-s}$

9-1-2

(a) $\dfrac{a}{s(s + a)}$

(b) $\dfrac{(s + 2a)}{(s + a)^2}$

(c) $\dfrac{(b - a)}{(s + a)(s + b)}$

(d) $\dfrac{b \cos \phi - s \sin \phi}{s^2 + b^2}$

(e) $\dfrac{\cos \phi \, (s + a - b \tan \phi)}{(s + a)^2 + b^2}$

(f) $\dfrac{b^2}{(s + a)^3 + b^2(s + a)}$

9-1-4

(a) $10/s$

(b) $10/(s + 1)$

(c) $10/(s + 10)$

(d) $10/s(s + 10)$

(e) Same as (d)

(f) $90/(s + 1)(s + 10)$

9-1-6

(Pole-zero diagram shows poles only.)

(a) $\dfrac{K}{s(s+1)(s+3)(s+6)}$

(b) Same as (a).

(c) $\dfrac{3s}{s^2+36} - \dfrac{100}{s^2+100} + \dfrac{7s}{s^2+100}$

(d) $\dfrac{3s}{s^2+36} - \dfrac{100}{s^2+100} + \dfrac{7(s+1)}{(s+1)^2+100}$

(e) $\dfrac{3s}{s^2+36} - \dfrac{100}{(s+1)^2+100} + \dfrac{6(s+1)}{(s+1)^2+100}$

(f) $\dfrac{3(s+1)}{(s+1)^2+36} - \dfrac{100}{(s+2)^2+100}$

$+ \dfrac{6(s+1)}{(s+1)^2+100}$

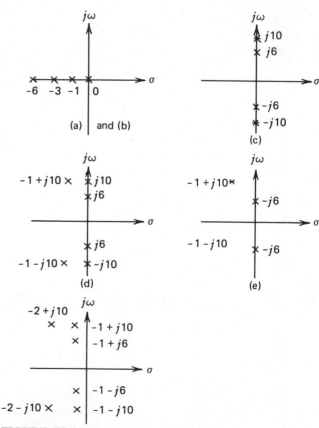

FIGURE F9-1-6
Pole-zero diagrams for Exercise 9-1-6.

9-2-3

(a) $v(t) = (-4/3) e^{-4t} + (4/3) e^{-t}$

(b) $v(t) = 4t e^{-2t}$

(c) $v(t) = (4/\sqrt{3}) e^{-t} \sin \sqrt{3}\, t$

(d) $v(t) = 4 e^{-4t}$

(e) $v(t) = 4(1 - t) e^{-2t}$

(f) $v(t) = 4 e^{-t} \cos \sqrt{3}\, t$

(g) $v(t) = 2 \sin 2t$

(h) $v(t) = -2 \sin 2t + \delta(t)$

(i) $v(t) = 2 \sin 2t - \cos 2t + 1$

9-2-4

(a) $v(t) = -e^{-2t} - 9e^{-4t} + 16e^{-5t}$

(b) $v(t) = -e^{-4t} + 2e^{-5t}$

(c) $v(t) = 2 + e^{-2t} \sin 2t$

(d) $v(t) = 2e^{-t} + 2t\, e^{-t} - 2 \cos t$

(e) $v(t) = 2 + 6 \cos 2t$

9-3-1

(a) $x(t) = 5e^{-10t}$

(b) $x(t) = 10 - 20\, e^{-10000t}$

(c) $x(t) = 1 - e^{10t}$

(d) $x(t) = -1 - e^{-10000t}$

(e) $x(t) = -e^{-5t} + \cos 5t + \sin 5t$

(f) $x(t) = 2.9\, e^{-10000t} + 0.1 \cos 100t + 10 \sin 100t$

(g) $x(t) = 0.915\, e^{-100t} - 0.915 \cos 100t +$
$3.145 \sin 100t$

9-3-2

C-1 $\quad v(t) = \left[\dfrac{R_2}{(R_1 + R_2)}\right] V_o \left[1 - e^{-t/R_T C}\right]$

C-2 $\quad i(t) = (V_o/R_1)[1 - e^{-R_T t/L}]$

C-3 $\quad v(t) = V_T + (V_o - V_T)[e^{-t/R_T C}]$

C-4 $\quad i(t) = V_T/R_T + (V_T/R_T - V_o/R_1)e^{-R_T t/L}$

9-3-4

C-1 $\quad v_o(t) = \left(\dfrac{R_2 V_o}{R_1 + R_2}\right) \left[1 - e^{-(R_1 + R_2)t/L}\right]$

C-2 $\quad v_o(t) = (I_o R_1) \left[e^{-(R_1 + R_2)t/L}\right]$

C-3 $\quad v_o(t) = \left[\dfrac{V_o R_2}{(R_1 + R_2)}\right] \left[e^{-t/(R_1 + R_2)C}\right]$

C-4 $\quad v_o(t) = (I_o R_1)[1 - e^{-t/(R_1 + R_2)C}]$

C-5 $\quad v_o(t) = V_o - \left(\dfrac{V_o R_1}{(R_1 + R_2)}\right) \left[1 - e^{-(t/R_T C)}\right]$

C-6 $\quad v_o(t) = \left(\dfrac{R_1 I_o R_2}{(R_1 + R_2)}\right) \left[1 - e^{-t(R_1 + R_2)/L}\right]$
$+ I_o R_1 e^{-t(R_1 + R_2)/L}$

9-3-6

C-1 $v(t) = 15.12\, e^{-3750t} \cos (3300t - 48.6°)$ V

C-2 $v(t) = 0.015 - 117\, e^{-3750t} \sin 3300t$ V

C-3 $i(t) = 0.0138\, e^{-1662t} \cos (4712t + 19.8°)$ A

C-4 $i(t) = 8.88 - 290\, e^{-11000t} - 47.5\, e^{-2250t}\, \mu$A

9-3-9

C-1 $v_o(t) = (-1/RC)\, e^{-t/RC}$

C-2 $v_o(t) = -A(1 - e^{-t/RC})$

C-3 $v_o(t) = A/(1 - aRC)(e^{-t/RC} - e^{-at})$

C-4 $v_o(t) = -\dfrac{A}{1 + \beta^2 R^2 C^2} (\cos \beta t + \beta RC \sin\beta t)$

$+ \dfrac{A\, e^{-t/RC}}{1 + \beta^2 R^2 C^2}$

Problems

9-1

(a) $V(s) = 1/(s + a)^2$

(b) $V(s) = 2bs/(s^2 + b^2)^2$

(c) $V(s) = 2/s3$

(d) $V(s) = 2/(s + a)^2$

9-3

In all answers $v(t) = e^{-t}$.

(a) For $K = 1$.

(b) For $K = 2$.

(c) For $K = 9$.

9-6

(a) $\displaystyle\lim_{s\to 0} sV(s) = \frac{1}{a}$

$\displaystyle\lim_{t\to\infty} v(t) = \frac{1}{a}$ for $a > 0$

(b) $\displaystyle\lim_{s\to 0} sV(s) = 0$

$\displaystyle\lim_{t\to\infty} v(t) = 0$ for a and $b > 0$

(c) $\displaystyle\lim_{s\to 0} sV(s) = 0$

$\displaystyle\lim_{s\to 0} v(t) = 0$ for $a > 0$

(d) $\displaystyle\lim_{s\to 0} sV(s) = 1$

$\displaystyle\lim_{t\to\infty} v(t) = 1$ for $a > 0$

9-8

$v_2(t) = 1 - 0.733\, e^{-0.382t} - 0.277\, e^{-2.62t}$

$v_1(t) = 1 - 0.447\, e^{-0.382t} - 0.477\, e^{-2.62t}$

CHAPTER 10

Exercises

10-1-2

C-1 $\quad v_c(t) = V_T\, e^{-t/R_2C}$

C-2 $\quad i_L(t) = (V_T/R_T)\, e^{-R_2t/L}$

C-3 $\quad v_C(t) = V_o + (V_T - V_o)\, e^{-t/R_1C}$

C-4 $\quad i_L(t) = (V_o/R_1) + [(V_T/R_T) - (V_o/R_1)]\, e^{-R_1t/L}$

10-1-4

C-1 $\quad v_C(t) = 97.6 + 7.7\, e^{-7.45 \times 10^4 t} - 105$
$e^{-0.55 \times 10^4 t}$

C-2 $\quad v_L(t) = 107.9\, e^{-7.45 \times 10^4 t} - 7.97\, e^{-0.55 \times 10^4 t}$

C-3 $\quad v_C(t) = 6.25\, e^{-2.5 \times 10^3 t} \sin (20 \times 10^3 t)$

C-4 $\quad v_R(t) = 100 + 6.198\, e^{-2.5 \times 10^3 t} \cos (20 \times 10^3 t$
$- 82.9°)$

10-1-5

C-1 $\quad Z_{\mathrm{EQ}} = \dfrac{sCR^2 + 2R}{sCR + 1}$

C-2 $\quad Z_{\mathrm{EQ}} = \dfrac{[s^2CRL + s(L + CR^2) + 2R]}{sCR + 1}$

C-3 $\quad Z_{\mathrm{EQ}} = \dfrac{(s^2C^2R^2 + 3sCR + 1)}{(s^2C^2R + 2sC)}$

C-4 $\quad Z_{\mathrm{EQ}} = \dfrac{[s^2CRL + s(L + CR^2) + 2R]}{s^2CL + 2sCR}$

10-1-6

C-1 $\quad I_{zs} = ARC/(2sCR + 1)$
$I_{zi} = -V_o sC/s(2sCR + 1)$

C-2 $\quad V_{zs} = \dfrac{RLsA}{(s + \alpha)(R^2 + 2RLs)}$
$V_{zi} = -R^2LI_o/(R^2 + 2sLR)$

C-3 $\quad I_{zs} = \dfrac{AsL}{(s + \alpha)(2R + sL)}$
$I_{zi} = -I_o L/(2R + sL)$

C-4 $\quad V_{zs} = A/s(RsC + 2)$
$V_{zi} = V_o CR/(sCR + 2)$

10-2-3

C-1 $\quad V_2(s) = -Z_{\mathrm{EQ}} V_1(s)/R$

where $\dfrac{1}{Z_{\mathrm{EQ}}} = \dfrac{1}{R_1R_3C_2s} + \dfrac{1}{R_2R_3C_2s} + \dfrac{1}{R_3} + C_1s$
$+ \dfrac{1}{R_2}$

C-2 $\quad V_2(s) = -V_1(s)/(1 + sC_1R_1)$

10-2-4

C-1 $V_2(s) = V_1(s) \left[\dfrac{1}{sCR + 1} - \dfrac{R}{sL + R} \right]$

C-2 $V_2(s) = \dfrac{V_1(s)}{2RCs + 1}$

C-3 $V_2(s) = \dfrac{V_1(s)RCs}{RCs + 2}$

C-4 $V_2(s) = \dfrac{V_1(s)Ls}{2R + Ls}$

10-2-6

C-1 $V_2(s) = \dfrac{V_1(s)\beta R_o R_L}{[R_1 + R_\pi][R_o + R_L + (1/Cs)]}$

C-2 $V_2(s) = \dfrac{V_1(s)(\beta + 1) Z_{EQ}}{[R_1 + R_\pi + (\beta + 1) Z_{EQ}]}$

where $Z_{EQ} = R_E/(sR_E C_E + 1)$

C-3 $V_2(s) = V_1(s) Z_{EQ}/R_1$

where $1/Z_{EQ} = [(K_1/R_1) + (K_1/R_\pi) - C_\mu s + C_\mu s k_1]$

10-3-1

$\dfrac{V_2(s)}{V_1(s)} = \dfrac{R_2}{R_o R_2 C_2 s + R_o + R_2}$

10-3-3

$\dfrac{I_2(s)}{I_1(s)} = \dfrac{R_o C_1 s}{s(C_1 R_o + C_1 R_1) + 1}$

10-3-5

$\dfrac{V_2(s)}{V_1(s)} = \dfrac{[sC_1(R_o + R_1) + 1]}{R_1 C_1 s + 1}$

10-3-7

$\dfrac{I_2(s)}{I_1(s)} = \dfrac{R_2}{R_2 C_2 L_3 s^2 + (L_3 + R_2 C_2 R_3)s + (R_2 + R_3)]}$

10-3-9

$\dfrac{V_2(s)}{V_1(s)} = \dfrac{R_4 L_4 C_1 s^2}{(R_1 + R_4)L_4 C_1 s^2 + (R_1 R_4 C_1 + L_4)s + R_4}$

10-3-11

$\dfrac{V_2(s)}{V_1(s)} = \dfrac{C_5 R_o s}{(s^2 L_5 C_5 + sR_o C_5 + 1)}$

10-2-13

$\dfrac{V_2(s)}{V_1(s)} = \dfrac{(s^2 L_5 C_5 + 1)}{(s^2 L_5 C_5 + sR_o C_5 + 1)}$

10-3-16

$\dfrac{I_2(s)}{I_1(s)} = \dfrac{(s^2 R_2 L_6 C_6 + R_2)}{[s^2 R_2 L_6(C_6 + C_2) + sL_6 + R_2]}$

10-4-1
See Exercise 10-3-3 for $H(s)$.
For $i_1(t) = u(t)$ we have two poles, one zero.
Natural real pole at $1/C_1(R_o + R_1)$.
Forced real pole at origin.
Natural real zero at origin.
(The forced pole is unobservable.)
System is stable.

10-4-4
See Exercise 10-3-13 for $\underline{H(s)}$.
For $v_1(t) = A \sin (t/\sqrt{L_5C_5})$.
Four poles, two zeros.
Two forced imaginary poles at $\pm j/\sqrt{L_5C_5}$.
Two natural imaginary zeros at $\pm j/\sqrt{L_5C_5}$.
(The above poles are unobservable.)
Two natural poles, form not determinable from given data.
System is stable.

10-4-5
$$V_2 = \frac{(s^2L_5C_5 + 1)}{[s^2(L_5C_5 + L_3C_5) + s(C_5R_3) + 1]s}$$
Three poles, two zeros.
One forced pole at origin.
Two natural poles of undetermined form.
Two zeros at $\pm j/\sqrt{L_5C_5}$.
System is stable.

10-4-8
$$C\text{-}1 \quad V_2 = \frac{(5R^2C^2s^2)}{(s^2C^2R^2 - s2RC + 1)s}$$
Three poles, two zeros.
Two natural zeros at origin.
One forced pole at origin (unobservable).
Two natural poles of undetermined form.
System is unstable because $-s(2RC)$ term will cause poles to fall on right-hand plane.
$$C\text{-}2 \quad V_2 = \frac{5sCR}{s(-8sCR + 2)}$$
Two poles, one zero.
One natural zero at origin.
One forced pole at origin (unobservable).
One natural pole at $1/4CR$.
System is unstable.
$$C\text{-}3 \quad V_2 = \frac{-5CRs}{[s^2(4C^2R^2) + s(7CR) + 4]s}$$
Three poles, one zero.
One natural zero at origin.
One forced pole at origin (unobservable).
Two natural poles of undetermined form.
System is stable.

Problems

10-1

$$T(s) = \frac{1}{[L_1 L_2 C_1 C_2 s^4 + (L_1 C_1 + L_2 C_2 + L_1 C_2)s^2 + 1]}$$

$$Z_{in} = \frac{L_1 L_2 C_1 C_2 s^4 + (L_1 C_1 + L_2 C_2 + L_1 C_2)s^2 + 1}{L_2 C_1 C_2 s^3 + (C_1 + C_2)s}$$

10-3

$$T(s) = \frac{Y_2 - Y_1}{Y_2 - Y_1 + Y_3 - Y_4}$$

This OP AMP circuit is called a *negative impedance converter*.

10-5

$T_{CM} = RCs/2(RCs + 1)$

$T_{DM} = -(RCs + 4)/2(RCs + 1)$

$\text{CMRR} = -(RCs + 4)/RCs$

10-7

(a) $T(s) = \dfrac{RCs + 1}{RCs + 2}$

(b) Zero at $s = -1/RC$, poles at $s = -2/RC$, and $s = -100$. To cancel forced pole, select $RC = 10^{-2}$.

(c) Zero at $s = -1/RC$, poles at $s = -2/RC$, and $s = -10 \pm j200$. To have no natural component, the zero at -10 must cancel the natural pole at $s = -2/RC$. Hence $RC = 0.2$

10-9

$$T(s) = \frac{1}{R^2 C^2 s^2 + RCs + 1}$$

Poles are located at $(1/2RC)(-1 \pm j\sqrt{3})$ and are always complex.

CHAPTER 11

Exercises

11-1-1

$$\frac{I_2}{I_1} = \frac{R_o}{R_o + R_3 + sL} = T(s) = H(s)$$

$$h(t) = (R_o/L)\, e^{-(R_o + R_3)t/L}$$

10-4-1

See Exercise 10-3-3 for $H(s)$.

For $i_1(t) = u(t)$ we have two poles, one zero.

Natural real pole at $1/C_1(R_o + R_1)$.

Forced real pole at origin.

Natural real zero at origin.

(The forced pole is unobservable.)

System is stable.

10-4-4

See Exercise 10-3-13 for $H(s)$.

For $v_1(t) = A \sin (t/\sqrt{L_5 C_5})$.

Four poles, two zeros.

Two forced imaginary poles at $\pm j/\sqrt{L_5 C_5}$.

Two natural imaginary zeros at $\pm j/\sqrt{L_5 C_5}$.

(The above poles are unobservable.)

Two natural poles, form not determinable from given data.

System is stable.

10-4-5

$$V_2 = \frac{(s^2 L_5 C_5 + 1)}{[s^2(L_5 C_5 + L_3 C_5) + s(C_5 R_3) + 1]s}$$

Three poles, two zeros.

One forced pole at origin.

Two natural poles of undetermined form.

Two zeros at $\pm j/\sqrt{L_5 C_5}$.

System is stable.

10-4-8

C-1 $$V_2 = \frac{(5R^2 C^2 s^2)}{(s^2 C^2 R^2 - s2RC + 1)s}$$

Three poles, two zeros.

Two natural zeros at origin.

One forced pole at origin (unobservable).

Two natural poles of undetermined form.

System is unstable because $-s(2RC)$ term will cause poles to fall on right-hand plane.

C-2 $$V_2 = \frac{5sCR}{s(-8sCR + 2)}$$

Two poles, one zero.

One natural zero at origin.

One forced pole at origin (unobservable).

One natural pole at $1/4CR$.

System is unstable.

C-3 $$V_2 = \frac{-5CRs}{[s^2(4C^2 R^2) + s(7CR) + 4]s}$$

Three poles, one zero.

One natural zero at origin.

One forced pole at origin (unobservable).

Two natural poles of undetermined form.

System is stable.

Problems

10-1

$$T(s) = \frac{1}{[L_1L_2C_1C_2s^4 + (L_1C_1 + L_2C_2 + L_1C_2)s^2 + 1]}$$

$$Z_{in} = \frac{L_1L_2C_1C_2s^4 + (L_1C_1 + L_2C_2 + L_1C_2)s^2 + 1}{L_2C_1C_2s^3 + (C_1 + C_2)s}$$

10-3

$$T(s) = \frac{Y_2 - Y_1}{Y_2 - Y_1 + Y_3 - Y_4}$$

This OP AMP circuit is called a *negative impedance converter*.

10-5

$T_{CM} = RCs/2(RCs + 1)$

$T_{DM} = -(RCs + 4)/2(RCs + 1)$

$CMRR = -(RCs + 4)/RCs$

10-7

(a) $T(s) = \dfrac{RCs + 1}{RCs + 2}$

(b)   Zero at $s = -1/RC$, poles at $s = -2/RC$, and $s = -100$. To cancel forced pole, select $RC = 10^{-2}$.

(c)   Zero at $s = -1/RC$, poles at $s = -2/RC$, and $s = -10 \pm j200$. To have no natural component, the zero at -10 must cancel the natural pole at $s = -2/RC$. Hence $RC = 0.2$

10-9

$$T(s) = \frac{1}{R^2C^2s^2 + RCs + 1}$$

Poles are located at $(1/2RC)(-1 \pm j\sqrt{3})$ and are always complex.

CHAPTER 11

Exercises

11-1-1

$$\frac{I_2}{I_1} = \frac{R_o}{R_o + R_3 + sL} = T(s) = H(s)$$

$h(t) = (R_o/L)\, e^{-(R_o + R_3)t/L}$

11-1-4

(1)

(a) $h(t) = u(t)$ $\quad H(s) = \dfrac{1}{s}$

$$\int_0^t h(t)dt = g(t) = \int_0^t dt = t - 0 = t$$

(b) $h(t) = r(t)$ $\quad H(s) = \dfrac{1}{s^2}$

$$\int_0^t h(t)dt = \int_0^t t\,dt = \left.\frac{t^2}{2}\right|_0^t = \frac{t^2}{2} = g(t)$$

(c) $h(t) = e^{-t}$ $\quad H(s) = \dfrac{1}{s+1}$

$$\int_0^t e^{-t}dt = g(t) = \left.-e^{-t}\right|_0^t = 1 - e^{-t}$$

(d) $h(t) = 1 - e^{-t}$ $\quad H(s) = \dfrac{1}{s} - \dfrac{1}{s+1}$

$$\int_0^t (1 - e^{-t})dt = g(t) = \left.t - e^{-t}\right|_0^t$$
$$= t - e^{-t} - 0 + 1$$
$$= (t + 1) - e^{-t}$$

(e) $h(t) = \sin t$ $\quad H(s) = \dfrac{1}{s^2 + 1}$

$$\int_0^t \sin t\, dt = g(t) = \left.-\cos t\right|_0^t = 1 - \cos t$$

(f) $h(t) = \cos t$ $\quad H(s) = \dfrac{s}{s^2 + 1}$

$$\int_0^t \cos t\, dt = g(t) = \left.\sin t\right|_0^t = \sin t$$

(g) $h(t) = 1 - \cos t$ $\quad H(s) = \dfrac{1}{s} - \dfrac{s}{s^2 + 1}$

$$\int_0^t (1 - \cos t)dt = g(t) = \left.t - \sin t\right|_0^t = t - \sin t$$

(2)

(a) $g(t) = u(t)$ $\quad G(s) = \dfrac{1}{s}$

$$sG(s) = H(s) = 1$$
$$\frac{dg(t)}{dt} = h(t) = \delta(t)$$

(b) $g(t) = r(t)$ $\quad G(s) = \dfrac{1}{s^2}$ $\quad$ then $\quad H(s) = \dfrac{1}{s}$

$$\frac{dg(t)}{dt} = h(t) = \frac{d(t)}{dt} = 1$$

(c) $g(t) = e^{-t}$ $G(s) = \dfrac{1}{s+1}$ then $H(s)$

$= \dfrac{s}{s+1}$

$\dfrac{dg(t)}{dt} = h(t) = -e^{-t}$

(d) $g(t) = 1 - e^{-t} \dfrac{1}{s} - \dfrac{1}{s+1}$ then $H(s) = 1$

$- \dfrac{s}{s+1} = \dfrac{1}{s+1}$

$\dfrac{dg(t)}{dt} = h(t) = e^{-t}$

(e) $g(t) = \sin t \dfrac{1}{s^2+1}$ then $H(s) = \dfrac{s}{s^2+1}$

$\dfrac{d \sin t}{dt} = h(t) = \cos t$

(f) $g(t) = \cos t \dfrac{s}{s^2+1}$ then $H(s) = \dfrac{s^2}{s^2+1}$

$\dfrac{d \cos t}{dt} = -\sin t = u(t)$

(g) $g(t) = 1 - \cos t \dfrac{1}{s} - \dfrac{s}{s^2+1}$ then $H(s) = 1$

$- \dfrac{s^2}{s^2+1} = \dfrac{1}{s^2+1}$

$\dfrac{d(1 - \cos t)}{dt} = h(t) = \sin t$

11-2-2
$H(s) = R_o/[sL + (R_o + R_3)]$
$G(s) = H(s)/s$
$g(t) = K_o - K_o\, e^{-(R_o + R_3)t/L}$

11-2-3
$H(s) = \dfrac{sR_oC_1}{[s(C_1R_o + C_1R_1) + 1]}$
$G(s) = H(s)/s$
$g(t) = \dfrac{(R_oC_1)}{(C_1R_0 + C_1R_1)}\, e^{-t/(C_1R_1 + C_1R_0)}$

11-2-5
$H(s) = \dfrac{[s(C_1R_0 + C_1R_1 + 1]}{(sC_1R_1 + 1)}$
$G(s) = H(s)/s$
$g(t) = 1 + (R_0/R_1)\, e^{-t/R_1C_1}$

11-3-1
$FV = 1/2 = K_0$
$IV = 0$, therefore $K_1 = -1/2$
$g(t) = 1/2 - e^{-1000t} \cos 1000t$

11-3-3
$g(t) = (2/\sqrt{3})\, e^{-500t} \sin (\sqrt{3}/2)\, 10^3 t$

11-3-6

$$H(s) = \frac{(s^2 + 2000s + 10^6)}{(s^2 + 3000s + 10^6)}$$

$$g(t) = \frac{H(s)}{s} = \frac{K_0}{s} + \frac{K_1}{s + 170} + \frac{K_2}{s + 5820}$$

$$- \frac{K_2}{s + 170}$$

$K_0 = 1 = FV$, $K_0 + K_1 = IV = 1$,
therefore $K_1 = 0$
$K_2 = 0.12(10^{-3})$
$g(t) = 1 + 0.12(10^{-3})e^{-5820t} - 0.12(10^{-3})e^{-170t}$

11-4-1

$\omega_0^2 = 10^6$, $\zeta = 0.5$
Let $R_1 = R_2 = 1$ kΩ and let $C_2 = 1$ μF so that
$L_3 = 1$ H.
Other values are possible.

11-4-3

$\omega_0^2 = 10^6$, $\zeta = 1.0$
Let $R_3 = 1$ kΩ and $L_3 = 1$ H. Solve and get
$R_4/L_4 = 10^3$.
Pick $L_4 = 1$ H, R_4 then equals 1 kΩ.

11-4-5

$\omega_0^2 = 1 = 1/R^2C^2$. Pick $R = 1$ MΩ,
$C = 1$ μF.
$\zeta = 1.5$

11-4-7

C-1 $g(t) = 1 - e^{-t/RC}$
$T_1 = 0.1$ ms, $T_2 = 2.3$ ms
$T_R = 2.2$ ms
$T_S = 5$ ms
$T_D = 0.69$ ms
PO = 0 percent
C-2 $g(t) = 1 + 1.15 \, e^{-500t} \cos(866t - 150°)$
$T_1 = 0.1$, $T_2 = 1$ ms
$T_R = 0.9$ ms
$T_S = 10$ ms
$T_D = 0.5$ ms
PO = 30 percent

Problems

11-1

(1) $f_1(t) = t - \dfrac{b}{(a^2 + b^2)} + \left[\dfrac{e^{-at}}{(a^2 + b^2)}\right]$ [a sin bt + b cos bt]

(2) $f_2(t) = t - \dfrac{a}{(a^2 + b^2)} - e^{-at}$ (a cos bt + b sin bt)

(3) $f_3(t) = t - (1/a^2) + (e^{-at}/a^2) + (t\, e^{-at}/a)$

(4) $f_4(t) = t - \left[\dfrac{(a + 1)}{a^2}\right] + \left[\dfrac{(a + 1)\, e^{-at}}{a^2}\right] + \left(t\, \dfrac{e^{-at}}{a}\right)$

11-3

For a step input, the diode is "forward biased" and acts as a short circuit. Hence

$$g(t) = 1 - e^{-t/RC},\ t \geq 0$$

Without the diode, the impulse response is

$$h(t) = (e^{-t/RC})/RC,\ t \geq 0$$

and the current

$$i_C(t) = C\, dh(t)/dt = \frac{\delta(t)}{R} - \frac{e^{-t/RC}}{R^2 C}$$

With the diode in the circuit, the negative component of this current is blocked (diode behaves as an open circuit), and hence the capacitor does not discharge,

$$h(t) = u(t)/RC$$

11-4

Circuit	IV	FV
1	0	1
2	1	1
3	0	0
4	0	K

11-5

Common mode:

$v_o = e^{-t/RC}/2$

Difference mode:

$v_o = -1 + [e^{-t/RC}/2]$

11-7

$$T(s) = \frac{2G}{(C - C_1)s + G}$$

If $C_1 = C$, $T(s) = 2$ and response time is zero.

If $C_1 > C$ by any amount, circuit is unstable.

11-10

$g(t) = 1 - e^{-\alpha t}$ could satisfy the requirement provided a suitable value of α can be found.
Constraint 1 at 1 ms requires $\alpha < 693$.
Constraint 2 at 2 ms requires $\alpha > 346$.
Constraint 3 at 10 ms requires $\alpha > 461$.
Hence $461 < \alpha < 693$. So pick $\alpha = 500$.
$G(s) = (1/s) - (1/s + 500)$
$T(s) = sG(s) = 500/(s + 500)$

11-12

If $\mathcal{L}\{u(t - T)\} = e^{-Ts}/s$, then $T(s) = e^{-Ts}$.

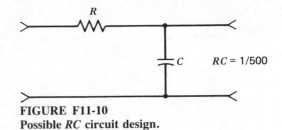

FIGURE F11-10
Possible *RC* circuit design.

CHAPTER 12

Exercises

12-1-2

(a) $v_1(t) = 10 \cos(\omega t + 30°)$
(b) $v_2(t) = 60 \cos(\omega t - 270°)$
(c) $i_1(t) = 5 \cos(\omega t + 180°)$
(d) $i_2(t) = 15 \cos(\omega t + 70°)$

12-1-3

(a) $v_1(t) = 3.92 \cos(\omega t + 101.3°)$
(b) $v_2(t) = 1.7 \cos(\omega t - 9.4°)$
(c) $i_1(t) = 3.16 \cos(\omega t - 71°)$
(d) $i_2(t) = \cos(\omega t + 142°)$

12-2-1

C-1 $V_{\text{OUT}} = \dfrac{RV_o\angle 0°}{R - (j/\omega C)}$

C-2 $I_{\text{OUT}} = \dfrac{RI_o\angle 0°}{R + j\omega L}$

C-3 $I_{\text{OUT}} = \dfrac{V_o\angle 0°}{R + j2\omega L}$

C-4 $V_{\text{OUT}} = \dfrac{-RjI_o\angle 90°}{2R\omega C - j}$

12-2-3

C-1 $I = 0.59\angle 26.5° \text{ A}$
C-2 $I = 0.62\angle -83° \text{ A}$

12-2-5

C-1 $I = 0.024\angle -53° \text{ A}$
C-2 $V = 64.84\angle 248° \text{ V}$

12-3-2

Gain at

Circuit	$\omega = 0$	$\omega = \infty$
C-1	0	0
C-2	-1	0
C-3	∞	1

12-3-4

At $RC = L/R$, $Z_{EQ} = R$.

12-4-1

C-1 $P_{AVE} = 52.8$ W

C-2 $P_{AVE} = 7.5$ W

12-4-2

Circuit	P_{AVEMAX}	I (A)	$P_{L(DEL)}$	$Z_L(\Omega)$
C-1	450 mW	$0.024\angle -53°$	144 mW	$1000 - j2000$
C-2	8.4 W	$0.287\angle 185°$	4.1 W	$75 + J25$

12-4-3

Circuit	P_{AVEMAX}	Z_T (Ω)
C-1	0.021 W	$36 + j418$
C-2	2.2 mW	176
C-3	4.6 W	$156 - j72$
C-4	13.24 W	$1248 + j504$

12-5-1

$V_P = 120$ V

$I_A = 10.71\angle -26°$ A

$I_B = 10.71\angle -146°$ A

$I_L = 10.71\angle -266°$ A

$P_A = 1155$ W

$P_T = 3.46$ kW

12-5-2

$I_1 = 40\angle 0°$ A

$I_2 = 40\angle -120°$ A

$I_3 = 40\angle -240°$ A

$I_A = 70\angle -30°$ A

$I_B = 70\angle -150°$ A

$I_C = 70\angle -270°$ A

$P_A = 140$ kW

$P_T = 420$ kW

$\cos \theta = 0.866 =$ power factor

12-5-4
$P_T = 10$ kW
$P_A = 3.33$ kW
$V_L = 440$ V
power factor $= 0.75$ lagging, $\theta = -41.4°$
$I_L = 17.5$ A
$I_A = 17.5\angle -41.4°$ A
$I_B = 17.5\angle -161.4°$ A
$I_C = 17.5\angle -281.4°$ A

Problems

12-1
$R = 125\ \Omega, L = 60.5$ mH
$V_0 = 10\angle 0°$ V
$V_1 = 2.6 - j3.04$ V
$V_2 = 7.4 + j3.04$ V

12-2
$$T(j\omega) = \frac{1}{[1 - 5\omega^2R^2C^2 + j\omega RC(6 - \omega^2R^2C^2)]}$$
When $\omega = 1/\sqrt{5}\,RC$, the real part of the denominator vanishes and $T(j/\sqrt{5}\,RC) = (5\sqrt{5}/29)\angle -90°$.
When $\omega = \sqrt{6}/RC$, the imaginary part of the denominator vanishes and $T(j\sqrt{6}/RC) = (1/29)\angle -180°$.

12-5
$Z_L = 10.3 + j7.70\ \Omega$
$Y_L = 0.0623 - j0.0466\ \Omega$
$Y_C + Y_L = 0.0623 + j(\omega C - 0.0466)\ \Omega$
To achieve unity power factor, imaginary part must vanish. Hence pick $C = 123\ \mu$F.
The increase in the power factor reduces the line current and the wire losses.

12-7
(a) $P_T = 288$ W, $P_A = P_B = 144$ W, $P_C = 0$ W
(b) $P_T = 288$ W, $P_A = 141.6$ W, $P_L = 146.4$ W, $P_C = 0$ W

12-9
(a) $I_P = 1.2$ A; $I_L = 2.08$ A; $I_A = 2.08\angle 30°$ A
$I_B = 2.08\angle -90°$ A, $I_C = 2.08\angle -210°$ A
$P_T = 3\,P_P = 432$ W
(b) $V_A = V_{AB} = 120\angle 0°$ V
$V_C = V_{CB} = 120\angle -120°$ V

CHAPTER 13

Exercises

13-1-1

$a_0 = A/3$

$a_n = (3A/\pi^2 n^2)(1 - \cos 2\pi n/3)$

$b_n = 0$

$BW = 8\pi/3T$

13-1-2

$a = A/4$

$a_n = (A/\pi n) \sin(\pi n/2)$

$b_n = (3A/\pi n) - (A/\pi n) \cos(\pi n/2)$

$BW = $ up to 21st harmonic.

13-1-4

$a_o = 0$

$a_n = 0$ except $a_1 = A$

$b_n = 0$

13-1-5

$a_0 = A[1 - T]$

$a_n = \dfrac{-4AT}{\pi^2 n^2} \qquad n_{odd}$

$a_n = + \dfrac{4AT}{\pi^2 n^2} \qquad n_{even}$

$b_n = 0$

$BW = 4\pi/T$

13-2-1
$|H(0)| = R_2/(R_o + R_2), \angle H(0) = 0°$
$\omega_C = (R_o + R_2)/R_o R_2 C_2$

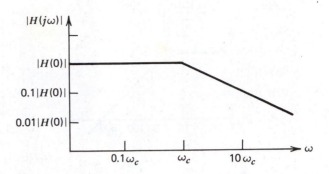

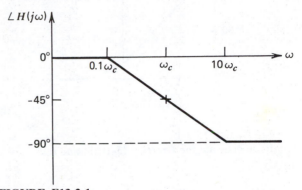

FIGURE F13-2-1
Bode diagram for Exercise 13-2-1.

13-2-5

$$H(j\omega) = \frac{(1/CR)}{[j\omega + (1/CR)]}$$

$|H(0)| = 1, \angle H(0) = 0°, \omega_C = 1/RC$
(Same plots as 13-2-1.)

13-2-8
$H(j\omega) = (-1/CR)/[j\omega + (1/CR)]$
$|H(0)| = 1, \angle H(0) = +180°, \omega_C = 1/RC$
Magnitude plot same as 13-2-1, phase plot same shape
as 13-2-1 except that it starts at 180° and ends at 90°.

13-2-13
$V_2/V_1 = K/(s + K)$
Single-pole low-pass filter with $\omega_C = K$.

13-2-14
$V_2/V_1 = Ks/(s + K)$
Single-pole high-pass filter with $\omega_C = K$.

13-3-1

(a) $|H(0)| = 2, \angle H(0) = 0°$

(c) $|H(j.1)| = 0.02, \angle H(j0) = 90°$

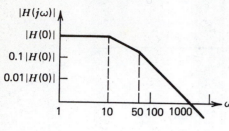

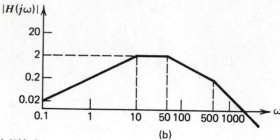

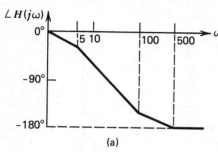

(a)

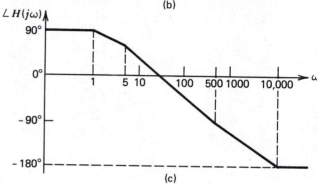

(c)

FIGURE F 13-3-1

Bode diagram for Exercise 13-1-1(a) and (c).

13-3-3

$$H(s) = \frac{62500}{(s + 50)(s + 250)}$$

13-3-4

$$H(s) = \frac{10(s + 10^3)^2}{(s + 100)(s + 10^4)}$$

13-4-1

$$H(s) = \frac{10^6}{[s^2 + 2(10^3)\,s + 2(10^6)]}$$

$|H(0)| = 0.5$, $\omega_o = 1.414(10^3)$, $\zeta = 0.707$

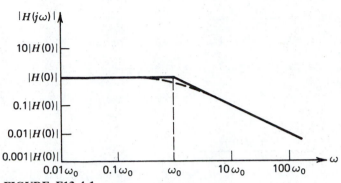

FIGURE F13-4-1
Magnitude plot for Exercise 13-4-1.

13-4-4

$$H(s) = \frac{10^3 s}{[s^2 + s(10^3) + 10^6]}$$

$\omega_{C1} = 618$ rad/second, $\omega_{C2} = 1620$ rad/second

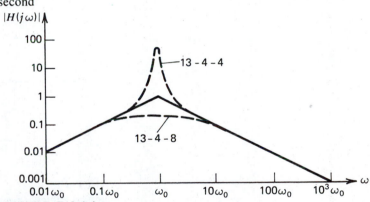

FIGURE F13-4-4
Magnitude plots for Exercises 13-4-4 and 13-4-8.

13-4-8

$H(s) = (s/CR)/[s^2 + (2/RC)\,s + (1/R^2C^2)]$
$\omega_{C1} = 0.414/RC$, $\omega_{C2} = 2.414/RC$

13-4-12

$$H(s) = \frac{-RCs}{(s^2C^2R^2 + 2RCs + 1)}$$

Everything the same as 13-4-8. Minus sign only affects phase plot.

13-4-14

Same as 13-4-8.

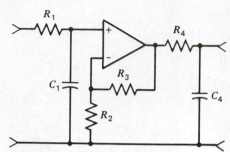

13-5-1

$$H(s) = \frac{62500}{(s + 50)(s + 250)}$$

$\omega_{C1} = 1/R_1C_1 = 50$ Let $R_1 = 10$ kΩ, then $C_1 = 2$ μF.

$\omega_{C2} = 1/R_4C_4 = 250$ Let $R_4 = 10$ kΩ, then $C_4 = 0.4$ μF.

Want gain = 5, hence let $R_2 = 1$ kΩ, then $R_3 = 4$ kΩ.

FIGURE F13-5-1
Possible OP AMP design for Exercise 13-5-1.

13-5-3

$$H(s) = \frac{1.6(10^4)s}{(s + 100)(s + 8000)}$$

$\omega_{cH} = 1/R_1C_1$ Let $R_1 = 1$ kΩ, then $C_1 = 10$ μF.

$\omega_{cL} = 1/R_4C_2$ Let $R_4 = 1$ kΩ, then $C_2 = 0.125$ μF.

Want overall gain = 2, pick $R_2 = 2$ kΩ, $R_3 = 1$ kΩ.

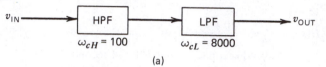

(a)

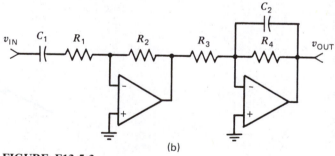

(b)

FIGURE F13-5-3
Bandpass filter design for Exercise 13-5-3. (a)Block diagram. (b) Actual circuit.

13-5-7

Suitable design is standard passive RC high-pass followed by non inverting OP AMP. Select cutoff of standard high-pass $1/RC = 100$ rad/seconds. Suitable values are $R = 1$ kΩ, $C = 10$ μF. Select gain of non inverter to equal 10. $R_2 = 1$ kΩ, $R_1 = 9$ kΩ. Using standard values pick $R_1 = 9.1$ kΩ (slight errror but O.K.). Cost—thru resistors at $0.15, one capacitor at $0.20, one OP AMP at $0.60, one power supply at $15.00 equals $15.95.

Problems

13-2

$$H(s) = \frac{-5(s + 10,000)(s + 100)}{(s + 1000)(s + 500)}$$

Bode diagram is described as follows:
$|H(0)| = 10$. Slope increases to "$+1$" at $\omega = 100$, levels off to "0" at $\omega = 500$. Slope decreases to "-1" at $\omega = 1000$ and levels off again at 10,000. $|H(\infty)| = 5$.

13-3
Circuit (d) is best design since no loading on source occurs.

13-4
$v_o = (2C^3)(s + 1/RC)(s^2 + K/R^2C^2)/$denominator
zero at $-1/RC$, 2 zeros at $\pm j\sqrt{K}/RC$

13-5
$K = (R_2/R_1) + (C_1/C_2) + 1$

13-6
$|H(j\omega)| = 1/2$ (independent of ω) $\angle H(j\omega)$
$= \angle -2 \tan^{-1}(\omega/R_xC)$

APPENDIX A

Block I

B1-1
(a)
$$v_0 = \underbrace{-R_6/(R_1 + R_5)v_1}_{K_1} +$$
$$\underbrace{[R_4(R_1 + R_5)/(R_2 + R_3 + R_4)(R_1 + R_5 + R_6)]\, v_2}_{K_2}$$

(b)   For
$R_1 = 1\ k\Omega$, $R_2 = 2\ k\Omega$, $R_5 = 1\ k\Omega$
$K_1 = 10$, $K_2 = 10$
so that
$R_1 = R_3 = R_5 = 1\ k\Omega$, $R_4 = R_6 = 20\ k\Omega$, $R_2 = 2\ k\Omega$
(c) $v_o = 10(v_2 - v_1)$
(d)   C-2 requires two OP AMPs but does not load the input source.

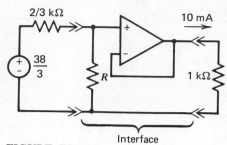

FIGURE FB1-3
Interface design for Block problem B1-3e.

B1-3

$V_T = (38/3)V$, $R_{TH} = (2/3)$ kΩ
(a) For P_{MAX}, $R_L = (2/3)$ kΩ.
(b) For V_{MAX}, $R_L = \infty \Omega$.
(c) For $I_L = 10$ mA, $R_L = 0.6$ kΩ.
(d) For $R_L = 1$ kΩ, $I_L = 7.6$ mA.
Therefore the interface circuit is just an R in series
with R_L. $R_{ADDED} = 4.33$ kΩ.
(e) Must add OP AMP buffer to obtain necessary
current. $R = 6.25$ kΩ

B1-5

Two possible solutions are shown in Figure F B1-5
circuit (a) dissipates 5 mW; circuit (b) 25 mW

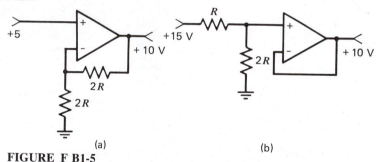

(a)   (b)

FIGURE F B1-5
Two possible design solutions for Block problem B1-5.

B1-7

(a) $V_{OC} = V_1$, $I_{SC} = V_1/R$ common mode
$V_{OC} = V_1$, $I_{SC} = V_1/R$ difference mode
(b) $V_{OUT} = -2V_D$ difference mode
$V_{OUT} = 0$ common mode

B1-9

(a) $v = V_B$, $i = (v - V_B)/R$
$P = V_B(v - V_B)/R$
   If $V_B = v/2$, $P = v^2/4R = P_{MAX}$.
(b) Note the maximun power theorem says that *if*
the load is a resistor, *then* $R_L = R_{TH}$. It does not say
if maximum power is delivered, *then* the load must
be a resistor. There are infinitely many loads that will
draw maximum power from the source.

Block II

B2-1

(a) $v_{\text{forced}} = (250/9) \, e^{-100t}$ V

$v_{\text{natural}} = (-250/9) \, e^{-1000t}$ V

$v_{\text{zero state}} = (250/9) \, (e^{-100t} - e^{-1000t})$ V

$v_{\text{zero input}} = 0$V

(b) Select C so that $R_{TH}C = 0.01$, that is, $C = 2$ μF.

B2-3

(a) $v_{\text{zero state}} = u(t) \, [10 - 11.5 \, e^{-5000t} \sin 8660t]$ V

$v_{\text{zero input}} = u(t) \, [5.81 \, e^{-5000t} \sin 8660t]$ V

(b) $v_C(0) = 10$V

(c) $v_1(t) = \sin (t/\sqrt{LC})$

B2-5

(a) C-1 $T(s) = \dfrac{-R_2 C_1 s}{R_1 C_1 s + 1}$

Zero at origin, pole at $-1/R_1 C_1$.

C-2 $T(s) = \dfrac{[(R_1 + R_2)Cs + 1]}{R_1 C_1 s + 1}$

Zero at $-1/(R_1 + R_2)C$, pole at $-1/R_1 C_1$.

(b) For $T_1(s)$ use two C-1's in cascade. First C-1 R's $= 10$ kΩ, $C = 1$ μF. Second C-1 R's $= 10$ kΩ, $C = 0.25$ μF.

For $T_2(s)$ use one each C-1 and C-2 in cascade. C-1 R's $= 10$ kΩ, $C = 1$ μF. C-2 R's $= 10$ kΩ, $C = 0.25$ μF.

For $T_3(s)$ use two C-2's in cascade. First C-2 R's $= 10$ kΩ, $C = 1$ μF. Second C-2 R's $= 10$ kΩ, $C = 0.25$ μF.

(c) The zero at $s = -500$ is outside all the poles and cannot be achieved.

B2-8

(a) $T(s) = \dfrac{1}{R^2 C^2 s^2 + 2RCs + 2}$

$= \dfrac{1}{(RCs + 1)^2 + 1}$

Poles are at $(-1 \pm j)/RC$. At $\omega = 0$ gain is 1/2. At $\omega = \sqrt{2}/RC$ gain is $(1/2)/\sqrt{2}$; BW is verified.

(b) $T(s) = \dfrac{RCs}{(RCs + 1)^2 + 1}$ same poles as (a). $\omega_0 = \sqrt{2}/RC$, BW $= 2/\sqrt{RC}$ from bandpass form.

Block III

B3-1

(a) $T(s) = \dfrac{15(10^6)}{(s + 1000)(s + 4000)}$

(b) $v_o(t) = 0.586 \cos(4000t - 141°)$

(c) $g(t) = (15/4) = 5e^{-1000t} + 1.25\,e^{-2000t},\ t \geq 0$

(d) See Figure F B3-1d.

(e) A suitable design would be two low-pass filters in cascade: first with $RC = 0.001$ and gain of 2; second with $RC = 0.00025$ and gain of 1.875.

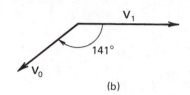

(b)

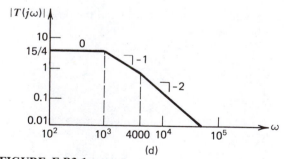

(c)

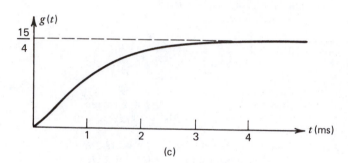

(d)

FIGURE F B3-1
**Solution for Block problem B3-1. (b) Phasor diagram. (c) Step resp
(d) Bode plot.**

B3-3

(a) Filter no. 1 $T(s) = \dfrac{1}{1 + (s/10^4)}$

Bandwidth is too wide.

Filter no. 2 $T(s) = \dfrac{0.6}{1 + (s/3000)}$

This is best design.

Filter no. 3 $T(s) = \dfrac{0.4}{1 + (s/10^3)}$

Bandwidth is too narrow.

(b) Let $T(s) = \dfrac{3000}{(s + 3000)}$, then RC
$= 1/3000$.
Standard passive RC low-pass filter followed by OP AMP buffer will work.

B3-5

(a) $P_L = V_0^2/8R_0$

(b) $P_L = \dfrac{V_0^2}{8R_0 \, [1 + (\omega L/2R_0)^2]}$

(c) Insertion ratio $= 1 + (\omega L/2R_0)_2$

(d) When $\omega L = 0$ insertion ratio is 1 because inductor acts as a short circuit. When $\omega L = \infty$ insertion ratio is ∞ because inductor acts as an open circuit.

(e) At 3 dB the insertion ratio equals 2. Hence $(\omega L/2R_0)^2 = 1$ or $\omega = 2R_0/L$

B3-7

The function $T(s) = 10s^2/(s + 10^4)^2$ meets the requirements. The design can be achieved using two standard high-pass active filters in cascade. First filter has $RC = 10^{-4}$, gain 2. Second has $RC = 10^{-4}$, gain 5. One solution lets $R_S = 10 \text{ k}\Omega$ and $C_s = 0.01 \ \mu\text{F}$ for both filters. R_F for first filter is 20 kΩ and R_F for second filter is 50 kΩ.

Appendix E

E-1

(a) $43.01\angle-1.3°$
(b) $2.3 + j0.5$
(c) $6.2\angle99.3°$
(d) $-14.9 - j3$
(e) $27.9 - j2.8$

E-2

(a) $H(\omega) = \dfrac{10}{1 + j\omega}$

@ $\omega = 0.1$, $H(0.1) = 10\angle-5.7°$
@ $\omega = 1.0$, $H(1) = 7.07\angle-45°$
@ $\omega = 10.0$, $H(10) = 1.0\angle-84.3°$

(b) $H(\omega) = \dfrac{10 + j\omega}{j\omega}$

@ $\omega = 1$, $H(1) = 10\angle-84.3°$
@ $\omega = 10$, $H(10) = 1.414\angle-45°$
@ $\omega = 100$, $H(100) = 1.0\angle-5.7°$

(c) $H(\omega) = \dfrac{(10 + j\omega)}{1 + j\omega}$

@ $\omega = 0.1$, $H(.1) = 11\angle-4.6°$
@ $\omega = 1$, $H(1) = 7.5\angle-36.8°$
@ $\omega = 10$, $H(10) = 1.414\angle-39°$
@ $\omega = 100$, $H(100) = 1.0\angle-5.1°$

(d) $H(\omega) = \dfrac{(10 + j\omega)}{1 + j\omega}$

@ $\omega = 0.1$, $H(.1) = 11\angle-4.6°$
@ $\omega = 1$, $H(1) = 7.5\angle-36.8°$
@ $\omega = 10$, $H(10) = 1.414\angle-39°$
@ $\omega = 100$, $H(100) = 1.0\angle-5.1°$

Historical Index

Subject Index

Two-Pole Low-Pass Case

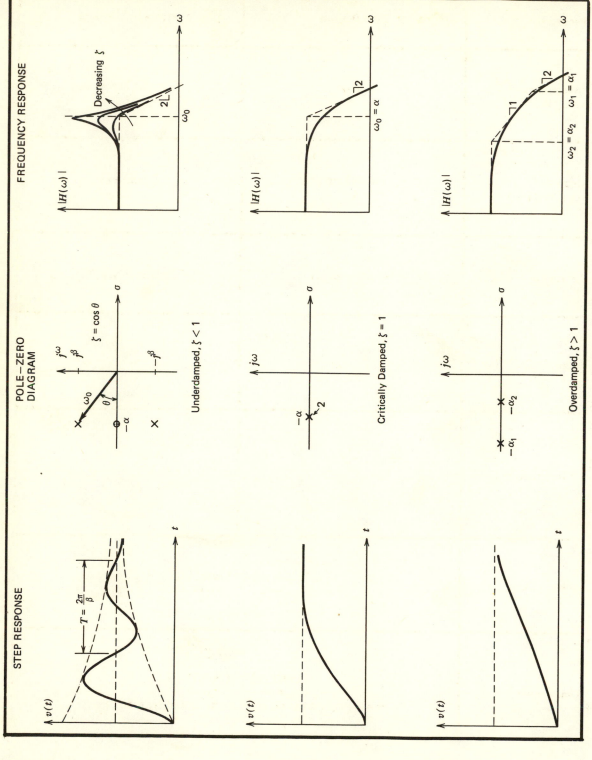